PLASTIC LIMIT ANALYSIS OF PLATES, SHELLS AND DISKS

NORTH-HOLLAND SERIES IN

APPLIED MATHEMATICS AND MECHANICS

VOLUME 43

ELSEVIER
AMSTERDAM • LAUSANNE • NEW YORK • OXFORD • SHANNON • SINGAPORE • TOKYO

PLASTIC LIMIT ANALYSIS OF PLATES, SHELLS AND DISKS

M.A. SAVE
Faculté Polytechnique de Mons
Mons, Belgium

C.E. MASSONNET
Institut du Génie Civil
Liège, Belgium

G. de SAXCE
Université des Sciences et Technologies de Lille
Ville-Neuve-D'Ascq, France

1997
ELSEVIER
AMSTERDAM • LAUSANNE • NEW YORK • OXFORD • SHANNON • SINGAPORE • TOKYO

ELSEVIER SCIENCE B.V.
Sara Burgerhartstraat 25
P.O. Box 211, 1000 AE Amsterdam, The Netherlands

This book is the 2nd edition (largely revised and updated) of "Plastic Analysis and Design of Plates, Shells and Disks" by M.A. SAVE and C.E. MASSONNET, published in 1972 by North-Holland Publishing Company, Amsterdam-London, in its Series in Applied Mathematics and Mechanics, vol. 15.

This first edition was itself the translation into English (by M.A. SAVE) of the book "Calcul plastic des Constructions", vol. 2, 2nd edition published by the Centre Belgo-Luxemburgeois d'Information de l'Acier, 47, rue Montoyer, Brussels, in 1972.

ISBN: 0-444-89479-9

This book is printed on acid-free paper.

PRINTED IN THE NETHERLANDS

Preface to the first edition.

Plastic design of steel beams and frames has been included in building codes since 1948 in Great Britain and since 1959 in the United States. The increased use of plastic design and the present trend toward applying its methods to reinforced and prestressed concrete indicate that the basic advantages of plastic analysis - accurate estimate of the collapse load, simplicity of application, and economy of structure - are more and more appreciated by practicing engineers. Codes on reinforced concrete slabs (European Committee for Concrete) and on pressure vessels (ASME) are on the way to adopt plastic limit state as one of the design criterions. Hence, it was felt urgent to provide all interested people with a synthesis work on the subject.

The present book is the English version of " Calcul plastique des Constructions ", Volume 2, 2nd edition, published by the Centre Belgo-Luxembourgeois d'Information de l'Acier in Brussels, Belgium.

Together with this second edition in French, which is to appear nearly contemporarily, not only has it been brought up to date but also appreciably improved with respect to the first edition in French :

- new experimental results have been introduced, to better clarify the real physical significance of the theoretical limit load, particularly in Chapters 6 and 8.

- the treatment of circular plates has been re-written in a more systematic manner

- Section 6-8 on minimum-weight design has been re-written and amplified

- in Chapter 7 on reinforced concrete plates, a more satisfactory derivation of the yield surface has been achieved, the delicate question of nodal forces has been faced in a clearer manner, statically admissible and complete solutions have been amplified, economy of reinforcement has been treated in am more rational manner, as well as the influence of the axial forces

- Chapters 8 and 9 contain many new additions of new theoretical solutions and experimental information

- only Chapter 10 is practically unchanged except for the new section on notched bars in tension

On the other hand, the main features of the first edition in French have been maintained, namely :

- the book is directed toward engineering applications ; consequently, we have limited the mathematics to what was strictly needed, and we have avoided any formalism that could be unfamiliar to some engineers (as tensor notations for example). Physical significance, with reference to experimental evidence, is emphasized throughout.

- the book should serve the engineering student in plastic limit analysis : therefore, the theory is rigorously developed, as a rule from simpler to more complicated problems, though it was felt necessary to base applications to particular structures on the firm ground of a general theory.

Though it can be read completely independently, with the sole pre-requisite basis of some knowledge of the classical theory of structures, the present book can be regarded as the companion volume of " Plastic Analysis and Design of Beams and Frames " Published in 1965 by Blaisdell Publishing Co., Waltham, Mass., U.S.A. , which is the adaptation in English to the first volume of " Calcul plastique des Constructions ", devoted by the same authors to the simpler problems of beams and frames. The knowledge of the subjects treated in " Plastic Analysis and Design of Beams and Frames " is by no means necessary to undertake studying the present work. However, references may be useful. Whenever this is done, the cited book is referred to as " Companion Volume ", abbreviated as Com. V. throughout the text.

We believe that the false conflict between elastic design and plastic design is now superseded. Our hope is that the present book will contribute to enlarge the designer's ability by giving him one pore tool, namely the theory of limit analysis and design, together with the definition of its range of applicability. Though extremely powerful, this theory remains nothing but a tool, to be used by the design engineer within the frame of a wider and more sophisticated design philosophy.

Last but not least, we are pleased to express here our gratitude to the numerous authors whose original works have been used in this book. We have tried to give due acknowledgement to all of them. We apologize for any, involuntary, omission.

The junior author (M.A.S.), who wrote the adaptation in English, is particularly grateful to Professor W. Prager, who read the manuscript and made many valuable suggestions that resulted in appreciable improvement of the text.

Preface to the second edition.

Though the present edition has retained the general organization of the first, it differs substantially on various aspects, as could be expected at the end of two decades.

It was first decided to delete completely the parts dealing with optimal design, for the following reasons :

on the one hand this subject has developed enormously during these last 10 or 15 years, and it would have been impossible to give an acceptable summary without adding an important number of pages. On the other hand, several books now exist on the subject, among which a three-volume treatise [1] edited by one of the authors (M.S.) and late professor William Prager .

In the updating process, it was felt preferable to introduce the fundamental concepts on the basis of Drucker's postulate than using Prager's assumptions on the dissipation function, because the first presentation is more widely used. This was done keeping in mind that this " postulate " is, as emphasized by prof. Druker himself, nothing but the definition of a class of material.

Problems of cyclic loading have been given some more extensive treatment, both in the general theory and in applications. We wish to stress here that we kept the original terminology in which shake-down is, by definition, a return to elastic behaviour, the alternative being either ratchetting (incremental collapse) or alternating plasticity that we avoided to call plastic shake-down. Indeed, though displacements may be stabilized at constant values, alternating plasticity is likely to generate fracture by exhaustion of ductility and thus is, from the point of view of safety, quite distinct from shake-down.

Because a very large number of new solutions were published in these last twenty years, it would have been impossible to quote all (or even most) of them without increasing excessively the number of pages of the book. Such an accumulation would also have been very likely more negative than positive for the clarity and understandability of the theory. The ideal solution was found by the quasi-simultaneous publishing by Elsevier of an " Atlas of Limit Loads of Metal Plates, Shells and Disks ", edited by one of the authors (M.S.). Not only does this Atlas contain most (if not all) presently existing solutions of the title subject, but also extensive lists of references and texts of introduction to the various parts (plates, shells and disks) which summarize the theory and are, to this respect, complementary to the corresponding texts in the present book. They are not substitute however because they had no didactical goal and hence are much too concise, whereas we kept trying, in the present book, to develop the theory progressively and go from simple to more complicated problems. It is worth pointing out that this Atlas will be enriched regularly.

For reinforced concrete plates and shells, though new solutions have been developed since 1972, we also avoided to overload the book without benefit to the understanding of the theory, and limited ourselves to general indications and references. On the contrary, a more general presentation of the yield condition for both plates and shells was introduced in chapter 5 and used later in chapters 7 and 9. The section on the influence of axial force in plates also was nearly completely re-written, with the help, here gratefully acknowledged, of Dr. Marek Janas of the Polish Academy of Sciences in Warsaw.

Last but certainly not least, the present text differs from the preceding by the introduction of an important chapter 6 exclusively devoted to the numerical approach to limit load and shake-down load evaluation, and by its illustration in the various chapters of the second part dealing with applications to plates, shells and disks. The matrix notations, already introduced as an alternative in the preceding chapters, was extensively used in chapter 6 as usual for computer-oriented subjects.

Concerning units, we decided to remain close to engineering habits, keeping the original (even old) units in figures related to experiments, and using millimeters (mm), centimeters (cm), meters (m), Newton (N), decanewtons (dN which is approximately one kg), and kilonewton (kN).

We hope that, in its present form, this book will serve both the engineering student in plastic limit analysis and, coupled with the Atlas, or analogous documents for reinforced concrete structures (*), the engineers of design offices.

[1] Structural Optimization, Plenum Press, New-York and London, SAVE and PRAGER editors.
Vol. 1. Optimality criteria, 1985
Vol. 2. Mathematical programming, 1990
Vol. 3. Metal and concrete structures : to appear.

(*) A catalog of this kind can be found for example in the book " Limit Analysis of Plates " by A. Sawczuk and J. Sokol-Supel, Polish Scientific Publisher, Warsaw, 1993.

List of Symbols

a	length, radius, diameter of a hole, thickness of a web, coefficient
b	length, breadth, radius, coefficient
c	parameter, radius, coefficient
d	specific power of stresses on strain rates, coefficient, distance
e	average axial strain, coefficient, base of natural logarithms
$\dot{\mathbf{e}}$	unit strain rate vector
f	function, reduced force, coefficient
g	gravity acceleration, distance between centers of holes
$\mathbf{g}$	nodal force vector
h	height, distance
$\mathbf{h}$	stress parameter vector
i	strength parameter
k	coefficient, orthotropy coefficient, economy coefficient, parameter
$\mathbf{k}$	constant vector
k'	orthotropy coefficient (in negative bending)
l, m, n	direction cosines of an outward pointing normal
l	span, length, small side of a rectangle
m	reduce bending moment
m_x, m_y	bending components of the reduced moment tensor in rectangular cartesian x, y axes
m_{xy}	twisting component of the reduced moment tensor in rectangular cartesian x, y axes
m_t	reduced twisting moment
n	reduced axial force, number
$\mathbf{n}$	outward normal vector
p, q, r	direction parameters
p	pressure, distributed load, number of parameters
$\bar{p}$	line load per unit of length
p_s	shake-down pressure
p^*	reduced pressure
p^M	limit pressure of a membrane
q	reduced load
$\dot{q}$	generalized strain rate
$\dot{\mathbf{q}}$	nodal velocity vector
r	generic radius

r_1, r_2	principal radii of curvature of a shell
s	safety factor, curvilinear abscissa, distance, reduced principal stress
s_x, s_y, s_z	normal components of the stress deviator
t	time, thickness
u, v, w	components of the displacement vector
V	volume
$\dot{\mathbf{v}}$	strain rate parameter vector
w	transversal displacement of a plate
x, y, z	orthogonal cartesian coordinates, reduced forces
A	area, boundary of the body, interface between finite elements, coefficient
A_S	sheared area in punching, steel reinforcement cross-sectional area
$\mathbf{A}_u$	portion of the boundary where velocities are prescribed
$\mathbf{A}_o$	portion of the boundary where surface forces are imposed
B	breadth of a stiffener, coefficient
C	statical connection matrix
D	specific power of dissipation, diameter
D	matrix of elastic coefficients
E	Young's modulus, thickness
F	function, intensity of a force, bending stiffness of a plate
F	body-force vector, compliance matrix
G	elastic Coulomb's modulus
G	static connection matrix
H	thickness, thickness of a core, height of a pressure vessel head
H	stiffness matrix
I	stress invariant
K	parameter, coefficient, constant, nodal force
K	stiffness matrix
L	length, long side of a rectangle, lagrangian
L	matrix multiplying the stress vector to obtain the surface traction vector
M	bending moment
M_x, M_y	bending components of the moment tensor
M_{xy}	twisting component of the moment tensor
M_e	maximum elastic bending moment of a cross-section
M_p	plastic moment
M'_p	plastic moment for negative bending
M_t	twisting moment
M_{tp}	plastic twisting moment

$\mathbf{M}$	velocity shape function matrix
N	axial force
N_p	plastic axial force
N_s	strength force of the unit layer of yield-line
$\mathbf{N}$	constant matrix
O	origin of axes
P	load parameter, concentrated load, point, stress point
P_-	licit statical load parameter
P_+	licit kinematical load parameter
P_l	limit load
P_i	limit load for inscribed yield surface
P_c	limit load for circumscribed yield surface, carrying capacity
P_s	service load
P_V	punching load
P_v	live load
Q	generalized stress component, "equivalent" shear in plates, line load intensity, point
R	radius, pole of Mohr's circle, joint, region, corner force in plates
$\mathbf{R}$	strain rate shape function matrix
S	surface
$\mathbf{S}$	stress field shape function matrix
T	magnitude of edge force in rotating disk
$\mathbf{T}$	surface traction vector, Lagrange's multiplier field on interface
U	elastic strain energy per unit volume
$\mathbf{U}$	displacement vector
V	volume, shear force
V_e	finite element
$\mathbf{V}$	velocity vector
V_x, V_y, V_z	components of the velocity vector
$(\mathbf{V})$	velocity field
W	weight
X,Y,Z	components of the body-force vector
$\mathbf{X}$	body-force vector
$\overline{\mathbf{X}}$	surface-traction vector
	lagrangian
	weight ratio, fictitious residual elastic energy

$\overline{X}, \overline{Y}, \overline{Z}$	components of the surface traction vector
B	power of body forces
D	total dissipation in a structure
E	strain energy
I	stress intensity
J	strain intensity
$\dot{J}$	strain rate intensity
P	power of applied forces
R	weight ratio
V	reinforcement volume
V_M	moment volume
W	work of applied forces
div.	divergence of a vector
α, β	angle, coefficient, parameter, ratio, load parameter
γ	shear strain, affinity coefficient, angle, parameter, weight per unit volume
γ_p	coefficient of plate-beam interaction for square plate
γ_b	coefficient of plate-short beams interaction for rectangular plates
γ_B	coefficient of plate-long beams interaction for rectangular plates
$\gamma_{xy}, \gamma_{yz}, \gamma_{zx}$	shear strains
$\dot{\gamma}_{xy}, \dot{\gamma}_{yz}, \dot{\gamma}_{zx}$	shear strain rates
δ	deflection, elongation, variation, parameter
Δ	modified dissipation rate, variation
Δ	symbolic matrix representing the tensor divergence operator
ε	axial strain
$\dot{\varepsilon}$	axial strain rate
(ε)	strain tensor
$\varepsilon_1, \varepsilon_2, \varepsilon_3$	principal strains
$(\dot{\varepsilon})$	strain rate tensor
$\boldsymbol{\varepsilon}^e$	elastic strain rate vector
$\boldsymbol{\varepsilon}^p$	plastic strain rate vector
$\boldsymbol{\varepsilon}^*$	strain vector in the corresponding fictitious indefinitely elastic body
$\dot{\boldsymbol{\varepsilon}}$	strain rate vector in $\varepsilon_x, \ldots, \gamma_{xy}, \ldots$, space
ζ	parameter

η	line, reduced load, reduced abscissa
$\boldsymbol{\eta}$	residual strain vector
η, ξ	coordinates
θ	angle, angle of rotation
θ_{ij}	angle of relative rotation of i and j
$\dot{\theta}$	rotation rate
$\dot{\boldsymbol{\theta}}$	rotation vector
κ	curvature
$\dot{\kappa}_x$, $\dot{\kappa}_y$	curvature components of the curvature rate tensor
$\dot{\kappa}_{xy}$	twist component of the curvature rate tensor
λ	non negative scalar, displacement, load parameter, Lagrange's multiplier
$\dot{\lambda}$	reduced strain rate
μ	slenderness, reinforcement ratio
$\boldsymbol{\mu}$	Lagrange multiplier vectoron interfaces
ν	Poisson's ratio
ξ	reduced abscissa, parameter
π	3.1416
ρ	radius of curvature, radius, reduced radius, cutout factor, efficiency, ratio
$\boldsymbol{\rho}$	residual stress vector
σ	normal component of a stress vector
σ_x, σ_y, σ_z	normal components of the stress tensor
(σ)	stress tensor
	σ_1, σ_2, $\sigma_3$principal stress
$\boldsymbol{\sigma}$	stress vector in σ_x ,..., τ_{xy} ,... space
$\boldsymbol{\sigma}^*$	stress vector in the corresponding fictitious indefinitely elastic body
σ_R	reference stress
σ_Y	yield stress in tension or compression
σ_c, $\sigma'_{r(cyl)}$	crushing stress of concrete in compression, on cylinders
σ_r	tensile rupture stress of concrete
$(\sigma\_)$	licit stress field
Σ	summation
τ	shear stress

$\tau_{xy}, \tau_{yz}, \tau_{zx}$ shear components of the stress tensor

τ_Y yield stress in pure shear

τ_{oct} octahedral shear stress

φ angle, function, shape factor

Φ Markov's functional, function

$\dot{\Phi}$ reduced strain rate

Π Hill's functional

ψ function, angle

ω geometrical parameter of a circular cylindrical shell, parameter, angular velocity

Ω total internal dissipation

Ω_s reinforcement ratio

x] discontinuity on x

∇ laplacian

$(\dot{\varepsilon}_+)$ licit strain rate field

$(\dot{u}_+)$ licit velocity field

Remark : when used in matrix formulation, bold-face lower case letters indicate, except otherwise explicitely stated, *column* vectors.

Table of contents

PART TWO : APPLICATIONS TO PLATES, SHELLS AND DISKS

Part one : General theory

1 . Stress and Strain.

1.1. Stress, strain, and strain-rate tensors.

1.1.1. *Stress tensor.*

Consider an elementary parallelepiped at a generic point O of a continuum referred to orthogonal cartesian axes x, y, z (fig. 1.1). Each of the three faces in the reference planes is in general subjected to one normal stress component and two shearing stress components. The state of stress at O is thus characterized by nine components. From one face to the neighbouring parallel face, these components experience a small variation.

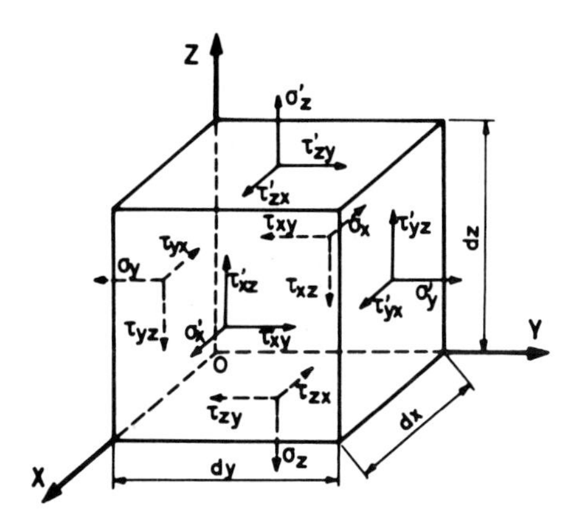

Fig. 1.1.

This fact is indicated by the primes shown in fig.1.1. Rotational equilibrium of the parallelepiped about its axes yields the equality of the shearing stresses :

$$\tau_{xy} = \tau_{yx} \qquad \ldots (x,y,z)\ (*)\ . \tag{1.1}$$

There thus remain six independent stress components, namely the normal stresses σ_x, σ_y, σ_z and the shearing stresses τ_{xy}, τ_{yz}, τ_{zx}. The *stress tensor*, which will be symbo-

(*) The three dots, followed by (x,y,z), at the end of a line, mean that similar expressions or equations are obtained by cyclic permutation of x,y,z.

lically denoted by (σ), si completely defined, in the chosen system of coordinates, by these six components. Translational equilibrium in the directions of the coordinate axes furnishes the three *differential equations of equilibrium* :

$$\frac{\partial \sigma_x}{\partial x} + \frac{\partial \tau_{xy}}{\partial y} + \frac{\partial \tau_{xz}}{\partial z} + X = 0 \qquad ...(x,y,z), \tag{1.2}$$

where X,Y,Z are the components of the body force per unit volume.

For later use in numerical methods implemented on computers, it proves suitable to re-write the system of equations (1.2) in the matrix form

$$\mathbf{\Delta}\boldsymbol{\sigma} + \mathbf{X} = \mathbf{o} \tag{1.2'}$$

where Δ is the operator matrix

$$\mathbf{\Delta} = \begin{vmatrix} \frac{\partial}{\partial x} & 0 & 0 & \frac{\partial}{\partial y} & 0 & \frac{\partial}{\partial z} \\ 0 & \frac{\partial}{\partial y} & 0 & \frac{\partial}{\partial x} & \frac{\partial}{\partial z} & 0 \\ 0 & 0 & \frac{\partial}{\partial z} & 0 & \frac{\partial}{\partial y} & \frac{\partial}{\partial x} \end{vmatrix}$$

and $\boldsymbol{\sigma}$ and $\mathbf{X}$ are one column matrices (that is column vectors) with components (σ_x ,.., τ_{xy} ,...) and (X,Y,Z) respectively.

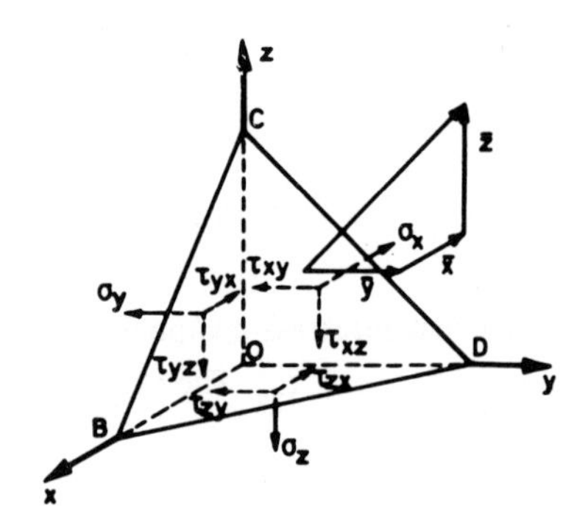

Fig. 1.2.

Consider now the elementary tetrahedron of fig. 1.2, with the plane BCD infinitely close to the origin O. Let l,m,n, be the direction cosines of the exterior normal to face BCD of the tetrahedron. Denote by $\overline{X}$, $\overline{Y}$, $\overline{Z}$ the components of the stress vector acting on this face. From equilibrium considerations we obtain

$$\overline{X} = \sigma_x l + \tau_{xy} m + \tau_{xz} n \qquad \ldots(x,y,z) \qquad \ldots(l,m,n). \tag{1.3}$$

Or, equivalently,

$$\mathbf{L}\boldsymbol{\sigma} = \overline{\mathbf{X}} \tag{1.3'}$$

where $\overline{\mathbf{X}}$ denotes the column vector with components $(\overline{X}, \overline{Y}, \overline{Z})$, l, m, n are the direction cosines to the outward pointing normal and

$$\mathbf{L} = \begin{vmatrix} l & 0 & 0 & m & 0 & n \\ 0 & m & 0 & l & n & 0 \\ 0 & 0 & n & 0 & m & l \end{vmatrix}$$

Note that $\overline{\mathbf{X}}$ is given on the loaded boundary. A stress field satisfying both relations (1.2) and (1.3) is called *statically admissible* for the given loads.

It can be shown [1.1] that, among all plane elements such as BCD at point O, three are subjected solely to normal stresses σ_1, σ_2 and σ_3, respectively. The direction of each of these stresses is orthogonal to the plane determined by the other two directions. The stresses σ_1, σ_2, σ_3 are called *principal stresses* and their directions, *principal directions*. By definition, the shearing stresses associated with the principal directions vanish :

$$\tau_{12} = \tau_{23} = \tau_{31} = 0 .$$

The stress tensor is completely characterized by the principal directions and the algebraic values of the principal stresses (the sign convention being defined on fig. 1.1 where all components are positive).

There exist various geometrical representations of the stress tensor [1.1], [1.2], [1.3]. Mohr's classical plane representation is particularly useful [1.4]. We assume it to be known.

Functions of the components of (σ) that are not altered by rotation of coordinate axes are called invariants. The three fundamental invariants are

$$I_1 \equiv \sigma_x + \sigma_y + \sigma_z, \tag{1.4}$$

$$I_2 \equiv \sigma_x\sigma_y + \sigma_y\sigma_z + \sigma_z\sigma_x - \tau_{xy}^2 - \tau_{xz}^2 - \tau_{zx}^2, \tag{1.5}$$

$$I_3 \equiv \sigma_x\sigma_y\sigma_z + 2\tau_{xy}\tau_{yz}\tau_{zx} - \sigma_x\tau_{yz}^2 - \sigma_y\tau_{zx}^2 - \sigma_z\tau_{xy}^2. \tag{1.6}$$

States of *plane stress*, where one principal stress vanishes, are frequently encountered. Let σ_3 be the vanishing principal stress. All stress vectors (with components $\overline{X}, \overline{Y}, \overline{Z}$) then lie in the plane (O, σ_1, σ_2) of the two other principal directions. With the special sign convention of fig. 1.3 where positive stress components are drawn, the equality of shearing stresses is expressed by $\tau_{xy} = -\tau_{yx}$. Consider a plane surface element with the exterior normal lying in the (O, σ_1, σ_2) plane and making an angle α (clockwise positive) with respect to the x-axis (fig. 1.3). Let σ and τ be the normal and shear components of the stress vector acting on this plane element. From equilibrium considerations, we find :

$$\sigma = \sigma_x\cos^2\alpha + \sigma_y\sin^2\alpha + \tau_{xy}\sin 2\alpha, \tag{1.7}$$

$$\tau = \frac{1}{2}(\sigma_y - \sigma_x)\sin 2\alpha + \tau_{xy}\cos 2\alpha. \tag{1.8}$$

The principal directions are given by

$$\tan 2\alpha = \frac{2\tau_{xy}}{\sigma_x - \sigma_y}. \tag{1.9}$$

When these directions are used for coordinate axes (0, 1) and (0, 2), formulas (1.7) and (1. 8) become

$$\sigma = \sigma_1\cos^2\alpha + \sigma_2\sin^2\alpha, \tag{1.10}$$

$$\tau = \frac{\sigma_2 - \sigma_1}{2}\sin 2\alpha. \tag{1.11}$$

Mohr's circle for a generic state of plane stress is shown on fig. 1.4.

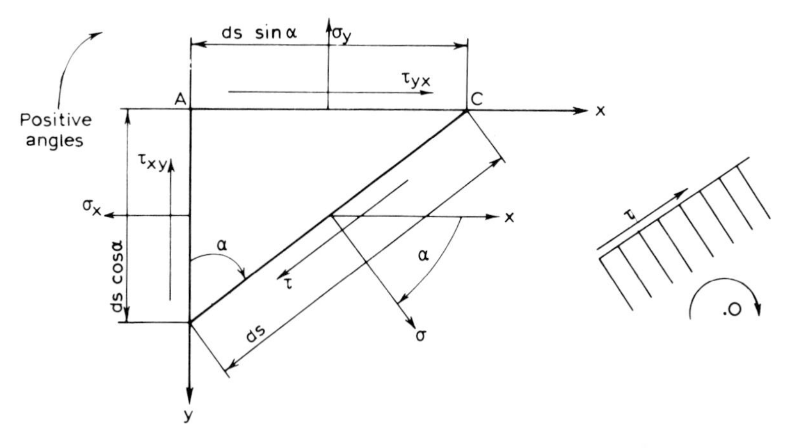

Fig. 1.3.

Let D and D_1 be the points representing the stress vectors acting on surface elements normal to the axes Ox and Oy, respectively. Through D and D_1, draw straight lines parallel to Ox and Oy, respectively (that is, parallel to the traces, in the stress plane, of the surface elements on which the respective stress vectors act). Their intersection determines a point R on the circle. This point R is called the *pole* of the stress circle. The direction of the trace of the surface element subjected to a given stress vector is obtained by drawing the straight line through the pole and the point of the circle representing the stress vector.

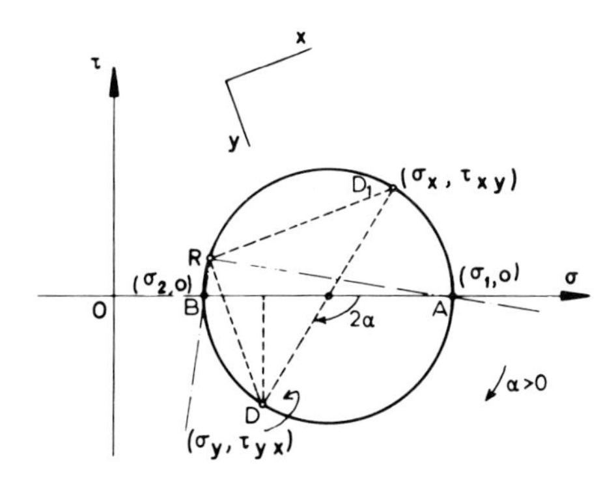

Fig. 1.4.

For example, RA is parallel to the trace of the surface element subjected to the principal stress represented by point A, and RB is parallel to that of σ_2 (represented by point B). The principal stresses are related to the stress components σ_x, σ_y, $\tau_{yx} = -\tau_{yx}$ by (*)

$$\sigma_1 = \frac{\sigma_x + \sigma_y}{2} + \left(\frac{1}{4} (\sigma_x - \sigma_y)^2 + \tau_{xy}^2 \right)^{1/2}, \tag{1.12}$$

$$\sigma_2 = \frac{\sigma_x + \sigma_y}{2} - \left(\frac{1}{4} (\sigma_x - \sigma_y)^2 + \tau_{xy}^2 \right)^{1/2} \tag{1.13}$$

and σ_1 is inclined, with respect to positive direction Ox, by an angle α given by

$$\tan 2\alpha = - \frac{2\,\tau_{xy}}{\sigma_x - \sigma_y}. \tag{1.14}$$

1.1.2. *Strain tensor.*

When a continuum is deformed, a generic point experiences a displacement **U** with components u, v, w with respect to cartesian orthogonal axes x, y, z, respectively. For very small strains, relations between axial strains ε_x, ε_y, ε_z, shear strains γ_{xy}, γ_{yz}, γ_{zx}, and displacement components are [1.1] :

$$\varepsilon_x = \frac{\partial u}{\partial x} \qquad \ldots (x, y, z) \qquad \ldots (u, v, w), \tag{1.15}$$

$$\gamma_{xy} = \frac{\partial u}{\partial y} + \frac{\partial v}{\partial x} \qquad \ldots (x, y, z) \qquad \ldots (u, v, w). \tag{1.16}$$

In matrix form, we write

$$\mathbf{\Delta}^{T'}\,\dot{\mathbf{u}} = \dot{\boldsymbol{\varepsilon}}$$

where

(*) Formulas (1.12) and (1.13) also hold when $\sigma_z \equiv \sigma_3 \neq 0$ provided σ_z is a principal stress.

where Δ is the operator matrix

$$\Delta^T = \begin{vmatrix} \frac{\partial}{\partial x} & 0 & 0 \\ 0 & \frac{\partial}{\partial y} & 0 \\ 0 & 0 & \frac{\partial}{\partial z} \\ \frac{\partial}{\partial y} & \frac{\partial}{\partial x} & 0 \\ 0 & \frac{\partial}{\partial z} & \frac{\partial}{\partial y} \\ \frac{\partial}{\partial z} & 0 & \frac{\partial}{\partial x} \end{vmatrix}$$

Because the strain field derives by equations (1.15) and (1.16) from the field of displacements, the six functions of position $\varepsilon_x ,..., \gamma_{xy} ,...,$ are not independent. They must satisfy three linearly independent differential equations, called *compatibility equations*, obtained by elimination of u, v, w from (1.15) and (1.16). The displacement field must also satisfy the *kinematic boundary conditions* at the supports. Fields of displacement and strains exhibiting these properties are called *kinematically admissible*.

The six components $\varepsilon_x ,..., \gamma_{xy} ,...,$ completely describe the state of strain at the considered point, and it is found [1.1, 1.3] that they satisfy relations formally identical to those obtained in the analysis of the stress tensor. $\varepsilon_x, \varepsilon_y, \varepsilon_z$ are substitutes for $\sigma_x, \sigma_y, \sigma_z$ respectively, and $\frac{\gamma_{xy}}{2}, \frac{\gamma_{yz}}{2}$, and $\frac{\gamma_{zx}}{2}$ for $\tau_{xy}, \tau_{yz}, \tau_{zx}$. The state of strain is thus a tensor, with three principal directions subjected to vanishing shear strain and with three fundamental strain invariants.

Especially important is the state of *plane strain*, with one vanishing principal strain, say ε_3. In this case, we have [see equation (1.12) and (1.13)]

$$\varepsilon_1 = \frac{\varepsilon_x + \varepsilon_y}{2} + \left(\frac{1}{4}(\varepsilon_x - \varepsilon_y)^2 + (\frac{\gamma_{xy}}{2})^2 \right)^{1/2} \tag{1.17}$$

$$\varepsilon_2 = \frac{\varepsilon_x + \varepsilon_y}{2} - \left(\frac{1}{4}(\varepsilon_x - \varepsilon_y)^2 + (\frac{\gamma_{xy}}{2})^2 \right)^{1/2} . \tag{1.18}$$

Principal directions are given by

$$\tan 2\alpha = -\frac{\gamma_{xy}}{\varepsilon_x - \varepsilon_y}. \tag{1.19}$$

Formulas (1.17) to (1.19) still hold when $\varepsilon_z \equiv \varepsilon_3 \neq 0$, provided ε_z is a *principal strain.*

1.1.3. *Strain-rate tensor.*

When plastic flow is considered, the strain components depend on time t and it proves necessary to use *strain rates* defined as follows :

$$\dot{\varepsilon}_x = \frac{\partial \varepsilon_x}{\partial t} \qquad \text{...(x, y, z)}, \tag{1.20}$$

$$\dot{\gamma}_x = \frac{\partial \gamma_{xy}}{\partial t} \qquad \text{...(x, y, z)}. \tag{1.21}$$

If **V** is the velocity of a given point ($\mathbf{V} = \partial \mathbf{U} / \partial t$) with components V_x, V_y, V_z, obviously

$$\dot{\varepsilon}_x = \frac{\partial}{\partial t}\frac{\partial u}{\partial x} = \frac{\partial}{\partial x}\frac{\partial u}{\partial t} = \frac{\partial \dot{u}}{\partial x} = \frac{\partial V_x}{\partial x} \qquad \text{...(x, y, z)} \qquad \text{...(u, v, w)},$$

$$\dot{\gamma}_{xy} = \frac{\partial}{\partial t}\left(\frac{\partial u}{\partial y} + \frac{\partial v}{\partial x}\right) = \frac{\partial}{\partial y}\frac{\partial u}{\partial t} + \frac{\partial}{\partial x}\frac{\partial v}{\partial t}$$

$$= \frac{\partial \dot{u}}{\partial y} + \frac{\partial \dot{v}}{\partial x} = \frac{\partial V_x}{\partial y} + \frac{\partial V_y}{\partial x} \qquad \text{...(x, y, z)} \qquad \text{...(u, v, w)} \tag{1.22}$$,

because

$$V_x \equiv \dot{u} = \frac{\partial u}{\partial t} \qquad \text{...(x, y, z)} \qquad \text{...(u, v, w)}. \tag{1.23}$$

The state of strain rate is also a tensor. For example, when Oz is a principal direction, we have (with $\varepsilon_3 \equiv \varepsilon_z$)

$$\dot{\varepsilon}_1 = \frac{\dot{\varepsilon}_x + \dot{\varepsilon}_y}{2} + \left(\frac{1}{4} (\dot{\varepsilon}_x - \dot{\varepsilon}_y)^2 + \left(\frac{\dot{\gamma}_{xy}}{2} \right)^2 \right)^{1/2}, \tag{1.24}$$

$$\dot{\varepsilon}_2 = \frac{\dot{\varepsilon}_x + \dot{\varepsilon}_y}{2} - \left(\frac{1}{4} (\dot{\varepsilon}_x - \dot{\varepsilon}_y)^2 + \left(\frac{\dot{\gamma}_{xy}}{2} \right)^2 \right)^{1/2}. \tag{1.25}$$

For purely plastic strains occurring with zero volume change, we have

$$\dot{\varepsilon}_1 + \dot{\varepsilon}_2 + \dot{\varepsilon}_3 = 0. \tag{1.26}$$

1.2. Yield conditions.

When the state of stress is uniaxial tension or compression, the yield condition for most metals is

$$\sigma = \pm \sigma_Y . \tag{1.27}$$

In a multiaxial state of stress, yielding will occur when a certain physical condition related to the state of stress will be satisfied. In uniaxial tension or compression, this condition must reduce to Equation (1.27). For metals, and particularly for mild steel, it has been observed that plastic deformations basically consist of slip in crystals. Hence, it was thought that the maximum shearing stress determined the onset of yielding, which was always to occur for a *fixed* value of the maximum shearing stress. This is Tresca's yield condition.

The value of the maximum shear stress at yielding can be obtained, for instance, from a tensile test where

$$\tau_{max} = \frac{\sigma}{2},$$

and hence,

$$\tau_{max,Y} = \frac{\sigma_Y}{2}, \tag{1.28}$$

In the case of plane stress defined by $\sigma_x = \sigma, \ \ \sigma_y = 0, \ \ \tau_{xy} = \tau$ (bending of beams with shear) we have

$$\tau_{max} = \frac{1}{2}\left(\sigma^2 + 4\,\tau^2\right)^{1/2}, \tag{1.29}$$

and Tresca's yield condition

$$\tau_{max} = \tau_{max,Y} \tag{1.30}$$

becomes, using Equation (1.28)

$$\sigma^2 + 4\,\tau^2 = \sigma_Y^2\,. \tag{1.31}$$

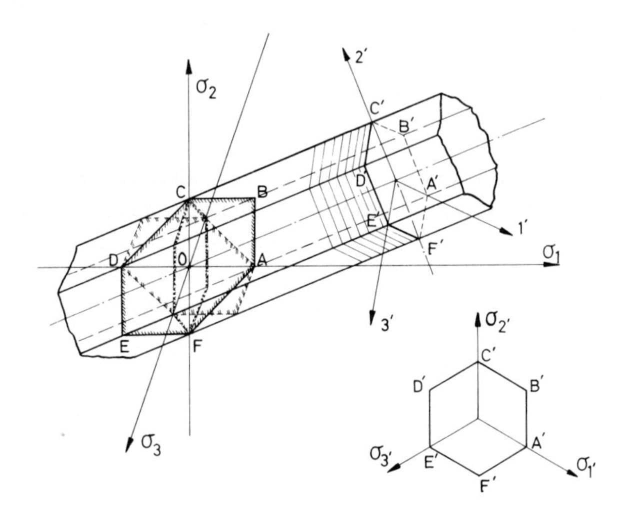

Fig. 1.5.

In plane orthogonal cartesian coordinates (O σ, O τ), relation (1.31) is represented by an ellipse.

In a multiaxial state of stress with principal stresses $\sigma_1, \sigma_2, \sigma_3$, the magnitude of the maximum shearing stress is the largest among the three absolute values

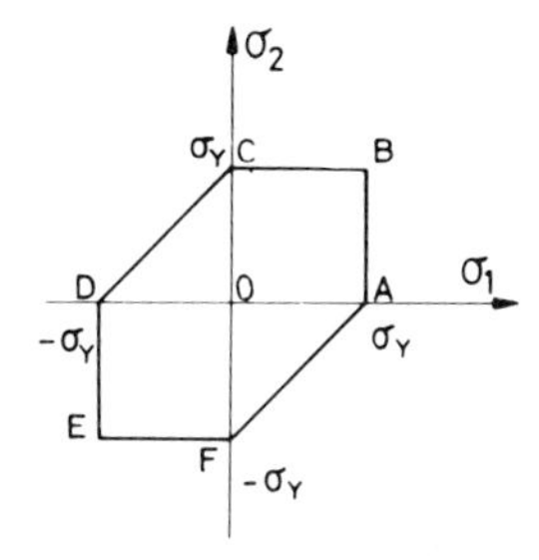

Fig. 1.6.

$$\frac{|\sigma_1 - \sigma_2|}{2}, \frac{|\sigma_2 - \sigma_3|}{2}, \frac{|\sigma_3 - \sigma_1|}{2}.$$

Consequently, condition (1.30) is represented, in cartesian orthogonal axes $(\sigma_1, \sigma_2, \sigma_3)$ by the hexagonal prism formed by the planes with equations

$$\sigma_1 - \sigma_2 = \pm \sigma_Y, \quad \sigma_2 - \sigma_3 = \pm \sigma_Y, \quad \sigma_3 - \sigma_1 = \pm \sigma_Y.$$

This prism is shown on fig. 1.5. Its axis is equally inclined with respect to the coordinate axes. When one of the principal stresses vanishes, say σ_3, the surface reduces to the hexagon obtained by intersecting the prism with the plane $\sigma_3 = 0$ (fig. 1.6). The yield condition becomes

$$\max\left[\,|\sigma_1|, |\sigma_2|, |\sigma_1 - \sigma_2|\,\right] = \sigma_Y. \tag{1.32}$$

Note finally that because the magnitude of the maximum shear stress is half the (algebraic) difference of the extreme principal stresses, the intermediate principal stress plays no role in Tresca's yield criterion.

More refined tests have however shown ([1.2, 1.4, 1.5, 1.6] and fig. 1.7) that the circular cylinder circumscribed to the considered hexagonal prism was a more exact "yield surface" for most metals. This surface represents the yield condition of Maxwell, Huber, Hencky, and von Mises, and will be simply called "the von Mises yield condition" in the following. The equation of this surface is

$$\sigma_1^2 + \sigma_2^2 + \sigma_3^2 - \sigma_1 \sigma_2 - \sigma_2 \sigma_3 - \sigma_3 \sigma_1 = \sigma_Y^2 \tag{1.33}$$

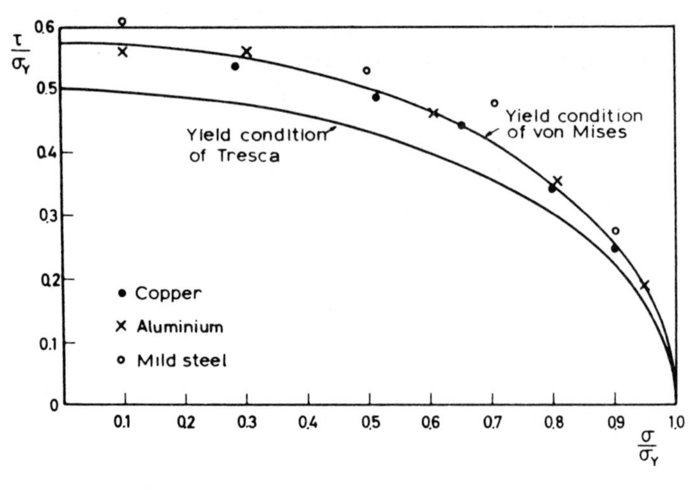

Fig. 1.7.

When the components σ_x ,...,τ_{xy} ,... of the stress tensor are used, condition (1.33) becomes

$$\sigma_x^2 + \sigma_y^2 + \sigma_z^2 - \sigma_x \sigma_y - \sigma_y \sigma_z - \sigma_z \sigma_x + 3\tau_{xy}^2 + 3\tau_{yz}^2 + 3\tau_{zx}^2 = \sigma_Y^2 \tag{1.34}$$

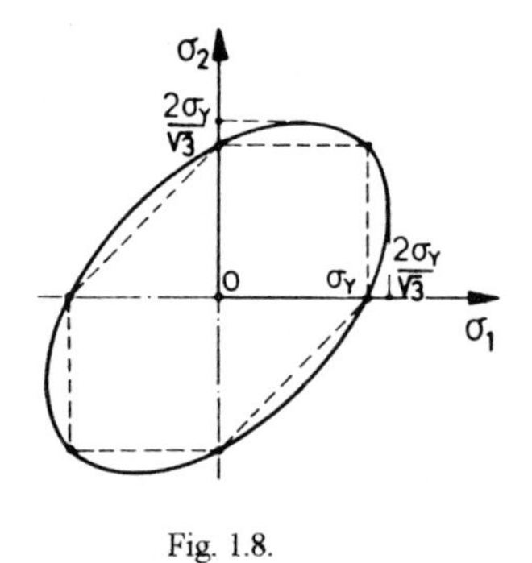

Fig. 1.8.

For a state of plane stress ($\sigma_3 = 0$) condition (1.33) represented by the ellipse of fig. 1.8 with equation :

$$\sigma_1^2 + \sigma_2^2 - \sigma_1 \sigma_2 = \sigma_Y^2 \tag{1.35}$$

In the particular state of plane stress where $\sigma_x = \sigma$, $\sigma_y = 0$, $\sigma_z = 0$ (bending of beams with shear), relation (1.34) transforms into

$$\sigma^2 + 3\,\tau^2 = \sigma_Y^2. \tag{1.36}$$

Therefore, the yield stress in pure shear is

$$\tau_Y = \frac{\sigma_Y}{\sqrt{3}}, \tag{1.37}$$

whence it had the magnitude $\frac{\sigma_Y}{2}$ according to Tresca's condition.

The yield condition of von Mises sometimes is called the criterion of the octahedral shear stress. The square root of the left-hand side of Equation (1.33) [or of Equation (1.34)] is indeed proportional to the magnitude of the octahedral shear stress τ_{oct} which acts on the octahedron, the faces of which make equal intercepts on the coordinate axes (fig. 1.9).

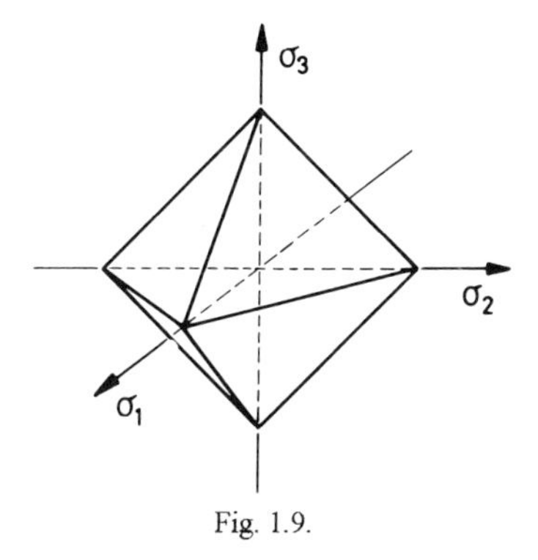

Fig. 1.9.

From a more general point of view, we call the reference stress σ_R the function of the stress components that must reach the value of the yield stress in simple tension (or compression) for yielding to occur. For the von Mises yield condition, σ_R is the square root of the left-hand sides of Equation (1.33) or (1.34), and the octahedral shearing stress is $\tau_{oct} = (\frac{\sqrt{2}}{3})\,\sigma_R$.

The two yield conditions of Tresca and of von Mises are the most commonly accepted for metals, but other conditions could be valid for other materials. Experiments will show which condition best describes a given material. Anyway, yielding can be expected to occur at a given particle when the reference stress at this particle attains the value σ_Y. This reference stress obviously is not a physical stress component but simply a reference value to compare with σ_Y. The general form of the yield condition thus is

$$\sigma_R(\sigma_x, \sigma_y, \sigma_z, \tau_{xy}, \tau_{yz}, \tau_{zx}) = \sigma_Y, \tag{1.38}$$

or, for an isotropic material,

$$\sigma_R(\sigma_1, \sigma_2, \sigma_3) = \sigma_Y. \tag{1.39}$$

1.3. Hooke's law.

For later use in elastic-plastic problems, Hooke's law of linear elasticity will be needed. We recall it in its classical form

$$\begin{aligned} \varepsilon_x &= \frac{1}{E}\left[\sigma_x - \nu\left(\sigma_y + \sigma_z\right)\right] \qquad \ldots (x, y, z) \\ \gamma_{xy} &= \frac{\tau_{xy}}{G} \qquad \ldots (x, y, z) \end{aligned} \tag{1.40}$$

whereas its matrix form can be written as

$$\boldsymbol{\varepsilon} = \mathbf{De}\ \boldsymbol{\sigma} \tag{1.40'}$$

where

$$\mathbf{D} = \frac{1}{E}\begin{vmatrix} 1 & -\nu & -\nu & 0 & 0 & 0 \\ -\nu & 1 & -\nu & 0 & 0 & 0 \\ -\nu & -\nu & 1 & 0 & 0 & 0 \\ 0 & 0 & 0 & 2(1+\nu) & 0 & 0 \\ 0 & 0 & 0 & 0 & 2(1+\nu) & 0 \\ 0 & 0 & 0 & 0 & 0 & 2(1+\nu) \end{vmatrix} \tag{1.41}$$

is the matrix of elasticity coefficients for an isotropic homogeneous body.

1.4. Problems.

1.4.1. show that the square of the octahedral shearing-stress magnitude is proportional to the elastic-distortion energy density (that is the difference between the total elastic energy density and the elastic energy density associated with volume change).

Answer :

$$\varepsilon_{dist} = \frac{3}{2}\left[\frac{(1+\nu)}{E}\right]\tau_{oct}^2.$$

1.4.2. For a closed thin tube with radius R and thickness t subjected to internal pressure p, find the maximum shearing stress and corresponding surface elements.

Answer :

$$\tau_{max} = \frac{pR}{2t}.$$

The corresponding surface elements are parallel to the axis of the tube and form angles of 45° with the radial direction.

1.4.3. Show that, in a state of plane stress, the expression $\left(\sigma_x - \sigma_y\right)^2 + 4\,\tau_{xy}^2$ is an invariant.

References.

[1.1] S.P. TIMOSHENKO, J.N. GOODIER, Theory of Elasticity, Mc Graw-Hill, New-York, 1951.

[1.2] A.A. ILIOUCHINE, Plasticité, Eyrolles, Paris, 1956.

[1.3] L. BAES, Résistance des Matériaux, vol. 1, Lamertin, Bruxelles, 1934.

[1.4] A. NADAI, Theory of Flow and Fracture of Solids, vol. 1, Mc Graw-Hill, New-York, 1950.

[1.5] W. SAUTER, A. KOCHENDÖRFER, V. DEHLINGER, Über die Gesetzmässigkeiten der plastischen Verformung von Metallen unter einem mehrachsigen Spannungszustand, (a dissertation) Stuttgart, 1952.

[1.6] J. MARIN, A.B. WISEMAN, Plastic Stress-Strain Relations for Combined Tension and Torsion, NACA Technical Note 2737, 1952.

[1.7] G.I. TAYLOR, H.QUINNEY, The Plastic Distortions of Metals, Phil. Trans. Roy. Soc. London, **230** : 323, 1931.

2. Fundamental Concepts and Laws.

2.1. Perfectly plastic solid.

We are now able to define a perfectly plastic solid as follows : a solid will be called perfectly plastic if it can undergo unlimited plastic deformations under constant reference stress when it is subjected to a homogeneous state of stress with $\sigma_R = \sigma_Y$.

The value σ_Y is well defined for each material in a given environment, and is the limiting value that σ_R cannot exceed. States of stress with $\sigma_R > \sigma_Y$ are not possible.

In the following we shall frequently be interested in *purely plastic* strains and strain rates and in the corresponding displacements and velocities, and we shall disregard the corresponding elastic elements (except when the contrary is explicitly stated).

When dealing with *incipient plastic flow*, we assume that strains remain very small. Hence, strains and displacement are related through eqs. (1.15) and (1.16), whereas strain rates derive from displacement rates (or velocities) through relations (1.22) and (1.23).

2.2. Power of dissipation.

During incipient plastic flow at a given particle, where the state of stress is described by (σ_x ,..., τ_{xy} ,...) and the state of strain rate by ($\dot{\varepsilon}_x$,..., $\dot{\gamma}_{xy}$,...), the power of the stresses per unit volume of material is

$$d = \sigma_x \dot{\varepsilon}_x + \ldots + \tau_{xy} \dot{\gamma}_{xy} + \ldots, \qquad (x, y, z). \qquad (2.1)$$

For purely plastic strain rates, this power is dissipated in heat during plastic flow. Therefore, it is called "power of dissipation". It is essentially *positive*.

2.3. Geometrical representation.

In a six-dimensional Euclidean space, consider a rectangular cartesian coordinate system with the origin O. The vectors $\boldsymbol{\sigma}$ and $\dot{\boldsymbol{\varepsilon}}$ that have the components σ_x, σ_y, σ_z, τ_{xy}, τ_{yz}, τ_{zx} and $\dot{\varepsilon}_x$, $\dot{\varepsilon}_y$, $\dot{\varepsilon}_z$, $\dot{\gamma}_{xy}$, $\dot{\gamma}_{yz}$, $\dot{\gamma}_{zx}$, respectively, with respect to this coordinate system, will be called the *stress vector* and the *strain rate vector*, and the point P with the radius vector **OP** = $\boldsymbol{\sigma}$ will be called the *stress point*. (An extension of these definitions to generalized stresses and strains will be discussed in Section 5.1.) Eq. (2.1) indicates an important property of this geometrical representation : the specific power of the stress on the strain rate is given by the scalar product of the vectors $\boldsymbol{\sigma}$ and $\dot{\boldsymbol{\varepsilon}}$:

$$d = \boldsymbol{\sigma} \,.\, \dot{\boldsymbol{\varepsilon}}. \tag{2.2}$$

If elastic strain rates are neglected so that $\dot{\boldsymbol{\varepsilon}}$ represents the plastic strain rate, this scalar product is the *specific rate of dissipation*, which will be denoted by D.

The surface with the equation

$$\sigma_R(\sigma_x,...,\tau_{xy},...) - \sigma_Y = 0 \tag{2.3}$$

is called the *yield surface*, because states of stress at the yield limit are represented by stress points on this surface. Note that for the perfectly plastic materials considered here, the reference stress depends only on the state of stress but not on the state of strain because these materials do not exibit work hardening. The yield surface is therefore a fixed surface in our six-dimensional space. The yield surface divides this space into two regions : the region $\sigma_R \leq \sigma_Y$ which consists of stress points representing attainable states of stress, and the region $\sigma_R > \sigma_Y$ which corresponds to states of stress that cannot be attained in the considered perfectly plastic material. For convenient reference, interior points of the attainable region will be described as lying inside the yield surface, while stress points representing unattainable states of stress will be described as being *outside the yield surface*. The coordinate origin, which represents the stress free state, must lie inside the yield surface because the material will not yield in the absence of stress.

2.4. Drucker's postulate, normality, and convexity of the yield surface.

Starting from general consideration about the behaviour of inelastic materials, Drucker [2.1, 2.2] has presented a general definition of a so-called standard stable material which distinguishes this type of material from the others, called non-standard. Figure 2.1 represents the various behaviours of inelastic material under positive increment of stress (fig. a and c) or strain (b). In Drucker's sense, material (a) is stable at point A because, deformed by the positive stress increment $\Delta\,\sigma$, it absorbs a positive work $\Delta\,\sigma \,.\, \Delta\,\varepsilon$. On the contrary, materials (b) and (c) are considered as instable because the corresponding work $\Delta\,\sigma \,.\, \Delta\,\varepsilon$ is negative. In general, Drucker's definition of a standard material is that *the net work performed during any arbitrary complete cycle in stress space is non-negative.*

Let us consider (figure 2.2) a particular loading cycle ABCDA of a perfectly plastic material.

The initial stress state $\boldsymbol{\sigma}^*$ si represented by a certain point A located inside the fixed load surface defined by equation (2.3).

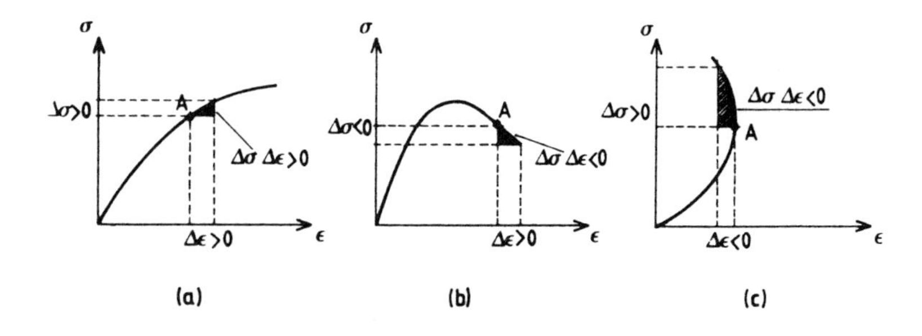

Fig. 2.1.

Under increasing external loads the stresses reach the value $\boldsymbol{\sigma}$ represented by the point C on the yield surface. Now suppose that the external loading keeps the stress state $\boldsymbol{\sigma}$ on the yield surface for a short time. Plastic flow must occur and plastic work takes place during this process. Then the loading releases and the stresses return to the state of stress $\boldsymbol{\sigma}^*$ along a certain path CDA inside the yield surface. As the total elastic work of a closed

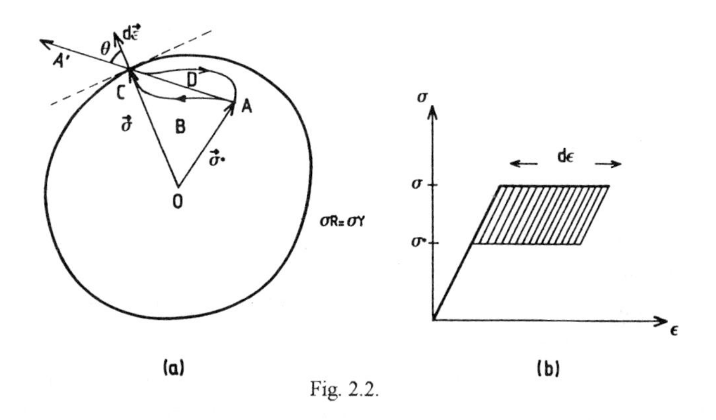

Fig. 2.2.

cycle vanishes the net work performed by the complete cycle is the scalar product of the stress vector $\boldsymbol{\sigma} - \boldsymbol{\sigma}^*$ by the plastic increment d $\boldsymbol{\varepsilon}$ as can be understood in the uniaxial case of fig. 2.2 b where the net work is represented by the shaded area.

Drucker's requirement (called sometimes postulate) is :

$$\left(\boldsymbol{\sigma} - \boldsymbol{\sigma}^*\right) d\,\boldsymbol{\varepsilon} \geq 0. \tag{2.4}$$

This inequality is also called Hill's principle of maximum local work [2.3]. If A is chosen infinitely near to C then

$$d\boldsymbol{\sigma} \, . \, d\boldsymbol{\varepsilon} \geq 0 \, . \tag{2.5}$$

The fundamental inequalities above have the following important consequences :

a) **convexity** : the yield surface is convex

b) **normality** : the yield vector of plastic strain increments is normal to the loading surface and directed toward the exterior of this surface. This property is expressed by

$$d\boldsymbol{\varepsilon} = d\lambda \, . \, \frac{\partial \sigma_R}{\partial \boldsymbol{\sigma}} \, . \tag{2.6}$$

We may write relation (2.6) under engineering form by replacing $d\lambda$ by $\lambda \, dt$:

$$\dot{\boldsymbol{\varepsilon}} = \lambda \, \frac{\partial \sigma_R}{\partial \boldsymbol{\sigma}} \, , \text{ or } \begin{cases} \dot{\varepsilon}_x = \lambda \, \dfrac{\partial \sigma_R}{\partial \sigma_x} , \ldots \, (x,y,z) \\ \dot{\gamma}_{xy} = \dfrac{\partial \sigma_R}{\partial \tau xy} , \ldots \, (x,y,z) \end{cases} \, . \tag{2.7}$$

where λ is a positive scalar factor. The normality laws (2.6), (2.7) are also called plastic potential flow law. It is widely accepted, though introduced in various manners [2.4, 2.5, 2.6]. Let us proceed to prove these two properties a) and b). Starting from a stress state A on the loading surface, inequality (2.4) is required for any additional stress $(\boldsymbol{\sigma} - \boldsymbol{\sigma}^*)$.

Therefore, the angle made by any vector $\mathbf{AC} = \boldsymbol{\sigma} - \boldsymbol{\sigma}^*$ with the vector $d\boldsymbol{\varepsilon}$ must be necessarily acute (figure 2.2). As $d\boldsymbol{\varepsilon}$ depend on the position of C but not on the position of A, this requirement is satisfied for all points A on the yield surface if and only if $d\boldsymbol{\varepsilon}$ is normal to the loading surface and this surface is convex (figure 2.2). Convexity means that a tangent plane at any point is on one side of the surface.

If both conditions of normality and convexity are not simultaneously satisfied, it is always possible to find a vector $\mathbf{AC} = \boldsymbol{\sigma} - \boldsymbol{\sigma}^*$ which forms with $d\boldsymbol{\varepsilon}$ an obtuse angle (figure 2.3 and 2.4).

1. If the yield surface is smooth, that is it may have a continuously turning normal (as does the von Mises yield surface), there then is one-to-one correspondence between the stress point and the flow mechanism $\dot{\boldsymbol{\varepsilon}}$ (figure 2.5) because the gradient of the yield surface is defined everywhere.

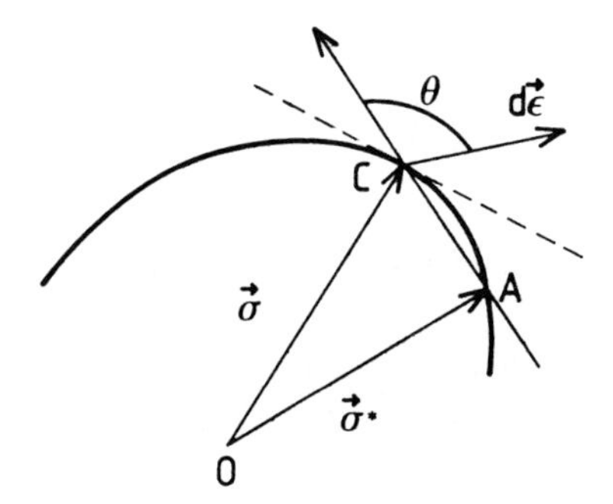

Fig. 2.3. Convexity satisfied by not normality, $(\boldsymbol{\sigma} - \boldsymbol{\sigma}^*) \cdot d\boldsymbol{\varepsilon} < 0$

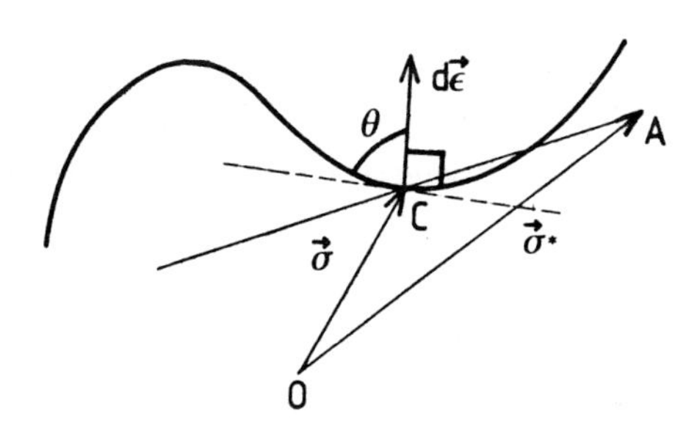

Fig. 2.4. Normality is satisfied but not convexity, $(\boldsymbol{\sigma} - \boldsymbol{\sigma}^*) \cdot d\boldsymbol{\varepsilon} > 0$

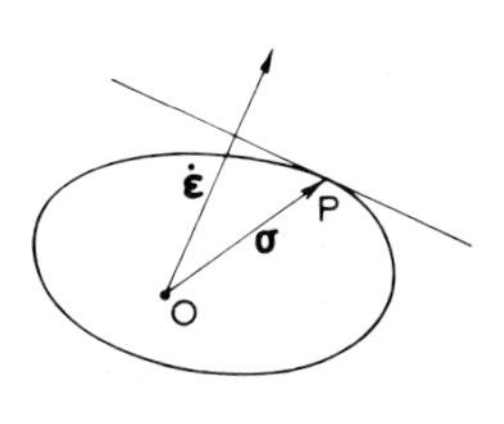

Fig. 2.5.

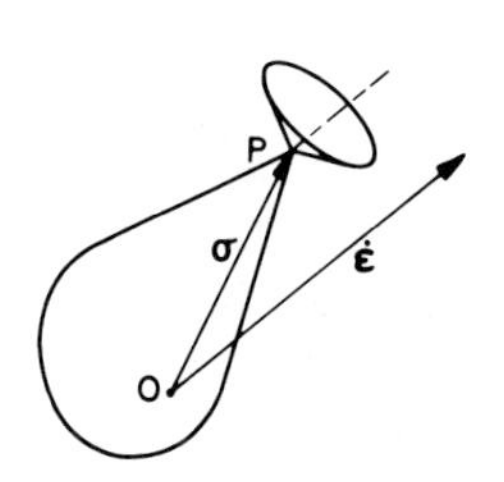

Fig. 2.6.

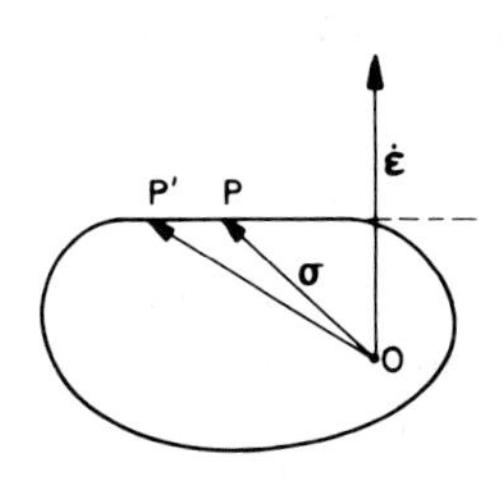

Fig. 2.7.

2. If the yield surface exhibits a vertex P, as in figure 2.6, all outward pointing vectors that lie on or within the cone of normals define possible flow mechanisms. The vector $\dot{\boldsymbol{\varepsilon}}$ of a flow mechanism still determines the stress point P but the converse is no longer true.

3. If the yield surface exhibits flat parts as in figure 2.7 (as does the Tresca yield surface), the stress point then determines the flow mechanism but the converse is not true when $\dot{\boldsymbol{\varepsilon}}$ is normal to a flat part of the yield surface.

2.5. Some properties of the specific power of dissipation.

1. Because the plastic deformation is an irreversible phenomenon the specific power of dissipation (2.2) where the stress vector $\boldsymbol{\sigma}$ and the strain rate vector $\dot{\boldsymbol{\varepsilon}}$ are now associated by the normality law, is non negative.

$$D = \boldsymbol{\sigma} \,.\, \dot{\boldsymbol{\varepsilon}} = \lambda\, \boldsymbol{\sigma} \,.\, \frac{\partial \sigma_R}{\partial \boldsymbol{\sigma}} \geq 0. \tag{2.8}$$

Since the scalar product of the vector $\boldsymbol{\sigma}$ on the yield surface and the exterior normal of the yield surface $\frac{\partial \sigma_R}{\partial \boldsymbol{\sigma}}$ is non negative (convexity of the yield surface), *the factor λ must always be non negative*. Note that the non negativeness of the plastic scalar factor λ is of thermodynamic origin, in opposition to the convexity of the yield surface.

2. *The specific power of dissipation is homogeneous of order one in the component of* $\dot{\boldsymbol{\varepsilon}}$.

In fact, the yield function σ_R is homogeneous. Assuming that the order is n, it follows from equation (2.8) and (2.3) that

$$D = \lambda\, n\, \sigma_Y. \tag{2.9}$$

As the specific power of dissipation depends linearly of λ as do the strain rate components (2.7), then

$$D\,(\lambda \dot{\boldsymbol{\varepsilon}}) = \lambda\, D\,(\dot{\boldsymbol{\varepsilon}}). \tag{2.10}$$

3. *The specific power of dissipation is a single-valued function of the strain rate vector.*

As λ may always be expressed as a single valued function of $\dot{\boldsymbol{\varepsilon}}$, by (2.9) it is the same for D.

$$D = D(\dot{\varepsilon}). \qquad (2.11)$$

In other terms, it is clear from figures 2.5 to 2.7 that whatever the type of the yield surface, the scalar product $\boldsymbol{\sigma} \cdot \dot{\boldsymbol{\varepsilon}}$ is single valued function of $\dot{\boldsymbol{\varepsilon}}$.

4. *The specific power of dissipation is a function which provides* $\boldsymbol{\sigma}$ *by gradient law.*

In fact, from (2.2) and (2.11)

$$\frac{\partial D}{\partial \dot{\varepsilon}} = \sigma + \lambda \, \frac{\partial \sigma}{\partial \dot{\varepsilon}} \cdot \frac{\partial \sigma_R}{\partial \sigma} = \sigma + \lambda \frac{\partial \sigma_R}{\partial \dot{\varepsilon}}.$$

Since in perfect plasticity the yield surface is independent of the strain rate $\dot{\boldsymbol{\varepsilon}}$, it follows that

$$\sigma = \frac{\partial D}{\partial \dot{\varepsilon}}. \qquad (2.12)$$

5. *The specific power of dissipation is a convex function in terms of the strain rate components.*

Consider two associated couples of vectors $(\sigma_1, \dot{\varepsilon}_1)$, $(\sigma_2, \dot{\varepsilon}_2)$ Drucker's postulate (2.4) gives :

$$\left(\sigma_2 - \sigma_1\right)\dot{\varepsilon}_2 \geq 0.$$

Subtracting $\sigma_1 \, \dot{\varepsilon}_1$ from each side of this inegality gives

$$\sigma_2 \, \dot{\varepsilon}_2 - \sigma_1 \, \dot{\varepsilon}_1 \geq \left(\dot{\varepsilon}_2 - \dot{\varepsilon}_1\right)\sigma_1.$$

Then, using (2.2) and (2.12)

$$D(\dot{\varepsilon}_2) - D(\dot{\varepsilon}_1) - \left(\dot{\varepsilon}_2 - \dot{\varepsilon}_1\right)\frac{\partial D}{\partial \dot{\varepsilon}_1} \geq 0$$

which proves the convexity.

If the strain rates at a particle are specified to within a common positive factor, they are said to determine a *local flow mechanism* at this particle. The strain rate vectors $\dot{\boldsymbol{\varepsilon}}$ and

$\lambda \dot{\boldsymbol{\varepsilon}}$ in eq. (2.10), where λ is positive, thus determine the same local flow mechanism, and this mechanism is completely define by the unit vector $\dot{\mathbf{e}}$ along $\dot{\boldsymbol{\varepsilon}}$.

In the following, we call *flow mechanism of a body* (or structure) a field of plastic strain rate vectors $\dot{\boldsymbol{\varepsilon}}$ whose magnitudes are defined to within a common positive scalar factor. Note that a given stress field may be related by the normality law to several fields of strain rate vectors $\dot{\boldsymbol{\varepsilon}}$. These fields have in common the field of unit vectors $\dot{\mathbf{e}}$.

When the solid is isotropic, at least as far as its yield condition is concerned, the normality law, expressed in the $(\sigma_x ,... \tau_{xy} ,...)$ and $(\dot{\varepsilon}_x ,... \dot{\gamma}_{xy} ,...)$ spaces, ensures that principal directions of the stress tensor and of the strain rate tensor coincide. Hence, when principal directions of these two tensors are known, or otherwise determinable, spaces of principal stresses and strain rates may as well be used. As a rule all spaces in which the dissipation is unambiguously determined may be used. All properties obtained above remain valid.

Finally, it should be noted that, the convexity of the yield surface and the normality rule can be derived from basic assumptions onthe specific dissipation, as introduced by Prager [2-8] [2-9]. Indeed, let us assume that :

1. *The specific power of dissipation is a single-valued function of the strain rate components.*

2. *The dissipation function $D\,[(\dot{\varepsilon})]$ is homogeneous of the order one.*

According to this second assumption, multiplying every strain rate component by a positive scalar λ means multiplying the power of dissipation by the factor λ :

$$D\,[(\lambda\dot{\varepsilon})] = \lambda\, D\,[(\dot{\varepsilon})]\,,\ \lambda\ \ 0.$$

Using the geometrical representation introduced above, we rewrite the preceding relation in the form

$$D\,[(\lambda\dot{\boldsymbol{\varepsilon}})] = \lambda\, D\,[(\dot{\boldsymbol{\varepsilon}})]\,,\ \lambda\ \ 0.$$

This second assumption expresses the inviscid nature of the considered perfectly plastic material.

Let a local flow mechanism $\dot{\mathbf{e}}$ be given. According to assumption 1 above, this flow mechanism determines a unique specific rate of dissipation $D\,(\dot{\mathbf{e}})$.

It follows from eq. (2.2) that a state of stress $\boldsymbol{\sigma}$ for which

$$\boldsymbol{\sigma}\dot{\mathbf{e}} < \mathrm{D}\ (\dot{\mathbf{e}})$$

cannot produce the flow mechanism $(\dot{\mathbf{e}})$.

Now, the stress points with radius vectors $\boldsymbol{\sigma}$ satisfying this last relation are interior points of the half-space that contains the origin 0 and is bounded by a plane normal to $\dot{\mathbf{e}}$ at the distance D $(\dot{\mathbf{e}})$from O. As we let the vector $\dot{\mathbf{e}}$ of the flow mechanism rotate about the origin, the interior points that all corresponding half-spaces have in common are the points inside the yield surface, and the bounding planes of the half-spaces envelope the yield surface.

As the boundary of the domain common to all half-spaces, the yield surface is convex.

2.6. Illustrating examples.

Let us first express the power of dissipation for the von Mises yield condition. We use the principal stresses and strain rates as components of the vectors $\boldsymbol{\sigma}$ and $\dot{\boldsymbol{\varepsilon}}$ (fig. 2.8). The yield surface is the circular cylinder with the eq. (1.33). Let P be a generic stress point on the surface and C the foot of the perpendicular from P on the axis of the cylinder. The line CP is normal to the cylinder at P. According to relation (2.3) the power of dissipation corresponding to the stress vector $\boldsymbol{\sigma} = \mathbf{OP}$ will be given by the modulus $|\dot{\boldsymbol{\varepsilon}}|$ of the strain rate vector $\dot{\boldsymbol{\varepsilon}} = \mathbf{OQ}$ (parallel to line CP) multiplied by the projection of $\mathbf{OP}$ on the direction of line $\mathbf{OQ}$. This projection has the length of CP, that is the magnitude $\left(\frac{2}{3}\right)^{1/2} . \sigma_Y$ of the radius of the cylinder. Hence,

$$D = \left(\frac{2}{3}\right)^{1/2} \sigma_Y \, |\mathbf{OQ}| = \left(\frac{2}{3}\right)^{1/2} \sigma_Y \left(\dot{\varepsilon}_1^2 + \dot{\varepsilon}_2^2 + \dot{\varepsilon}_3^2\right)^{1/2}. \tag{2.13}$$

Consider next Tresca's yield condition. When the three principal stresses are taken as rectangular cartesian coordinates in stress space, this yield condition is represented by a hexagonal prism (fig. 1.5). Fig. 2.9 shows the normal cross section through a point P of this prism that does not lie on an edge. The normal to the prismatic surface at P is the perpendicular η in the cross-sectional plane to the corresponding side of the regular hexagon. The projection of $\boldsymbol{\sigma} = \mathbf{OP}$ on η is the segment NP. Its magnitude is that of the distance from O to any side of the hexagon, namely $\frac{\sigma_Y}{\sqrt{2}}$. Consequently, we have

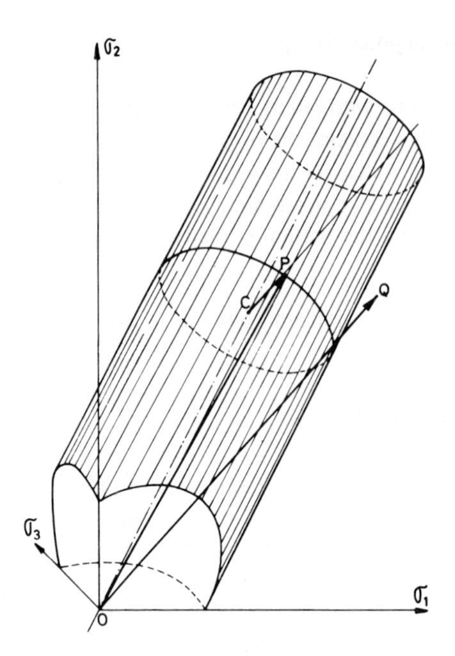

Fig. 2.8.

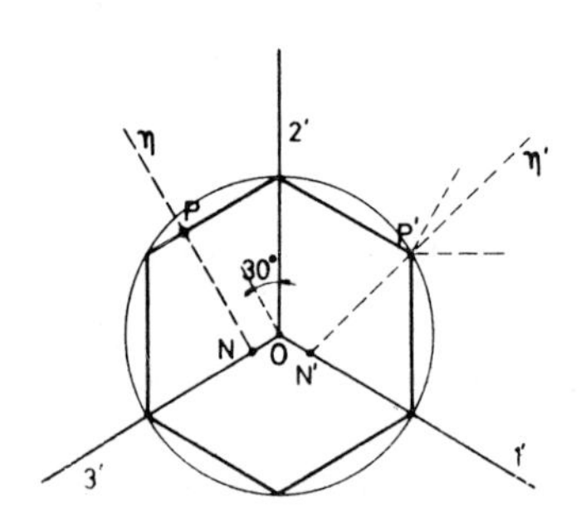

Fig. 2.9.

$$D = \boldsymbol{\sigma} \cdot \dot{\boldsymbol{\varepsilon}} = \frac{\sigma_Y}{\sqrt{2}} \, |\dot{\boldsymbol{\varepsilon}}|. \tag{2.14}$$

When the stress point lies on an edge of the yield prism (e.g., P' in fig. 2.9) and when the direction of the strain rate vector is intermediate between those of the normals to the adjacent planes (as η' on fig. 2.9), relation (2.14) is no longer valid. A strain rate vector with direction η' can be considered as resulting from the linear combination of strain-rate vectors corresponding to adjacent planes [2.6, 2.7]. Hence, we have the general expression

$$\dot{\boldsymbol{\varepsilon}} = \alpha\, \dot{\mathbf{e}}_a + \beta\, \dot{\mathbf{e}}_b, \tag{2.15}$$

where $\dot{\mathbf{e}}_a$ and $\dot{\mathbf{e}}_b$ are unit strain vectors for two adjacent planes, and α and β are arbitrary nonnegative scalar factors. From eqs. (2.3), (2.14) and (2.15) we obtain

$$D = \frac{\sigma_Y}{2}(\alpha + \beta). \tag{2.16}$$

Clearly, from relation (2.16) it is seen that the dissipation depends on the orientation of $\dot{\boldsymbol{\varepsilon}}$, except for vanishing α or β [when relation (2.14) is valid].

The preceding results can be more directly obtained when remembering that, as shown in Section 2.4, the dissipation for a unit vector $\dot{\mathbf{e}}$ is measured by the distance from the origin to the tangent plane to the yield surface at the stress point P. This distance is $\sigma_Y\left(\frac{2}{3}\right)^{1/2}$ for von Mises'condition. It is $\frac{\sigma_Y}{\sqrt{2}}$ for Tresca's except at points on the edged where it varies with the inclination of the tangent plane.

We finally want to show that, for the yield conditions of both von Mises and Tresca, *plastic flow occurs with no volume change*. To this purpose, it is sufficient to prove that

$$\dot{\varepsilon}_1 + \dot{\varepsilon}_2 + \dot{\varepsilon}_3 = 0. \tag{2.17}$$

As we have seen, the yield condition of von Mises is represented by a cylinder in the space of principal stresses, whereas Tresca's yield condition is represented by an hexagonal prism. But in both cases, the generatrices are normal to the plane with equation

$$\sigma_1 + \sigma_2 + \sigma_3 = 0. \tag{2.18}$$

Hence, any strain rate vector must be parallel to this plane and, consequently, its components must satisfy eq. (2.17).

For the last example, consider a state of plane stress such that the yield function is given by relation (1.35). The normality law (2.7) gives :

$$\dot{\varepsilon}_1 = \lambda(2\sigma_1 - \sigma_2), \qquad \dot{\varepsilon}_2 = \lambda(2\sigma_2 - \sigma_1)$$

so that

$$\sigma_1 = \frac{1}{3\lambda}(2\dot{\varepsilon}_1 + \dot{\varepsilon}_2), \qquad \sigma_2 = \frac{1}{3\lambda}(2\dot{\varepsilon}_2 + \dot{\varepsilon}_1).$$

Replacing these stress components into (1.35), one finds :

$$\lambda = \frac{1}{\sqrt{3}\cdot\sigma_Y}\left(\dot{\varepsilon}_1^2 + \dot{\varepsilon}_1\,\dot{\varepsilon}_2 + \dot{\varepsilon}_2^2\right)^{1/2}.$$

Using (2.9) one obtains the following specific power of dissipation

$$D\left(\dot{\varepsilon}_1, \dot{\varepsilon}_2\right) = \frac{\sigma_Y}{\sqrt{3}}\left(\dot{\varepsilon}_1^2 + \dot{\varepsilon}_1\,\dot{\varepsilon}_2 + \dot{\varepsilon}_2^2\right)^{1/2}. \tag{2.19}$$

We recognize that :

$$\sigma_1 = \frac{\partial D}{\partial \dot{\varepsilon}_1}, \qquad \sigma_2 = \frac{\partial D}{\partial \dot{\varepsilon}_2}.$$

2.7. Problems.

2.7.1. A perfectly plastic, incompressible solid obeying the von Mises yield condition is in a state of both plane stress ($\sigma_3 = 0$) and plane strain ($\dot{\varepsilon}_3 = 0$). Determine its yield limit σ_1.

Answer :

$$\sigma_1 = \pm\left(\frac{2}{\sqrt{3}}\right)\sigma_Y.$$

2.7.2. Apply the result of Problem 2.7.1 to find the ultimate plastic bending moment of a very long rectangular plate with thickness t, subjected to cylindrical bending.

Answer :

$$M_p = 1.155\,\sigma_Y\,\frac{t^2}{4}.$$

2.7.3. A very thin cylindrical shell with radius R and thickness t (under membrane stresses only) obeys the von Mises yield condition. It is subjected to uniform internal pressure p. Find its yield pressure

1. if it possesses heads,

2. if it is in plane strain (vanishing axial strain, incompressible material),

3. if the axial force vanishes.

Answer :

1. $$p=\left(\frac{2}{\sqrt{3}}\right)\left(\frac{\sigma_Y t}{R}\right)$$

2. $$p=\left(\frac{2}{\sqrt{3}}\right)\left(\frac{\sigma_Y t}{R}\right)$$

3. $$p=\frac{\sigma_Y t}{R}$$

2.7.4. Give the graphical representation of the yield conditions of von Mises and Tresca, respectively, with reference to principal axes O, σ_1, σ_2 for a perfectly plastic incompressible solid in plane strain.

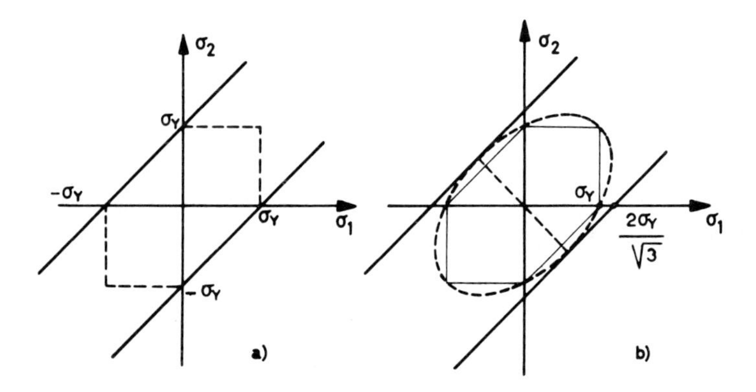

Fig. 2.10. (a) and (b)

Answer :

See fig. 2.10.

2.7.5. Same problem as in Problem 2.7.4, but for von Mises' condition only and with reference to cartesian orthogonal axes x, y, z such that $\dot{\varepsilon}_z = \dot{\gamma}_{xz} = \dot{\gamma}_{yz} = 0$.

Answer :

$$\left(\sigma_x - \sigma_y\right)^2 + 4\,\tau_{xy}^2 = \frac{4}{3}\,\sigma_Y^2, \qquad \text{(fig. 2.11).}$$

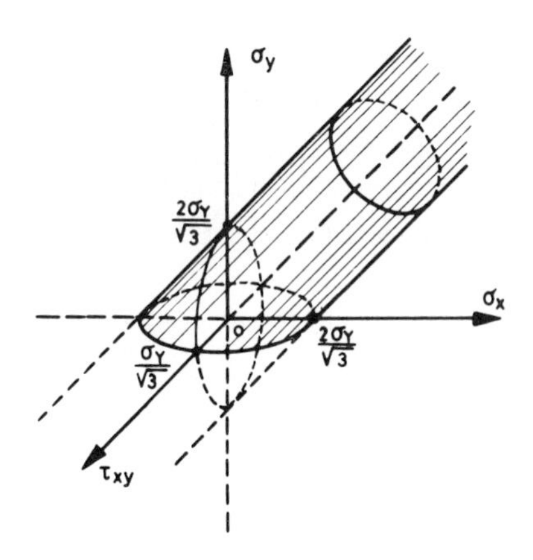

Fig. 2.11.

References.

[2.1] D. C. DRUCKER, "On uniqueness in the theory of plasticity", Quart. Appl. Math., **14** : 35, 1956.

[2.2] D. C. DRUCKER, "On the postulate of stability of material in mechanics of continua", J. de Mécanique, **3** : 235, 1964.

[2.3] R. HILL, "A variational principle of maximum plastic work in classical plasticity", Quant. J. Mech. Appl. Math., **1**, 18, 1948.

[2.4] R. von MISES, "Mechanik der plastichen Formänderung von Kristallen",

Zeitschrift angew, Math. Mech., **8** : 161, 1928.

[2.5] D.C. DRUCKER, "A more Fundamental Approach to Plastic Stress-Strain Relations", Proc.. Ist U.S. Nat. Congr. Appl. Mech., Chicago, 1951, 487 (J.W. Edwards, Ann Arbor, Mich., 1952).

[2.6] W.T.KOITER, "Stress-Strain Relations, Uniqueness and Variational Theorems for Elastic-Plastic Materials with a Singular Yield Surface", Quart. of Appl. Math., **11** : 350, 1953. See also J.L. SANDERS, "Plastic Stress-Strain Relations Based on Infinitely Many Plane Loading Surfaces", Proc. 2nd U.S. Nat. Congr. Appl. Mech., Ann Arbor, Mich., 1954, 455, A.S.M.E., NewYork, 1955.

[2.7] W. PRAGER, "On the Use of Singular Yield Conditions and Associated Flow Rules", J. of Appl. Mech., **20** : 317, 1953.

[2.8] W. PRAGER, "Théorie générale des états limites d'équilibre", J. de mathématiques pures et appliquées, **34** : 4 , 1955.

[2.9] W. PRAGER, "The General Theory of Limit Design", Proc. 8th Int. Cong. Appl. Mech., Istanbul, 1952, **2** : 65, 1956.

3. Fundamental Theorems.

3.1. Basic definitions and statements of theorems.

Let a structure be subjected to a system of loads that are increased *quasi-statically* and *in proportion*, starting from zero. Here the term "quasistatic" indicates that the loading process is sufficiently slow for all dynamic effects to be disregarded. The term "in proportion" signifies that the ratio of the intensities of any two loads remains constant during the loading process(*). The points of application and the lines of action of the loads, and the constant ratios of their intensities will be said to determine the *type of loading*. Choosing one of the loads, we use its magnitude P as a measure for the *intensity of loading*. The variable P will also be called the *loading parameter*.

For beams and frames the transition from purely elastic behaviour to contained plastic deformation and to unrestricted plastic flow is readily studied (see Com. V, Section 3.2). For more complex structures, however, this becomes quite difficult, and the emphasis is on the direct determination of the *limit state* in which the plastic deformation in the plastic zones is no longer contained by the adjacent nonplastic zones and the structure begins to flow under constant loads. The intensity of loading for this limit state is called the *limit load* ; it will be denoted by P_l

Limit analysis is concerned with these limits states. The limit state of incipient unrestrained plastic flow is characterized by three facts :

1. The stress field is

a) statically admissible : internal equilibrium, eq. (1.2), and equilibrium with the applied loads of parameter P, eq. (1.3), are satisfied ;

b) plastically admissible : The stresses nowhere violate the yield inequality

$$\sigma_R \leq \sigma_Y.$$

A stress field of this kind, both statically and plastically admissible, is called *licit* (**)

2. the strain rate field is

(*) The terms "proportional loading", "simple loading", and "radial loading" are used as alternatives to "loading in proportion".

(**) according to prof. Jean MANDEL's terminology.

a) kinematically admissible : it derives, by eq. 1.2.2, from a velocity field which satisfies the boundary conditions of the body ;

b) plastically admissible : The strain rate vectors belong to the set of normals to the yield locus and the external power of the load is positive.

A flow mechanism of this kind is called *licit.*

3. The stress vectors $\boldsymbol{\sigma}$ and the strain rate vectors $\dot{\boldsymbol{\varepsilon}}$ are *associated* by the normality law at every plastified point, and, consequently, the total power P of the applied loads is equal to the total dissipation D.

For a given type of loading, there is an infinity of licit stress fields. Each of these fields corresponds to a certain intensity of loading, which will be denoted by P_-.

Similarly, for a given licit mechanism and a given type of loading, an intensity of loading P_+ can be defined in such a manner that the power of the loads at this intensity of loading equals the power of dissipation in the yield mechanism.

The fundamental theorems of limit analysis can then be stated as follows :

Theorem 1 : *Statical (or lower bound) theorem : The limit load* P_l *is the largest of all loads* P_- *corresponding to licit stress fields.*

Theorem 2 : *Kinematical (or upper bound) theorem : The limit load* P_l *is the smallest of all loads* P_+ *corresponding to licit mechanisms.*

3.2. Proofs of the theorems.

We shall prove (*) the two fundamental theorems for the rigid perfectly plastic body (which can be regarded as an elastic perfectly plastic body with infinitely large elastic modulus). We shall then show that as far as limit loads are concerned, this idealization of the actual material is as satisfactory as the elastic perfectly plastic scheme. We assume proportional loading. Proofs are based on two preliminary theorems :

Theorem 3 : Virtual work : *If a continuum is in equilibrium, the virtual work of external forces (surface tractions and body forces) on any (infinitesimal) virtual displacement field compatible with the kinematic boundary conditions is equal to the virtual work*

(*) The following proofs are completely general, contrary to those given in Com. V., Section 3.4, for the particular case of beams and frames - .

of the internal forces (stresses) on the virtual strains corresponding to the considered virtual displacements.

The proof of this classical theorem is given in Section 3.8. A few remarks are worth noting :

1. If the structure under consideration is rigid, it must have at least one degree of freedom to admit nonvanishing virtual displacements. The work of the stresses on the rigid body displacements is zero, and the total work of the external forces is also equal to zero since they form an equilibrium system.

2. If the continuum is deformable, virtual deformations need only be compatible and infinitesimal ; that is, to derive from a sufficiently continuous (*) field of infinitesimal displacements satisfying the kinematic boundary conditions.

3. For incipient flow, displacements and velocities are proportional to each other and the theorem may be rephrased as a theorem of virtual power.

Theorem 4 : *Maximum dissipation : the power of dissipation D ($\dot{\boldsymbol{\varepsilon}}$) associated by the normality law to a given local flow mechanism $\dot{\boldsymbol{\varepsilon}}$ is larger than or at least equal to the (fictitious) power dissipated in this mechanism by any attainable state of stress.*

When referring to a local flow mechanism $\dot{\boldsymbol{\varepsilon}} = \mathbf{OQ}$, (fig. 3.1), this theorem expresses the inequality

$$\mathbf{OQ} \cdot \mathbf{OP} \geq \mathbf{OQ} \cdot \mathbf{OP'} \tag{3.1}$$

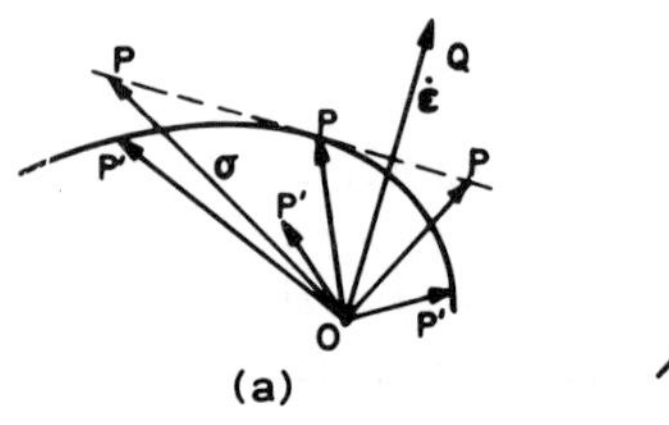

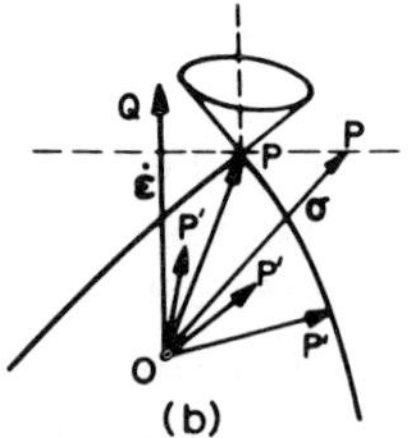

Fig. 3.1.

all possible stress points P' being on or within the yield surface. Inequality (3.1) is a direct consequence of convexity of the yield surface, coupled with the normality law. As inequality (3.1) is valid at every particle, it may be integrated over the entire volume of the body.

(*)Continuity conditions will be discussed in Section 5.6.

Denoting by $(\dot{\boldsymbol{\varepsilon}})$ the strain rate field, $(\boldsymbol{\sigma})$ the stress field corresponding to $(\dot{\boldsymbol{\varepsilon}})$ by the normality law (*) and $(\boldsymbol{\sigma}_*)$ any attainable stress field, we have

$$\text{Power of } (\boldsymbol{\sigma}) \text{ on } (\dot{\boldsymbol{\varepsilon}}) \geq \text{power of } (\boldsymbol{\sigma}_*) \text{ on } (\dot{\boldsymbol{\varepsilon}}). \quad (3.2)$$

For the sake of brevity, we shall symbolically write

$$P\,(\boldsymbol{\sigma}, \dot{\boldsymbol{\varepsilon}})$$

to mean : power of the field $(\boldsymbol{\sigma})$ on the field $(\dot{\boldsymbol{\varepsilon}})$. As the left-hand side of the inequality (3.2) is the power of dissipation $\int_V D[\dot{\boldsymbol{\varepsilon}}]\, dv$ of the mechanism, we finally have

$$\int_V D\,[\dot{\boldsymbol{\varepsilon}}]\, dv \geq P\,(\boldsymbol{\sigma}_*, \dot{\boldsymbol{\varepsilon}}). \quad (3.3)$$

We now proceed to prove the fundamental theorems.

Theorem 1 : Statical theorem : *Any load* P_- *corresponding to a licit stress field is smaller than or at most equal to the limit load* P_l.

Let $(\boldsymbol{\sigma}_-)$ be a licit stress field and P_- the corresponding load. In the limit state of incipient unrestrained plastic flow, let $(\boldsymbol{\sigma})$ be the actual stress field, and $(\mathbf{V})$ and $(\dot{\boldsymbol{\varepsilon}})$ the velocity and strain rate fields. Apply the theorem of virtual power to the stress field $(\boldsymbol{\sigma}_-)$ and $(\boldsymbol{\sigma})$ with the deformation state described by $(\mathbf{V})$ and $(\dot{\boldsymbol{\varepsilon}})$. We have, respectively,

$$P\,(\boldsymbol{\sigma}_-, \dot{\boldsymbol{\varepsilon}}) = P \text{ of } P_- \text{ on } (\mathbf{V}), \quad \text{(a)}$$

$$P\,(\boldsymbol{\sigma}, \dot{\boldsymbol{\varepsilon}}) = P \text{ of } P_l \text{ on } (\mathbf{V}). \quad \text{(b)}$$

As $(\boldsymbol{\sigma})$ corresponds to $(\dot{\boldsymbol{\varepsilon}})$ by the normality law, the left-hand side of eq. (b) is the dissipation $\int_V D[\dot{\boldsymbol{\varepsilon}}]\, dv$. We substract from the two sides of eq. (b) the corresponding sides of eq. (a), note that $(\boldsymbol{\sigma}_-)$ is an attainable stress field and hence apply eq. (3.2) to obtain

$$P \text{ of } P_l \text{ on } (\mathbf{V}) \geq P \text{ of } P_- \text{ on } (\mathbf{V}). \quad \text{(c)}$$

(*)It might not be unique, but the dissipation would remain uniquely determined by the $(\dot{\boldsymbol{\varepsilon}})$ field.

Because the velocity field (**V**) is common and the systems of loads differ only by a positive scalar factor, the inequality (c) furnishes immediately

$$P_{-} \leq P_{l}. \tag{3.4}$$

Theorem 2 : *Kinematical theorem : Any load* P_{+} *corresponding to a licit admissible mechanism is larger than or at least equal to the limit load.*

Let $(\mathbf{V}_{+})$ and $(\dot{\boldsymbol{\varepsilon}}_{+})$ be the considered licit velocity and strain rate fields. The corresponding load P_{+} is given, by definition, by the relation

$$\mathcal{P} \text{ of } P_{+} \text{ on } (\mathbf{V}_{+}) = \int_{V} D\left[(\dot{\boldsymbol{\varepsilon}}_{+})\right] d\,v, \tag{a}$$

where the dissipation D is a (well defined) single valued function of $(\dot{\boldsymbol{\varepsilon}}_{+})$ through the yield condition and the normality law. Let $(\boldsymbol{\sigma})$ be the actual stress field at the limit state, in equilibrium with P_{l}. The theorem of virtual powers, applied to the field $(\boldsymbol{\sigma})$ and the fields $(\mathbf{V}_{+})$ and $(\dot{\boldsymbol{\varepsilon}}_{+})$, gives

$$\mathcal{P} \text{ of } P_{l} \text{ on } (\mathbf{V}_{+}) = \mathcal{P}(\boldsymbol{\sigma}, \dot{\boldsymbol{\varepsilon}}_{+}). \tag{b}$$

Because $(\boldsymbol{\sigma})$ is an attainable stress field that as a rule does not correspond to the field $(\dot{\boldsymbol{\varepsilon}}_{+})$ by the normality law, relation (3.3) gives

$$\int_{V} D\left[\dot{\boldsymbol{\varepsilon}}_{+}\right] d\,v \geq \mathcal{P}(\sigma, \dot{\boldsymbol{\varepsilon}}_{+}). \tag{c}$$

Comparison of eqs. (a), (b) and (c) shows that

$$\mathcal{P} \text{ of } P_{l} \text{ on } (\mathbf{V}_{+}) \leq \mathcal{P} \text{ of } P_{+} \text{ on } (\mathbf{V}_{+}). \tag{d}$$

Because the field $(\mathbf{V}_{+})$ is common and the systems of loads differ only by a positive factor, the inequality (d) immediately furnishes

$$P_{l} \leq P_{+}. \tag{3.5}$$

Note finally that it is not necessary that the sets of values of P_{-} and P_{+} be continuous. The load P_{l} corresponding to a state that is simultaneously statically and kinematically licit belongs to both sets and is thus their unique common bound.

3.3 Important remarks.

3.3.1. *Exact value of the limit load (complete solution).*

Assume that we have found a licit stress field and a licit mechanism that correspond to the same load P. According to the two fundamental theorems we have $P \leq P_l$ and $P \geq P_l$. Hence, $P = P_l$, and we have obtained the exact limit load. This situation occurs most often when it is possible to associate a licit stress field and a licit mechanism by the plastic potential flow law. The work equation defining P_+ can then be regarded as a virtual work equation expressing the equilibrium of the associated stress field. Consequently $P_+ = P_-$, and denoting by P this common value, we have $P = P_l$.

The combined theorem is thus as follows :

Theorem 5 : *When it is possible to associate by the plastic potential flow law a licit stress field and a licit mechanism, the load* P *corresponding simultaneously to both fields is the exact limit load* P_l.

The two fields above form what is called a "complete solution" of the limit analysis of the structure. In practical problems, one starts either from a mechanism or from a licit stress field, and tries to obtain the other field. Examples for this technique are given in Part II of this volume.

3.3.2. *Addition or subtraction of material or of kinematic constraints.*

a) If the dimensions of a perfectly plastic structure are increased without changing the nature of the material, and if the dead weight of the additional material is neglected, the stress field consisting of vanishing stresses in the additional material and of the stress field at the limit state in the original structure is licit for the modified structure. According to the statical theorem we can hence state that *increasing the dimensions of a perfectly plastic structure cannot result in a lower limit load.*

Similarly, it could be shown that *decreasing the dimensions of a perfectly plastic structure cannot result in a larger limit load.*

b) *Addition of kinematic constraints (external or internal) cannot result in a lower limit load* because the flow mechanism of the modified body is licit for the original one. Conversely, decreasing the kinematical constraints cannot result in a larger limit load.

3.3.3. *Residual stresses.*

In the preceding sections, no assumption has been made concerning an initial stress-free state. The possible presence of residual stresses does not interfere with the proofs of the theorems, nor the existence of slight initial deformations, provided these deformations do not significantly change the geometry of the structure so that the equilibrium conditions can be set up without taking account of theses deformations. Hence we can state that *unknown initial stresses and deformations have no effect on the limit load provided they do not significantly alter the geometry of the structure.*

For example, a slight settling of the supports of a continuous beam, or a small permanent twist of a beam subjected to bending, or residual stresses generated by rolling or welding do not influence the limit load but only the load at which the behaviour of the structure ceases to be elastic (provided the plasic properties of the material are preserved).

Numerous experiments (see Com. V., Section 3.3.8 and 3.3.9) have supported this property, which has been implicitly used by practicing engineers for a long time.

3.3.4. *Constancy of the stresses during plastic flow.*

Considering an elastic perfectly plastic body we want to prove that *if all changes of geometry are disregarded, all stresses remain constant during the unrestrained plastic flow.*

Disregarding changes of geometry is justified because we are concerned exclusively with the *incipient* unrestrained plastic flow, of infinitesimal magnitude. This flow occurs under constant load P_l. Assume some changes ($\dot{\boldsymbol{\sigma}}$) in the stress field during the flow. They are in equilibrium with vanishing load changes, ($\dot{\mathbf{P}}$). Apply the theorem of virtual powers to these variations of loads and stresses with the velocity and strain rate fields at the limit state, (V) and ($\dot{\boldsymbol{\varepsilon}}$), respectively. We have

$$P \text{ of } \quad (\dot{\mathbf{P}}) \quad \text{ on } \quad (\mathbf{V}) = P\,(\dot{\boldsymbol{\sigma}}, \dot{\boldsymbol{\varepsilon}}). \tag{3.6}$$

Because all $\dot{\mathbf{P}}$ vanishes, the left-hand side of eq. (3.6) is equal to zero, and we obtain

$$P\,(\dot{\boldsymbol{\sigma}}, \dot{\boldsymbol{\varepsilon}}) = 0. \tag{3.7}$$

We divide the total strain rates into their elastic and plastic parts as follows :

$$\dot{\varepsilon}_x = \dot{\varepsilon}_x^e + \dot{\varepsilon}_x^p \ldots; \quad \dot{\gamma}_{xy} = \dot{\gamma}_{xy}^e + \dot{\gamma}_{xy}^p \ldots (x, y, z). \tag{3.8}$$

Relations (3.8) enables us to write eq. (3.7) in the following manner :

$$\int_V \left(\dot{\sigma}_x \dot{\varepsilon}_x^e + \ldots + \dot{\tau}_{xy} \dot{\gamma}_{xy}^e + \ldots \right) d\,v + \int_V \left(\dot{\sigma}_x \dot{\varepsilon}_x^p + \ldots + \dot{\tau}_{xy} \dot{\gamma}_{xy}^p + \ldots \right) d\,v = 0. \tag{3.9}$$

The second integral in eq. (3.9) vanishes because, for continuing flow, we must either have constant stress or a stress variation vector $\Delta \boldsymbol{\sigma}$ in the tangent plane to the yield surface normal to the strain rate vector $\dot{\boldsymbol{\varepsilon}}$, and therefore $\dot{\sigma}_x \dot{\varepsilon}_x^p + ... + \dot{\tau}_{xy} \dot{\gamma}_{xy}^p + ... = 0$. Because the elastic strain rates are related to the stress rates by Hooke's law, $\left(\dot{\sigma}_x \dot{\varepsilon}_x^e + ... + \dot{\tau}_{xy} \dot{\gamma}_{xy}^e + ...\right)$ will be positive unless,

$$\dot{\sigma}_x = 0 \qquad ... (x, y, z),$$

$$\dot{\tau}_{xy} = 0 \qquad ... (x, y, z), \tag{3.10}$$

(see [3.1]* for proof). As the first integral must vanish, conditions (3.10) are necessarily satisfied at all points.

3.3.5. *Final remark.*

In certain circumstances, nonassociated flow rules are used that derive the strain rate components from a potential function not identical with the yield condition (for example, the plastic potential of von Mises with the yield condition of Tresca). Obviously, the fundamental theorems proved in Section 3.2 and their corollaries, particularly the theorem of Section 3.3.4 [3.2], are no longer valid.

3.4. Elastic-plastic and rigid-plastic bodies.

The two fundamental theorems were introduced by Gvozdev [3.3], Hill [3.4] and Prager [3.5], in the case of the rigid perfectly plastic body, and by Drucker, Prager and Greenberg [3.6] for the elastic perfectly plastic material. Both idealizations are appropriate for the purposes of limit analysis [3.5].

When the elastic-plastic idealization is used, the limit state corresponds to incipient unrestrained plastic flow. The graph of a relevant displacement δ versus the applied load P [fig. 3.2 (a)] is initially a ray OA (elastic range), then a curve AB (elastic-plastic range : restricted plastic flow) and finally, when plastification has spread sufficiently through the body, the graph becomes a parallel to the axis, representing the unrestrained plastic flow.

When using the rigid-plastic idealization, the rigid part of the body prevents all deformations up to the onset of unrestrained plastic flow at the limit load P_l, sometimes also called the *yield-point load* [fig. 3.2 (b)].

On the other hand, *the fundamental theorems of limit analysis are absolutely identical for both idealizations*. They are based exclusively on the concepts of licit stress fields and licit plastic strain rate fields, with no reference to the elastic or rigid nature of the material not at yield. Lower bound P_-, upper bound P_+, and complete solutions (as described in Section 3.3.1) are valid for both idealizations.

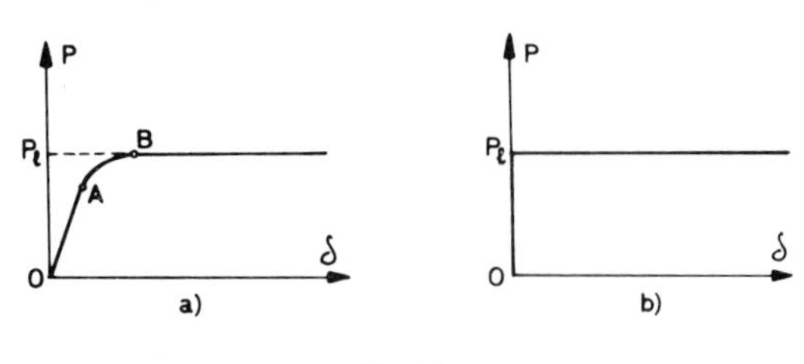

Fig. 3.2.

Why the elastic-plastic idealization is not better than the rigid-plastic idealization is easily understood, if we remember that the former must assume vanishingly small elastic and elastic-plastic strains so that the known undeformed geometry of the structure can be used up to the limit state.

3.5. Influence of changes of geometry.

Consider a rigid perfectly plastic structure at incipient unrestrained flow. The flow mechanism is supposed to be known. The velocity field is thus determined to within a positive scalar factor. Assuming that the flow mechanism does not change (*), the displacement field has the same form as the velocity field, the successive deflected shapes transforming one into the other by similarity. The magnitudes of all displacement vectors are monotonically increasing functions of a scalar parameter that can be regarded as measuring the time t. The work W of the applied loads and the energy E dissipated in plastic deformations are functions of t only, because the mechanism is given. Let $t = 0$ at impending plastic flow and consider the deflected shape at the end of the first time interval, taken as the unit of time. If this interval is small enough, we need only retain the first term in the series expansion of W (t).

From the theorem of virtual work $W = E$, we obtain

$$1 \,.\, k \,.\, P_1 = E\,(1), \tag{3.11}$$

(*)Note however that, in the course of charges of geometry, bifurcation of collapse mechanis ms may occur

where k is a proportionality factor characteristic of the mechanism.

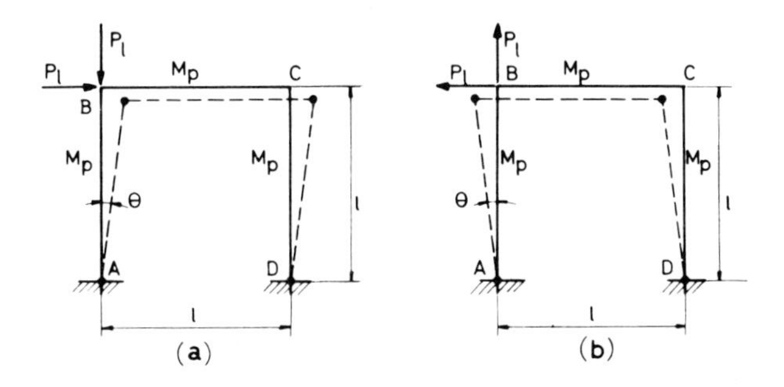

Fig. 3.3.

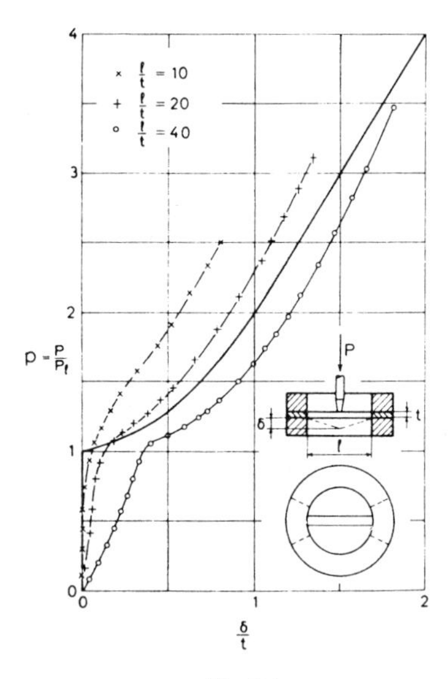

Fig. 3.4.

On the other hand, after a time t large enough to force us to retain two terms in the expansion of W (t), we have

$$W(t) = (t + A\,t^2)\,k\,.\,P_l \tag{3.12}$$

whereas, according to eq. (2.4),

$$E(t) = t\,E(1). \tag{3.13}$$

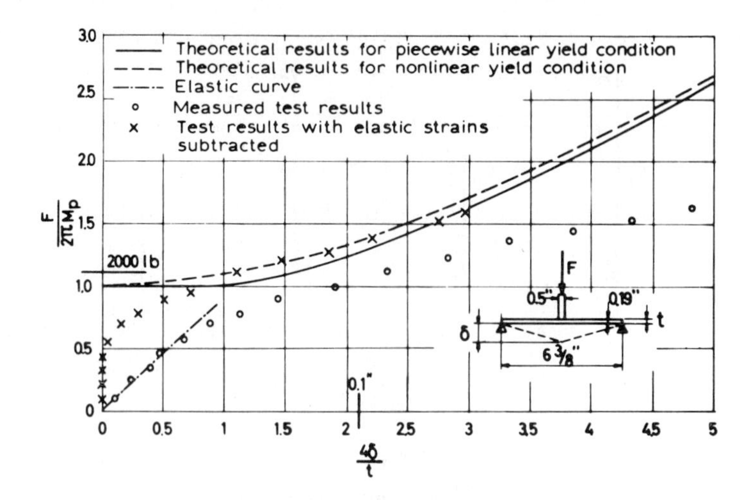

Fig. 3.5. from Plastic Analysis of Structures by P.G. Hodge Jr. Copyright 1959, Mc Graw-Hill Book Company. Used by permission of Mc Graw-Hill Book Company, Inc.

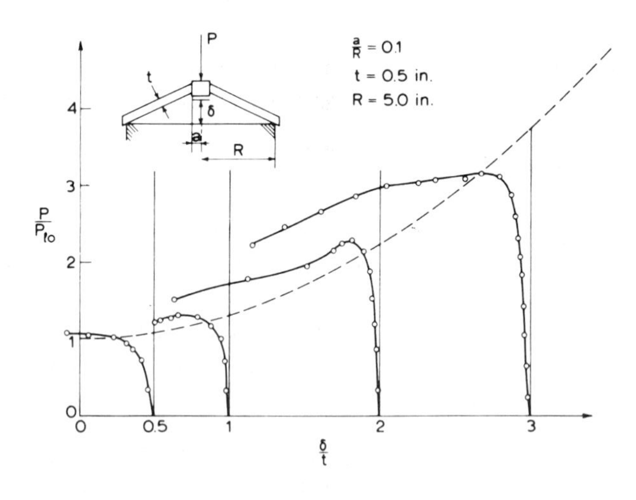

Fig. 3.6.

Substracting from eq. (3.12) the corresponding sides of eq. (3.13) and using eq. (3.11), we obtain

$$W(t) - E(t) = At^2 k p_1 \tag{3.14}$$

When A is positive, as in fig. 3.3 (a), the work of the loads is larger than the work dissipated and the limit state is an unstable (collapse) state [3.7]. In the case of fig. 3.3 (b), A is negative, the energy dissipated is larger than the work of the loads, and the limit state is stable. Increasing loads are needed to maintain plastic flow.

Fig. 3.4 and 3.5 are the graphs of load versus deflection for a built-in beam [3.8] and a circular simply supported plate [3.9], respectively.

Fig. 3.6 gives similar curves for a truncated conical shell subjected to a uniform load at the upper edged [3.10]. The dashed curved showing the theoretical load versus the deflection is obtained by taking account of the changes of geometry [3.11].

In fig. 3.4 and 3.5 the changes of geometry have a favorable effect on the strength of the structure whereas in fig. 3.6 the limit state is unstable and results in plastic collpase of the structure. Experimental and theoretical curves differ for two main reasons :

1. The solid is not rigid-plastic but elastic-plastic. If the structure is elastically very deformable and highly redundant, deformations for loads smaller than P_1 may already influence the effects of the forces : examples are string and membrane effects in beams and plates primarily subjected to bending, and instability by divergence of structures under compression.

2. When large plastic deformations occur, the real material exhibits some work-hardening that was neglected in the theory.

3.6. Solutions given by the deformation theory.

Whereas limit analysis is only concerned with the determination of the intensity of loading at the onset of unrestrained plastic flow, we may be interested in studying the elastic and elastic-plastic states that precede this onset of flow. When the potential-flow law (2.6) is applied to the yield condition of von Mises,

$$\sigma_R^2 \equiv \sigma_x^2 + \sigma_y^2 + \sigma_z^2 - \sigma_x\sigma_y - \sigma_y\sigma_z - \sigma_z\sigma_x + 3\left(\tau_{xy}^2 + \tau_{yz}^2 + \tau_{zx}^2\right) \equiv \sigma_Y^2 \qquad (3.15)$$

we find

$$\dot{\varepsilon}_x \equiv \lambda(2\,\sigma_x - \sigma_y - \sigma_z) \qquad ...(x, y, z),$$

$$\dot{\gamma}_{xy} \equiv 6\,\lambda\,\tau_{xy} \qquad ...(x, y, z). \qquad (3.16)$$

Indeed, we have

$$\frac{\partial \sigma_R}{\partial \sigma_x} = \frac{\partial \left(\sigma_R^2 \right)^{1/2}}{\partial \sigma_x} = \frac{1}{2\sigma_Y} \frac{\partial \sigma_R^2}{\partial \sigma_x} \qquad \ldots(x,y,z),$$

$$\frac{\partial \sigma_R}{\partial \tau_{xy}} = \frac{1}{2\sigma_Y} \frac{\partial \sigma_R^2}{\partial \tau_x} \qquad \ldots(x,y,z); \tag{3.17}$$

and the constant $\frac{1}{2\,\sigma_Y}$ may be included in the positive scalar factor λ.

Introducing the stress deviator with normal components

$$s_x = \sigma_x - \frac{\sigma_x + \sigma_y + \sigma_z}{3} = \frac{2\,\sigma_x - \sigma_y - \sigma_z}{3} \qquad \ldots(x, y, z), \tag{3.18}$$

and the same shear components as the stress tensor, eqs. (3.16) become

$$\dot{\varepsilon}_x = 3\,\lambda\, s_x \qquad \ldots(x, y, z),$$

$$\dot{\gamma}_{xy} = 6\,\lambda\, \tau_{xy} \qquad \ldots(x, y, z). \tag{3.19}$$

With relations (3.18), the yield condition (3.15) can be written as

$$\frac{3}{2}\left(s_x^2 + s_y^2 + s_z^2 \right) + \left(\tau_{xy}^2 + \tau_{yz}^2 + \tau_{zx}^2 \right) = \sigma_Y^2 . \tag{3.20}$$

Substitution of expression (3.19) for s_x and τ_{xy} in (3.20) gives

$$\lambda^2 = \frac{1}{18\,\sigma_Y^2}\left[3\left(\dot{\varepsilon}_x^2 + \dot{\varepsilon}_y^2 + \dot{\varepsilon}_z^2 \right) + \frac{3}{2}\left(\dot{\gamma}_{xy}^2 + \dot{\gamma}_{yz}^2 + \dot{\gamma}_{zx}^2 \right)\right]. \tag{3.21}$$

With the condition of plastic incompressibility,

$$\dot{\varepsilon}_x + \dot{\varepsilon}_y + \dot{\varepsilon}_z = 0,$$

the bracket in eq. (3.21) can be written, after subtracting $(\dot{\varepsilon}_x + \dot{\varepsilon}_y + \dot{\varepsilon}_z)^2$, as

$$\left(\dot{\varepsilon}_x - \dot{\varepsilon}_y\right)^2 + \left(\dot{\varepsilon}_y - \dot{\varepsilon}_z\right)^2 + \left(\dot{\varepsilon}_z - \dot{\varepsilon}_x\right)^2 + \frac{3}{2}\left(\dot{\gamma}_{xy}^2 + \dot{\gamma}_{yz}^2 + \dot{\gamma}_{zx}^2\right).$$

Introducing the strain-rate intensity

$$\dot{J} = \frac{\sqrt{2}}{3}\left[\left(\dot{\varepsilon}_x - \dot{\varepsilon}_y\right)^2 + \left(\dot{\varepsilon}_y - \dot{\varepsilon}_z\right)^2 + \left(\dot{\varepsilon}_z - \dot{\varepsilon}_x\right)^2 + \frac{3}{2}\left(\dot{\gamma}_{xy}^2 + \dot{\gamma}_{yz}^2 + \dot{\gamma}_{zx}^2\right)\right]^{1/2} \tag{3.22}$$

we write eq. (3.21) in the form

$$\lambda = \frac{\dot{J}}{2\,\sigma_Y}. \tag{3.23}$$

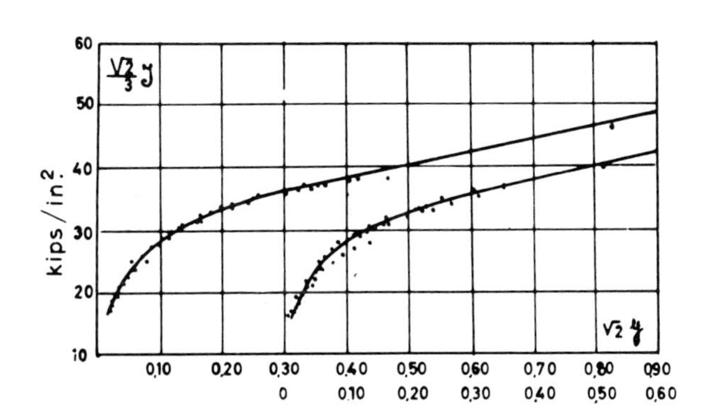

fig. 3.7.

Hence, eqs. (3.19) become

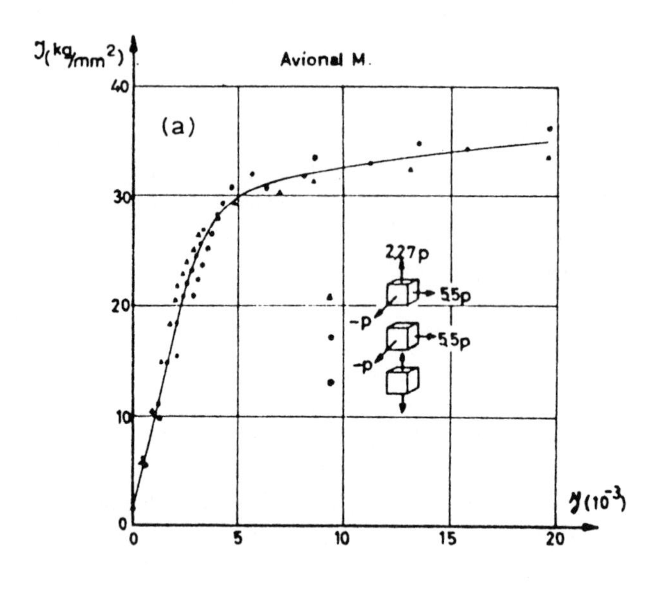
Avional M.
(a)
2,27p
5,5p
-p
5,5p
-p
0
10
20
30
40
0
5
10
15
20

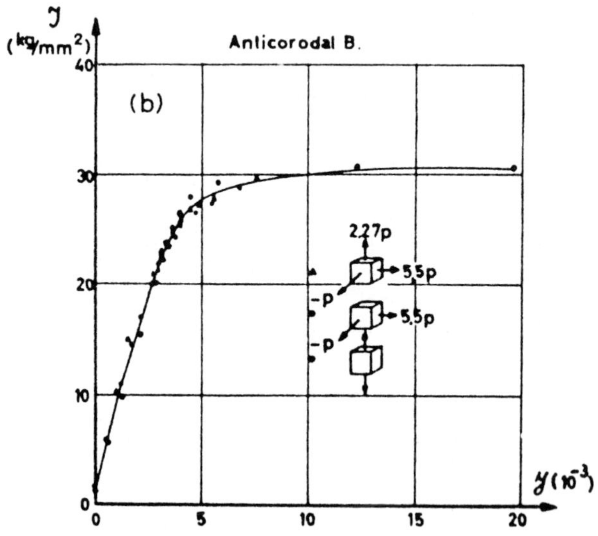
Anticorodal B.
(b)
2,27p
5,5p
-p
5,5p
-p
0
10
20
30
40
0
5
10
15
20

fig. 3.8.

$$\dot{\varepsilon}_x = \frac{3\dot{J}}{2\sigma_Y} s_x \qquad \ldots(x, y, z),$$

$$\dot{\gamma}_{xy} = \frac{3\dot{J}}{\sigma_Y} \tau_{xy} \qquad \ldots(x, y, z). \tag{3.24}$$

Eqs. (3.24) are the classical Levy-von Mises equations. Note that they involve purely plastic strain rates (rigid-plastic body). To account for elastic strains, one must divide total strain rates into their elastic and plastic parts,

$$\dot{\varepsilon}_x = \dot{\varepsilon}_x^e + \dot{\varepsilon}_x^p \qquad \ldots(x, y, z),$$

$$\dot{\gamma}_{xy} = \dot{\gamma}_{xy}^e + \dot{\gamma}_{xy}^p \qquad \ldots(x, y, z), \tag{3.25}$$

as in Section 3.3.4. Eqs. (3.24) furnish the plastic parts and differentiation of Hooke'law with respect to time the elastic parts. In this manner, one obtains the Prandtl-Reuss equations (see [3.13], pp. 28-29). Solutions of most practical problems with the Mises or Prandtl-Reuss equations unfortunately turn out to be very difficult [3.12, 3.13].

Some authors approach the problem from a different point of view [1.2, 3.14]. Their starting point is the experimental evidence that for any loading such that all the components of the stress tensor increase in proportion, there is a unique relation between the "intensity",

$$J = \frac{\sqrt{2}}{3}\left[\left(\varepsilon_x - \varepsilon_y\right)^2 + \left(\varepsilon_y - \varepsilon_z\right)^2 + \left(\varepsilon_z - \varepsilon_x\right)^2 + \frac{3}{2}\left(\gamma_{xy}^2 + \gamma_{yz}^2 + \gamma_{zx}^2\right)\right]^{1/2}$$

of the *total* strains, and the stress intensity,

$$I = \frac{1}{\sqrt{2}}\left[\left(\sigma_x - \sigma_y\right)^2 + \left(\sigma_y - \sigma_z\right)^2 + \left(\sigma_z - \sigma_x\right)^2 + 6\left(\tau_{xy}^2 + \tau_{yz}^2 + \tau_{zx}^2\right)\right]^{1/2} \tag{3.27}$$

as shown on figs. 3.7 taken from references [1.2] and [3.15], respectively. Consequently, the adopted fundamental equations are([1.2], p. 98)

$$\varepsilon_x - e = \frac{3}{2}\frac{J}{I}s_x \qquad \ldots(x, y, z),$$

$$\gamma_{xy} = 3\frac{J}{I}\tau_{xy} \qquad \ldots(x, y, z), \tag{3.28}$$

where

$$e = \frac{\varepsilon_x + \varepsilon_y + \varepsilon_z}{3}, \tag{3.29}$$

and

$$J = J(I).$$

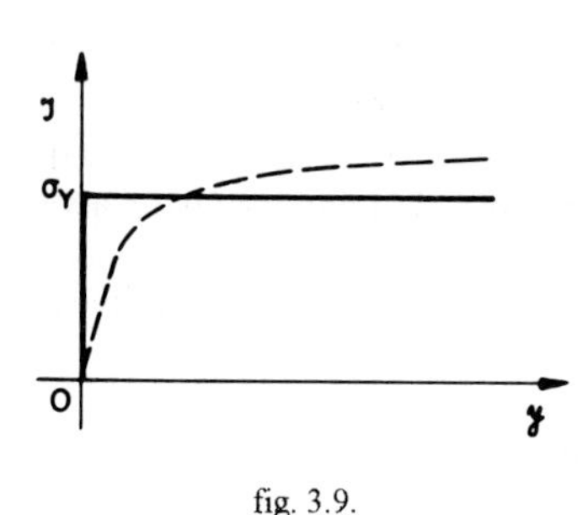

fig. 3.9.

The function $J(I)$ is characteristic of the material. The range of applicability of this second theory of plasticity, called *deformation theory*, is narrow compared to that of the former *flow theory* ([3.16, 1.2]), but because of the simple form of eq. (3.28), many practical problems have been solved by the deformation theory. If we consider a perfectly plastic structure from the point of view of the deformation theory, and if we want to determine its "carrying capacity", we may neglect the elastic strains ([1.2], pp. 170 and 232). Hence we let e=0 because plastic volume change vanishes. For perfect plasticity the relation $J=J(I)$ simply becomes (fig. 3.9)

$$J = \sigma_Y. \tag{3.31}$$

From the definition (3.27) of I, eq. (3.31) is yield condition of von Mises. Since the strains are purely plastic, eqs. (3.28) reduce to

$$\varepsilon_x = \frac{3}{2}\frac{J}{\sigma_Y}s_x \qquad \ldots(x, y, z),$$

$$\gamma_{xy} = 3 \frac{J}{\sigma_Y} \tau_{xy} \qquad \ldots(x, y, z), \tag{3.32}$$

Comparing eqs. (3.32) with eqs(3.24), we note that *they are formally identical except that the former deal with plastic strains and the latter with plastic strain rates.*

Now imagine that the "carrying capacity" of a structure has been determined from eqs. (3.32). This means that there has been found :

1. an equilibrium stress field satisfying $\sigma_R \leq \sigma_Y$,

2. a field of plastic strains at impending unrestrained plastic flow.

From the point of view of limit analysis, the stress field is licit. The strain field, *when regarded as a strain rate field* defined except for a constant scalar factor, *specifies a licit mechanism that corresponds to the stress field by the plastic potential flow law* (3.24), because of the formal identity of eqs. (3.32 and 3.24). Consequently, *a "carrying capacity" determined by the deformation theory is an exact limit load for limit analysis.*

This result justifies including in our solutions those obtained by the deformation theory(*)

Two final points are worth nothing :

1. The "carrying capacity" is often obtained solely from a statically admissible stress field. It must then be regarded as a lower bound P_- to P_l. Nevertheless it is in some instances reasonable to expect that one obtains the "best" bound (or at least a very good bound) in this manner.

   This is the case in plate problems where plastic flow occurs at all points.

2. The "carrying capacity" is sometimes obtained with the yield condition of Tresca and eqs. (3.32). That carrying capacity must be regarded exclusively as a lower bound P_- as long as a mechanism has not been found that corresponds to the stress field by the plastic potential flow law *applied to the yield condition of Tresca.*

(*) The formation analogy of eqs. (3.24) and (3.32) also explains that the yield surfaces obtained by Ilouchine [1.2] are identical to those obtained on the basis of plastic potential ([5.8, 5.9],...).

3.7. Uniqueness.

As already noted at the end of Section 3.2, the limit load for proportional loading P_l is unique. Indeed, consider a load P^* that simultaneously corresponds to a licit stress field $(\boldsymbol{\sigma})$ and a licit strain rate field $(\dot{\boldsymbol{\varepsilon}})$. Let us assume that there exist several limit loads. The proofs of the fundamental theorems show that P^* must be equal to any one of them. Hence, P_l is unique and coincides with P^*.

Are the fields $(\boldsymbol{\sigma})$ and $(\dot{\boldsymbol{\varepsilon}})$ also unique? As was emphasized at the end of Section 2.4, to the field $(\boldsymbol{\sigma})$ there may correspond by the normality law several mechanisms, all of them furnishing the same load P by their work equation. Moreover, it was also noted in Com. V. for beams and frames that different mechanisms could result in the same exact limit loads (e.g., Section 4.2.4). Hence, *the flow mechanism clearly need not be unique*.

When two complete solutions $(\boldsymbol{\sigma})_1$, $(\dot{\boldsymbol{\varepsilon}})_1$ and $(\boldsymbol{\sigma})_2$, $(\dot{\boldsymbol{\varepsilon}})_2$ are known, it can be proved [3.5, 3.17] that the stress fields of the two solutions are identical except possibly

1. in the common rigid regions,

2. where both states of stress are represented by stress points on the same flat part of the yield surface.

3.8. Appendix.

Proof of the theorme of virtual works.

Consider a continuous body in equilibrium under the applied forces and the reactions at the supports (fig. 3.10). Imagine a virtual field of displacements $\mathbf{U}$ of the points of the continuum. Except that $\mathbf{U}$ must be a continuous function of position satisfying the kinematic boundary conditions of the body, this displacement field may otherwise be chosen arbitrarily.

Evaluate the work done on these displacements by the surface tractions $\mathbf{T}$ per unit area (including the reactions of the supports if they do work) and the body forces $\mathbf{F}$ per unit volume. This work W_e is given by

$$W_e = \int_A \mathbf{T} \cdot \mathbf{U} \, dA + \int_V \mathbf{F} \cdot \mathbf{U} \, dV , \tag{a}$$

where A and V are the area and the volume of the body, respectively.

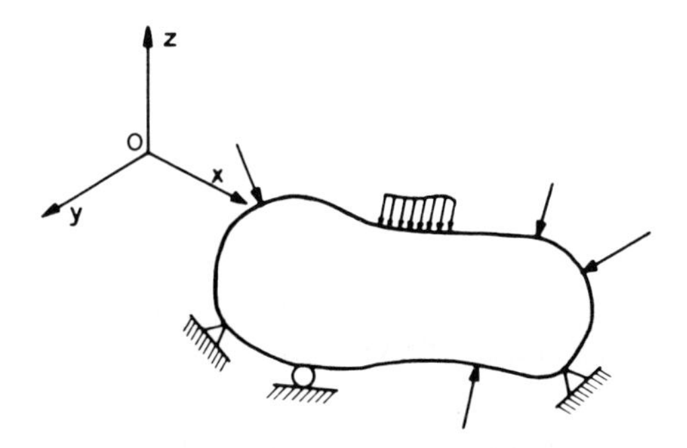

Fig. 3.10.

The assumed equilibrium (internal and at the boundary) is expressed by the following eqs. (b) and (c), respectively :

$$\frac{\partial \sigma_x}{\partial x} + \frac{\partial \tau_{xy}}{\partial y} + \frac{\partial \tau_{zx}}{\partial z} + X = 0 \qquad \ldots(x,y,z), \tag{b}$$

$$l\sigma_x + m\tau_{xy} + n\tau_{xz} = \overline{X} \qquad \ldots(x,y,z), \tag{c}$$

In eqs. (b) and (c), l, m, n are the direction cosines of the exterior normal at a point of the surface A of the body subjected to force T with components $\overline{X}$, $\overline{Y}$, $\overline{Z}$ and X, Y, Z are the components of **F**.

We rewrite the first terms of eq. (a) as

$$\int_A \mathbf{T} \cdot \mathbf{U}\, dA = \int_A (\overline{X}\, u + \overline{Y}\, v + \overline{Z}\, w)\, dA\,, \tag{d}$$

where u,v,w are the components of U. Using eq. (c) in eq. (d) we obtain

$$\int_A \mathbf{T} \cdot \mathbf{U}\, dA = \int_A (lu\sigma_x + \mu\tau_{xy} + v\tau_{zx} + lv\tau_{xy} + mv\tau_{yz} + lw\tau_{zx} + mw\tau_{yz}$$

$$+ nw\sigma_z)\, dA$$

$$= \int_A \bullet[\, l\,(\sigma_x\, u + \tau_{xy}\, v + \tau_{zx}\, w) + m\,(\tau_{xy}\, u + \sigma_y\, v + \tau_{yz}\, w)$$

$$+ n\,(\tau_{zx}\, u + \tau_{yz}\, v + \sigma_z\, w)\,]\; dA. \qquad \text{(d')}$$

We apply the Green-Ostrogradsky formula

$$= \int_A \mathbf{P} \cdot \mathbf{n}\, dA = \int_V \operatorname{div} \mathbf{P}\, dV. \qquad \text{(e)}$$

Because l, m, n are the projections of the outward normal unit vector **n**, relation (d') becomes

$$\int_A \mathbf{T} \cdot \mathbf{U}\, dA = \int_V [\frac{\partial}{\partial x}\,(\sigma_x\, u + \tau_{xy}\, v + \tau_{zx}\, w) + \frac{\partial}{\partial y}\,(\tau_{xy}\, u + \sigma_y\, v + \tau_{yz}\, w)$$

$$+ \frac{\partial}{\partial z}\,(\tau_{zx}\, u + \tau_{yz}\, v + \sigma_z\, w)\,]\, dV,$$

or more explicitly,

$$\int_A \mathbf{T} \cdot \mathbf{U}\, dA = \int_V [\, u\left(\frac{\partial \sigma_x}{\partial x} + \frac{\partial \tau_{xy}}{\partial y} + \frac{\partial \tau_{zx}}{\partial z}\right) + \left(\frac{\partial \tau_{xy}}{\partial x} + \frac{\partial \sigma_y}{\partial y} + \frac{\partial \tau_{yz}}{\partial z}\right)$$

$$+ w\left(\frac{\partial \tau_{zx}}{\partial x} + \frac{\partial \tau_{yz}}{\partial y} + \frac{\partial \sigma_z}{\partial z}\right) + \sigma_x \frac{\partial u}{\partial x} + \sigma_y \frac{\partial v}{\partial y} + \sigma_z \frac{\partial w}{\partial z}$$

$$+ \tau_{xy}\left(\frac{\partial v}{\partial x} + \frac{\partial u}{\partial y}\right) + \tau_{yz}\left(\frac{\partial w}{\partial y} + \frac{\partial v}{\partial z}\right) + \tau_{zx}\left(\frac{\partial u}{\partial z} + \frac{\partial w}{\partial x}\right)]\, dV. \qquad \text{(f)}$$

For very *small* deformations we have

$$\varepsilon_x = \frac{\partial u}{\partial x}, \qquad \varepsilon_y = \frac{\partial v}{\partial y}, \varepsilon_z = \frac{\partial w}{\partial z},$$

$$\gamma_{xy} = \frac{\partial u}{\partial x} + \frac{\partial v}{\partial y}, \quad \gamma_{yz} = \frac{\partial w}{\partial y} + \frac{\partial v}{\partial z}, \quad \gamma_{zx} = \frac{\partial u}{\partial z} + \frac{\partial w}{\partial x}. \tag{g}$$

Using eq. (b) and (g) in eq. (f) we obtain

$$\int_A \mathbf{T} \cdot \mathbf{U} \, dA = \int_V [-(X u + Y v + Z w) + \sigma_x \varepsilon_x + \sigma_y \varepsilon_y + \sigma_z \varepsilon_z$$

$$+ \tau_{xy} + \gamma_{xy} + \tau_{yz} + \gamma_{yz} + \tau_{zx} + \gamma_{zx}] \, dV .$$

Because Xu + Yu + Zu = **F** . **U**, we have

$$\int_A \mathbf{T} \cdot \mathbf{U} \, dA + \int_V \mathbf{F} \cdot \mathbf{U} \, dV = \int_V \sigma_x \varepsilon_x + \sigma_y \varepsilon_y + \sigma_z \varepsilon_z$$

$$+ \tau_{xy} + \gamma_{xy} + \tau_{yz} + \gamma_{yz} + \tau_{zx} + \gamma_{zx} \; dV . \tag{h}$$

The right-hand side of eq. (h) is the virtual work of the stresses σ_x, ..., τ_{xy}, ... On the strains ε_x, ..., γ_{xy}, ..., the latter being as a rule independent of the former because they are derived from an arbitrary displacement field by relations (g). We thus have proved the theorem :

If a continuum is in equilibrium under the influence of given stresses, loads, and reactions, the virtual work of the loads and reactions on an arbitrary continuously differentiable displacement field equals the virtual work of the stresses on the corresponding strain field.

When the displacement field corresponds to a rigid-body displacement, we have

$$\varepsilon_x = \varepsilon_y = \varepsilon_z = \gamma_{xy} = \gamma_{yz} = \gamma_{zx} = 0 \ ,$$

and hence

$$\int_A \mathbf{T} \cdot \mathbf{U} \, dA + \int_V \mathbf{F} \cdot \mathbf{U} \, dV = 0$$

We thus have the following theorem :

For a continuum that is in equilibrium under the influence of given stresses, loads, and reactions, the virtual work of the loads and reactions on any virtual rigid-body displacement vanishes.

3.8.1. *Remarks.*

1. The theorems of virtual works above are simply a different way of saying that the body is in equilibrium.

2. The preceding proof only assumes that (a) continuity conditions for the applicability of the Green formula transforming a surface integral into a volume integral are fulfilled ; (b) the strains are sufficiently small for eq. (g) to be valid (this condition is regarded to be included in the term " virtual "). With these two assumptions, the theorem applies to bodies of any nature : elastic, plastic, viscous, and so forth.

3.9. Problems.

3.9.1. Show that, for the yield condition of von Mises the dissipation per unit volume is $D = \sigma_Y \dot{J}$.

3.9.2. Prove the following theorems (in analogy with those of Section 3.3.2).

(a) When kinematic boundary restraints are tightened, the limit load does not decrease.

(b) When kinematic boundary restraints are relaxed, the limit load does not increase. *Hint : discuss kinematical admissibility of actual collapse mechanisms for original and modified structures.*

3.9.3. Prove that, when the statical boundary conditions are unchanged (nature of the reactions of the supports) the limit load does not depend on the kinematic boundary conditions. For example, simple supports may be (slightly) deformable.

Answer : see Section 9.5.10.

3.9.4. Show that unrestrained plastic flow under constant load is impossible for an elastically compressible perfectly plastic solid with the von Mises yield condition, when it is subjected to plane strain conditions. *Hint* : study the variation of the principal stress acting in the direction of vanishing principal strain.

References.

[3.1] W. PRAGER, Introduction to mechanics of Continua, Ginn and Co. Boston, 1961.

[3.2] W.T.KOÏTER, "On the Stress-Strain Relations and the General Theorem of Plasticity", Lab. van toegepaste Mech., Techn.Hoge-school, Delft, 1953.

[3.3] A.A GVOZDEV, "The determination of the Value of the Collapse Load for Statically Indeterminate Systems Undergoing Plastic Deformation", Proc. of the Conf. on Plastic Deformations, December 1936, Akademia Nauk S.S.S.R., Moscow ; translated by R.M. HAYTHORNTHWAITE, Int. J. of Mech. Sci., **1** : 322, 1960.

[3.4] R. HILL, "On the State of Stress in a Plastic-Rigid Body at the Yield Point", Phil. Mag, **42** : 868, 1951.

[3.5] W. PRAGER, "Problèmes de plasticité théorique", Dunod, Paris, 1958.

[3.6] D.C. DRUCKER, W. PRAGER, and H.J. GREENBERG, "Extended Limit Design Theorems for Continuous Media", Quart. Appl. Math., **9** : 381, 1952.

[3.7] E.T. ONAT, "On Certain Second Order Effects in the Limit Design of Frames", J. of Aero. Sci., **22** : 681, 1955.

[3.8] R.M. HAYTHORNTHWAITE, "Beams with Full End Fixity", Engineering, **183** : 110, 1957. See also by the same author, "Plastic Behaviour of Beams with Elastic End Constraints", Comptes rendus 9è Cong. Int. Mec. Appl., Bruxelles, 1956.

[3.9] P.G. HODGE jr, Plastic Analysis of Structures, Mc Graw-Hill, 1959.

[3.10] E.T. ONAT, "Plastic Analysis of Shallow Conical Shells", Paper presented at the Xth Int. Congr. Appl. Mech., Stresa, 1960 (diagrams reproduced by kind permission).

[3.11] E.T. ONAT "On the plastic Analysis of Shallow Conical Shells", Proc. A.S.C.E., J. Eng. Mech. Div., December, 1960.

[3.12] R. HILL, The Mathematical Theory of Plasticity, Clarendon Press, Oxford, 1950.

[3.13] W. PRAGER, and P.G. HODGE jr., Theory of Perfectly Plastic Solids, J. Wiley, New York, 1951.

[3.14] V.V. SOKOLOVSKI, Theorie der Plastizität, V.E.B. Verlag Technik, Berlin, 1955

[3.15] Swiss Fed. Lab. for Testing Materials, Report 126, Zürich, Feb. 1940.

[3.16] B. BUDIANSKY, “Reassessment of Deformation Theories of Plasticity”, J. Appl. Mech., **26 :** 259, June 1959.

4. General Loading Case.

4.1. Structures with nonnegligible dead load.

Under the assumption of proportional loading made in the previous sections, the safety of a structure with respect to incipient unrestrained plastic flow is measured by the quotient $s_p = \frac{P_l}{P_s}$ of limit load P_s. It must be noted that the dead load is not likely to vary (except for corrosion, or uncertainties on specific weight and dimensions), whereas live load may vary in a wide range. This fact must be taken into account when the dead load is an important part of the total loads. Assuming the dead load to be given and fixed and the live loads to consist of a oneparameter loading with magnitude P_v, one defines statically and kinematically licit loads P_{v-} and P_{v+} as follows :

1. P_{v-} is any intensity of the live load that, together with the fixed dead load, corresponds to a licit stress field.

2. P_{v+} is any intensity of the live load that, together with the fixed dead load, corresponds by the work equation to a kinematically admissible mechanism.

If the limit live load P_{vl} is defined as the live load that produces unrestrained plastic flow when associated with the dead load, it is easily shown, along the lines of Section 3.2, that

$$P_{v-} \leq P_{vl} \leq P_{v+}. \tag{4.1}$$

Note that *the point of view adopted above implies that the dead load alone can not cause the collapse of the structure.*

4.2. Loading depending on several parameters.

When the loading depends on several *independent* scalar loading parameters P_1, P_2, ..., P_k, There does not exist as a rule one limit state - but a infinity of possible combinations of the "loads" P_1, P_2, ..., P_k may produce collapse.

The whole loading path must then be examined to determine *at every loading stage* whether there is unrestrained plastic flow or not.

The theorems of limit analysis that can be established along a lime similar to that of Section 3.3, are then as follows :

1. If a licit stress field with $\sigma_R < \sigma_Y$ can be found *at every loading stage*, plastic collapse will not occur on this loading path.

2. If at a certain loading stage, a licit mechanism can be found in which the power of applied loads is not smaller than the power of dissipation, plastic collapse must have occurred on the considered loading path or must be impending.

3. As long as plastic collapse has not occurred, it is possible to find a statically admissible stress field *at every loading stage.*

4.3. Shakedown analysis.

4.3.1. *Introduction.*

To use the preceding theorems, the loading path must be completely described and every loading stage must be studied. This analysis is in most cases a very long and difficult task.

On the other hand, the loading parameters very often vary in an unknown manner and the problem is to determine *the permissible range of variation of each parameter* to avoid some kind of plastic collapse.

4.3.2. *One-parameter loading.*

When the load does not increase monotonically but varies arbitrarily between prescribed limits, the structure can fail either by accumulation of plastic deformations of the same sign or by alternating plastic deformations that eventually result in fracture.

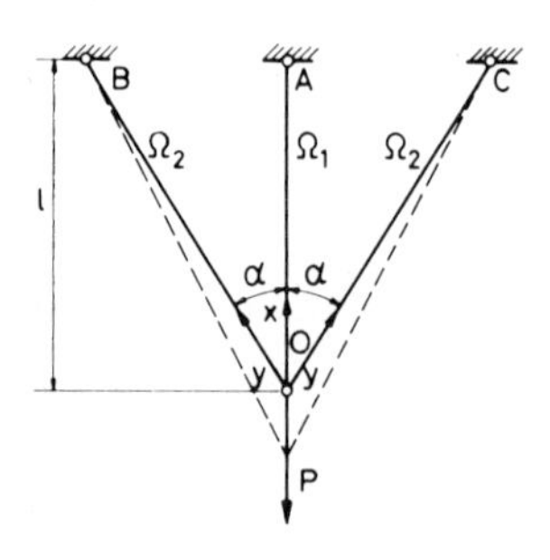

fig. 4.1.

These types of failure occur, as a rule, with the highest value of the load smaller than the limit load for proportional loading. This problem was already studied in Com. V., Chapter 7.

We consider here in more detail the example of the three bar truss of fig. 4.1, in a slightly more general form than in Section 1.1 of Com. V. because the angle α is arbitrary end the cross sections of the bars OB and OC, though indentical and denoted by A_2 may differ from the dross section A_1 of the var OA. All the bars are made of the same elastic perfectly plastic material with an elastic modulus E and the stress-strain diagram shown in fig. 4.2.

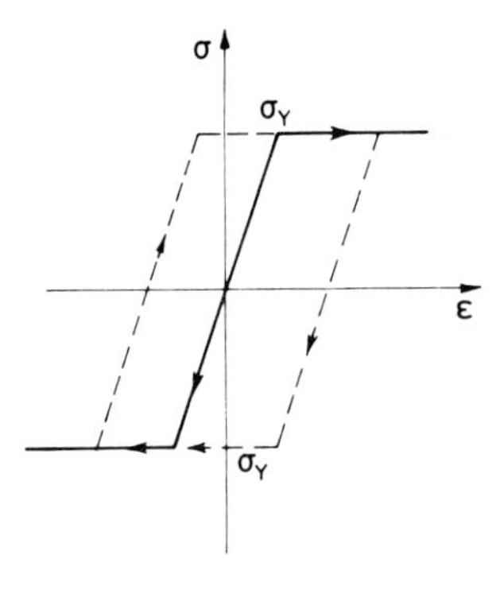

fig. 4.2.

Let

$$\frac{1}{EA_1} = C_1, \qquad \frac{1}{EA_2 \cos\alpha} = C_2, \tag{4.2}$$

and denote by Δ the total elongation of a bar and Δ_p its plastic elongation. We then have

$$\Delta_1 = C_1 X + \Delta_1 p, \qquad \Delta_2 = C_2 Y + \Delta_2 p, \tag{4.3}$$

where X and Y are the axial forces in the bars OA and OB (or OC), respectively. With the following definitions of nondimensional forces and elongations

$$x = X\left(\frac{C_1}{2}\right)^{1/2}, \qquad y = Y\sqrt{C_2},$$

$$\delta_1 = \frac{\Delta_1}{\sqrt{2C_1}}, \qquad \delta_2 = \frac{\Delta_2}{\sqrt{C_2}},$$

$$\delta_{1p} = \frac{\Delta_{1p}}{\sqrt{2C_1}}, \qquad \delta_{2p} = \frac{\Delta_{2p}}{\sqrt{C_2}}, \tag{4.4}$$

the preceding expressions for the total elongations become

$$\delta_1 = x + \delta_{1p}, \qquad \delta_2 = y + \delta_{2p}. \tag{4.5}$$

Equilibrium of joint O requires that

$$X + 2Y \cos\alpha = P. \tag{4.6}$$

With notations (4.4), eq. (4.6) can be rewritten :

$$\frac{x}{\sqrt{C_1}} + \frac{\sqrt{2}}{\sqrt{C_2}}\, y \cos\alpha = \frac{P}{\sqrt{2}}. \tag{4.7}$$

Compatibility at joint O furnishes

$$\Delta_1 = \frac{\Delta_2}{\cos\alpha},$$

or, equivalently,

$$\delta_1 \sqrt{C_1} - \frac{\sqrt{C_2}\; \delta_2}{\sqrt{2}\,\cos\alpha} = 0. \tag{4.8}$$

The yield condition of the bars are $|X| = \sigma_Y A_1$, $|Y| = \sigma_Y A_2$, or, in nondimensional form,

$$|x| = \sigma_Y A_1 \left(\frac{C_1}{2}\right)^{1/2} \equiv x_p,$$

$$|y| = \sigma_Y A_2 \sqrt{C_2} \equiv y_p. \tag{4.9}$$

In a system of cartesian orthogonal axes x and y the yield condition is represented by a rectangle (fig. 4.3). The forces in the bars are the coordinates of *the force point* that, in order to satisfy equilibrium, must, for each value of P, remain on the corresponding straight line with eq. (4.7), called the equilibrium line.

Superimpose on the x, y axes the ξ, η axes, respectively, with

$$\xi = -\delta_{1p},$$

$$\eta = -\delta_{2p}, \tag{4.10}$$

and call the plastic deformation point the point with coordinates (xi, η). The compatibility equation (4.8) can be written

$$(x - \xi)\sqrt{C_1} - (y - \eta)\frac{\sqrt{C_2}}{\sqrt{2}\cos\alpha} = 0. \tag{4.11}$$

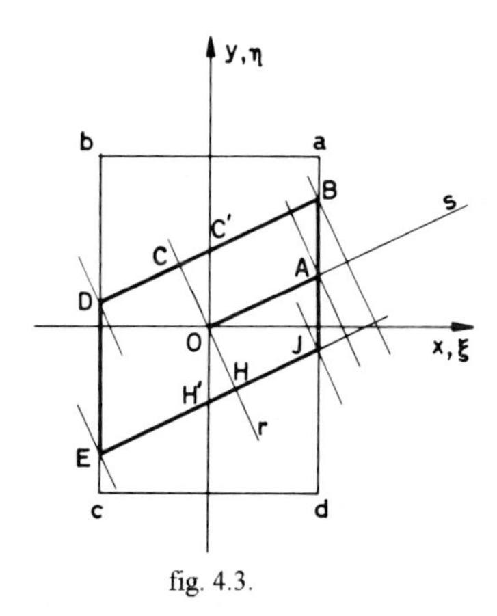

fig. 4.3.

It is represented by the line of compatibility which is normal to the equilibrium line through the plastic deformation point, as is easily verified from inspection of the coefficients of the variables in eqs. (4.7) and (4.11). The intersection of the line of compatibility with the equilibrium line is the force point P representing the state of stress of the truss. The position of the force point thus depends on both the value of the load P and the existing plastic deformation.

Note that the force point has a smaller distance from the plastic deformation point than any other point on the equilibrium line. This minimum distance,

$$E^{*} = (x+\delta_{1p})^2 + (y+\delta_{2p})^2 = \delta_1^2 + \delta_2^2 = \Delta_1^2 \frac{EA_1}{2l} + 2\,\Delta_2^2 \frac{EA_2}{2l} \cos\alpha,$$

is the fictitious strain energy computed from the total elongations as if they were purely elastic. This result is a particular form of a more general theorem of Colonnetti [4.1, 4.2].

In a stress-free initial state, the force and permanent deformation points both coincide with the origin, and the equilibrium and compatibility lines are rays Or and Os respectively. When the truss is loaded, the point P moves on Os whereas the line Or experiences a translation. When P has the position A (fig. 4.3) we have

$$\xi = \eta = 0,$$

$$x = x_p. \tag{4.12}$$

With use of eqs. (4.12) and (4.9), eq. (4.11) becomes

$$y = y_p \cos^2 \alpha. \tag{4.13}$$

It is seen that the yield limit cannot be reached simultaneously in all three bars for nonvanishing α. The central bar always yields first.

With increasing load intensity P, the force point moves from A towards the corner a where $P = P_l$ (fig. 4.3). Suppose we stop increasing P when we have reached the force point B. The plastic deformation point has moved from O to C' along axis Oy. We then unload and eventually reverse the sign of P. The force point first moves on BC down to D where the central bar yields in compression, and then on bc towards c, whereas the plastic deformation point moves down the Oy axis. If we again reverse the sign of P when we have reached E, the force point climbs on EJ and the plastic deformation point stays at H'. If we cycle the force point along EHKBD (zero load P corresponding to points H and C) the central bar yields alternatively in tension (JB) and compression (DE), though the load is bounded by $P_1 < P_l$ (point B) and $P_2 > -P_l$ (point E).

Alternating plasticity of this type is likely to rapidly produce fracture (*).

On the other hand, if loading to point B is followed by load variations restricted to values of P corresponding to the points B and D, the system will shake-down to purely elastic behaviour after the first plastic deformation OC'. This situation occurs in particular for repeataed loading (along BC) for all $P < P_1$. As emphasized in Com. V., chapter 7, shake-down results from a favourable state of selfstress (represented by the coordinates of C in repeated loading).

4.4. Loading depending on several parameters : basic definitions and statements of theorems.

4.4.1. *Elastic-perfectly plastic and fictitious elastic bodies.*

Let a structure made of an elastic-perfectly plastic material be subject to loads applied *quasi statically* and allowed to vary *independently* between given limit. It is clear from preceding sections that the elastic strain must be taken into account. Hence, the vector of total strains $\boldsymbol{\varepsilon}$ divided into its elastic and plastic parts :

$$\boldsymbol{\varepsilon} = \boldsymbol{\varepsilon}^e + \boldsymbol{\varepsilon}^p \qquad (4.14)$$

Elastic strains are related to the stresses through Hooke's law. Plastic strain increments are given by the normality law (2.6). The stresses $\boldsymbol{\sigma}$ must satisfy the yield inequality $\sigma_R(\boldsymbol{\sigma}) \leq \sigma_Y$.

Continued application of the variable loads in time may result in the following possible situations :

a - after possibly some elastic-platic strains, the structure *shakes down* to purely elastic behaviour : after some sequency of loading the plastic flow ceases completely. Hence, the total work dissipated is bounded.

b - plastic flow continues with continued loading variations : if no reversal in the signs of the strains occurs, accumulation of plastic strains gives birth to an incremental collapse mechanism, with monotonically increasing permanent deflections (a phenomenon also called ratchetting)

(*) Loading cycles as above may also produce accumulating plastic deformations of the same sign (see [4.3], p. 28), though this cannot occur in the simple structure considered here.

if strains reversals occur, ratchetting is prevented but a real structure with limited ductility will fail by fracture due to *exhaustion of ductility.*

With our elastic-perfectly plastic (with unlimited ductility) both failures are pointed out by the fact that the *total plastic work dissipated is unbounded.*

4.4.2. *Residual strains and stresses.*

It will prove convenient to associate to the given body a *fictitious indefinitely elastic body, for* which no yield condition exists. We denote σ^* and ε^* the instantaneous values of the stresses and strains in the corresponding fictitious elastic body.

The residual stresses and strains are defined by :

$$\boldsymbol{\rho} = \boldsymbol{\sigma} - \boldsymbol{\sigma}^* \tag{4.15}$$

$$\boldsymbol{\eta} = \boldsymbol{\varepsilon} - \boldsymbol{\varepsilon}^*$$

As the field $\boldsymbol{\varepsilon}^*$ is purely elastic, we have

$$\boldsymbol{\eta}^p = \boldsymbol{\varepsilon}^p, \qquad \boldsymbol{\eta} = \boldsymbol{\eta}^e + \boldsymbol{\varepsilon}^p \tag{4.16}$$

Note that, in general, the residual stress field (also called selfstress field) is a function of time. But, as we have seen in the previous example, its limit is time independent. So, we now suppose that we have found some time independent field of fictitious residual stresses ρ. This field can be taken as any solution of the homogeneous equilibrium equations

$$\Delta\,\boldsymbol{\rho} = 0 \tag{4.18}$$

and in equilibrium with vanishing loads. Fields of this type are used to define a statical condition of shakedown. Indeed, we can prove the following theorem :

Theorem 6 : Melan's, shakedown theorem. *If a time independent field of selfstress can be found that does not violate the yield condition when superimposed on the (fictitious) purely elastic stress fields produced by the load cycles, then shade-down will occur.*

Suppose that such a field of residual stresses $\bar{\boldsymbol{\rho}}$ exists. We shall show that the structure shakes down. Let us consider the fictitious residual elastic energy of the stress difference $(\boldsymbol{\rho} - \bar{\boldsymbol{\rho}})$

$$R = \frac{1}{2}\int_V (\boldsymbol{\rho} - \bar{\boldsymbol{\rho}})^T(\boldsymbol{\eta}^e - \bar{\boldsymbol{\eta}}^e)\, dv \tag{4.19}$$

The stress difference $(\boldsymbol{\rho} - \bar{\boldsymbol{\rho}})$ is connected with the strain difference $(\boldsymbol{\eta}^{e} - \bar{\boldsymbol{\eta}}^{e})$ by the linear relations of Hooke's law, and therefore the derivative of the energy with respect to time is

$$\dot{R} = \int_V (\boldsymbol{\rho} - \bar{\boldsymbol{\rho}})^T \dot{\boldsymbol{\eta}}^e \, dv. \tag{4.20}$$

According to (4.17), we have :

$$\dot{R} = \int_V (\boldsymbol{\rho} - \bar{\boldsymbol{\rho}})^T \dot{\boldsymbol{\eta}} \, dv - \int_V (\boldsymbol{\rho} - \bar{\boldsymbol{\rho}})^T \dot{\boldsymbol{\varepsilon}}^p \, dv. \tag{4.21}$$

By the theorem of virtual powers, applied to the field $(\boldsymbol{\rho} - \bar{\boldsymbol{\rho}})$ and the field $\boldsymbol{\eta}$, the first term in (4.21) vanishes.

Because of (4.15) and the definition

$$\bar{\boldsymbol{\rho}} = \bar{\boldsymbol{\sigma}} - \boldsymbol{\sigma}^*. \tag{4.22}$$

we have :

$$\dot{R} = -\int_V \boldsymbol{\sigma}^T \dot{\boldsymbol{\varepsilon}}^p \, dv + \int_V \bar{\boldsymbol{\sigma}}^T \dot{\boldsymbol{\varepsilon}}^p \, dv. \tag{4.23}$$

The first term is the rate of dissipation and the second the power produced by the attainable stresses $\bar{\boldsymbol{\sigma}}$ acting on the plastic strain rates $\dot{\boldsymbol{\varepsilon}}^p$

$$\dot{R} = P\int_V (\boldsymbol{\rho}, \dot{\boldsymbol{\varepsilon}}^p) - \int_V D(\dot{\boldsymbol{\varepsilon}}^p) \, dv \tag{4.24}$$

Because of the theorem of maximum dissipation, we have :

$$\dot{R} \leq 0 \tag{4.25}$$

Since the residual elastic energy is non negative, a time will be reached when plastic flow ceases (i.e. $\dot{\boldsymbol{\varepsilon}}^p = 0$, $\dot{R} = 0$). The resisual stresses will no longer change with time and the theorem is proved.

Finally, to characterize completely these residual stress fields, we shall say that the field $\bar{\rho}$ is admissible if, besides, the stress field $\bar{\sigma}$ associated to $\bar{\rho}$ by (4.22) satisfies the yield inequality $\sigma_R \leq \sigma_Y$ at any time.

4.4.3. *Admissible cycle of plastic strains.*

Consider some arbitrary firld of palstic strains ε^p. We call this field admissible, if the field of plastic strain increments over some loading cycle

$$\Delta\,\varepsilon^p = \oint \dot{\varepsilon}^p \, dt \tag{4.26}$$

is kinematically admissible (cf. Sect. 1.1.2). These fields are used in kinematical condition of shakedown. We may associate by the normality law (2.6) a stress field σ and determinate a residual stress field ρ as the difference between this field of fictitious elastic stresses σ^*.

Because the plastic strain $\Delta\varepsilon^p$ over the loading cycle are kinematically sdmissiblen so are the associated elastic strains $\Delta\varepsilon^e$ together with the total strains. The residual stresses ρ at the end of the cycle return to their initial values. Then :

$$\oint \varepsilon^e \, dt = 0 \tag{4.27}$$

4.4.4. *Loading domain.*

Consider a set of loads P_1, P_2,..., P_k. Any independent load is varying in a given interval

$$P_{mi} < P_m < P_{ms}. \tag{4.27 bis}$$

In addition, this load domain is supposed to increase in a homothetic way qith a parameter noted α. We would like to know the smallest value α_s for which failure occurs. We call α_s the shakedown multiplier.

We may remark at once that it is aquivalent to give the loading domain or the domain of the stresses field σ^* in the fictitious elastic body. So that we may put

$$\sigma^* = \alpha\sigma^{*o} \tag{4.28}$$

where the elastic stress field σ^{*o} belongs to the domain of references.

4.4.5. *Fundamental theorems of shakedown analysis.*

The fundamental theorems can be stated as follows :

Theorem 7 : Lower bound theorem or statical theorem.

The load parameter α_- *corresponding to admissible residual stress fields are lower bounds to the shakedown multiplier* α_s.

Theorem 8 : Upper bound theorem or kinematical theorem.

The load parameter α_+ *corresponding to admissible fields of plastic strain increments are upper bounds to the shakedown multiplier* α_s.

The statical approach, with the key concept of time-independent residual stress field, was introduced historically the first. Papers by Bleich [4.11], Melan [4.12 to 4.16] and Symonds [4.6], are considered as the starting point of the method. The kinematical approach and the basic concept of admissible plastic strain increment was introduced by Koiter [4.10, 4.17 and 4.18] and developed by Neal [4.19], Gokhfeld [4.20, 4.21] and Sawczuk [4.22, 4.23]. An improved formulation including the concept of loading multiplier was proposed by Martin [4.45].

4.5. Proofs of the theorems.

Theorem 7 : Lower bound theorem or statical theorem.

The load parameter α_- corresponding to admissible residual stress field are lower bounds to the shakedown multiplier α_s.

Let α_- be the stress field associated to some admissible residual stress field $\bar{\rho}_-$.

Using (4.15) and (4.21), we have :

$$\sigma_- = \bar{\rho}_- + \alpha_- \sigma^{*o} . \tag{4.29}$$

Let σ be the stress field and $\Delta\varepsilon^p$ the plastic strain indrement field for which failure occurs. Then, we can put :

$$\sigma = \bar{\rho} + \alpha_s \sigma^{*o} \tag{4.30}$$

Taking account that $\bar{\rho}$ is a time independent field, it resuts immediately from (4.26) and (4.29) that we have :

$$\oint P(\sigma_-, \dot{\varepsilon}^p)\, dt = P\, (\bar{\rho}_-, \Delta\dot{\varepsilon}^p)\, dt = \alpha_- \oint P(\sigma^{*o}, \dot{\varepsilon}^p)\, dt. \tag{4.31}$$

As $\Delta\varepsilon^p$ satisfies the homogeneous kinematical boundary conditions of the body and $\bar{\rho}_-$ is a selfstress field, we diduced from the theorem of virtual powers that the first term of the right hand member of (4.31) vanishes. Hence,

$$\oint P(\sigma_-, \dot{\varepsilon}^p)\, dt = \alpha_- \oint P(\sigma^{*o}, \dot{\varepsilon}^p)\, dt. \tag{4.32}$$

In a similar way, we have

$$\oint P(\sigma, \dot{\varepsilon}^p)\, dt = \alpha_s \oint P(\sigma^{*o}, \dot{\varepsilon}^p)\, dt. \tag{4.33}$$

Because the power in the left hand member of (4.33) can be written $\int_V D\, dv$, subtraction of the two previous relations result in

$$\oint P(\sigma_-, \dot{\varepsilon}^p)\, dt - \oint \int_V D\, (\dot{\varepsilon}^p)\, dV\, dt = (\alpha_- - \sigma_s) \oint P(\sigma^{*o}, \dot{\varepsilon}^p)\, dt \tag{4.34}$$

The left hand member of (4.34) is non positive according to the theorem of maximum dissipation. Besides, we have deduced from (4.33) that the integral in the right hand member of (4.34) is positive. Therefore :

$$\alpha_s \leq \alpha_- . \tag{4.35}$$

Theorem 8 : Upper bound theorem or kinematical theorem.

The load parameter α_+ corresponding to admissible fields of plastic strain increments are upper bounds to the shakedown multiplier α_s. (*)

(*) This theorem may be considered as a special form of Koiter's shakedown theorem [4.10, 4.17 and 4.18].

Let $\Delta\, \varepsilon^p_+$ be some admissible increment of the plastic strain field.

Let σ be the time dependent stress field from which failure occurs.

The load parameter α_+ corresponding to the admissible field ε^p_+ is defined by the equality

$$\alpha_+ \oint P\,(\sigma^{*o}, \dot{\varepsilon}^p_+)\, dt = \oint \int_V D\,(\dot{\varepsilon}^p_+)\, dV\, dt \tag{4.36}$$

From (4.26) and (4.30), we obtain :

$$\alpha_s \oint (\sigma^{*o}, \dot{\varepsilon}^p_+)\, dt + (\bar{\rho}, \Delta\varepsilon^p) = \oint P\,(\sigma, \dot{\varepsilon}^p_+)\, dt$$

As $\dot{\varepsilon}^p_+$ is an admissible field, and $\bar{\rho}$ a selfstress field, we deduce from the theorem of virtual powers that the second term of the left hand member of the previous relation vanishes.

$$\alpha_s \oint P\,(\sigma^{*o}, \dot{\varepsilon}^p_+)\, dt = \oint P\,(\sigma, \dot{\varepsilon}^p_+)\, dt \tag{4.37}$$

Because σ is the stress field as failure, it is licit, but it does not correspond generally with the field $\Delta\varepsilon^p_+$.

Hence, we can apply relation (3.3) to $\dot{\varepsilon}^p_+$.

If we subtract (4.36) from (4.37), we have, taking account of (3.3) :

$$(\alpha_s - \alpha_+) \oint P\,(\sigma^{*o}, \dot{\varepsilon}^p_+)\, dt \leq 0. \tag{4.38}$$

By (4.36), the integral in (4.38) is strictly positive, and we have

$$\alpha_s \leq \alpha_+ . \tag{4.39}$$

4.6. Application of the shakedown theorems.

4.6.1. *Combined theorem.*

It is obvious that the fundamental theorems of shakedown theory are similar to those of limit analysis..

Hence, we similarly deduce from (4.35) and (4.39) a combined theorem :

Theorem 9 : Combined theorem.

When it is possible to associate to an admissible increment of plastic strain field an admissible residual stress field, the corresponding common multiplier α *is exactly the shakedown multiplier* α_s.

4.6.2. *Thermal loads and other extensions.*

These theorems can easily be generalized to bodies with thermal stresses by simply regarding σ^* as the field of thermal stresses in a fictitious perfectly elastic body.

Readers interested by more details can consult references such as [4.4 to 4.10, 4.21, 4.24 to 4.28, 4.47]. Different other extensions of the precious theory were also investigated, including dynamical effects [4.29 to 4.34], hardening [4.35 to 4.43], nonstandard materials [4.28], and geometrical second order effects [4.28, 4.35, 4.36, 4.44].

4.6.3. *Example of one-parameter loading.*

Consider [4.7] a thick-wall tube with radii a and b, which is in plane strain under the influence of uniform internal pressure (fig. 4.4). In chapter 10 we shall prove that, for monotonically increasing pressure, the values p_e and p_l of the pressure for first yielding and complete plastification are respectively given by

$$p_e = \tau_Y \left(1 - \frac{a^2}{b^2} \right), \tag{4.40}$$

and

$$p_l = 2\,\tau_Y \ln \frac{b}{a}. \tag{4.41}$$

The two formulas above are valid for the yield conditions of both Tresca and von Mises, τ_Y being the yield limit in pure shear (greater by a factor of $\frac{2}{\sqrt{3}}$ in von Mises'condition than in Tresca's).

Beyond the elastic limit pressure p_e, a rig-shaped plastic region ($a \leq r \leq \rho$) appears which is surrounded by an elastic ring ($\rho \leq r \leq b$) (fig. 4.4). In chapter 10, we shall prove that the radial stress distribution for Tresca's condition in the plastic ring is given by

$$\sigma_r = \sigma_Y \ln r + C_1 \qquad \text{for} \qquad a \leq r \leq \rho . \tag{4.42}$$

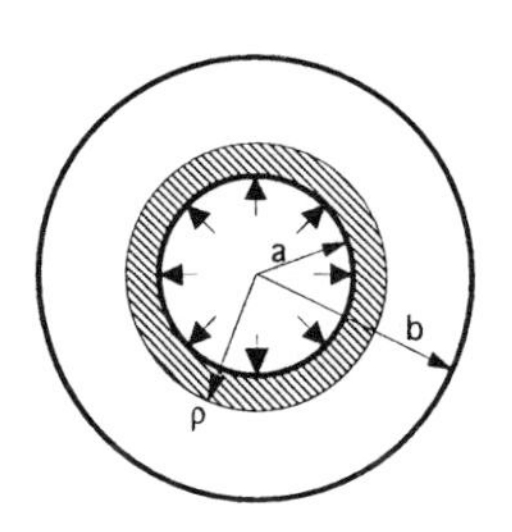

Fig. 4.4.

The constant C_1 is determined by $\sigma_r = -p$ at $r = a$. Eq. (4.42) is rewritten

$$\sigma_r = \sigma_Y \ln \frac{r}{a} - p . \tag{4.43}$$

From Tresca's condition :

$$\sigma_\theta - \sigma_r = \sigma_Y, \tag{4.44}$$

we deduce the distribution of σ_θ in the plastic ring

$$\sigma_\theta - \sigma_Y (\ln \frac{r}{a} + 1) - p \qquad \text{for} \qquad a \leq r \leq \rho . \tag{4.45}$$

To find the relation connecting p and ρ, we must express the continuity $\sigma_r = -p'$ through the surface $r = \rho$, where p' is the limit elastic pressure of the elastic ring-shaped region.

If a is replaced by ρ in (4.40), we obtain :

$$p = \sigma_Y \left[\ln \frac{\rho}{a} + \frac{1}{2}\left(1 - \frac{\rho^2}{b^2}\right)\right]. \tag{4.46}$$

The stress field in the elastic region is given by the well known Lame equations :

$$\sigma_r = \frac{p\rho^2}{b^2 - \rho^2}\left(1 - \frac{b^2}{r^2}\right) \qquad \text{for} \qquad \rho \le r \le b$$

$$\sigma_\theta = \frac{p\rho^2}{b^2 - \rho^2}\left(1 + \frac{b^2}{r^2}\right) \tag{4.47}$$

These equations give also the elastic stresses in the fictitious elastic body if ρ is replaced by a :

$$\sigma_r^* = \frac{p a^2}{b^2 - a^2}\left(1 - \frac{b^2}{r^2}\right) \qquad \text{for } a \le r$$

$$\sigma_\theta^* = \frac{p a^2}{b^2 - a^2}\left(1 + \frac{b^2}{r^2}\right). \tag{4.48}$$

The field of residual stresses is obtained by subtracting the elasto-plastic field (4.43) (4.45) from the fictitious elastic field (4.48) in the plastic ring

$$\rho_r = \sigma_Y \ln \frac{r}{a} - p - \frac{p a^2}{b^2 - a^2}\left(1 - \frac{b^2}{r^2}\right) \qquad \text{for} \qquad a \le r \le \rho$$

$$\rho_\theta = \sigma_Y\left(\ln \frac{r}{a} + 1\right) - p - \frac{p a^2}{b^2 - a^2}\left(1 + \frac{b^2}{r^2}\right). \tag{4.49}$$

and by subtracting the elastic field (4.47) from the fictitious elastic field (4.48) in the elastic ring

$$\rho_r = \frac{p\, b^2 (\rho^2 - a^2)}{(b^2 - \rho^2)(b^2 - a^2)} \left(1 - \frac{b^2}{r^2} \right)$$

$$\text{for} \qquad \rho \leq r \leq b$$

$$\rho_\theta = \frac{p\, b^2 (\rho^2 - a^2)}{(b^2 - \rho^2)(b^2 - a^2)} \left(1 + \frac{b^2}{r^2} \right). \tag{4.50}$$

The hatched ordinates of figure 4.5 give the residual stresses in the cylinder.

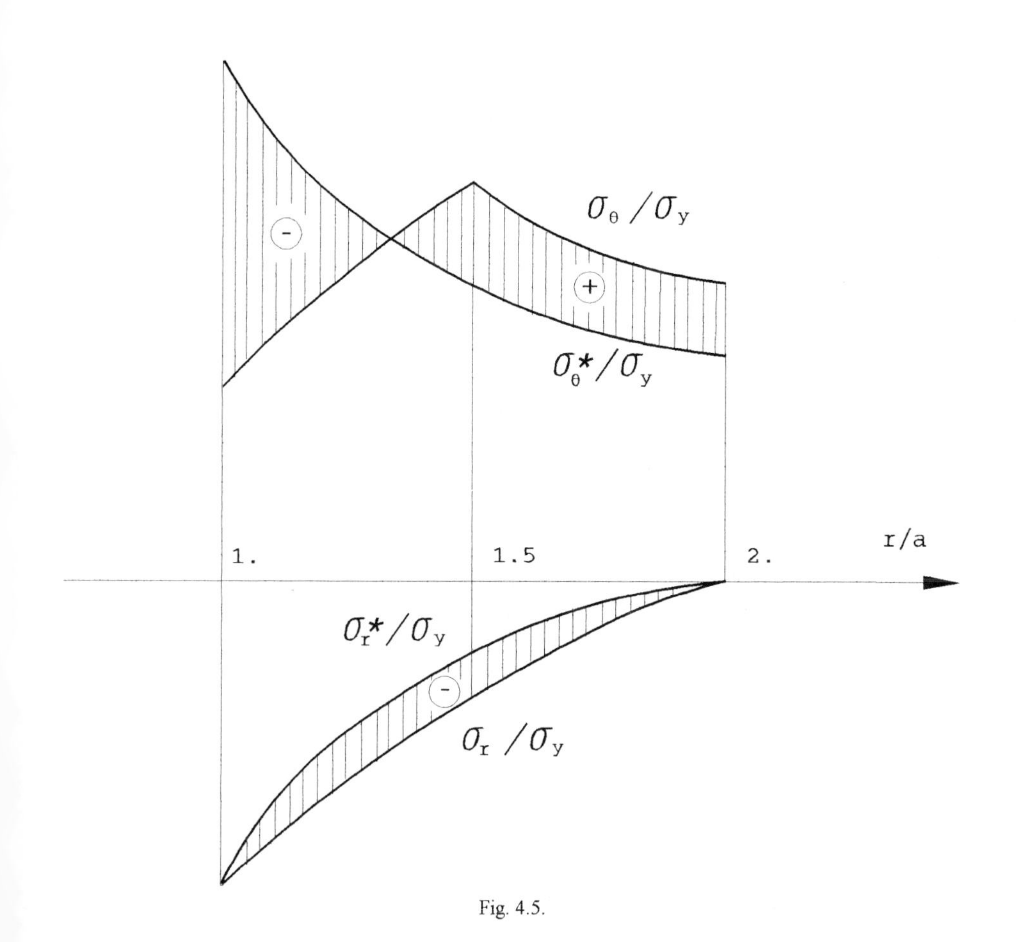

Fig. 4.5.

The results (4.14) and (4.50) show that, at r=a we have :

$$\rho_{ra} = 0\,,\ \rho_{\theta a} = \sigma_Y\left(1 - \frac{p}{p_l}\right). \tag{4.51}$$

Melan's theorem states that, for repeated loading to p_1, the cylinder shakes down as long as the circumferential residual stress $\rho_{\theta a}$ at the internal boundary satisfies the relation

$$\rho_{\theta a} - \rho_{ra} \geq -\sigma_Y\,. \tag{4.52}$$

This situation occurs for $p_1 \leq 2p_e$ with the obvious supplementary condition $p_1 < p_l$ to avoid collapse. Hence, the shake-down load p_s for repeated loading is

$$p_s = \text{min of}\left[\,2p_e\,,\ p_l\right]. \tag{4.53}$$

Eq. (4.53) is represented graphically on fig. 4.6. For $p_1 > p_s$, failure occurs from alternated or cumulative yielding.

When p is allowed to vary from p'' to p', the shake-down conditions are

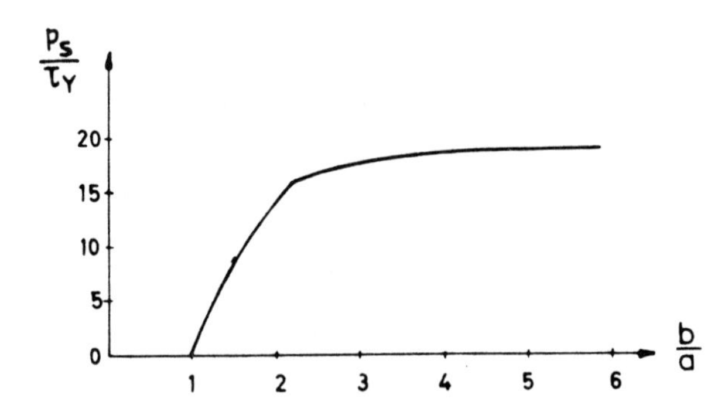

Fig. 4.6.

$$-p_l \leq p'' \leq p' \leq p_l$$
$$p' - p'' \leq 2p_e . \tag{4.54}$$

Let us now consider the kinematical approach.

Taking account of Tresca's condition (4.44) and the normality law (2.6),we have :

$$\dot{\varepsilon}_r^p = -\dot{\varepsilon}_\theta^p \tag{4.55}$$

and the dissipation is equal to

$$D = \sigma_Y |\dot{\varepsilon}_\theta^p| . \tag{4.56}$$

Using relations (4.48) of the stress in the fictitious elastic body and (4.55), we have :

$$\oint P\left(\sigma^{*o}, \dot{\varepsilon}^p\right) dt = \frac{4\pi a^2 b^2}{b^2 - a^2} \int_a^b \left[\oint p^o \dot{\varepsilon}_\theta^p \, dt\right] \frac{dr}{r}, \tag{4.57}$$

where the loading domain of reference is

$$0 \leq p^o \leq 1. \tag{4.58}$$

From eq. (4.56), we have immediately :

$$\oint \int_V D\left(\underline{\dot{\varepsilon}^p}\right) dV \, dt = 2\pi \sigma_Y \int_a^b \left[\oint |\dot{\varepsilon}_\theta^p| \, dt\right] r \, dr . \tag{4.59}$$

Hence, the pressure p_+ corresponding to the field $\dot{\varepsilon}^p$ is given by (4.36) :

$$p_+ = \frac{b^2 - a^2}{2b^2 a^2} \sigma_Y \frac{\int_a^b \left[\oint |\dot{\varepsilon}_\theta^p| \, dt\right] r \, dr}{\int_a^b \left[\oint p^o \dot{\varepsilon}_\theta^p \, dt\right] \frac{dr}{r}}. \tag{4.60}$$

At this point, we can make several assumptions concerning the field of plastic strain. First, we suppose that :

- the plastic yielding occurs only when $p^o = 1$

- the plastic stains are cumulative : $\dot{\varepsilon}^p_\theta \geq 0$.

Accordingly we have :

$$p_+ = \frac{b^2 - a^2}{2a^2 b^2} \sigma_Y \frac{\int_a^b \Delta \varepsilon^p_\theta \, r \, dr}{\int_a^b \Delta \varepsilon^p_\theta \frac{dr}{r}} . \tag{4.61}$$

Because the plastic strain field is admissible, we must have

$$\Delta \varepsilon^p = \varepsilon_o \left(\frac{b}{r} \right)^2 . \tag{4.62}$$

The value of the expression (4.61) for the field (4.62) is :

$$p_+ = p_l . \tag{4.63}$$

This type of failure in which a collapse mechanism appears progressively is a case of incremental collapse or ratchetting (see Sect. 4.4.1). We can also consider another type of collapse for which

- plastic yielding occurs in a small ring at the radius r=c, that we can represent by means of the DIRAC distribution denoted $\delta(r - c)$ and such that we have

$$\int_a^b \delta(r - c) \, f(r) \, 2\pi r \, dr = f(c).$$

A physical significance of such a singular distribution is given in Ref. [4.46]

- the plastic strains are alternated :

$$\dot{\varepsilon}^p_\theta |_{r=c} = \dot{\varepsilon}_o > 0 \quad \text{when} \quad p^o = 1$$

$$\dot{\varepsilon}^p_\theta |_{r=c} = -\dot{\varepsilon}_o \qquad \text{when} \quad p^o = 0$$

so that we have

$$\Delta \varepsilon_\theta^p = 0 \tag{4.64}$$

With this admissible plastic strain rate cycle, eq. (4.60) gives :

$$p_+ = \frac{b^2 - a^2}{2a^2 b^2} \sigma_Y \frac{\int_a^b \dot{\varepsilon}_0 \, \delta(r-c) \, r \, dr}{\int_a^b \dot{\varepsilon}_0 \, \delta(r-c) \frac{dr}{r}} = 2 \, p_e \left(\frac{c}{a}\right)^2 . \tag{4.65}$$

Such type of collapse is called *plastic fatigue* or failure by alternating plasticity (see Sect. 4.4.1). Finally, from (4.63) and (4.65), we obtain :

$$p_s = \inf \left[\inf_{a \le c \le b.} \; 2 \, p_e \left(\frac{c}{a}\right)^2 , p_l \right] = \inf \left[2 \, p_e , p_l \right] \tag{4.66}$$

In conclusion, the statical and kinematical solutions are completely associated, according to the combined theorem.

4.7. Problems.

4.7.1. For structures with nonnegligible dead load, prove that $p_{v-} \le P_{vl} \le P_{v+}$ where the statically and kinematically admissible loads p_{v-} and p_{v+} are defined as in Sect. 4.1.

4.7.2. We consider a simply supported circular sandwich plate of radius a loaded by a uniform pressure q(t). The plate material obeys Tresca's yield condition (see fig. 6.13 b).

The load domain is defined by :

$$-p\alpha \le a \, (t) \le \alpha .$$

Show by the statical theorem that the shakedown multiplier is :

$$\alpha_s = \frac{6M_p}{a^2} \qquad \text{if} \qquad p \le \frac{7 - 3\,\nu}{3\,(3 + \nu)},$$

$$\alpha_s = \frac{1}{1+p}\,\frac{32}{3+\nu}\,\frac{M_p}{a^2} \qquad \text{otherwise.}$$

Hint : the moment field in the fictitious elastic body is given by :

$$M_r^* = \frac{q(t)}{16}(3+\nu)\,(a^2 - r^2)$$

$$M_\theta^* = \frac{q(t)}{16}\left[\,a^2(3+\nu) - (1+3\nu)\,r^2\right]$$

4.7.3. Solve problem 4.7.2 using the kinematic theorem.

Hint : use a ratchet mechanism on one hand, and a mechanism with plastic fatigue at the centre of the plate on the other hand.

4.7.4. We consider the simply supported sandwich plate of problem 4.7.2 but loaded by uniform pressure q(t) and a uniform distribution of circumferential bending moments M(t), applied at r=a.

The loads vary independently within the limits :

$$0 \le q(t) \le \bar{q}\,, \qquad\qquad -\mathbf{M} \le M(t) \le 0.$$

Using the statical theorem, show that the plate shakes down when :

$$\mathbf{M} \le M_p\ \bar{q} \le \frac{6M_p}{a^2} \qquad \text{and} \qquad \mathbf{M} + \frac{(3+\nu)}{16}\,a^2\ \bar{q} \le 2M_p$$

4.7.5. Solve problem 4.7.4 using the kinematic theorem.

References.

[4.1] G. COLONNETTI, L'équilibre des corps déformables, Dunod, Paris, 1955.

[4.2] M. SAVE, "Une interprétation du théorème de Colonnetti", Z.A.M.P., XIII, 5, 1962.

[4.3] W. PRAGER, An Introduction to Plasticity, Addison-Wesley Publ. Co., Inc., Reading. Mass., 1959.

[4.4] P.S. SYMONDS, W. PRAGER, "Elastic-Plastic Analysis of structures Subjected to Loads Varying Arbitrary Between Prescribed Limits", J. of Appl. Mech. **17** : 315, Sept. 1950.

[4.5] Discussion of paper [4.4] by G. WINTER, T.M. CHARLTON, and the authors, J. Appl. Mech. **18** : 117, March 1951

[4.6] P.S. SYMONDS, "Shake-down in Continuous Media", J. Appl. Mech. **18** : 85, March 1951.

[4.7] P.G. HODGE Jr., "Shake-down of Elastic-Plastic Structures", Residual Stresses in Metals and Metal Constructions, W.R. OSGOOD, ed., Reinhold Pub. Corp., New York, 1954.

[4.8] W. PRAGER, "Shake-down in Elastic-Plastic Media Subjected to Cycles of Loads and Temperature", Symposium sulla plasticita nella scienza delle costruzioni, N. ZANICHELLI, Bologna, 1957.

[4.9] W. PRAGER, "Plastic Design and Thermal Stresses", British Welding J., Aug. 1956.

[4.10] W.T. KOÏTER, "General Theorems for Elastic-Plastic Solids", Progress in Solid Mechanics, vol. 1, I.N. SNEDDON, R. HILL, eds., North-Holland, Co., Amsterdam, 1960.

[4.11] H. BLEICH, "Über die Bemessug statish unbestimmter Ztahltragwerk unter Berücksichtigung des elastisch-plastichen Verhaltens des Baustoffes", Bauingenieur, **13** : 2610, 1932.

[4.12] E. MELAN, "Theorie statisch unbestimmter System", Berlin, Prelim. Publ. 2nd Congres, A.I.P.C., 43, 1936.

[4.13] E. MELAN, "Theorie statisch unbestimmter Systeme aus ideal plastischen Baustoff", Sitz. Ber. Akad. Wiss. Wien IIa, 195-218, 1936.

[4.14] E. MELAN, "Der Spannungszustand eines HENCKY-MISES' chen Kontinuums bei veranderlichen Belastung", Sits. Ber. Akad. Wiss. Wien IIa, 73-87, 1938.

[4.15] E. MELAN, "Plastizität des raümlichen Kontinuums", Ing. Archiv, **9** : 116-126, 1938.

[4.16] E. MELAN, "Theorie statisch unbestimmter System", Berlin, Prelim. Publ. 2nd Int. Congres Assoc. Bridge and Struct. Eng., 1936.

[4.17] W.T. KOITER, "A new General Theorem on Shakedown of Elastic-Plastic Structures", Proc. Kon. Nad. Ak. Wet. B59, **24**, 1956.

[4.18] W.T. KOÏTER, "Some Remarks on Plastic Shakedown Theorems", Istambul, Proc. 8th Int. Congr. Appl. Mech., **1**, 1952.

[4.19] B.G. NEAL, "The plastic methods of structural analysis", Londres, Chapman and Hall, 1956.

[4.20] D.A. GOKHFELD, "Some problems of plastic shakedown of plates and shells" (in Russian), Proc. 6th Soviet Conf. Plates and Shells Bakou 1966, Moscou, Izd. Nauka, 284-291, 1966.

[4.21] D.A. GOKHFELD, "Load carrying capacity of structures subject to uneven heating" (in Russian), Moscou, Izd. Mashinostr., 1970.

[4.22] A. SAWCZUK, "On incremental collapse of shell under cycling loading", Proc. I.U.T.A.M., Symp. Copenhagen 1967, Berlin, Springer, 328-340, 1969.

[4.23] A. SAWCZUK, "Evaluation of upper bounds to shakedown loads for shells", J. Mech. Phys. Solids, **17** : 291-301, 1969.

[4.24] J.A. KÖNIG, "Engineering Applications of shakedown theory", Udine, C.I.S.M., 1977.

[4.25] V.I. ROSEMBLUM, "Shakedown of elasto-platic bodies subject to a non uniform Temperature" (in Russian) I sv. ak. N. U.R.S.S., OTN n°7, 1957.

[4.26] J.A. KÖNIG, "Theory of shakedown of Elastic Plastic Structures", Arch. Mech. Stos., **18** : 227-238, 1966.

[4.27] J.A. KÖNIG, "A shakedown Theorem for Temperature dependent Elastic Moduli", Bull. Acad. Pol. Sci., Ser. Sci. Techn. **17** : 161-165, 1969.

[4.28] G. MAIER, "Shakedown theory in perfect elastoplasticity with associated and nonassociated flow-laws : A finite element linear programming approach", Meccanica, **4** : 250-260, 1969.

[4.29] L. CORRADI, G. MAIER, "Inadaptation Theorems in the Dynamics of Elastic-work Hardening Structures", Ing. Arch., **43** : 44-57, 1973.

[4.30] L. CORRADI, G. MAIER, "Dynamic Non-shakedown Theorem for the Elastic Perfectly Plastic Continua", J. Mech. Phys. Solids, **22** : 401-413, 1974.

[4.31] C. POLIZZOTTO, "Adaptation of Rigid-Plastic Continua under Dynamic Loadings", Tech. Rep. SISTA-77-OMS-2, Facolta di Architetura di Palermo, Palermo, 1977.

[4.32] HO HWA SHAN, "Shakedown in Elastic-Plastic Systems under Dynamic Loading", J. Appl. Mech., **39** : 416-421, 1972.

[4.33] G. CERADINI, "Sull'adattamento dei corpi elasto-plastici soggeti ad azione dinamiche", Giornale del Genio Civile, **107** : 239-250, 1969.

[4.34] G. CERADINI, C. GAVARINI, "Applicazione della programmazione ai problemi adattamento plastico statico o dinamico", Giornale del Genio Civile, **107** : 471-476, 1969.

[4.35] G.A. MAIER, "A matrix structural theory of piecewise linear elastoplasticity with interacting yield planes", Meccanica, **5** : 54-66, 1970.

[4.36] G.A. MAIER, "A shakedown matrix treory allowing for work-hardening and second-order geometric effects", Int. Symposium on Found. of Plast., Warsaw, 1972, Fondations of Plasticity, **1**, Ed. SAWCZUK, A. Leyden, Noordhoff Int. Pub., 417-433, 1973.

[4.37] J.A. KÖNIG, G.A. MAIER, "Adaptation of Rigide-Workhardening Discrete Structures subjected to Load and Temperature cycles and second-order geometric effects", Comp. Meth. Appl. Mech. Eng., **8** : 37-50, 1976.

[4.38] W. PRAGER, "Adaptation Baushinger d'un solide plastique à écrouissage cinématique", C. R. Acad. Sc., Paris, t. 280, Série B, 585-587, 1975.

[4.39] J. MANDEL, "Adaptation d'une structure plastique écrouissable et approximations", Mech. Res. Comm., **3** : 483-488, 1978.

[4.40] C. POLIZZOTTO, "Workhardening Adaptation of Rigid-Workhardening Structures", Meccanica, **10**, n°4, 280-288, 1975.

[4.41] J. MANDEL, J. ZARKA, B. HALPHEN, "Adaptation d'une structure élastoplastique à écrouissage cinématique", Mech. Res. Comm., vol. 4, 1977.

[4.42] B. HALPHEN, "Plastic fatigue and shakedown of elastoviscoplastic and plastic structures" (in French), Seminar about the materials and structures submitted to cyclic loading, Palaiseau, 1978. Ecole Nationale des Ponts et Chaussées, Ecole Polytechnique, 203-229, 1979.

[4.43] A.R.S. PONTER, " A General Shakedown Theorem for Elastic-Plastic Bodies with Workhardening", Proc. of the 3rd Int. Conf. Struct. Mech. Renc. Tech., **5**, part 2, 1-8, 1975.

[4.44] G. MAIER, "Shakedown of Plastic Structures with Unstable Parts", ASCE, J. Eng. Mech. Div., **98** : 1322, 1972.

[4.45] J.B. PARTIN, "Plasticity : fundamentals and general results", The M.I.T. Press, Cambridge (Massachusetts) and London, 1975.

[4.46] G. DE SAXCE, "About some Problems of the Mechanics of Solids Considered as Materials with Convex Potentials" (in French), Ph. D. Thesis, Faculty of Applied Sciences, University of Liège, 1986.

[4.47] O. DE DONATO, "Second Shakedown Theorem Allowing for Cycles of both Loads and Tempreatures", Meccanica delle Costruzioni, A104, 265-277, 1970.

5. Generalized Variables.

5.1. The concept of generalized variables.

5.1.1. *Introduction.*

Limit analysis of a rigid perfectly plastic continuum is based on the three following concepts : (1) yield condition and related flow rule, (2) licit stress field, and (3) licit flow mechanism. Great simplification is achieved when these concepts can be applied without the need to discuss three-dimensional stress and displacement fields. This situation arises in linear elasticity when the considered solid is a beam, plate, or shell. Assumptions regarding the deformations of these particular structural elements are accepted as direct consequences of the fact that these elements are "thin" in certain directions (normal to the axis of a beam or to the median surface of a plate or shell).

For beams, the hypothesis of Bernoulli states that plane cross sections remain plane and orthogonal to the deformed material axis. For plates and shells, straight segments normal to the median surface remain straight and normal to the deformed median surface. As long as Hooke's law of linear elasticity is applicable, it is possible to obtain all stress components at every point of a beam, a plate, or a shell when stress resultants and resultant moments are known. To do this, we need only apply the assumption that normals to the median surface are preserved, and use the equilibrium equations and Hooke's law ; the latter immediately furnish all strain components [5.1]. Hence, stress resultants and resultant moments are sufficient for a complete description of stresses and strains.

Theory [5.2] and experiments [5.3] show that Bernoulli's hypothesis and its generalization to plates and shells (normals preserved) are *equally valid in the elastic and plastic ranges*. Bernoulli's hypothesis will therefore be adopted in the following discussion of rigid-plastic beams, plates, and shells.

5.1.2. *Beams without axial force.*

A generic cross section is subjected to a bending moment M and a shear force V.

From Bernoulli's hypothesis, shear strains are seen to vanish and longitudinal strains ε_x are given by (*)

(*) Transverse strains are irrelevant.

$$\varepsilon_x = y\,\kappa\,, \tag{5.1}$$

where y is the distance from the neutral plane and κ is the curvature of the material axis (*). (Note that κ is the reciprocal of the radius of curvature). The strain rate is therefore given by

$$\dot{\varepsilon}_x = y\,\frac{\partial \kappa}{\partial t} = y\,\dot{\kappa}\,. \tag{5.2}$$

Because the state of stress is uniaxial (*), the power dissipated per unit of length of the beam in a plastic region is

$$D = \int_{-h/2}^{h/2} \sigma_Y |\dot{\varepsilon}_x|\, b(y)\, dy\,, \tag{5.3}$$

where h is the height of the section and b(y) the width at the level y.

With the use of eq. (5.2), eq. (5.3) can be written

$$D = \int_{-h/2}^{h/2} \sigma_Y |y\dot{\kappa}|\, b(y)\, dy = |\dot{\kappa}|\, M_p\,, \tag{5.4}$$

where M_p is the (ultimate) plastic moment (see Com. V, Section 2.2).

The total rate of dissipation D_t is then

$$D_t = \int_{\text{struct}} |\dot{\kappa}|\; M_p\, ds\,, \tag{5.5}$$

or for plastic hinges with rotation rates $\dot{\theta}$,

(*) It is assumed that shear stresses do not influence yielding.

$$D_t = \sum M_{pi} |\dot{\theta}_i| , \tag{5.6}$$

as was established in Com. V, Section 3.4. We see that

1. The yield condition reduces to $|M|=M_p$ and the flow rule to sign $\dot{\theta}_i$ = sign M_i , or $M_i \dot{\theta}_i \geq 0$;

2. The stress field reduces to the M diagram ;

3. The strain rate field reduces to the distribution of the rate of curvature.

Beam and frame problems have been extensively studied in Com. V.

5.1.3. *Arches.*

In arches, neither the axial strain ε_o nor the axial force N can be neglected, even when we are not concerned with instability phenomena (see Com. V., Chapters 6 and 10). The longitudinal strain rate at the level y with respect to the centroid consists of a part due to bending $\dot{\varepsilon}_y = y\,\dot{\kappa}$, and of a part due to axial strain $\dot{\varepsilon}_o$.

The total rate of dissipation is

$$D_t = \int_{struct} \left(M\,\dot{\kappa} + N\,\dot{\varepsilon}_o \right) ds , \tag{5.7}$$

where M and N combine to produce complete plastification of the section. Interaction curves M versus N of various sections are given in Com. V., Section 5.2.

Assuming that the shear force V does not influence yielding, the functions M, N, $\dot{\kappa}$ and $\dot{\varepsilon}_o$ of the abscissa s are sufficient for the problem at hand.

5.1.4. *Simple plate and shell examples.*

In both plates and shells, the thickness t must be small compared to the other dimensions. A plate has a plane median surface and is subjected solely to forces normal to this median plane (when the applied forces are parallel to this plane, the structure is called a disk). A shell has a median surface with at least one finite radius of curvature. A "membrane" is a shell with no bending rigidity.

On the median surface of one of the structures described above, and through a given point P of this surface, draw a line element ds that has P as its centre. The normals to the median surface through the points of ds form the "cut based on ds". The stresses transmitted across this cut are statically equivalent to certain forces and couples acting at P, which are proportional to the length of ds. The factors of proportionality are called the "stress resultants" for the considered cut. The "state of stress" at P is specified by the stress resultants for two orthogonal cuts.

1. *Circular plate with constant thickness and rotational symmetry in loading and supports :*

With cylindrical coordinates r, θ, z (fig. 5.1), rotational symmetry indicates that the radial and circumferential bending moments M_r and M_θ are the principal moments (the

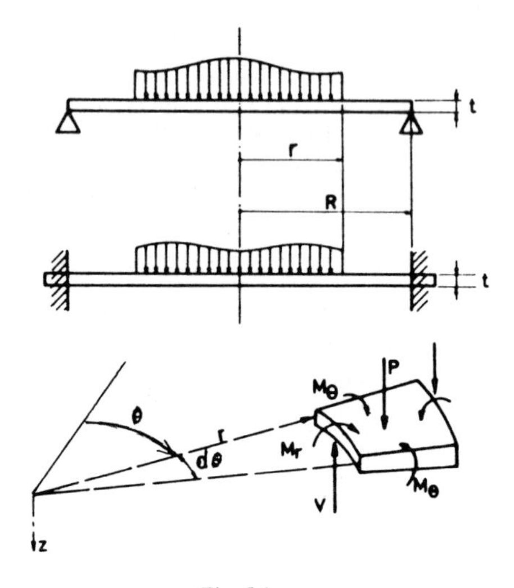

Fig. 5.1.

twisting moment $M_{r\theta}$ vanishes). These bending moments as well as the deflection rate depend solely on the coordinate r.

In accordance with the assumption that material normals remain normal to the deformed median surface, transverse shear strains are neglected. The strain rates are given by

$$\dot{\varepsilon}_r = z\dot{\kappa}_r ,$$

$$\dot{\varepsilon}_\theta = z\dot{\kappa}_\theta ,$$

where $\dot{\kappa}_r$ and $\dot{\kappa}_\theta$ are the radial and circumferential (that is the principal) rates of curvature.

In analogy with beams, the dissipation per unit area of the median plane is

$$D = M_r \dot{\kappa}_r + M_\theta \dot{\kappa}_\theta . \tag{5.8}$$

In relation (5.8), M_r and M_θ must combine to completely plastify the volume element $r\,dr\,d\theta$ at the considered point.

Since the yield condition can be expressed solely in therms of M_r and M_θ, the functions M_r, M_θ, $\dot{\kappa}_r$, $\dot{\kappa}_\theta$ of r are sufficient for the limit analysis of the plate.

2. Cylindrical shells subjected to rotationally symmetric internal pressure :

Internal resultant forces and moments that the symmetry does not oblige to vanish are shown on fig. 5.2. We immediately note that, because of the rotational symmetry, the circumferential curvature rate $\dot{\kappa}_\theta$ vanishes. Indeed there is no circumferential displacement. Any point of the shell displaces in the meridian plane in which it is contained.

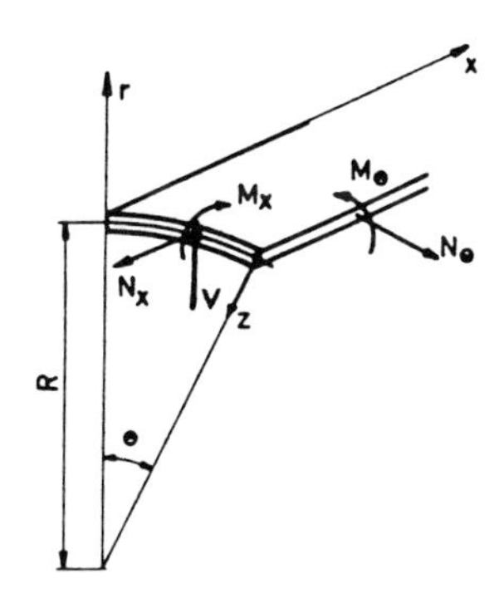

Fig. 5.2.

Hence, any two neighbouring meridian planes experience no relative rotation, and M_θ does not work. We thus have $\kappa_\theta = \dot{\kappa}_\theta = 0$, because our generalized variables are defined from the expression of the internal energy (they are the Lagrange variables). Although the radius of the median surface varies from R to R + w, where w is the radial displacement of the median surface, the circumferential strain,

$$\varepsilon_\theta = \frac{w}{R-z} \qquad (\frac{-t}{2} \leq z \leq \frac{t}{2}),$$

must be regarded as constant (and hence $\kappa_\theta = 0$) because z is negligible with respect to R from the very definition of a shell.

Rates of transversal shear vanish because we assume the material normals to remain normal to the deformed median surface. The dissipation rate is then

$$D = M_x \, \dot{\kappa}_x + N_x \, \dot{\varepsilon}_{xo} + N_\theta \, \dot{\varepsilon}_{\theta o}. \tag{5.9}$$

Expressing the yield condition solely in terms of M_x, N_x and N_θ (see Section 5.4), the functions M_x, N_x, N_θ, $\dot{\kappa}_x$, $\dot{\varepsilon}_{xo}$, $\dot{\varepsilon}_{\theta o}$ of x and θ will be sufficient for the limit analysis of the shell.

5.2. The general case : choice of the generalized variables.

Limit analysis of a structure will use collapse mechanisms of that structure. Denote by $\dot{q}_1$, $\dot{q}_2$, ... ,$\dot{q}_n$, *the generalized strain rates* suitable for describing these mechanisms. As just seen, the generalized strain rates will be rates of curvature and extension for beams, plates and shells. For a three-dimensional body, or in the case of plane stress and plane strain, they will be the components of the strain-rate tensor.

The generalized stresses are then, *by definition* [5.4], the stress-type variables Q_1, ... ,Q_n that must be associated with the generalized strain rates in order that the specific dissipation be given by

$$D = Q_1 \, \dot{q}_1 + \ldots + Q_n \, \dot{q}_n. \tag{5.10}$$

The variables Q_i and $\dot{q}_i$ may even be chosen nondimensional, and eq. (5.10) may be rewritten in the slightly more general form

$$D = C\left(Q_1 \, \dot{q}_1 + \ldots + Q_n \, \dot{q}_n\right) \tag{5.11}$$

where C is a dimensional constant.

We now call "reactions" the generalized stresses that do not *a priori* vanish for reasons of symmetry or equilibrium and that nevertheless do not appear in eq. (5.10) because they correspond to generalized strain rates that, in the considered problem, have been assumed to vanish throughout the structure. For example, in beams, plates, and shells, transversal shear forces are always reactions because normals are assumed to remain normal to the deformed median surface. For the shell of the second example of Section 5.14, M_θ is a "reaction" because $\dot{\kappa}_\theta$ vanishes.

Not only is it always possible to solve problems of limit analysis using only the generalized variables (with no reference to the reactions) but it is also the most efficient way for solving the problems. To that purpose, the reactions must be eliminated from the yield conditions. This will be discussed in Section 5.3.

Let us summarize as follows : The generalized stresses are the only stress-type variables that appear in the expression of the dissipation for the problem at hand. The yield condition is then expressed in terms of these generalized stresses only, by elimination of the reactions.

5.3. Eliminating the reactions.

5.3.1. *Introduction.*

We now remark that the preceding definitions of generalized stresses and strain rates preserve the validity of formula (2.2) if the stress space Q_i and the strain rate space $\dot{q}_i$ are superimposed. This fact is sufficient for all fundamental results of Chapter 3 to hold if one substitutes the generalized stresses Q_i for the components of the tensor (σ) and the generalized strain rates $\dot{q}_i$ for the components of the tensor $(\dot{\varepsilon})$. Fundamental properties (convexity of yield surface, plastic potential) and fundamental theorems (maximum dissipation, statical and kinematical theorems) are obtained in the very same manner, by mere modification of the terminology.

The only point to clarify is the elimination of the reactions, which, as a rule, initially appear in the most general yield condition.

Consider a structural element and denote by $Q_1,...,Q_i,...,Q_n$, the n stress-type variables acting on it. Suppose first that none is a reaction. The yield condition of this element can be written, in a normalized form :

$$F\left(Q_1,...,Q_i,...,Q_n\right)=1. \tag{5.12}$$

We assume for the time being that F is a known function. The normality law applies to surface with eq. (5.12) in the superimposed stress space $\left(Q_1,...,Q_i,...,Q_n\right)$ and strain-rate space $\left(\dot{q}_1,...,\dot{q}_i,...,\dot{q}_n\right)$

We now suppose that (n - k) relations

$$\begin{aligned}
\dot{q}_{k+1} &= 0,\\
\dot{q}_{k+2} &= 0,\\
&\dots\\
\dot{q}_n &= 0,
\end{aligned} \tag{5.13}$$

hold, expressing that, in the particular case under consideration, plastic flow can only occur with (n - k) vanishing generalized strain rates.

According to the normality law, eqs. (5.13) will select a set of points on the surface (5.12) where the projections of a normal vector on the axes k + 1, ..., n, vanish. This set of points form part of the original yield surface (5.12).

By projecting this part on the $\left(Q_1,...,Q_k\right)$ space one obtains the simplified yield condition :

$$\Phi\left(Q_1,...,Q_k\right) = 1, \qquad (5.14)$$

that contains only generalized stresses, and none of the reactions $Q_{k+1},...,Q_n$.

5.3.2. *Direct elimination of the reactions through the use of the dissipation function.*

Assuming that we know the dissipation function D $\left(\dot{q}_1,...,\dot{q}_k\right)$ for a given problem with generalized strain rates $\dot{q}_1,...,\dot{q}_k$, we can generate the yield surface in the stress space $Q_1,...,Q_k$ with the technique described at the en of Section 2.5. The normality law obviously applies to that surface. We shall show that the surface obtained in this manner is identical with the surface (5.14) obtained by projection, and that, consequently, the normality law applies to that latter surface.

The basic yield surface (5.12) is, by nature, unique. Hence the manner in which it is obtained is irrelevant. We suppose we construct it from our knowledge of the dissipation function D $\left(\dot{q}_1,...,\dot{q}_n\right)$ as described at the end of Section 2.5. We recall that, to every possible mechanism $\dot{\boldsymbol{\varepsilon}}$ with components $\dot{q}_1,...,\dot{q}_n$ (in the n dimensional space) there corresponds a plane tangent to the yield surface. This plane is normal to $\dot{\boldsymbol{\varepsilon}}$ and distant by D ($\dot{\mathbf{e}}$) from the origin (in the direction of $\dot{\boldsymbol{\varepsilon}}$), with $\dot{\mathbf{e}}$ the unit vector along $\dot{\boldsymbol{\varepsilon}}$. If we want to select, on the surface (5.12), the points where eqs. (5.13) are satisfied, we select a subset of tangent planes the normals of which have vanishing projections on axes k+1, ..., n.

The wanted simplified surface is the envelope of this subset of planes. Clearly this is identical to constructing the simplified surface directly from the knowledge of D $\left(\dot{q}_1,...,\dot{q}_k\right)$ because we so select all mechanisms with $\dot{q}_{k+1} = \dot{q}_{k+2} = ... = \dot{q}_n = 0$ among all possible mechanisms. But this is also identical with the projection procedure of Section 5.3.1 that merely consists of taking the intersection of the subset of planes above in the $\left(Q_1,...,Q_k\right)$ space.

To sum up, the same simplified yield condition (5.14) can be obtained either starting from the more general yield condition (5.12) and using conditions (5.13) or directly using the dissipation function $D\left(\dot{q}_1,...,\dot{q}_k\right)$

5.3.3. *Remark on the reactions.*

A distinction must be made between generalized strain rates that vanish because of the very definition of the structure and those that vanish because of special (symmetry) conditions. An example of the first class occurs when a shell or a plate or a beam is defined as a structure in which the direction normal to the median surface is a material direction. Hence, *there never is any transversal rate of shear*. Because of the normality law the yield surface is "cylindrical" with its axis parallel to the shear force axis (fig. 5.3). Consequently, *the shear forces may take any value*. They cannot be determined from mechanism and normality law, but may possibly be otained from equilibrium conditions.

On the other hand, the stress-type variables that are reactions because of some special (symmetry) conditions are assigned given values by the normality law. For example, in plane stress, $\dot{\varepsilon}_2 = 0$ imposes $\sigma_2 = \frac{\sigma_1}{2} = \frac{1.15\,\sigma_Y}{2}$ for the Mises yield condition [fig. 5.4 (a)], or $0 \leq \sigma_2 \leq \sigma_Y$ [fig. 5.4 (b)] for the Tresca condition.

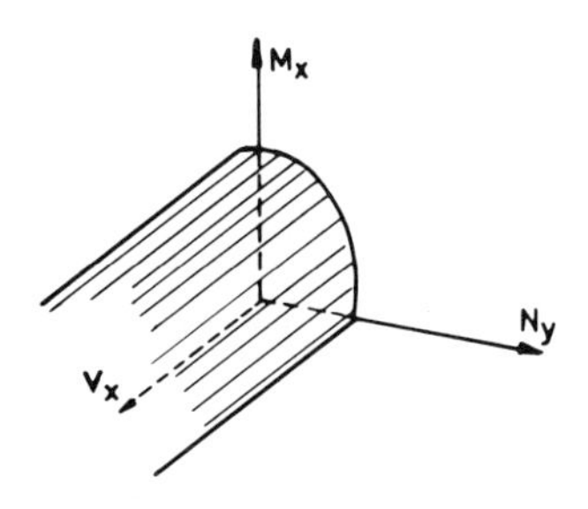

Fig. 5.3.

But, at the same time, the equilibrium equations do not contain these reactions because of the special symmetry conditions above. Hence, when equilibrium and yield conditions are satisfied in terms of generalized stresses only, they can also always be satisfied when reactions are considered. Moreover, note that the equilibrium equations can be obtained from the theorem of virtual work by a variational procedure. No reaction will enter the virtual work equation. This remark proves that it is always possible to eliminate the reactions from the equations of equilibrium. Obviously, any definition of a structure will have a certain range of validity. Shear forces will have no effects on yielding of shells in most cases, as the thickness-to-span ratio, that ranges from 10 to 30 for beams, goes from

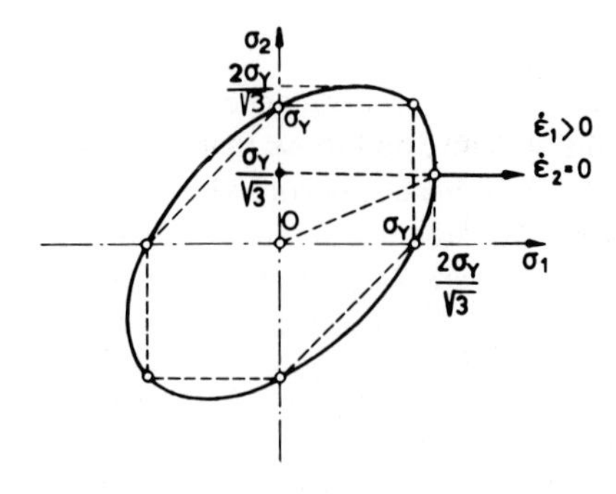

Fig. 5.4 (a)

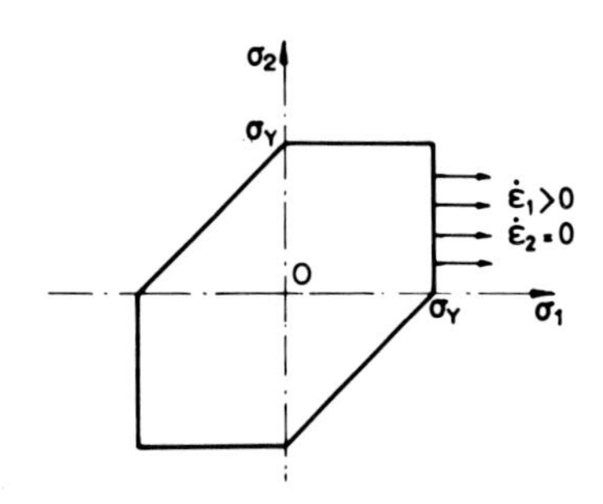

Fig. 5.4 (b).

20 to 50 for most plates and even to 500 for some shells [5.5]. Large concentrate forces may change the situation.

5.4. Obtaining yield conditions in generalized stresses.

5.4.1. *Method by integration.*

Instead of establishing the most general yield condition (5.12) in order to obtain the simplified yield surfaces by section or projection [5.7], it is often desired to obtain the simplified yield condition directly.

To this purpose, the first method is an integration method. On the basis that normals remain normals, generalized strain rates are related to the components of the strain rate tensor at every level in the thickness. The strain rate tensor is related to the stress tensor by the normality law. Integration over the thickness furnishes the yield condition in generalized stresses. We illustrate this method with the example of a *plate*.

Consider a plate of constant thickness, transversally loaded and subjected to arbitrary boundary conditions. Orthogonal cartesian coordinate axes x and y are located in the median plane, and the positive z-axis has the direction of the loads (fig. 5.5). We assume not only that material normals remain normal to the deformed median surface but also that the transversal displacements w are small with respect to the constant thickness t, which, in turn, is small with respect to in-plane dimensions and do not vary with the deformation [5.1].

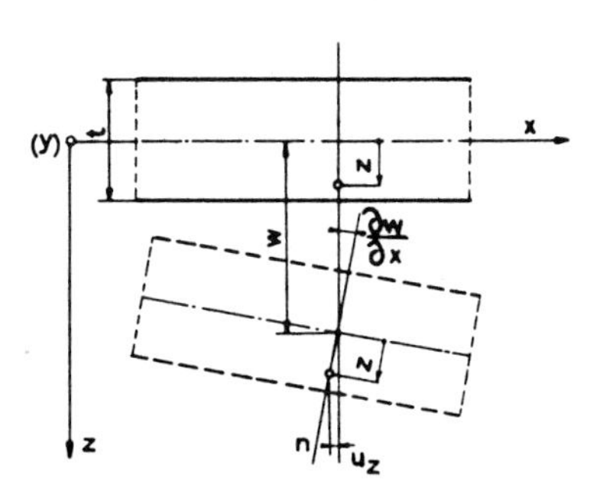

Fig. 5.5.

The deflected shape of the plate is then completely described by the single function w(x,y) because we have (fig.5.5) :

$$u = -z\,\frac{\partial w}{\partial x}, \qquad v = -z\,\frac{\partial w}{\partial y}, \qquad w = w\,(x,y), \tag{5.15}$$

where u,v,w, are the components of the displacements of the points at the distance z of the midplane on the x-, y-, and z-axes, respectively.

Using eqs. (1.15) and (1.16), we obtain

$$
\begin{aligned}
\varepsilon_x &= -z\,\frac{\partial^2 w}{\partial x^2},\\
\varepsilon_y &= -z\,\frac{\partial^2 w}{\partial y^2},\\
\varepsilon_z &= 0,\\
\tau_{xy} &= z\,\frac{\partial^2 w}{\partial x\,\partial y},\\
\tau_{xz} &= \tau_{yz} = 0.
\end{aligned}
\tag{5.16}
$$

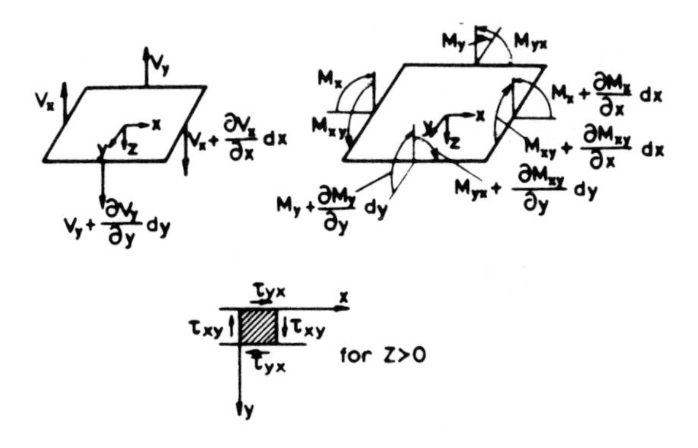

Fig. 5.6.

Because the midplane is deformation-free, we consider that the resultant forces parallel to this plane always vanish.

Hence, the remaining resultant forces and moments are shown in fig. 5.6, where all forces and moments are positive. Moments are related to stresses as follows :

$$M_x = \int_{-t/2}^{t/2} \sigma_x \, z \, dz,$$

$$M_y = \int_{-t/2}^{t/2} \sigma_y \, z \, dz,$$

$$M_{xy} = -M_{yx} = \int_{-t/2}^{t/2} \tau_{xy} \, z \, dz. \tag{5.17}$$

Note that τ_{xy} is positive as shown in fig. 5.6.

Now, the virtual energy per unit area of the median plane is

$$E = \int_{-t/2}^{t/2} \left(\sigma_x \, \varepsilon_x + \sigma_y \, \varepsilon_y + \tau_{xy} \, \gamma_{xy} \right) dz \, . \tag{5.18}$$

Using eqs. (5.16) in eq. (5.18), we obtain

$$E = \left(- \frac{\partial^2 w}{\partial x^2} \right) \int_{-t/2}^{t/2} \sigma_x \, z \, dz + \left(- \frac{\partial^2 w}{\partial y^2} \right) \int_{-t/2}^{t/2} \sigma_y \, z \, dz$$

$$+ \left(-2 \frac{\partial^2 w}{\partial x \, \partial y} \right) \int_{-t/2}^{t/2} \tau_{xy} \, z \, dz. \tag{5.19}$$

Within the framework of small deflection theory, the factors in parenthesis are the curvatures κ_x, κ_y and twice the torsion κ_{xy} of the deflected surface, respectively :

$$
-\frac{\partial^2 w}{\partial x^2} = \frac{1}{\rho_x} \equiv \kappa_x ,
$$

$$
-\frac{\partial^2 w}{\partial y^2} = \frac{1}{\rho_y} \equiv \kappa_y ,
$$

$$
-2\frac{\partial^2 w}{\partial x \partial y} = \frac{2}{\rho_{xy}} \equiv 2\kappa_{xy} . \tag{5.20}
$$

With the definitions (5.19) of the moments, relation (5.19) can hence be written $E = M_x \kappa_x + M_y \kappa_y + 2 M_{xy} \kappa_{xy}$, and the specific power of dissipation is

$$
D = M_x \dot{\kappa}_x + M_y \dot{\kappa}_y + 2 M_{xy} \dot{\kappa}_{xy} . \tag{5.21}
$$

The generalized stresses are M_x, M_y, M_{xy} and the corresponding generalized strain rates are $\dot{\kappa}_x$, $\dot{\kappa}_y$ and $2\dot{\kappa}_{xy}$, respectively.

Because we actually consider each layer of thickness dz to be in plane stress, the yield condition is

$$
\sigma_R\left(\sigma_x , \sigma_y , \tau_{xy}\right) = \sigma_Y . \tag{5.22}
$$

On the other hand, inspection of relations (5.16) reveals that the strain-rate vector has components proportional to z. Hence, the corresponding stress point is the same for all z with same sign. If we now assume that the yield surface with eq. (5.22) is symmetric with respect to the origin (as for the von Mises and Tresca conditions), the stress point goes to a position symmetric with respect to the origin when z changes sign ()(fig. 5.7). If the yield state of stress is $(\sigma_x, \sigma_y, \tau_{xy})$ for positive z, it is $(-\sigma_x , -\sigma_y , -\tau_{xy})$ for negative z, and one obtains

$$
M_x = \sigma_x \frac{t^2}{4} , \qquad M_y = \sigma_y \frac{t^2}{4} , \qquad M_{xy} = \tau_{xy} \frac{t^2}{4} . \tag{5.23}
$$

Because moments are seen to be propositional to the stress components, *the yield surface in the space of moments will have the same form as in the space of stress components.*

It often proves convenient to use nondimensional (also called "reduced") variables. Stress components are rendered nondimensional by division by σ_Y, and relation (5.22) takes the "canonic" form

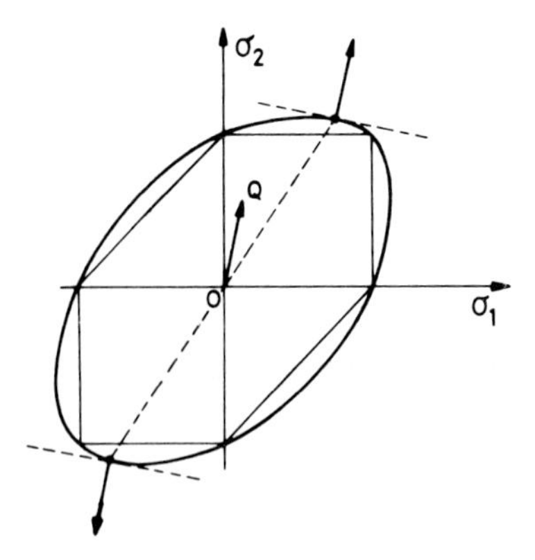

Fig. 5.7.

$$\Phi\left(\frac{\sigma_x}{\sigma_Y}, \frac{\sigma_y}{\sigma_Y}, \frac{\tau_{xy}}{\sigma_Y}\right) = 1. \tag{5.24}$$

Similarly, we define *reduced moments*

$$m_x = \frac{M_x}{M_p}, \qquad m_y = \frac{M_y}{M_p}, \qquad m_{xy} = \frac{M_{xy}}{M_p}, \tag{5.25}$$

where

$$M_p = \sigma_Y \frac{t^2}{4} \tag{5.26}$$

is the yield moment for uniaxial bending. From definitions (5.25) and relations (5.23), we obtain

$$m_x = \frac{\sigma_x}{\sigma_Y}, \qquad m_y = \frac{\sigma_y}{\sigma_Y}, \qquad m_{xy} = \frac{\tau_{yx}}{\sigma_Y}. \tag{5.27}$$

With relations (5.27), condition (5.24) becomes

$$\Phi\left(m_x, m_y, m_{xy}\right) = 1. \tag{5.28}$$

We see that the yield condition (5.28) in reduced moments is identical to that in reduced stresses.

For example, von Mises'condition (1.34) for plane stress, using reduced stresses, becomes

$$\left(\frac{\sigma_x}{\sigma_Y}\right)^2 + \left(\frac{\sigma_y}{\sigma_Y}\right)^2 - \frac{\sigma_x}{\sigma_Y} \cdot \frac{\sigma_y}{\sigma_Y} + 3\left(\frac{\tau_{xy}}{\sigma_Y}\right)^2 = 1 .$$

Hence, the corresponding yield condition for a plate is simply

$$m_x^2 + m_y^2 - m_x\, m_y + 3\, m_{xy}^2 = 1 . \tag{5.29}$$

Similarly, Tresca's condition gives

$$\max\left[\, |m_1|,\, |m_2|,\, |m_1 - m_2| \right] = 1 . \tag{5.30}$$

5.4.2. *Use of the power of dissipation.*

Consider the power of dissipation

$$D\left(\dot{q}_1,\ldots,\dot{q}_k\right) = Q_1\, \dot{q}_1 + \ldots + Q_k\, \dot{q}_k , \tag{5.31}$$

where $Q_1,\ldots,Q_k$ and $\dot{q}_1,\ldots,\dot{q}_k$ are the generalized variables.

Because the function $D\left(\dot{q}_1,\ldots,\dot{q}_k\right)$ is homogeneous with the order one, Euler's theorem on homogeneous functions gives

$$D\left(\dot{q}_1,\ldots,\dot{q}_k\right) = \frac{\partial D}{\partial q_1}\dot{q}_1 + \ldots + \frac{\partial D}{\partial q_k}\dot{q}_k . \tag{5.32}$$

By comparing eqs. (5.31) and (5.32) we find

$$Q_1 = \frac{\partial D\left(\dot{q}_1,\ldots,\dot{q}_k\right)}{\partial \dot{q}_1} ,$$

. . .

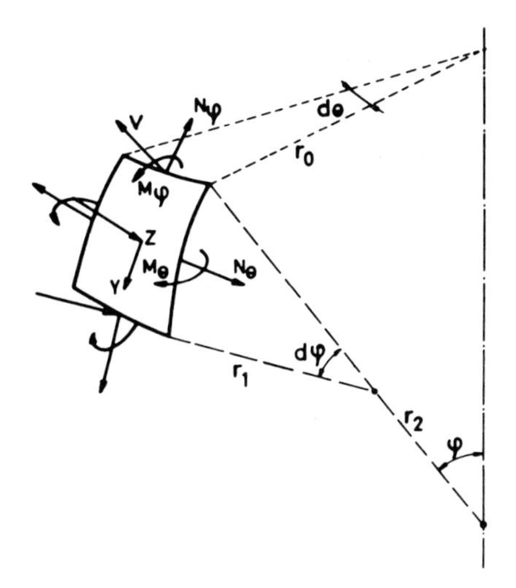

Fig. 5.8.

$$Q_k = \frac{\partial D\left(\dot{q}_1,\ldots,\dot{q}_k\right)}{\partial \dot{q}_k}. \tag{5.33}$$

Relations (5.33) are the parametric equations of the yield surface, with parameters $\dot{q}_1,\ldots,\dot{q}_k$. Actually, because a yield mechanism at a point defines the generalized strain rates except for a common positive factor, there are only k-1 parameters : for example the ratios

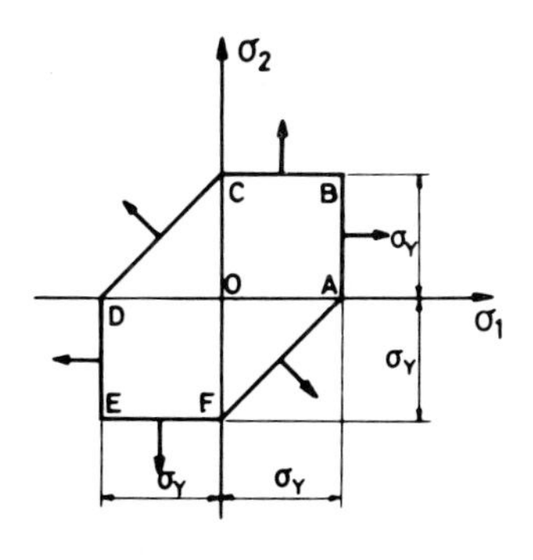

Fig. 5.9.

of the $\dot{q}_i$ to one of them.

We illustrate the method in the example of the yield condition of a shell of revolution with axisymmetric loading [5.8]. Fig. 5.8 shows an element of the shell with the nonvanishing resultant forces and moments (per unit of length) acting on it. Principal directions are φ and θ because of the symmetry. We denote by $\dot{\varepsilon}_\theta$ and $\dot{\varepsilon}_\varphi$ the principal rates of strain of the midsurface, and by $\dot{\kappa}_\theta$ and $\dot{\kappa}_\varphi$ the rates of curvature of that surface.

Because material normals remain normal to the deformed median surface, the dissipation per unit area of this surface is

$$D = M_\varphi \dot{\kappa}_\varphi + M_\theta \dot{\kappa}_\theta + N_\varphi \dot{\varepsilon}_\varphi + N_\theta \dot{\varepsilon}_\theta, \tag{5.34}$$

so that M_φ, M_θ, N_φ, N_θ are the generalized stresses.

We use Tresca's condition for plane stress ($\sigma_z \equiv \sigma_3 = 0$; see fig. 5.9).

To obtain the direction parameters of the outward pointing normal to the hexagonal cylinder at the various points of the plane hexagonal section of fig. 5.9, we simply note that the normal to the plane hexagon at one of its point is the projection in the $(O\sigma_1, O\sigma_2)$ plane of the normal to the hexagonal cvlinder at the same point. We further recall that the sum of the direction parameters p, q, and r, must vanish (see Section 2.6). We then can write table 5.1, valid for points on the hexagon other than vertices A, B, ..., F. From table 5.1 we conclude that, according to eq. (2.8), the dissipation per unit volume is given by

Table 5.1.

Plastic regimes	p	q	r	$\dot{\varepsilon}_1$	$\dot{\varepsilon}_2$	$\dot{\varepsilon}_3$	$\lvert\dot{\varepsilon}\rvert$
AB	1	0	-1	$\lvert\dot{\varepsilon}_1\rvert$	0	$-\dot{\varepsilon}_1$	$\sqrt{2}\ \lvert\dot{\varepsilon}_1\rvert$
BC	0	1	-1	0	$\lvert\dot{\varepsilon}_2\rvert$	$-\dot{\varepsilon}_2$	$\sqrt{2}\ \lvert\dot{\varepsilon}_2\rvert$
CD	-1	1	0	$-\dot{\varepsilon}_2$	$\lvert\dot{\varepsilon}_2\rvert$	0	$\sqrt{2}\ \lvert\dot{\varepsilon}_2\rvert$
DE	-1	0	1	$-\dot{\varepsilon}_3$	0	$\lvert\dot{\varepsilon}_3\rvert$	$\sqrt{2}\ \lvert\dot{\varepsilon}_3\rvert$
EF	0	-1	1	0	$-\dot{\varepsilon}_3$	$\lvert\dot{\varepsilon}_3\rvert$	$\sqrt{2}\ \lvert\dot{\varepsilon}_3\rvert$
FA	1	-1	0	$\lvert\dot{\varepsilon}_1\rvert$	$-\dot{\varepsilon}_1$	0	$\sqrt{2}\ \lvert\dot{\varepsilon}_1\rvert$

$$D_u = \sigma_Y |\dot{\varepsilon}_i|. \tag{5.35}$$

When the stress point is at a vertex, we may have a vertex of type A where only one stress component is not zero and hence has the value $\pm\sigma_Y$. At point A, $\sigma_1 = \sigma_Y$, $\sigma_2 = \sigma_3 = 0$. The directions of the vector with components $\dot{\varepsilon}_1, \dot{\varepsilon}_2, \dot{\varepsilon}_3$ are bounded by those of the vectors associated with the regimes FA and AB. We thus have :

$$\dot{\varepsilon}_1 \geq |\dot{\varepsilon}_2|, \qquad \dot{\varepsilon}_1 \geq |\dot{\varepsilon}_3| \qquad \text{and} \qquad \dot{\varepsilon}_1 > 0 .$$

Thus,

$$D_v = \sigma_Y \max |\dot{\varepsilon}_i|. \tag{5.36}$$

A second vertex is of type B where $\sigma_1 = \sigma_2 = \sigma_Y$, $\sigma_3 = 0$, $D_v = \sigma_Y(\dot{\varepsilon}_1 + \dot{\varepsilon}_2)$.

We also have $\dot{\varepsilon}_1 + \dot{\varepsilon}_2 = -\dot{\varepsilon}_3$ with $\dot{\varepsilon}_1 > 0$ and $\dot{\varepsilon}_2 > 0$.

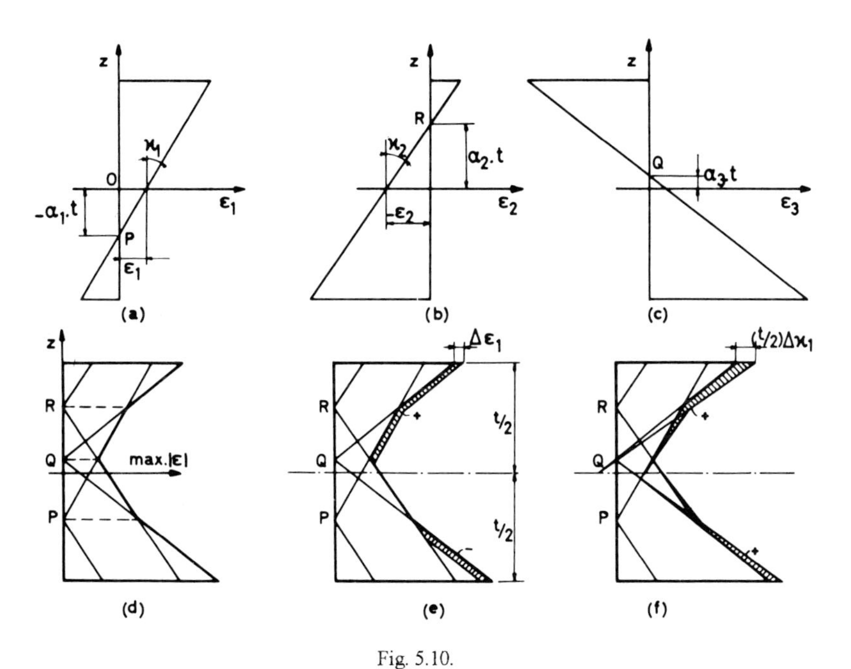

Fig. 5.10.

Thus, eq. (5.36) still holds. The same conclusion would be obtained for point E. As eq. (5.35) is a particular case of eq. (5.36) all cases are covered by eq. (5.36). By integration over the thickness t, we find the dissipation D per unit midsurface of the shell. The strain rates $\dot{\varepsilon}_i$ vary with z according to

$$\dot{\varepsilon}_{1z} = \dot{\varepsilon}_1 + \dot{\kappa}_1 z, \qquad \dot{\varepsilon}_{2z} = \dot{\varepsilon}_2 + \dot{\kappa}_2 z, \qquad \dot{\varepsilon}_{3z} = -\left(\dot{\varepsilon}_{1z} + \dot{\varepsilon}_{2z}\right). \tag{5.37}$$

The direction 3 is that of the z-axis, and the directions 1 and 2 coincide with the φ and θ directions. Note that the direction φ and θ are interchangeable as far as the yield condition is concerned.

A typical distribution of $\dot{\varepsilon}_{iz}$ is shown on fig. 5.10. The three parameters α_1, α_2, α_3, defined by

$$\alpha_1 = \frac{\dot{\varepsilon}_1}{t\,\dot{\kappa}_1}, \qquad \alpha_2 = \frac{\dot{\varepsilon}_2}{t\,\dot{\kappa}_2}, \qquad \alpha_3 = \frac{\dot{\varepsilon}_1 + \dot{\varepsilon}_2}{t\left(\dot{\kappa}_1 + \dot{\kappa}_2\right)}, \tag{5.38}$$

are sufficient to describe this distribution completely (see fig. 5.10 for notations). They locate the points P, Q, R of zero strain rate. The diagram of max $|\dot{\varepsilon}_i|$ is then constructed [fig. 5.10 (d)]. The dissipation D is given by the area of that diagram.

Now, relations (5.33) specialize to

$$N_1 = \frac{\partial D}{\partial \dot{\varepsilon}_1}, \qquad N_2 = \frac{\partial D}{\partial \dot{\varepsilon}_2}, \qquad M_1 = \frac{\partial D}{\partial \dot{\kappa}_1}, \qquad M_2 = \frac{\partial D}{\partial \dot{\kappa}_2}. \tag{5.39}$$

The derivatives in relations (5.39) are most readily evaluated from the variations of the areas of the diagram in fig. 5.10 (d) for small variations of $\dot{\varepsilon}_1$, $\dot{\varepsilon}_2$, $\dot{\kappa}_1$, $\dot{\kappa}_2$.

For example, limiting values of the ratios of the dashed areas in fig. 5.10 to the corresponding variations of the parameters give (see [5.8]) :

$$N_1 = \sigma_Y t \left[\frac{1}{2} - \alpha_3 - \left(\frac{1}{2} + \alpha_1 \right) \right] = -\sigma_Y t \left(\alpha_1 + \alpha_3 \right),$$

$$M_1 = \sigma_Y \frac{t^2}{4} \left[\frac{1}{4} - \alpha_3^2 + \left(\frac{1}{4} - \alpha_1^2 \right) \right] = \sigma_Y \frac{t^2}{4} \left[1 - 2\left(\alpha_1^2 + \alpha_3^2 \right) \right].$$

Similarly,

$$N_2 = -\sigma_Y \, t \left(\alpha_3 + \alpha_2\right),$$

$$M_2 = \sigma_Y \frac{t^2}{4}\left[1 - 2\left(\alpha_2^2 + \alpha_3^2\right)\right].$$

With the following definitions of the reduced generalized stresses,

$$n_1 = \frac{N_1}{N_p}, \qquad n_2 = \frac{N_2}{N_p}, \qquad m_1 = \frac{M_1}{M_p}, \qquad M_2 = \frac{M_2}{M_p}, \tag{5.40}$$

where $N_p = \sigma_Y t$ and $M_p = \sigma_Y \left(\frac{t^2}{4}\right)$, the preceding relations become

$$n_1 = -(\alpha_1 + \alpha_3), \qquad n_2 = -(\alpha_2 + \alpha_3),$$

$$m_1 = 1 - 2(\alpha_1^2 + \alpha_3^2), \qquad m_2 = 1 - 2(\alpha_2^2 + \alpha_3^2), \tag{5.41}$$

Eqs. (5.41) are parametric equations of the desired yield surface, in (n_1, n_2, m_1, m_2) space.

To obtain the complete surface, all relative positions of points P, Q, R of fig. 5.10 must be considered, with corresponding values of α_1, α_2, α_3.

table 5.2. Points P, Q, R are distinct.

Central point	$\pm n_1$	$\pm n_2$	$\pm m_1$	$\pm m_2$
P	$-(\alpha_1 + \alpha_3)$	$-(\alpha_3 - \alpha_2)$	$1 - 2(\alpha_1^2 + \alpha_3^2)$	$2(\alpha_2^2 - \alpha_3^2)$
Q	$-(\alpha_1 + \alpha_3)$	$-(\alpha_3 + \alpha_2)$	$1 - 2(\alpha_1^2 + \alpha_3^2)$	$1 - 2(\alpha_3^2 + \alpha_2^2)$
R	$-(\alpha_3 - \alpha_1)$	$-(\alpha_3 + \alpha_2)$	$2(\alpha_1^2 - \alpha_3^2)$	$1 - 2(\alpha_3^2 + \alpha_2^2)$

Table 5.3. Points P, Q, R are not distinct.

Coincidence	Yield surface
$P \equiv Q$	$m_1 = \pm(1 - n_1^2)$
$Q \equiv R$	$m_2 = \pm(1 - n_2^2)$
$R \equiv P$	$m_1 - m_2 = \pm[1 - (n_1 - n_2)^2]$

This discussion [5.8] gives the results shown in tables 5.2 and 5.3. Note that point P, Q, R must fall within the shell thickness. Hence α_1, α_2 and α_3 are bounded by $-\frac{1}{2}$ and $+\frac{1}{2}$.

Two important particular situations occur when either one of the axial forces or one of the moments can be eliminated. The yield surface is then an ordinary surface in three-dimensional space (and not a hypersurface in a space of a higher number of dimensions). As the yield surface is symmetric with respect to the origin, representation and specification of one half of it is sufficient.

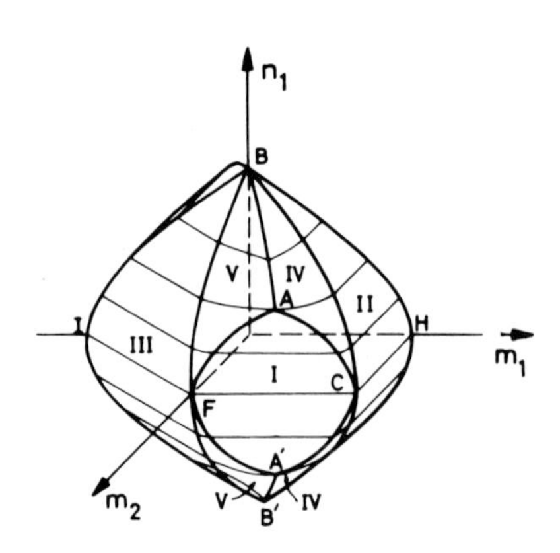

Fig. 5.11.

Consider first $n_2 = 0$ (see [5.8]). The yield surface is shown in fig. 5.11. Part I is a plane with the equation

$$m_2 = 1 , \tag{5.42}$$

corresponding to Q ≡ R (table 5.3), + sign, with $n_2 = 0$. Part II corresponds to P ≡ Q, + sign. Its equation is

$$m_1 = 1 - n_1^2 . \tag{5.43}$$

(cylinder with axis Om_2). Part III corresponds to P ≡ R, - sign. Its equation is

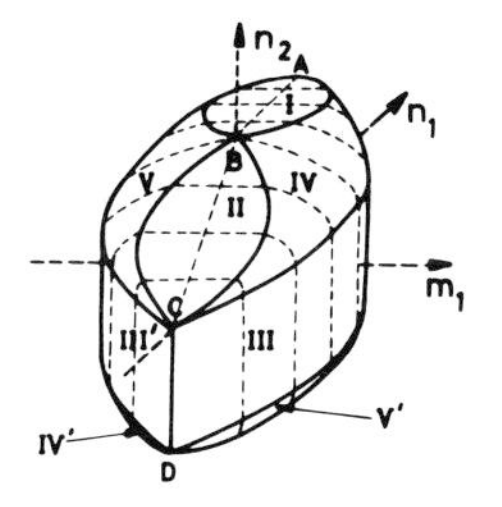

Fig. 5.12.

$$m_1 - m_2 = n_1^2 - 1 . \tag{5.44}$$

Eqs. of parts IV and V are obtained from table 5.1, taking Q and R for the central points, respectively. Condition $n_2 = 0$ gives $\alpha_3 = -\alpha_2$, and elimination of α_3 and α_2 from the remaining three relations furnishes, using the plus signs,

$$\text{for part IV,} \qquad m_1 = 1 - 2\left[\left(n_1 + \frac{\sqrt{1 - m_2}}{2}\right)^2 + \frac{1 - m_2}{4}\right], \tag{5.45}$$

for part V, $$m_1 = 2\left[\left(n_1 + \frac{\sqrt{1-m_2}}{2}\right)^2 - \frac{1-m_2}{4}\right], \tag{5.46}$$

The second interesting case occurs when m_2 is a reaction, because $\dot{\kappa}_2=0$ (cylindrical shells with axisimmetrical loading, $\dot{\kappa}_2$ denoting circumferential rate of curvature [5.9, 5.10 and 4.3]). Elimination of m_2 results in the yield surface of fig. 5.12. Part I is a plane with equation

$$n_2 = 1, \tag{5.47}$$

bounded by parabolic arcs

$$m_1 = \pm 2\, n_1\, (1 - n_1)\,. \tag{5.47'}$$

Part II is also a plane bounded by parabolic arcs. The relevant equations are

$$n_2 - n_1 = 1\,, \tag{5.48}$$

$$m_1 = \pm 2\, n_1 (1 + n_1)\,. \tag{5.48'}$$

Part III is a portion of the parabolic cylinder with equation

$$m_1 = 1 - n_1^2, \tag{5.49}$$

bounded by its intersection with the planes

$$m_1 = 0\,,$$

$$2\, n_2 - n_1 = \pm 1. \tag{5.49'}$$

Part IV and V belong to the paraboloids with equations

$$m_1 = \pm \frac{1}{2}\left[2 - (2\, n_2 - 1)^2 - (2\, n_2 - 2\, n_1 - 1)^2\right]. \tag{5.50}$$

Note that the yield surfaces are formed of parts with different analytic expressions, and the normality law can only be formulated explicitly when one knows on what part the stress point is located.

5.4.3. *Purely statical method : adaptation of the reactions.*

To begin with, suppose there is no reaction among the stress-type variables, the number of which is, say, three.

A part of the yield surface is shown schematically on fig. 5.13. Choose fixed values Q_1^0 and Q_2^0 such that the point with the coordinates Q_1^0, Q_2^0, 0 falls within the yield surface. Now let Q_3 vary from zero to the highest value Q_3 compatible with the yield condition of the material (expressed in terms of the ordinary stress components ($\sigma_x,...,\tau_{xy},...$). The coordinates Q_1^0, Q_2^0, Q_3 are that of a point P of the desired yield surface. In a general manner one attributes fixed values to all generalized stresses but one, which will be given the extreme magnitudes compatible with the yield condition of the material. In this way, the yield surface is generated point by point.

Now, if there exist reactions, the only nonfixed generalized stress depends not only on the yield condition of the material but remains a function of the reactions which, as a rule, are not fixed.

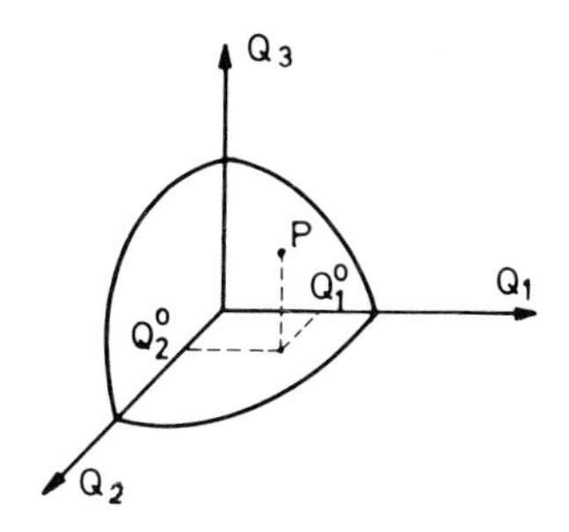

Fig. 5.13.

The following theorem has been proved [5.5] :

Adaptation : If one fixes all generalized stresses but one, the reactions adapt themselves to give the nonfixed generalized stress a maximum positive or minimum negative value.

Thus, the procedure just described still holds when reactions exist, and the reactions may be completely ignored.

We illustrate the results by these two examples :

1. *Bar with square cross section, subjected to two orthogonal bending moments* M_x *and* M_y (fig. 5.14). We treat this problem by a variational procedure [3.8]. When the cross section is completely plastic, we have at all points $|\sigma| = \sigma_Y$. If $y = \varphi(x)$ is the equation

of the boundary between the regions of tensions ($\sigma = \sigma_Y$) and compressions ($\sigma = -\sigma_Y$), we have

$$M_x = \int_{-t/2}^{t/2} \left[\int_{-t/2}^{\varphi(x)} x(-\sigma_Y)\, dy + \int_{\varphi(x)}^{t/2} x\sigma_Y\, dy \right] dx$$

$$= -2\,\sigma_Y \int_{-t/2}^{t/2} x\, \varphi(x)\, dx\,,$$

and, similarly,

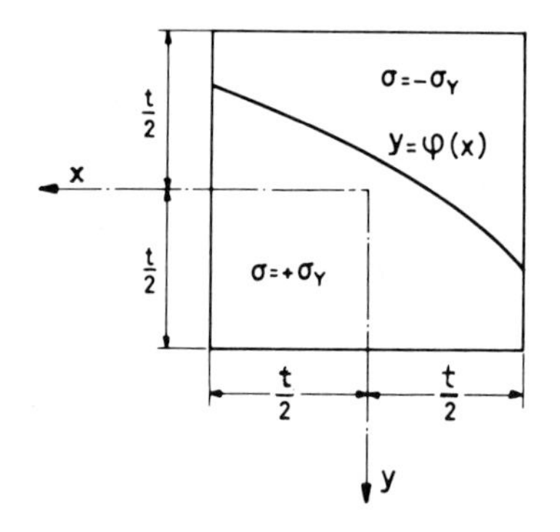

Fig. 5.14.

$$M_y = \sigma_Y \int_{-t/2}^{t/2} \left[\frac{t^2}{4} - \varphi(x)^2 \right] dx\,. \tag{5.51}$$

Consider a fixed value of M_x and assume M_y to be an analytic maximum. If the stress distribution is then varied by an arbitrary small amount $\delta\,\sigma$, we have $\delta\, M_x = 0$ because M_x is fixed ; and $\delta\, M_y = 0$ because M_y is a maximum. Hence, α being a parameter, we can write, using eq. (5.51),

$$\delta M_y + \alpha\,\delta M_x = -2\,\sigma_Y \int_{-t/2}^{t/2} [\,\varphi(x) + \alpha\,x]\;\delta\varphi(x)\,dx = 0\,,$$

for all $\delta\,\sigma$, that is for all $\delta\,\varphi(x)$. Consequently, $\varphi(x) + \alpha\,x = 0$.

We see from the preceding relation that the boundary between tensions and compressions is a ray emanating from the origin. Hence, we readily obtain

$$M_x = \frac{3}{4}\sigma_Y\,\frac{t^3}{8}\,\alpha\,, \qquad M_y = \frac{2}{3}\sigma_Y\,\frac{t^3}{8}\left(3 - \alpha^2\right).$$

Eliminating α from the two equations above, and letting

$$m_x = \frac{M_x}{M_p}\,, \qquad m_y = \frac{M_y}{M_p}\,, \qquad M_p = \sigma_Y\,\frac{t^2}{4}\,,$$

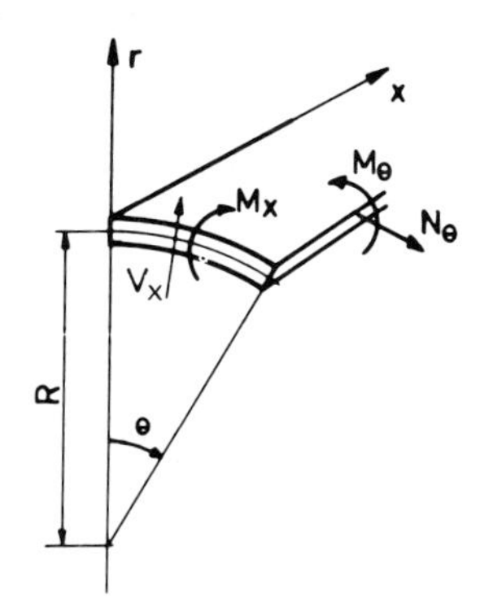

Fig. 5.15.

we finally obtain the desired equation

$$m_y + \frac{3}{4}m_x^2 = 1 \tag{5.52}$$

of the yield curve.

Note that eq. (5.52) is valid only for $|\alpha| \leq 1$, that is for $|m_x / m_y| \leq 1$. Because the yield curve is symmetric with respect to the rays $m_x / m_y = \pm 1$, the remaining part is obtained without difficulty.

2. *Circular cylindrical "sandwich" shell without axial force, subjected to axially symmetrical loading.* Because of the absence of axial force and of the symmetry of revolution, the only nonvanishing stress type variables are V_x, M_x, M_θ and N_θ (see fig. 5.15).

Shear forces V_x are reactions.

For a circular cylindrical shell, symmetry of revolution enforces $\dot{\kappa}_\theta = 0$. Hence, M_θ is a reaction and we must simply determine the yield condition in terms of M_x and N_θ.

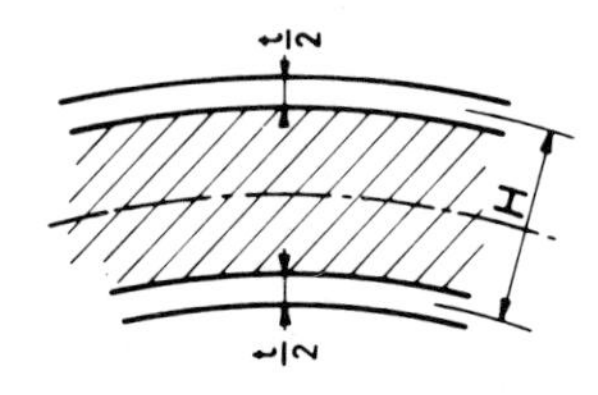

Fig. 5.16.

The "sandwich" shell is formed of a core with thickness H and two face sheets with thickness $t/2$ each (see fig. 5.16). The core carries exclusively shear forces V_x, to which it is always exceedingly resistant. The face sheets carry all other stresses and are assumed in a state of plane stress.

Denote by $\sigma_{\theta e}$ and σ_{xe} the principal normal stresses in the external face sheet, and by $\sigma_{\theta i}$ and σ_{xi} those in the internal sheet. We then have :

$$N_\theta = \frac{t}{2}\left(\sigma_{\theta\, i} + \sigma_{\theta\, e}\right),$$

$$M_\theta = \frac{tH}{4}\left(\sigma_{\theta\, i} - \sigma_{\theta\, e}\right),$$

$$M_x = \frac{tH}{4}\left(\sigma_{x\, i} - \sigma_{x\, e}\right). \tag{5.53}$$

Introducing reduced generalized stresses

$$n_\theta = \frac{N_\theta}{N_p}, \qquad m_\theta = \frac{M_\theta}{M_p}, \qquad m_x = \frac{M_x}{M_p},$$

where $N_p = \sigma_Y\, t$ and $M_p = \sigma_Y\ (tH/2)$, we have

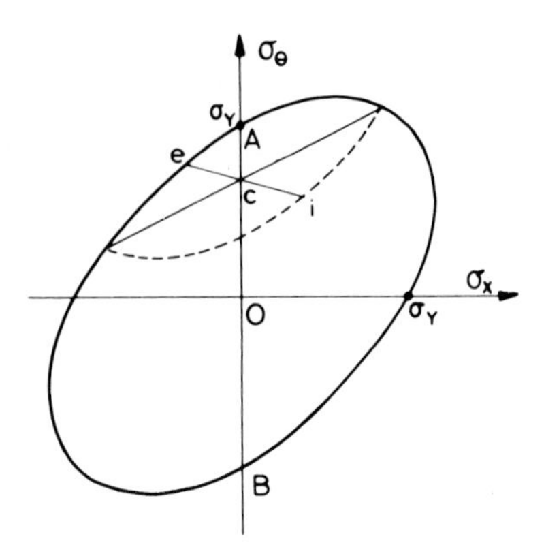

Fig. 5.17.

$$n_\theta = \frac{\sigma_{\theta\, i} + \sigma_{\theta\, e}}{2\,\sigma_Y},$$

$$m_\theta = \frac{\sigma_{\theta\, i} - \sigma_{\theta\, e}}{2\,\sigma_Y},$$

$$m_x = \frac{\sigma_{xi} - \sigma_{xe}}{2\,\sigma_Y}. \qquad (5.54)$$

Assume the material of the sheets obeys von Mises'yield condition :

$$\sigma_x^2 + \sigma_\theta^2 - \sigma_x\,\sigma_\theta = \sigma_Y^2, \qquad (5.55)$$

represented by the ellipse of fig. 5.17. The state of stress in the shell is represented by points e and i on fig. 5.17, with coordinates (σ_{xe}, $\sigma_{\theta\, e}$) and (σ_{xi}, $\sigma_{\theta\, i}$), respectively. According to relations (5.54), the coordinates of the midpoint c of segment ei give n_x and n_θ, whereas the projections of the segment ei on the axes gives m_x and m_θ (positive factors $1/\sigma_Y$ and $1/2\ \sigma_Y$ being irrelevant). Because $n_x = 0$, the point c must remain on the σ_θ axis. For a given position of point c (between points A and B) corresponding to some value of n_θ, plastification of the shell element requires that at least one of the two points e and i be on the yield locus.

Now, the adaptation theorem tells us that the slope of segment ei must be such that its projection on the σ_x axis is a maximum. This condition yields

$$m_\theta = \frac{m_x}{2}, \tag{5.56}$$

and points e and i both lie on the yield locus.

It is easily seen that when point c moved from A to B with condition (5.56) satisfied, the interaction relation is

$$\frac{3}{4} m_x^2 + n_\theta^2 = 1. \tag{5.57}$$

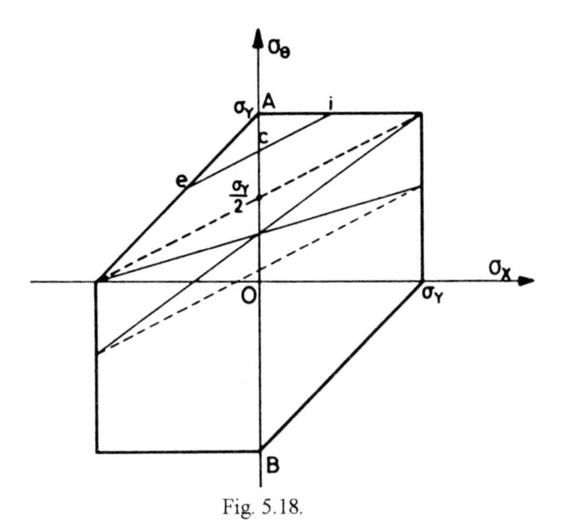

Fig. 5.18.

If the face-sheet material of the shell obeys Tresca's yield condition, represented on fig. 5.18, a similar analysis will furnish (a) for $0 \leq n_\theta \leq 1/2$,

$$n_\theta \leq \frac{m_\theta}{m_x} \leq 1 - n_\theta \qquad \text{and} \qquad m_x = 1, \tag{5.58}$$

and (b) for $1/2 \leq n_\theta \leq 1$,

$$m_\theta = \frac{m_x}{2} \qquad \text{and} \qquad \frac{m_x}{2} + n_\theta = 1. \tag{5.59}$$

The interaction curve is given by eqs. (5.58) and (5.59).

5.4.4. *Method of lower and upper bounds.*

Let us imagine an isolated element of a structure. For a beam, such an element may be specified by a line element of length dx along the undeformed axis. This element is bounded by the normal cross sections of the beam through the endpoints of the line element and by part of the lateral surface of the beam. For a plate or shell, an element may be specified by an infinitesimal rectangle of sides ds_1 and ds_2 on the undeformed median surface ; it is bounded by the normals of this surface through the points of the rectangle and by parts of the two surfaces of the shell, fig. 5.19.

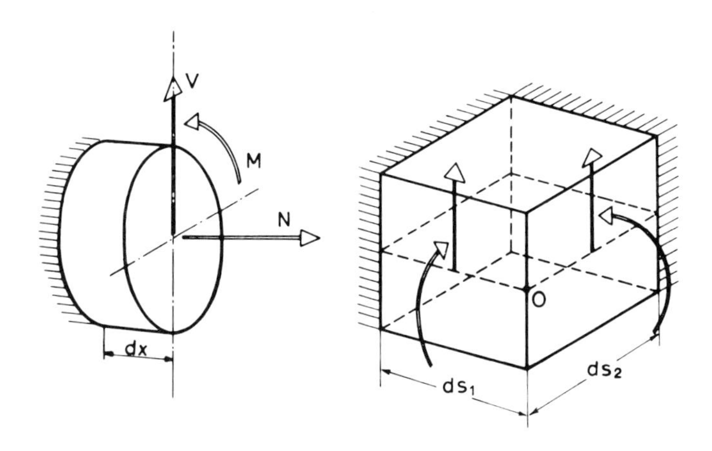

Fig. 5.19.

A structural element of this kind may be regarded as a free body subjected to the resultant forces and couples of the stresses transmitted by neighbouring elements and such loads as may be directly applied to the considered element.

Any combination of stress resultants that causes the element to yield specifies a point of the yield locus.

From this point of view, the fundamental theorems of limit analysis of Sections 3.1 and 3.2 can be used to obtain yield surfaces in generalized stress space :

1. Any licit stress distribution on the element will furnish generalized stresses that will be the coordinates of a point on or within the yield surface ;

2. Any licit strain rate distribution across the element will be associated, through the normality law, with generalized stresses that will be the coordinates of a point on or outside the yield surface.

The yield surface can thus be bounded from the interior and exterior.

The application of the lower-bound theorem is obviously identical to the statical method of Section 5.4.3. Indeed, reciprocity of shearing stresses is the only condition enforced by local equilibrium, and plastic admissibility of a stress distribution reduces to not violating the yield condition (in terms of stress components). Usually, the stress distribution is varied to maximize one generalized stress while fixing the others, in order to move the representing point from the inside onto the yield surface, as explained in Section 5.4.3. Note that the stress distributions will in general correspond to nonvanishing "reaction". For example, suppose that the curvature $\dot{\kappa}_x$ vanishes whereas the extension rate $\dot{\varepsilon}_x$ does not vanish. Hence, σ_x cannot be eliminated from the yield condition in terms of stress components, and the distribution of σ_x on the cross section will generally correspond to $M_x \neq 0$. However, according to the adaptation theorem, the stress distributions can be chosen without regard to the values of the reactions.

On the other hand, when the internal restraint concerns strain-rate components at every point - for example $\dot{\gamma}_{xy} = 0$ or $\dot{\varepsilon}_x = 0$ everywhere), the corresponding stress component (τ_{xy} or σ_x) can be eliminated from the yield condition (see Section 5.3.3 and Chapter 11) as well as the reactions they produce.

Consider now the application of the upper-bound theorem. Suppose that, within the frame work of the basic assumptions, the flow mechanism of a structural element is completely known (as in the shell example treated in Section 5.4.2). Application of the upper-bound theorem will then directly furnish the exact yield surface. Actually, the procedure is identical to using the dissipation power (Section 5.4.2), as we shall show.

Let $\dot{q}_1,...,\dot{q}_n$ be the generalized strain rates (curvature rates, extension rates of the median surface) used to describe the flow mechanism of an element. The corresponding strain-rate components are given by relations of the type

$$\dot{\varepsilon} = \dot{\varepsilon}\left(\dot{q}_1,...,\dot{q}_n,z\right), \tag{5.60}$$

where z is the distance of the considered point to the midsurface. Now, regard the $\dot{q}_i$ as the parameters of the problem. Through relations (5.60), each set of $\dot{q}_i$ gives a distribution of $\dot{\varepsilon}$ (z)to which the yield condition and the normality law relate a distribution σ (z). Hence, we can write

$$\sigma = \sigma (\dot{q}_i , \ldots \dot{q}_n , z) . \tag{5.61}$$

Next we obtain the corresponding generalized stress Q_i by integration over the thickness. The type of integration to be done is often obvious : if $\dot{q}_i$ is a curvature rate, Q_i is the corresponding moment, if $\dot{q}_i$ is an extension rate, Q_i is the corresponding axial force, etc. It must however be emphasized that, as a rule, the type of integration to achieve is determined by the *definition* of Q_i. This definition is related to the expression of dissipation power D (see relation 5.10). Integration over z will result in a function of $\dot{q}_1,\ldots,\dot{q}_n$ as shown by relation (5.61). This function Q_i must be such that $D = \sum Q_i \dot{q}_i$. Because D must be homogeneous of the order one in $\dot{q}_i$ (see Section 2.4) we have $Q_i = \frac{\partial D}{\partial \dot{q}_i}$. We conclude that, from the very definition of Q_i, integration over z must yield the same parametric form for Q_i as is obtained from relations (5.33).

It follows that the use of the lower-bound and upper-bound theorems to obtain yield surfaces in generalized stresses will differ from the other methods only if the yield mechanism *of the element* does not correspond by the normality law to the licit state of stress.

We illustrate this discussion by two examples.

1. *Beam with uniform solid cross section subjected to bending and torsion* [3.8].

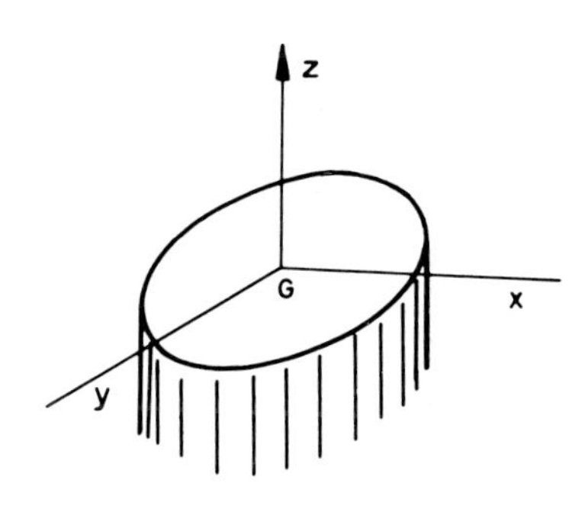

Fig. 5.20.

Let Gx and Gy be the two principal axes of inertia of the cross section of a beam (G being the centroid, fig. 5.20). Tresca's yield condition for pure torsion is $\tau^2 = \tau_{zx}^2 + \tau_{zy}^2 = \frac{\sigma_Y^2}{4}$. Assume a uniform distribution of shear stress τ similar to that corresponding to the limit torque M_{tp} but with $\tau < \frac{\sigma_Y}{2}$. The reduce torque m_t will be

$$m_t = \frac{M_t}{M_{tp}} = \frac{\tau}{\frac{\sigma_Y}{2}} . \tag{5.62}$$

To this shear stress distribution, we superimpose a distribution of normal stress σ to that corresponding to the fully plastic bending moment M_p about axis x, but with $\sigma < \sigma_Y$. The reduced bending moment is

$$m \equiv \frac{M}{M_p} = \frac{\sigma}{\sigma_Y} . \tag{5.63}$$

The two distributions of stresses will best combine to satisfy

$$\sigma^2 + 4\,\tau^2 = \sigma_Y^2 \tag{5.64}$$

everywhere, to furnish a fully plastified cross section. Substitution of expressions (5.62) and (5.63) for τ and σ, respectively, into eq. (5.64) yields

$$m^2 + m_t^2 = 1 . \tag{5.65}$$

Relation (5.65) is a lower bound for the interaction curve because the state of stress is licit but does not correspond to a kinematically admissible strain-rate distribution.

To obtain an upper bound, we arbitrarily *assume* that yielding results in a rate of curvature $\dot{\kappa}_y$ in the G_{zy} plane and a rate of twist $\dot{\kappa}_{xy}$ about G, but no warping of the cross section. The corresponding strain rates are

$$\dot{\varepsilon}_z = y\,\dot{\kappa}_y , \qquad \dot{\gamma}_{zx} = -y\,\dot{\kappa}_{xy} , \qquad \dot{\gamma}_{zy} = x\,\dot{\kappa}_{xy} . \tag{5.66}$$

The dissipation is

$$D = \int_A \left(\sigma_z \dot{\varepsilon}_z + \tau_{zx} \dot{\gamma}_{zx} + \tau_{xy} \dot{\gamma}_{zy} \right) dA$$

or

$$D = M \dot{\kappa}_y + M_t \dot{\kappa}_{xy} .$$

Because $\tau^2 = \tau_{zx}^2 + \tau_{zy}^2$, Tresca's condition (5.64) may be written

$$\sigma_z^2 + 4 \left(\tau_{zx}^2 + \tau_{zy}^2 \right) = \sigma_Y^2 . \tag{5.67}$$

The normality law (2.7) applied to condition (5.67) furnishes

$$\dot{\varepsilon}_z = 2 \lambda \sigma_z , \qquad \dot{\gamma}_{zx} = 8 \lambda \tau_{zx} , \qquad \dot{\gamma}_{zy} = 8 \lambda \tau_{zx} . \tag{5.68}$$

From comparison of relations (5.66) and (5.67) we have

$$\sigma_z = \frac{y \dot{\kappa}_y}{2 \lambda} , \qquad \tau_{zx} = \frac{y \dot{\kappa}_{xy}}{8 \lambda} , \qquad \tau_{zy} = \frac{x \dot{\kappa}_{xy}}{8 \lambda} . \tag{5.69}$$

Substituting expressions (5.69) for σ_z, τ_{zx}, τ_{zy} into relation (5.67), we obtain

$$2 \lambda = \left(\frac{\dot{\kappa}_y}{\sigma_Y} \right) \left\{ y^2 + \alpha^2 (x^2 + y^2) \right\}^{1/2} ,$$

where

$$\alpha = \frac{\dot{\kappa}_{xy}}{2 \dot{\kappa}_y} .$$

We now readily obtain

$$m \equiv \frac{1}{M_p} \int_A \sigma_x y \, dA = \frac{\sigma_Y}{M_p} \int_A \frac{y^2}{\left[y^2 + \alpha^2 (x^2 + y^2) \right]^{1/2}} dA ,$$

$$m_t \equiv \frac{1}{M_{tp}} \int_A \left(x\,\tau_{zy} - y\,\tau_{zx} \right) dA = \frac{\sigma_Y\,\alpha}{2\,M_{tp}} \int_A \frac{(x^2+y^2)\,dA}{\left[y^2 + \alpha^2 (x^2+y^2) \right]^{1/2}}, \tag{5.70}$$

Relations (5.70) are the parametric equations of the interaction curve m versus m_t, α being used as parameter. Different cross-sections will give rise to different coefficients $\frac{\sigma_Y}{M_p}$ and $\frac{\sigma_Y}{M_{tp}}$. Because the relations (5.70) are obtained from a kinematically admissible strain-rate field [eqs. (5.66)], they furnish an upper bound for the exact interaction curve. Fig. 5.21, shows the two bounds for a circular cross section, curves a ; and for square cross section, curves b.

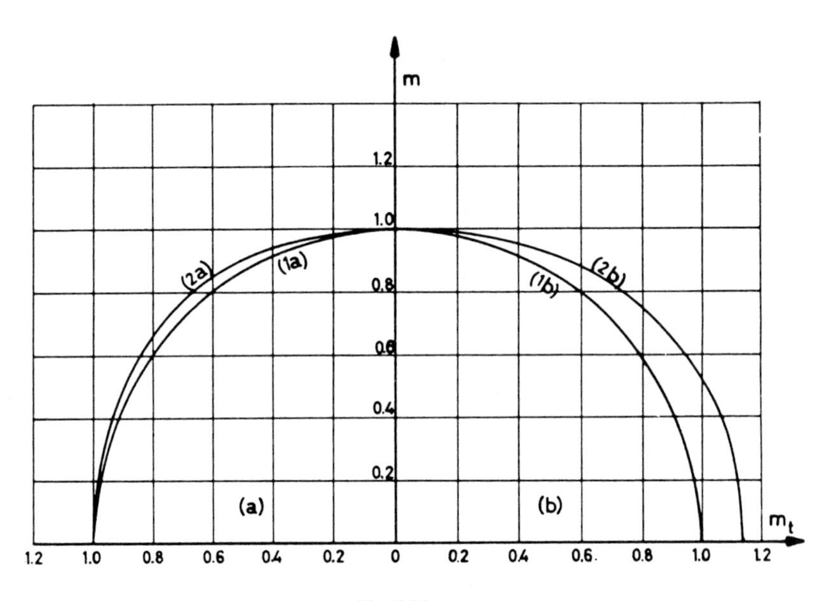

Fig. 5.21.

(From Plastic Analysis of Structures by P.G. Hodge Jr. Copyright 1959, McGraw-Hill Book Company, Used by permission of McGraw-Hill Book Company, Inc.)

2. *Cylindrical shell without axial force [3.9].*

Consider a circular cylindrical shell as shown in fig. 5.22. It is subjected to an internal pressure that may solely depend on the coordinate x. Because of the symmetry of revolution and the absence of axial load, the only nonvanishing stress resultants and moments are those shown in fig. 5.15. Shear force V_x is a reaction (see Section 5.3.3), as well as bending moment M_θ (see Section 5.1.4, example 2). Tresca's yield condition may be expressed as

$$\max\left[\,|\sigma_x|,|\sigma_\theta|,|\sigma_x-\sigma_\theta|\,\right]=\sigma_Y, \tag{5.71}$$

for all r. Plasticity occurs at a given level r of the shell thickness when one of the six relations

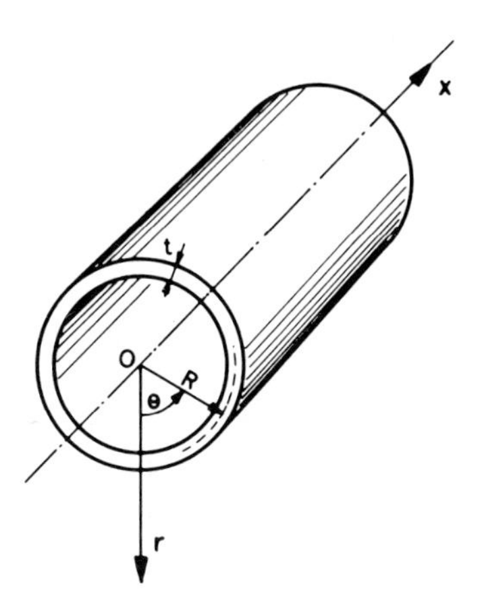

Fig. 5.22.

(5.71) is satisfied. Obviously, the relation to satisfy may change at some values of r.

Let

$$m_x=\frac{M_x}{M_p}=\frac{M_x}{\sigma_Y\,\frac{t^2}{4}},$$

$$n_\theta=\frac{N_\theta}{N_p}=\frac{N_\theta}{\sigma_Y\,t}, \tag{5.72}$$

be the reduced bending moment and axial force, respectively. They are the only generalized variables. We now distribute σ_x and σ_θ over the thickness in order to maximize n_θ for a given m_x, while satisfying the yield condition (5.71). We first try

$$m_x=1. \tag{5.73}$$

It is admissible as long as $n_\theta \leq 1/2$. Indeed, as shown on fig. 5.23, we have $n_\theta = 1/2 - \alpha$, and the reaction m_θ is $m_\theta = 1/2 - 2\alpha^2$, with $0 \leq \alpha \leq 1/2$. For $1/2 \leq n_\theta \leq 1$, we may not have $m_x = 1$ anymore. Fig. 5. 23 (b) shows the distribution that gives, for every m_x, the largest value of n_θ compatible with the yield condition (5.71). We obtain

$$m_x = 1 - 4\alpha^2,$$

$$n_\theta = \frac{1}{2} + \alpha \tag{5.74}$$

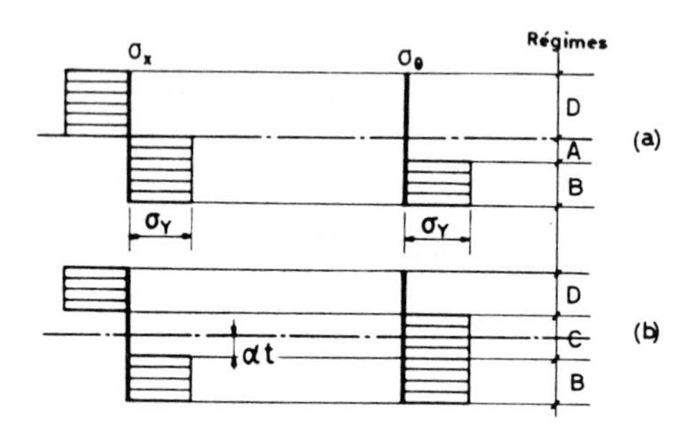

Fig. 5.23.

(and the reaction $m_\theta = \frac{1}{2} - 2\alpha^2$, with $0 \leq \alpha \leq \frac{1}{2}$). Eqs. (5.73) and (5.74) define the interaction curve shown on fig. 5.24.

To prove that this curve is not only a lower bound but the exact interaction curve, we must associate, to the stress distributions above, corresponding strain-rate distributions. The principal strain rates $\dot{\varepsilon}_x$, $\dot{\varepsilon}_\theta$ may be represented by the coordinates of a point (fig. 5.25). According to the normality law, all points of the first quadrant with $\dot{\varepsilon}_x > 0$ and $\dot{\varepsilon}_\theta > 0$ are associated with the vertex B of Tresca's yield hexagon (see the insert in fig. 5.25). Similarly, it is easily seen that the regions bounded by the axes and the diagonals of the second and fourth quadrants correspond to the six vertices, whereas these six rays correspond to the six sides of the hexagon.

We now recall (see Section 5.4.2) that

$$\dot{\varepsilon}_x = \dot{\varepsilon}_{xo} + z\dot{\kappa}_x \ ,$$
$$\dot{\varepsilon}_\theta = \dot{\varepsilon}_{\theta o} + z\dot{\kappa}_\theta = \dot{\varepsilon}_{\theta o} \ , \tag{5.75}$$

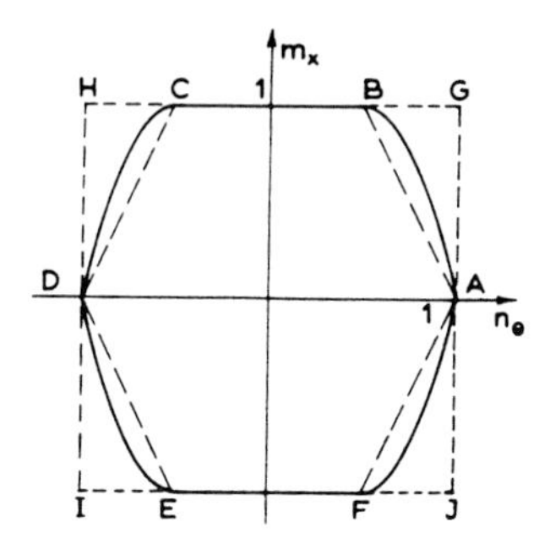

Fig. 5.24.

(From Plastic Analysis of Structures by P.G. Hodge Jr. Copyright 1959, McGraw-Hill Book Company, Used by permission of McGraw-Hill Book Company, Inc.)

where $\dot{\varepsilon}_{xo}$, $\dot{\varepsilon}_{\theta o}$, $\dot{\kappa}_x$, $\dot{\kappa}_\theta$ are rates of extension and of curvature of the midsurface in the longitudinal and circumferential directions, respectively. Curvature rate $\dot{\kappa}_\theta$ is known to vanish (Section 5.1.4) and z is the radial distance of a layer from the median surface, counted as positive when directed inwards. Hence, if a point M (fig. 5.25) represents the state of strain rate at a point P on the median surface, strain rates of the various points on the normal at P will be represented by the points of a segment parallel to the $\dot{\varepsilon}_x$-axis and with center M. But because the axial force n_x must vanish, there must be as many layers in regime D as in regime B (see the insert in fig. 5.25), or as many in regime A as in regime E. Consequently, the point M must fall on a certain dashed ray LON. Et is then easily seen that the kinematically admissible strain-rate distributions represented by the points of segments M' M_1 M' and M'' M_2 M'' in fig. 5.25 correspond to the stress distributions in fig. 5.23. The interaction curve shown in fig. 5.24 is therefore exact. Finally, we note that the parametric eqs. (5.73) and (5.74) could have been deduced from the results of Section

5.4.2 where the dissipation function was used. Indeed, if we relabel x and θ as 1 and 2, respectively, eqs. (5.75) show that the central point in fig. 5.10 is either P or Q because R goes to infinity. In both cases $\alpha_1 = -\alpha_3$ because $n_1 = 0$, and $m_1 = 1 - 4\alpha^2$, whereas $n_2 = \frac{1}{2} + \alpha_1$, because $\alpha_2 = \frac{1}{2}$ or $-\frac{1}{2}$, respectively (see [5.9]).

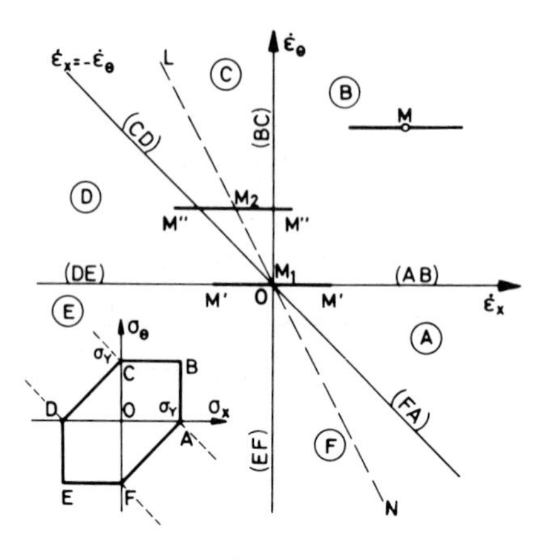

Fig. 5.25.

5.5. Simplified yield surface.

5.5.1. *Convenience of a simplified yield surface.*

A linear yield condition is very attractive from the mathematical point of view. Indeed, if the yield surface consists of plane facets, as long as the stress point remains on a given plane

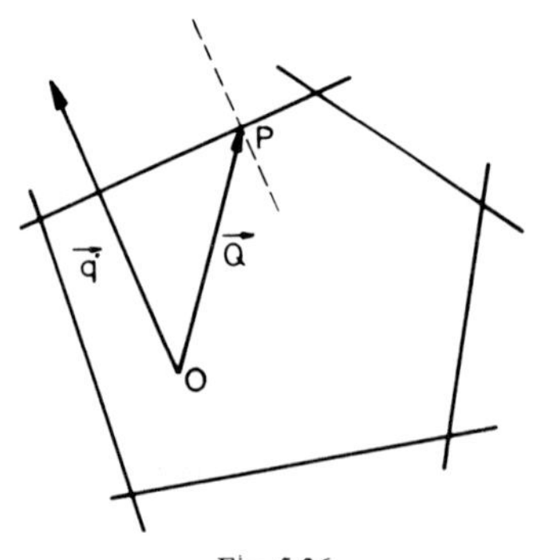

Fig. 5.26.

the yield vector retains the same direction and the yield "mechanism" does not change. All possible yield mechanisms can thus be classified into a finite number of plastic "regimes", each regime corresponding to the contact of the stress point with one plane, one edge, or one vertice (fig. 5.26). Limit analysis proves much easier in these circumstances than when the yield surface is curved, especially when the principal directions are known beforehand. There reasons explain the preference given to Tresca's yield condition over the condition of von Mises. But when generalized stresses are used, even Tresca's linear yield condition need not result in a piecewise linear yield surface, as shown in the second example of Section 5.4.4. The exact yield surface must then be replaced by a polyhedron, either inscribed (dashed lines in fig. 5.24) or circumscribed (dotted lines in fig. 5.24).

5.5.2. *Influence on the limit load of linearizing the yield condition.*

For the sake of simplification, consider a yield condition that involves only two generalized stresses Q_1 and Q_2. Let the exact yield curve be represented by the heavy line in fig. 5.27, and let P_l be the corresponding exact limit load for a given structure and a given loading scheme. If, instead of the exact yield curve e, the inscribed polygon i (dashed lines) is used, the state of stress at collapse for polygon i is licit for curve e. If the corresponding limit load is P_i, the static theorem furnishes the inequality

$$P_i \leq P_l . \tag{5.76}$$

If P_i^- is statical licit load for i, we have

$$P_i^- \leq P_i . \tag{5.77}$$

On the other hand, using a circumscribed polygon c would result in

$$P_l \leq P_c . \tag{5.78}$$

If P_c^+ is a kinematical licit load for c, the kinematic theorem asserts that

$$P_c \leq P_c^+ . \tag{5.79}$$

Inequalities (5.76) to (5.79) may be combined into one continued inequality,

$$P_i^- \leq P_i \leq P_l \leq P_c \leq P_c^+ , \tag{5.80}$$

that enables us to bound the error introduced by the linearization process as follows.

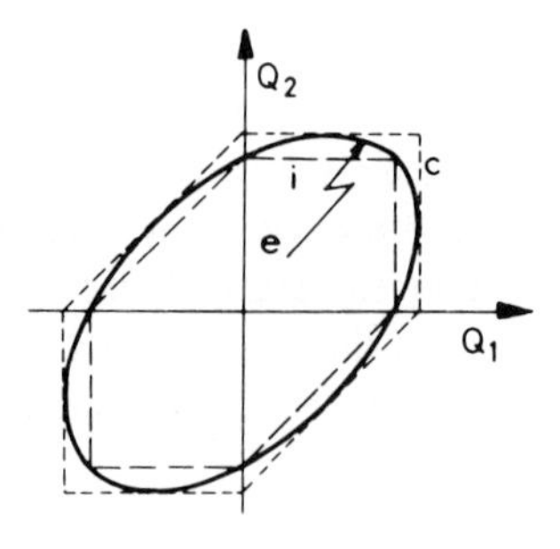

Fig. 5.27.

The limit load of a structure is directly proportional to the yield stress σ_Y of the material. If the yield stress is multiplied by a factor that is greater or smaller than unity, the yield surface is similarly expanded or contracted with respect to the origin. Hence, if polygons i and c are homothetical with factor k, we have

$$P_c = kP_i \, . \tag{5.81}$$

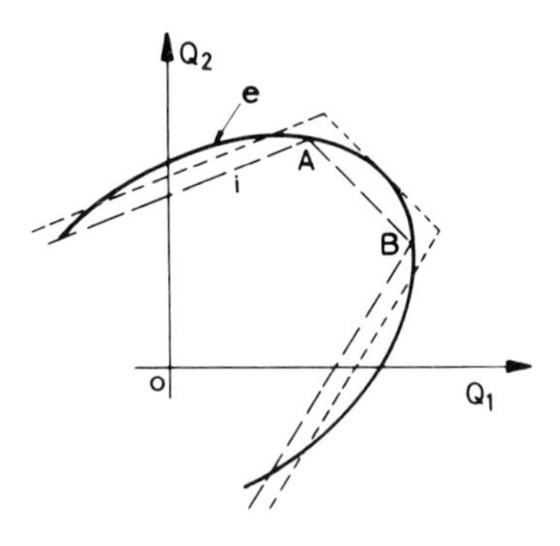

Fig. 5.28.

If we want to bound P_l from above and below, it is sufficient to know either P_i or P_c, say P_i. We then determine the lowest expansion factor k that makes polygon i become circumscribed to e. Note also that, if the stress point, for all plastic regions, remains on a certain part of curve e, as AB in fig. 5.28, the yield polygon obtained by expansion of polygon i must be external to curve e in that part only. This remark enables us to use the smallest possible expansion factor k.

5.5.3. *Linearization process.*

A first method consists of finding the exact yield surface and then inscribe (or circumscribe) more or less arbitrarily a polyhedron in order to simplify the subsequent analysis.

A second point of view consists of using an approximation not to the yield surface but to the structure itself [5.11].

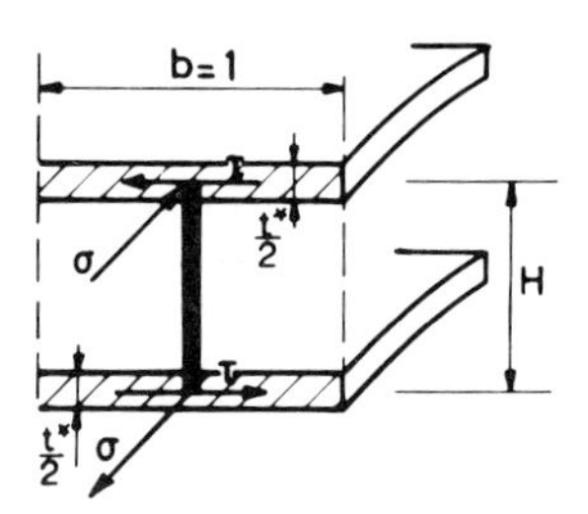

Fig. 5.29.

Consider (fig. 5.29) a structure (plate, shell) of the "ideal sandwich" type defined in Section 5.4.3, example 2. On a cross section, the axial force and bending moment per unit of length are [see eq. (5.53)]

$$N = \frac{t^*}{2}\left(\sigma_i + \sigma_e\right), \tag{5.82}$$

$$M = \frac{t^* H}{4}\left(\sigma_i - \sigma_e\right), \tag{5.83}$$

where σ_i and σ_e are the normal stresses in the internal and external sheets, respectively, $t^*/2$ the thickness of each sheet and H the core thickness. A twisting moment M_t (per unit of length) will produce shear stresses τ such that

$$M_t = \tau \frac{\overset{*}{t}}{2} H. \tag{5.84}$$

If we denote by $\overset{*}{\tau}_Y$ the yield shearing stress of the sandwich sheets and $\overset{*}{\sigma}_Y$ its yield stress in tension, the full plastic axial force, bending moment and twisting moment are

$$\overset{*}{N}_p = \overset{*}{\sigma}_Y \overset{*}{t},$$

$$\overset{*}{M}_p = \overset{*}{\sigma}_Y H \frac{\overset{*}{t}}{2},$$

$$\overset{*}{M}_{tp} = \overset{*}{\tau}_Y H \frac{\overset{*}{t}}{2},$$

For the sandwich structure to be substituted for a structure with uniform cross section of thickness t and yield stresses σ_Y and τ_Y, H and $\overset{*}{t}$ must be so chosen as to satisfy

$$\sigma_Y t = \overset{*}{\sigma}_Y \overset{*}{t} \equiv N_p, \tag{5.85}$$

$$\overset{*}{\sigma}_Y H \frac{\overset{*}{t}}{2} = \sigma_Y \frac{t^2}{4} \equiv M_p, \tag{5.86}$$

$$\overset{*}{\tau}_Y H \frac{\overset{*}{t}}{2} = \tau_Y \frac{t^2}{4} \equiv M_{tp}. \tag{5.87}$$

Relations (5.85) to (5.87) then give

$$\overset{*}{\sigma}_Y \overset{*}{t} = \sigma_Y t,$$

$$H = \frac{t}{2},$$

$$\frac{\overset{*}{\tau}_Y}{\overset{*}{\sigma}_Y} = \frac{\tau_Y}{\sigma_Y}. \tag{5.88}$$

The last relation of eqs.(5.88) just means that the same physical yield condition must hold for both structures.

Because the generalized stresses N, M, M_t are *linear* functions of the stress components (relation (5.82) to (5.84)), any linear yield condition in terms of the stress components will thus generate a linear yield condition in terms of generalized stresses. Hence, the preceding procedure directly furnishes an approximate yield polyhedron without recourse to the exact yield surface. This polyhedron is *inscribed* in the exact yield surface because the latter is convex and, according to relations (5.88), both have in common the points on the axes. Fig. 5.30 shows, for example, the yield polyhedron of a sandwich cylindrical shell axisymmetrically loaded and made of a Tresca material. Equations of the various planes are :

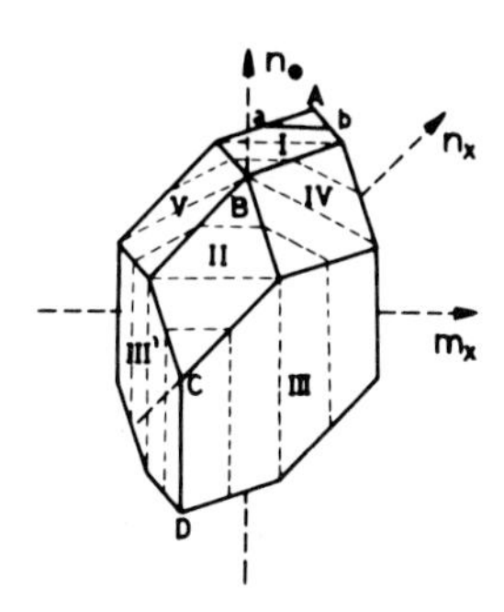

Fig. 5.30.

I : $n_\theta = 1,$

II : $n_\theta - n_x = 1,$

III : $n_x - m_x = -1,$

IV : $2\,n_\theta - n_x + m_x = 2$

V : $2\,n_\theta - n_x - m_x = 2,$ (5.89)

Coordinates x and θ are as indicated in fig. 5.22. This yield polyhedron is a linearization of the yield surface in fig. 5.12, with which it should be compared (subscripts 1 and 2 becoming x and θ, respectively).

5.5.4. *Example of application.*

Following Hodge and Sawczuk [5.12], consider an infinitely long cylindrical shell without axial force, loaded in a cross section by a radial uniform line load of total magnitude

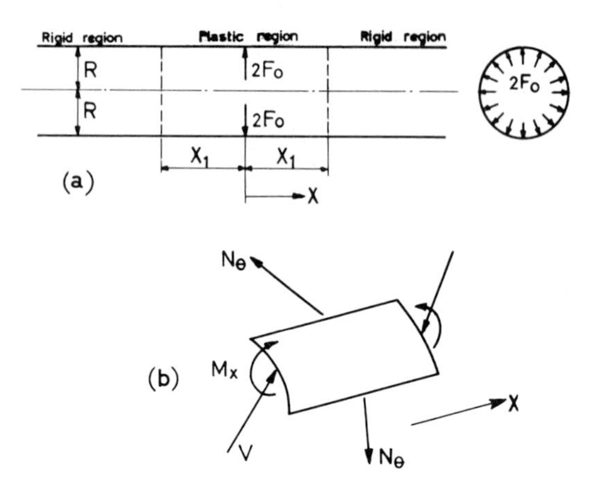

Fig. 5.31.

$2\,F_o$ (fig. 5.31).

Because of the symmetry, the plastic region at collapse will extend over an (unknown) length x, on each side of the loaded cross section where we locate the origin of the abscissae. The only generalized stresses are M_x and N_θ (fig. 5.31).

We assume the real shell to exhibit uniform thickness and satisfy von Mises' yield criterion. We compare it to these shells regarded as approximations to the former : (a) sandwich shell made of von Mises material ; (b) uniform shell made of Tresca material ; (c) sandwich shell made of Tresca material ; (d) shell with the (arbitrarily simple) "limited interaction" yield curve.

The various yield curves are shown in fig. 5.32. In the absence of twisting moment, conditions (5.88) reduce to

$$\sigma_Y^* t = \sigma_Y t ,$$

$$H = \frac{t}{2}.$$

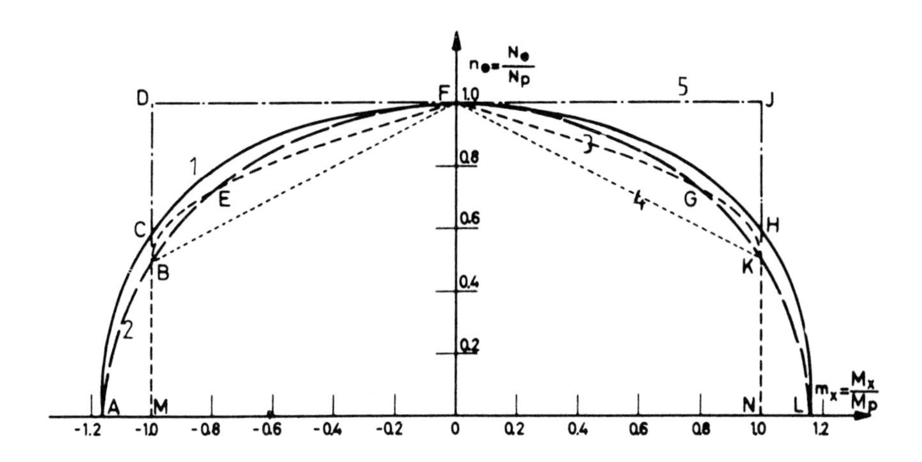

Fig. 5.32. Interaction curves.

The lines 2, 3 and 4 in fig. 5.32 were obtained in Section 5.4. Line 1 was given by Hodge [5.13] as a special case of the yield surface of a shell of revolution loaded with rotational symmetry. Line 5 was arbitrarily chosen for the sake of simplicity.

Line 4 is a linear approximation to lines 1, 2 and 3. Line 5 is an arbitrary linear approximation to lines 1 to 4.

The load was nondimensionalized as follow : $f_o = R^{1/2}\left(M_p N_p\right)^{-1/2} F_o$.

The various exact limit loads obtained by Hodge and Sawczuk are given in column 3 of table 5.4.

In column 4 we find the deviations of the preceding limit loads from those of the real shell (line 1 in fig. 5.32). Column 5 contains the relative deviations. The figures given in columns 6 and 7 were obtained as follows : each yield curve was similarly enlarged or reduced by a factor k (see Section 5.5.2) to become : (a) external to curve 1 for curves 2, 3, and 4 or (b) internal to curve 1 for curve 5 (see fig. 5.33). The exact yield curve is thus bounded, for each "approximation" shell, by two approximate curves that furnish lower and upper bounds for the limit load. The relative deviations of these bounds are indicated in columns 6 and 7. Note that the original "limited interaction" curve (line 5 in fig. 5.32) gives neither a lower bound nor an upper bound as it arbitrarily cuts across the exact yield curve. After geometrically similar expansion or reduction, it does furnish bounds to the limit load, which differ appreciably from the other bounds. Nevertheless, the original curve itself furnishes a fairly good approximate limit load.

Table 5.4. Comparison of reduced limit loads $f_o = R^{1/2} \left(M_p N_p\right)^{-1/2} F_o$.

1	2	3	4	5	6	7
Type of shell	Yield condition	f_o	Deviation from uniform von Mises	Deviation Actual	Lower bound	Upper bound
Uniform	von Mises	1.949	0	0	0	0
Sandwich	von Mises	1.905	- 0.044	- 2.3	- 2.3	2.6
Uniform	Tresca	1.826	- 0.123	- 6.3	- 6.3	7.9
Sandwich	Tresca	1.732	- 0.217	- 11.1	- 11.1	5.7
Limited interaction curve		2.000	+ 0.051	2.6	- 20.0	18.2

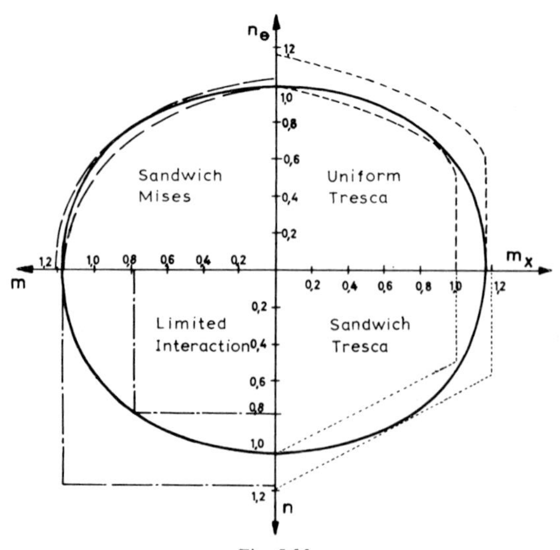

Fig. 5.33.

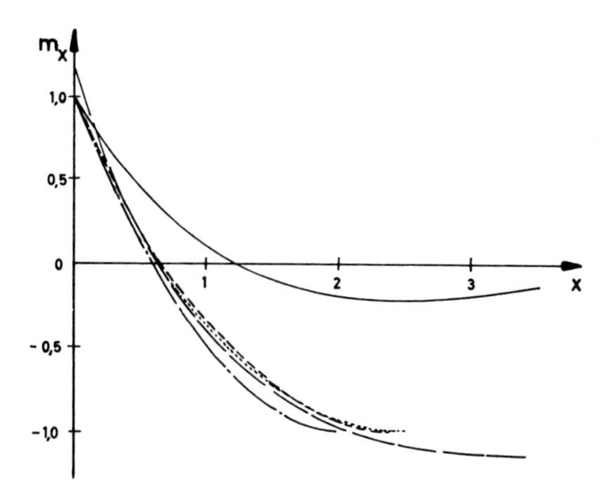

Fig. 5.34.

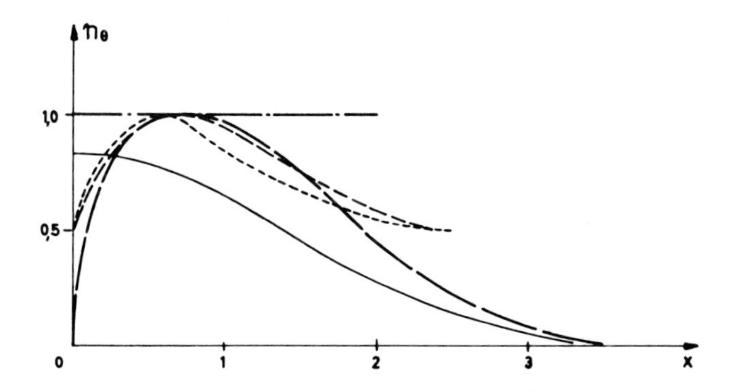

Fig. 5.35.

Diagrams of the reduced bending moment $m_x = \frac{M_x}{M_p}$ and axial force $n_\theta = \frac{N_\theta}{N_p}$ are shown in figs. 5.34 and 5.35, where elastic diagrams corresponding to the maximum load in elastic range are also given. It is easily seen that an important redistribution of both n_θ and m_x takes place prior to collapse.

5.6. Yield conditions of reinforced concrete plates and shells.

5.6.1. *General assumptions.*

Each layer of concrete is assumed to be in state of plane stress and regarded as a rigid-perfectly plastic material. In terms of the in-plane principal stresses σ_1 and σ_2, the yield condition of such a layer is given in fig. 5.36, where σ_c' is the compressive yield strength (crushing strength), whereas the tensile stress has a vanishing value because the concrete is cracked under any tensile stress.

Steel reinforcements are supposed to have a rigid-perfectly plastic behaviour, with the same tensile and compressive yield stress. Now, according to Johansen [5.14], we introduce the important concept of "yield lines" ("fracture lines" or "hinges lines"). Indeed, cracks patterns can be observed in some local regions of reinforced concrete slabs or shells (see fig. 9.2, 9.3 and 10.3). A yield-line is interpreted as a mathematical idealization of such a narrow zone.

The present yield condition, established by the junior author (G.S) by using the rate of dissipation (upper bound method, [5.15], [5.16]), extends those by Capurso [5.17] and Sawczuk-Olszak [5.18].

A layer of parallel reinforcing bars separated by a distance p and inclined of the angle α on the normal to the yield-line (fig. 5.3) is constituted by a number of bars equal to

$$n = \frac{|\cos\alpha|}{p} \qquad (5.90)$$

for a unit length of yield line.

If A denotes the cross-sectional area of any reinforcement, the strength force of the layer by unit length of yield-line is equal to :

$$N_s = \Omega_s \sigma_Y \qquad (5.91)$$

where the quantity :

$$\Omega_s = \frac{A}{p}\cos^2\alpha \tag{5.92}$$

involves the reinforcement ratio A/p. This quantity is additive if we consider several layers.

5.6.2. *The dissipation in a concrete layer.*

First, the rate of dissipation by unit of volume D_c associated with the yield curve of fig. 5.36 is given by :

$$D_c = \sigma_1 \dot{\varepsilon}_1 + \sigma_2 \dot{\varepsilon}_2. \tag{5.93}$$

Its value for each regime of plastic yielding is obtained by the use of the normality law and is given in table 5.5.

Table 5.5.

Regime	Constraints	Yielding law	Power of dissipation
O	$\sigma_1 = \sigma_2 = 0$	$\dot{\varepsilon}_1 \geq 0,\ \dot{\varepsilon}_1 \dot{\varepsilon}_2 \geq 0$	$D_c = 0$
OA	$\sigma_2 = 0$	$\dot{\varepsilon}_1 = 0,\ \dot{\varepsilon}_2 \geq 0$	$D_c = 0$
A	$\sigma_1 = -\sigma'_c\ \sigma_2 = 0$	$\dot{\varepsilon}_1 \leq 0,\ \dot{\varepsilon}_1 \dot{\varepsilon}_2 \leq 0$	$D_c = -\sigma'_c \dot{\varepsilon}_1$
AB	$\sigma_1 = -\sigma'_c$	$\dot{\varepsilon}_1 \leq 0,\ \dot{\varepsilon}_2 = 0$	$D_c = -\sigma'_c \dot{\varepsilon}_1$
B (other regimes by symmetry)	$\sigma_1 = \sigma_2 = -\sigma'_c$	$\dot{\varepsilon}_1 \leq 0,\ \dot{\varepsilon}_1 \dot{\varepsilon}_2 \geq 0$	$D_c = -\sigma'_c (\dot{\varepsilon}_1 + \dot{\varepsilon}_2)$

5.6.3. *Assumptions on discontinuities.*

Let $\dot{w}$ be the rate of transverse displacement, $\dot{u}$ and $\dot{v}$ the rates of the tangential displacements, respectively normal and parallel to the yield-line (fig. 5.37). $\dot{w}$ must obviously remain a continuous function, and consequently, $\frac{\partial \dot{w}}{\partial s}$ also :

$$\dot{w}\,] = \frac{\partial \dot{w}}{\partial s}\,] = 0\,. \tag{5.94}$$

Table 5.6.

Regime	Range of η	Plastic yielding in the reinforcement	Expression of n	Expression of m	Extreme points
CD'	$-1 \leq -\eta < \rho$	$\dot{u}_o + 2\rho\dot{\theta} < 0$	$n = -\gamma - \frac{1+\eta}{2}$	$m = -2\gamma\rho - \frac{1+\eta^2}{2}$	C $(-\gamma-1, -2\gamma\rho)$ D' $\left(-\gamma-\frac{1-\rho}{2}, -2\gamma\rho-\frac{1-\rho^2}{2}\right)$
AA'	$\rho < -\eta \leq 1$	$\dot{u}_o + 2\theta\dot{\rho} < 0$	$n = \gamma - \frac{1+\eta}{2}$	$m = 2\gamma\rho - \frac{1+\eta^2}{2}$	A $(\gamma, 2\gamma\rho)$ A' $\left(\gamma-\frac{1-\rho}{2}, 2\gamma\rho-\frac{1-\rho^2}{2}\right)$
A'D'	$\eta = -\rho$	$\dot{u}_o + 2\rho\dot{\theta} = 0$	$n = \gamma s - \frac{1-\rho}{2}$ $\leq$ $(\lvert s \rvert \leq 1)$	$m = 2\gamma\rho s - \frac{1+\rho^2}{2}$	A' $\left(\gamma-\frac{1-\rho}{2}, 2\gamma\rho-\frac{1-\rho^2}{2}\right)$ D' $\left(-\gamma-\frac{1-\rho}{2}, -2\gamma\rho-\frac{1-\rho^2}{2}\right)$

Table 5.7.

Regime	Range of η	Plastic yielding in the reinforcement	Expression of n	Expression of m	Extreme points
AA'	$-1 \le \eta < -\rho$	$\dot{u}_o + 2\rho\dot{\theta} > 0$ $\dot{u}_o - 2\rho\dot{\theta} > 0$	$n = -\frac{1+\eta}{2} - 2\gamma$	$m = \frac{1-\eta^2}{2}$	A $[2\gamma, 0]$ A' $\left[-\frac{1-\rho}{2} + 2\gamma, \frac{1-\rho^2}{2}\right]$
B'B''	$-\rho < \eta < \rho$	$\dot{u}_o + 2\rho\dot{\theta} > 0$ $\dot{u}_o - 2\rho\dot{\theta} < 0$	$n = -\frac{1+\eta}{2}$	$m = \frac{1-\eta^2}{2} + 4\gamma\rho$	B' $\left[-\frac{1-\rho}{2}, \frac{1-\rho^2}{2} + 4\gamma\rho\right]$ B" $\left[-\frac{1+\rho}{2}, \frac{1-\rho^2}{2} + 4\gamma\rho\right]$
C''C	$\rho < \eta \le 1$	$\dot{u}_o + 2\rho\dot{\theta} < 0$ $\dot{u}_o - 2\rho\dot{\theta} < 0$	$n = -\frac{1+\eta}{2} - 2\gamma$	$m = \frac{1-\eta^2}{2}$	C'' $\left[-\frac{1+\rho}{2} - 2\gamma, \frac{1-\rho^2}{2}\right]$ C $[-1-2\gamma, 0]$

The problem becomes simpler if we make the following assumptions :

a) no sliding of both sides of the yield line :

$$\dot{v}\,]=0 \tag{5.95}$$

b) plane cross-sections remain plane during deformation :

$$\dot{u}\,]=\dot{u}_o+z\,\dot{\theta} \tag{5.96}$$

with

$$\dot{u}_o=\dot{u}\,], \qquad z=0, \qquad \dot{\theta}=\frac{\partial \dot{w}}{\partial n}\,]. \tag{5.97}$$

Hence, the strain rates in the yield-line are given by :

$$\dot{\varepsilon}_n=\dot{u}_o+z\,\dot{\theta}, \qquad \dot{\varepsilon}_s=0, \qquad \dot{\gamma}_{ns}=0. \tag{5.98}$$

5.6.4. *Yield condition for shells with simple reinforcement.*

It is obvious that n and s (fig. 5.37) are the principal axes of strain rates.

As $\dot{\varepsilon}_s$ vanishes, the concrete yields according to regime OC when $\dot{\varepsilon}_n>0$, and to the regime AB when $\dot{\varepsilon}_n\leq 0$. First, let us assume that the upper layers of the concrete are in compression. Let η be the non-dimensional ordinate of the neutral axis so that $\dot{\varepsilon}_n$ is negative for $z<\eta\frac{t}{2}$ (fig. 5.38). According to table 5.5, the rate of dissipation is :

$$D_c=-\sigma'_c\left(\dot{u}_o+z\,\dot{\theta}\right) \quad \text{if} \quad z\leq\eta\frac{t}{2},$$

$$D_c=0 \quad \text{if} \quad z>\eta\frac{t}{2}.$$

Now, let $\rho\frac{t}{2}$ be the ordinate of the steel reinforcement (fig.5.38). The rate of dissipation in the reinforcement per unit of length of yield-line is :

$$D_s = \sigma_y\, \Omega_s |\dot{u}_o + \rho \frac{t}{2} \dot{\theta}|.$$

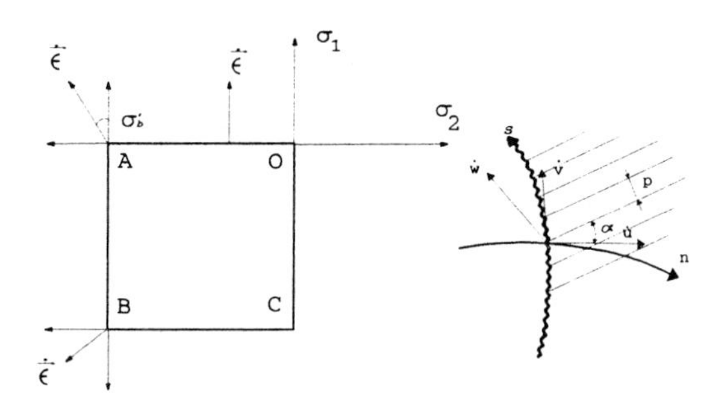

Fig. 5.36. Fig.5.37.

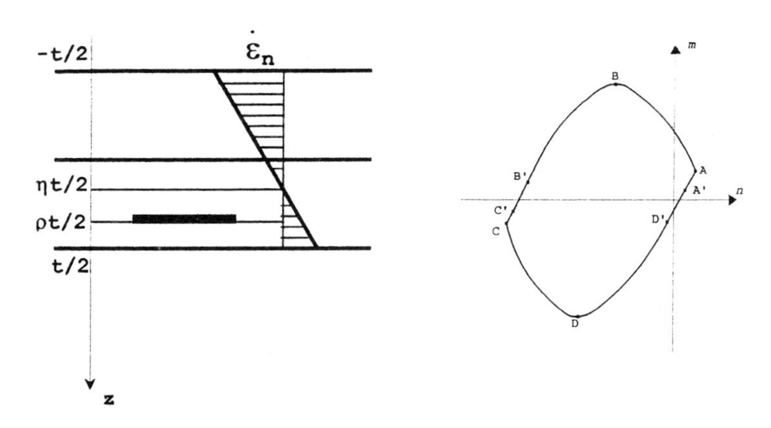

Fig. 5.38. Fig. 5.39.

The power of dissipation by unit length of yield-line is therefore :

$$\Phi = -\sigma'_c \int_{-t/2}^{\eta t/2} \left(\dot{u}_o + z\,\dot{\theta}\right) dz + \sigma_y\ \Omega_s\ |\dot{u}_o + \rho \frac{t}{2} \dot{\theta}|.$$

At yielding, the cross-section is subjected to a bending moment M and to a normal force N which are deduces by the normality law :

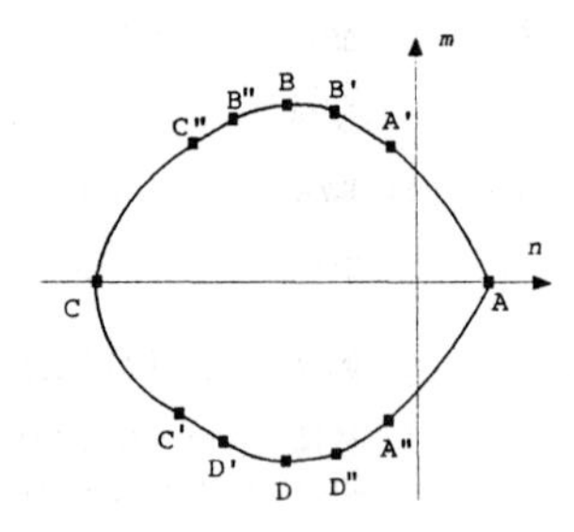

Fig. 5.40.

$$N = \frac{\partial \Phi}{\partial \dot{u}_o}, \qquad M = \frac{\partial \Phi}{\partial \dot{\theta}}, \qquad \Phi = N\,\dot{u}_o + M\dot{\theta}\,.$$

Introducing the following reduced variables:

$$n = \frac{N}{\sigma'_c t}, \qquad m = \frac{M}{\frac{\sigma'_c t^2}{4}}, \qquad \dot{\bar{u}}_o = \frac{\dot{u}_o}{\frac{t}{4}}, \qquad \varphi = \frac{\Phi}{\frac{\sigma'_c t^2}{4}}, \tag{5.99}$$

one has :

$$n = \frac{\partial \varphi}{\partial \dot{\bar{u}}_o}, \qquad m = \frac{\partial \varphi}{\partial \dot{\theta}}, \qquad \varphi = n\,\dot{\bar{u}}_o + m\dot{\theta}\,. \tag{5.100}$$

If γ denotes the reduced reinforcement ratio

$$\gamma = \frac{\sigma_y \Omega_s}{\sigma'_c t} \tag{5.101}$$

the dissipation is expressed, in terms of reduced variables, by

$$\varphi = -\frac{1+\eta}{2}\dot{\bar{u}}_o + \gamma|\dot{\bar{u}}_o + 2\rho\dot{\theta}| + \frac{1-\eta^2}{2}\dot{\theta}. \qquad (5.102)$$

This expression, which is not differentiable, has to be discussed with regard to the regime of plastic yielding :

a) *the reinforcement is fully in tension :*

$$-1 \leq \eta < \rho \; , \; \dot{\bar{u}}_o + 2\rho\dot{\theta} > 0 \, .$$

The normality law (5.100) gives :

$$n = \gamma - \frac{1+\eta}{2} \quad , \quad m = 2\gamma\rho + \frac{1-\eta^2}{2} \, . \qquad (5.103)$$

To obtain the yield condition, one has to eliminate the parameter η between both relations (5.103) and one gets easily the following condition :

$$m = 2n^2 + 2n(1-2\gamma) + 2\gamma^2 - 2\gamma(1+\rho) = 0 \qquad (5.104)$$

wich is indentical to that of Sawczuk and Olszak [5.18] :

Equation (5.104) represents a parabole, the apex of which is point B, (fig. 5.39) with coordinates ($\gamma - \frac{1}{2}$, $2\gamma\rho + \frac{1}{2}$) ; (5.104) is obtained by translation of (γ, $2\gamma\rho$) from the parabole corresponding to the plain concrete :

$$m = -2n(1+n). \qquad (5.105)$$

As the regime under consideration is allowed only when η is varying between -1 and ρ, we must preserve only the part of the parabole between points A (γ, $2\gamma\rho$) and B' $\left(\gamma - \frac{1+\rho}{2}, 2\gamma\rho + \frac{1-\rho^2}{2}\right)$ (fig. 5.39).

b) *the reinforcement is fully in compression :*

$$\rho < \eta < 1 \quad , \quad \dot{\bar{u}}_o + 2\rho\dot{\theta} < 0.$$

The normality law gives :

$$n = -\gamma - \frac{1+\eta}{2} \quad , \quad m = -2\gamma\rho + \frac{1-\eta^2}{2} \tag{5.106}$$

The stress state point belongs to a parabole obtained by translation of the parabole (5.105), but now in the opposite way $(-\gamma, -2\gamma\rho)$. As above, one must consider only the segment between the points C' $\left(-\gamma-\frac{1+\rho}{2}, -2\gamma\rho+\frac{1-\rho^2}{2}\right)$ and C $(-\gamma-1, -2\gamma\rho)$.

c) *the reinforcement is partly in compression :*

$$\eta = \rho, \quad \dot{\bar{u}}_o + 2\rho\dot{\theta} = 0.$$

There is a bifurcation of the stresses. The stress state point belongs to the straight segment between C' and B'. (*)

Il it is now assumed that the lower layers of the concrete are in compression $(z < \frac{\eta t}{2})$. Three new regimes of plastic yielding are obtained in a similar way ; they are shown in table 5.5.

Finally, the yield curve is ABB'C'CDD'A'A (fig. 5.39). We can chek easily the convexity of the rigid domain.

5.6.5 *Yield condition for shells with double reinforcements.*

By reasoning in a similar way, the yield curve for doubly reinforced shells is determined. The general case of the non-symmetrical reinforcement layers is done in [5.15]. The table 5.6 shows the quite interesting case of symmetrical reinforcements (at $z = \pm\rho\, dt$ with the same reinforcement ratio). These curves may be deduced in a way similar to that of Section 5.6.4 , from the following expression of the power of dissipation :

$$\varphi = -\frac{1+\eta}{2}\dot{\bar{u}}_o + \frac{1-\eta^2}{2}\dot{\theta} + \gamma\left(|\dot{\bar{u}}_o + 2\rho\dot{\theta}| + |\dot{\bar{u}}_o - 2\rho\dot{\theta}|\right). \tag{5.107}$$

(*) Remark that the function (5.102) is not differentiable at this point. The straight segment B'C' is the subdifferential to φ, in the sense of convex analysis [5.19].)

Obviously, the yield curve is symmetrical with respect to the n-axis, as shown in fig.5.40.

As a conclusion, a general approach for any number of reinforcement layers is thus established. Besides, as shown in [5.18], the present upper bound is the exact solution because, through the normality law, a licit stress distribution can be associated to the corresponding yield mechanisms.

5.7. Discontinuities.

5.7.1. *Introduction.*

When searching for fields of generalized stresses Q_i and strain rates $\dot{q}_i$, it often proves useful to introduce discontinuities. We already know some of these discontinuities from Com. V. : at a plastic hinge in a beam the slope jumps by a finite amount and in a plastically bent segment of the beam the stress σ jumps from σ_Y to $-\sigma_Y$ when the neutral layer is crossed. For a given theory, that is, for a given degree of idealization, these discontinuities are permissible because perfect plasticity relaxes some of the restraints of geometrical compatibility.

As a rule, displacements normal to the midsurface of a shell will be kept continuous if the shell is not to break, but slopes might be discontinuous, as well as displacements in the tangent plane. Obviously, these discontinuities must be regarded as limiting cases (that is, idealizations consistent with the level of the theory used) of very rapid variations over very narrow regions, in which compatibility remains fully satisfied. The stress fields will also be allowed to exhibit some discontinuities because stresses are no longer related in a unique manner to a continuous strain field.

It is important to know what the admissible discontinuities are and what relations they obey. We refer the reader to Prager [5.20] and Hill [5.21]. We restrict ourselves hereafter to the minimum amount of indications necessary for the applications in the coming chapters.

5.7.2. *Stress discontinuities.*

Consider (fig. 5.41) a point O on a surface of discontinuity S, and a elementary parallelepiped with center O. The general rule is as follows : *elementary forces and moments forces and moments that balance each other across the discontinuity surface must be continuous.*

On the other hand, elementary forces and moments that balance each other along the discontinuity surface may experience discontinuities across that surface.

If we examine, for example, a plate subjected to bending, the discontinuity surface reduces to a discontinuity line DD as shown in fig. 5.42. The action at O across the line is a moment vector with components M_x and M_{xy}. This vector must be continuous across the

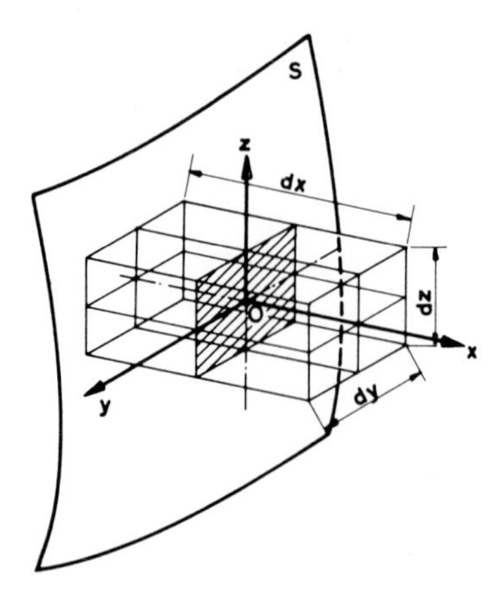

Fig. 5.41.

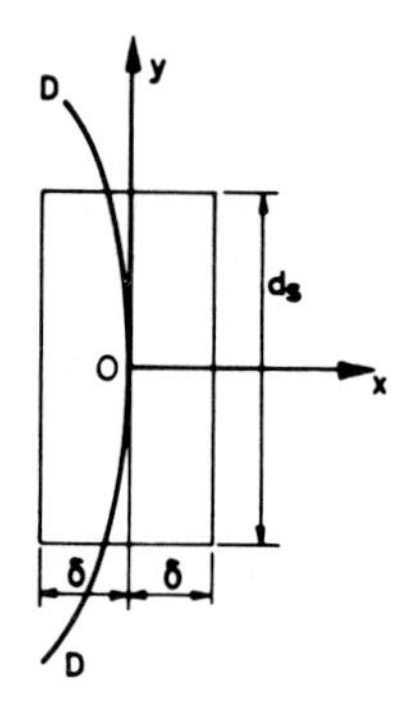

Fig. 5.42.

line, and hence there is no discontinuity admissible on M_x and M_{xy}. We express this result symbolically as

$$M_x] = 0, \qquad M_{xy}\,] = 0. \tag{5.108}$$

The remaining component M_y of the moment tensor at O may be discontinuous :

$$M_y] \neq 0. \tag{5.109}$$

The equilibrium of an element with center O furnishes the relations :

$$\frac{\partial M_x}{\partial x} + \frac{\partial M_{xy}}{\partial y} = V_x , \tag{5.110}$$

$$\frac{\partial M_y}{\partial y} + \frac{\partial M_{xy}}{\partial x} = V_y , \tag{5.111}$$

$$\frac{\partial V_x}{\partial x} + \frac{\partial V_y}{\partial y} = -p , \tag{5.112}$$

Eq. (5.110) shows that

$$V_x] = 0 \tag{5.113}$$

whereas eq. (5.111) does not exclude

$$V_y] \neq 0 \tag{5.114}$$

because $\frac{\partial M_y}{\partial y}$ may be discontinuous as well as M_y.

The preceding considerations will be used in Chapter 10 in dealing with reinforced concrete shells, as well as in Chapter 11 where plane stress and strain are studied.

5.7.3. *Strain-rate discontinuities.*

As noted in Section 5.6.1, *the field of transversal displacement rates* $\dot{w}$ *must be continuous*. If we examine the consequences of this statement for the first derivatives $\frac{\partial \dot{w}}{\partial x}$ and $\frac{\partial \dot{w}}{\partial y}$, we see that we must have

$$\frac{\partial \dot{w}}{\partial y}] = 0 \tag{5.115}$$

(to exclude different displacement rates for points with coordinates $(-\delta, ds/2)$ and $(+\delta, ds/2)$ with vanishingly small δ, fig. 5.42). But we may have

$$\frac{\partial \dot{w}}{\partial x}] \neq 0 \tag{5.116}$$

a situation where the discontinuity lime DD (fig. 5.42) is a "yield line" or "hinge line" that generalizes the plastic hinge of beams.

Obviously, a certain quantity of energy is dissipated in the strain-rate discontinuities, and due account must be taken of this fact. Indeed, some mechanisms will be made solely of discontinuities, as we shall see in the coming chapters.

5.8. Final remark.

Derivation of yield conditions is necessary not only for plates, shells and disks but also for beams subjected to combined loadings, as was discussed in Com. V in Section 5.2 for the simple case of simultaneous action of bending moment and axial force. Many papers have been devoted to this subject, among which we may quote the book of 1981 by Zyczkowski [5.22] containing detailed discussions of most (if not all) possible cases and an extremely extensive list of references.

5.9. Problems.

5.9.1. Show that the yield condition for a sandwich shell of revolution axisymmetrically loaded and made of a von Mises material is :

$$\left(n_\theta + m_\theta\right)^2 - \left(n_\theta + m_\theta\right)\left(n_\varphi + m_\varphi\right) + \left(n_\varphi + m_\varphi\right)^2 = 1,$$

$$\left(n_\theta - m_\theta\right)^2 - \left(n_\theta - m_\theta\right)\left(n_\varphi - m_\varphi\right) - \left(n_\varphi - m_\varphi\right)^2 = 1,$$

with $n = N/N_p$ and $m = M/M_p$.

5.9.2. Determine analytically the minimum amplification coefficient k to apply to the yield curve of the sandwich structure to have it circumscribe the exact yield curve (corresponding to uniform cross section of a Tresca material).

(a) For a rectangular beam subjected to bending and axial force the exact interaction curve being (see Com. V., relation (5.4)), $m = 1 - n^2$.

Answer : k = 1.25.

(b) For a cylindrical shell without axial force and axisymmetrically loaded.

Answer : k = 1.225.

5.9.3. Determine analytically the minimum reduction coefficient k to apply to the "square" yield condition (of the type of curve 5, fig. 5.32) to have it inscribed in the exact Tresca yield curve :

(a) When the yield curve m_1 versus n_1 is considered.

Answer : k = 0.50.

(b) When the yield curve m_1 versus n_2 is considered.

Answer : k = 0.75.

5.9.4. Determine the interaction curve for biaxial bending (Section 5.4.3, example 1) in the presence of a given nonvanishing twisting moment. *Hint* : use statical approach.

Answer :

$$m_y (1 - m_t^2)^{1/2} + \frac{3}{4} m_x^2 + m_t^2 = 1 \qquad (|m_x| < m_y), \text{ with } m_t = \frac{M_t}{M_{tp}} .$$

References.

[5.1] K. GIRKMAN, Flächentragwerke, Springer, 1959.

[5.2] C. MASSONNET, "Faut-il introduire l'hypothèse de Bernouilli en résistance des matériaux?", Bull. Soc.Roy. des Sci., **12** : 301, Liège, 1947.

[5.3] A.R. RIANITSYN, Calcul à la rupture et plasticité des constructions, p. 34, Eyrolles, Paris, 1959.

[5.4] W. PRAGER, "The General Theory of Limit Design", Proc. 8th Int. Congr. Appl. Mech., Istambul, 1952, **2** : 65, 1956.

[5.5] M. SAVE, "On Yield Conditions in Generalized Stresses", Quart. of Applied Math., **XIX** : 3, October 1961.

[5.6] "Structures", L'architecture d'aujourd'hui, March 1956.

[5.7] A. SAWCZUK and J. RYCHLEWSKI, "On Yield Surfaces for Plastic Shells", Archiwus Mechaniki Stosowanej, **1** : 12, 1960.

[5.8] E.T. ONAT and W. PRAGER, "Limit Analysis of Shells of Revolution", Koninki. Nederl. Akademie van Wetenschappen, Amsterdam, **57** : 5, 1954.

[5.9] P.G. HODGE Jr., "The Rigid-Plastic Analysis of Asymmetrically Loaded Cylindrical Shells", J. of Appl. Mech., **21** : 336, 1954.

[5.10] E.T. ONAT, "The Plastic Collapse of Cylindrical Shells under Axially Symmetrical Loading", Quart. Appl. Math. **13** : 63, 1955.

[5.11] P.G. HODGE Jr., "The Linearization of Plasticity Problems by Means of Nonhomogeneous Materials", Proc. I.U.T.A.M. Symp., 1958. Nonhomogeneity in Elasticity and Plasticity, W. OLSZAK, ed. pp. 147-156, Pergamon Press, 1959. See also W. PRAGER, "On the Plastic Analysis of Sandwich Structures", in Problems in Continuum Mechanics, pp. 342-349, Soc. for Ind. and Appl. Math., Philadelphia, 1961.

[5.12] A. SAWCZUK and P.G. HODGE Jr., "Comparison of Yield Conditions for Circular Cylindrical shells", J. of the Franklin Inst., **269** : n°5, May 1960.

[5.13] P.G. HODGE Jr., "The Mises Yield condition for Rotationally Symmetric Shells", Quart. of Appl. Math., **18** : 305, 1961.

[5.14] K.W. JOHANSEN, "Yield line Theory", (translated from the Danish), Cement and Concrete Association, London, 1962.

[5.15] G. DE SAXCE, "Extension de la méthode de JOHANSEN aux coques en béton armé", Laboratoire de Mécanique des Matériaux et Stabilité des Constructions, Int. Report, Univ. of Liège, Belgium, 1984.

[5.16] G. DE SAXCE, "Extension of the Yield-line Method to the Reinforced Concrete Shells", Proceedings Int. Conf., IASS-85, Moscow, September 1985.

[5.17] M. CAPURSO, "Sul calcolo a rottura delle volte in cemento armato", Giornale del genio civile, fasc. 2, pp. 83-100, February 1966.

[5.18] A. SAWCZUK and W. OLSZAK, "A Method of Limit Analysis of Reinforced Concrete Thank", Int. Coll. on Simplified Calculation Methods, report III/6, Brussels, IASS-ABEM, 1961.

[5.19] I. EKELAND and R. TEMAM, "Convex Analysis and Variational Problems", New-York North Holland, 1975.

[5.20] W. PRAGER, "Discontinuous Field of Plastic Stress and Flow", Proc. 2nd U.S. Nat. Congr. Appl. Mech., 21-32, A.S.M.E., Ann Arbor, 1954.

[5.21] R. HILL, "Discontinuity Relations in Mechanics of Solids", Progress in Solid Mechanics, vol. II, Sneddon, Hill, ed., North-Holland Publ. Co., Amsterdam, 1961.

[5.22] M. ZYCZKOWSKI, "Combined Loadings in the Theory of Plasticity", PWN-Polish Scientific Publishers, Warsaw, 1981.

6. Numerical Methods.

6.1. Introduction.

The theorems of limit analysis provide a powerful method for determination of collapse loads. However, although exact collapse loads have been determined analytically for various problems, complete solution rarely was obtained for individual problems with complex geometric forms and boundary conditions.

Even for simple forms such as rectangular plates, the bounds that have been obtained analytically may differ substantially (see for instance 7.6). In the range of shells, the analytical solutions are still very poor. For shakedown analysis, the analytical evaluation of α_s is much more difficult because of numerous possible combinations of the loads and have been performed only for very few continuous structures.

In view of the outstanding progress in numerical analysis and mathematical programming, numerical solutions of the problems of limit and shakedown analysis have been extensively developed by several investigators in the three past decades.

The first attempts to treat two-dimensional problems of limit analysis as mathematical programming problems were by Koopman and Lance [6.1], Ceradini and Gavarini [6.2] and Sacchi [6.3]. In their investigations, static admissibility was enforced by imposing *finite difference* formulation of the equilibrium equations at all points of a grid. The principal disadvantage of the type of formulation is that it is difficult to code this method with sufficient versatility to enable one program to handle a wide variety of geometries and loadings. Furthermore, the result obtained is an approximation to the limit load rather than a true lower bound. Indeed, there is no theoretical justification for assuming that a stress field which is statically admissible at a finite grid of points is truly statically admissible.

For these reasons, descretization by *finite element* has become a classical and indispensable tool. It can be noted from the literature that the problems of limit and shakedown analysis can be dealt with by two different approaches :

a) incrementaly analysis, whereby the entire history of the loading is examined ;

b) direct determination of the ultimate state.

The first method may be performed in two ways, Newton's method or mathematical programming techniques.

The principal promoters of the Newton's method are Argyris [6.4], Pope [6.5], Marcal and King [6.6], Yamada and his co-workers [6.7], and Zienkiewicz and his co-workers [6.8]. Another way using quadratic programming or linear complementary problems was introduced by Maier and De Donato [6.70 - 6.79]. It seems however that the step-by-step process may involve too much computation if the collapse load and its corresponding mechanism alone are required for design purposes.

The disadvantage becomes strongly accentuated for variable repeated loads, because the structure may shakedown only after an extensive number of cycles. A complete step-by-step computation of the overall history up to the ultimate state can be very time-consuming or even exceed the capacity of computer. Besides, if the loading path is unknown, then shakedown analysis *must* be used. Nevertheless, even if only step-by-step computer packages are available, it is possible to get partial informations about shakedown domain, as we shall see at Section 6.9.

For these reasons, we restrict the developments to the direct methods of limit and shakedown analysis. The main aspects that characterize these approaches are as follows :

1. the variational principle ;
2. the finite element discretization of the field ;
3. the analytical form of the yield condition ;
4. the way the yield condition is enforced ;
5. the mathematical programming algorithm.

In the present chapter we shall consistently use matrix notations (and rules) : general matrices will be indicated by capital bold face-letters, whereas a *column* vector (that is a one column matrix) will be noted by a lower-case bold face-letter, and the corresponding line vector indicated as the transpose of the preceeding. Transposition is, as usually, indicated by a superscript T.

We shall use vector and matrices (matrix operators) in terms of $\boldsymbol{\sigma}$, $\boldsymbol{\varepsilon}$ but it may be verified that they are valid for generalized variables $\mathbf{Q}$, $\mathbf{q}$ as in previous Chapter 5.

6.2. Variational principles.

The choice of the variational principle utilized and, therefore, the choice of the basing assumptions determines entirely the model of numerical approach with all its consequences, namely the accuracy and the speed of convergence. In present Section, we shall endeavour to present a general overview of the main variational principles which govern limit analysis theory. The various principles of limit analysis are presented with their filiation in fig. 6.1.

Now we consider a rigid-perfectly plastic body, noted V, with a boundary A which is composed of two disjointed portions :

- A_u, where displacement rates $\dot{\bar{\mathbf{u}}}$ are prescribed ;

- A_σ, where surface forces $\overline{\mathbf{X}}$ are imposed.

Finally, the notation $\mathbf{X}$ represents body forces.

6.2.1. *One-field principles.*

These preliminary definitions allow the ready introduction of the following variational principles :

Theorem 9 : *Markov's principle : among the licit strain rate fields, the true one makes the functional*

$$\Phi = \int_V D(\dot{\boldsymbol{\varepsilon}})\, dV - \int_V \mathbf{X}^T \dot{\mathbf{u}}\, dV - \int_{A_\sigma} \overline{\mathbf{X}}^T \dot{\mathbf{u}}\, dA_\sigma \tag{6.1}$$

an absolute minimum [6.9]

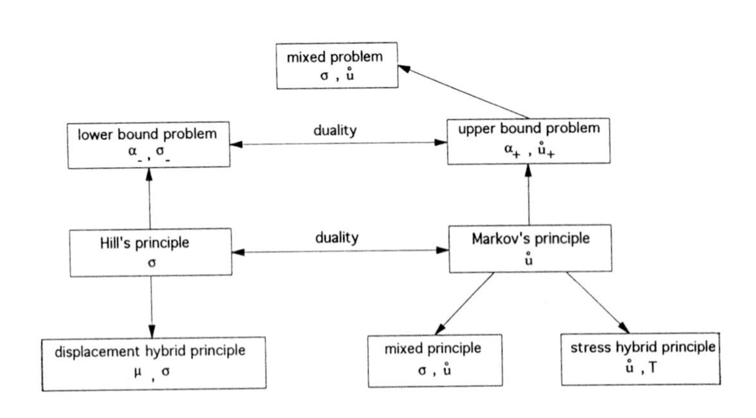

Fig.6.1.

Theorem 10 : *Hill's principle : among the licit stress fields, the true one makes the functional*

$$\Pi = -\int_{A_u} \boldsymbol{\sigma}^T \mathbf{L}^T \dot{\bar{\mathbf{u}}} \, dA_u \tag{6.2}$$

an absolute minimun [6.10]

First, let us consider Markov's principle. Let us prove that the true strain rate field $\dot{\boldsymbol{\varepsilon}}$ minimizes (6.1), among all licit fields $\dot{\boldsymbol{\varepsilon}}^*$. Indeed, we have :

$$\Phi(\dot{\boldsymbol{\varepsilon}}^*) - \Phi(\dot{\boldsymbol{\varepsilon}}) = \int_V \left[D(\dot{\boldsymbol{\varepsilon}}^*) - D(\dot{\boldsymbol{\varepsilon}}) \right] dV$$

$$-\int_V \mathbf{X}^T \left(\dot{\mathbf{u}}^* - \dot{\mathbf{u}} \right) dV - \int_{A_\sigma} \overline{\mathbf{X}}^T \left(\dot{\mathbf{u}}^* - \dot{\mathbf{u}} \right) dA_\sigma .$$

Now, from the definition (2.1) of specific rate of dissipation, it results :

$$D(\dot{\boldsymbol{\varepsilon}}^*) - D(\dot{\boldsymbol{\varepsilon}}) = \boldsymbol{\sigma}^T (\dot{\boldsymbol{\varepsilon}}^* - \dot{\boldsymbol{\varepsilon}}) + (\boldsymbol{\sigma}^* - \boldsymbol{\sigma})^T \dot{\boldsymbol{\varepsilon}}^* .$$

Using the *theorem of maximum dissipation* (see (3.3)), it follows :

$$D(\dot{\boldsymbol{\varepsilon}}^*) - D(\dot{\boldsymbol{\varepsilon}}) \geq \sigma^T (\dot{\boldsymbol{\varepsilon}}^* - \dot{\boldsymbol{\varepsilon}}) . \tag{6.3}$$

Finally, taking account of *theorem of virtual powers* and (6.3), we have :

$$\Phi(\dot{\varepsilon}^*) - \Phi(\dot{\boldsymbol{\varepsilon}}) \geq \int_V \boldsymbol{\sigma}^T (\dot{\varepsilon}^* - \dot{\boldsymbol{\varepsilon}}) \, dV - \int_V \mathbf{X}^T (\dot{\mathbf{u}}^* - \dot{\mathbf{u}}) \, dV$$

$$-\int_{A_\sigma} \overline{\mathbf{X}}^T (\dot{\mathbf{u}}^* - \dot{\mathbf{u}}) \, dA_\sigma = 0$$

that proves theorem 9.

On a similar way, we can prove that the true stress field $\boldsymbol{\sigma}$ minimizes (6.2) among all statically admissible fields σ^*.

Indeed, we have :

$$\Pi(\sigma^*) - \Pi(\sigma) = \int_{A_u} (\boldsymbol{\sigma} - \sigma^*)^T \mathbf{L}^T \dot{\bar{u}} \, dA_u \tag{6.5}$$

i.e., by *theorems of virtual powers and maximum dissipation.*

$$\Pi(\sigma^*) - \Pi(\boldsymbol{\sigma}) = \int_{V_u} (\boldsymbol{\sigma} - \sigma^*)^T \dot{\boldsymbol{\varepsilon}} \, dV = \int_{V_u} D(\dot{\boldsymbol{\varepsilon}}) \, dV - P(\sigma^*, \dot{\boldsymbol{\varepsilon}}) \geq 0$$

Now, we may deduce the Euler-Lagrange equations of the principle (6.1). Therefore, let us take the variation of Markov's functional, with respect to the displacement, taking account of the definition (2.1) of the specific rate of dissipation and the compatibility conditions (1.22) :

$$\delta \Phi = \int_V \boldsymbol{\sigma}^T \boldsymbol{\Delta}^T \delta \dot{\mathbf{u}} \, dV - \int_V \mathbf{X}^T \delta \dot{\mathbf{u}} \, dV - \int_{A_\sigma} \overline{\mathbf{X}}^T \delta \dot{\mathbf{u}} \, dA_\sigma .$$

We apply the Green-Ostrogradsky formula :

$$\delta \Phi = \int_{A_\sigma} \delta \dot{\mathbf{u}}^T (\mathbf{L} \boldsymbol{\sigma} - \overline{\mathbf{X}}) dA_\sigma - \int_V \delta \dot{\mathbf{u}}^T (\Delta \boldsymbol{\sigma} + \mathbf{X}) \, dV$$

So, the Euler-Lagrange equations are *equilibrium equations* (1.2) and (1.3) :

$$\Delta \boldsymbol{\sigma} + \mathbf{X} = 0 \qquad \text{in} \qquad V, \tag{6.6}$$

$$\mathbf{L} \boldsymbol{\sigma} = \overline{\mathbf{X}} \qquad \text{on} \qquad A_\sigma . \tag{6.7}$$

To deduce in a simple way the Euler-Lagrange equations of Hill's principle, it is necessary to remark that the stress field should enforce the yield condition (1.38). So that we consider, instead of the functional Π, the corresponding lagrangian :

$$L = \int_V \lambda (\sigma_R^2 - \sigma_Y^2) \, dV - \int_{A_u} \boldsymbol{\sigma}^T \mathbf{L}^T \dot{\bar{u}} \, dA_u . \tag{6.8}$$

Now, take the variation, with respect to the normality law (2.6) :

$$\delta L = \int_V \dot{\boldsymbol{\varepsilon}}^T \, \delta\boldsymbol{\sigma} \, dV - \int_{A_u} \delta\boldsymbol{\sigma}^T \, L^T \, \dot{\bar{u}} \, dA_u . \tag{a}$$

On the other hand, by applying the Green-Ostrogradsky formula, we have :

$$\int_V \delta\boldsymbol{\sigma}^T \, \Delta^T \, \dot{\mathbf{u}} \, dV = -\int_V \dot{\mathbf{u}}^T \, \Delta \, \boldsymbol{\delta\sigma} \, dV$$

$$+ \int_{A_\sigma} \delta\sigma^T \, \mathbf{L}^T \, \dot{\mathbf{u}} \, dA_\sigma + \int_{A_u} \delta\sigma^T \, \mathbf{L}^T \, \dot{\mathbf{u}} \, dA_u . \tag{b}$$

As we consider only the variations of licit stress fields, the two first terms of right-hand member of (b) vanish. From the addition of (b) to (a), it results :

$$\delta L = \int_V \delta\sigma^T \left[\dot{\boldsymbol{\varepsilon}} - \Delta^T \, \dot{\mathbf{u}} \right] dV + \int_{A_u} \delta(\mathbf{L}\,\boldsymbol{\sigma})^T \, (\dot{\mathbf{u}} - \dot{\bar{\mathbf{u}}}) \, dA_u .$$

So we see that the Euler-Lagrange equations of Hill's principle are the compatibility equations (1.22) and the corresponding boundary conditions :

$$\dot{\boldsymbol{\varepsilon}} = \Delta^T \, \dot{\mathbf{u}} \qquad \text{in} \qquad V \tag{6.9}$$

$$\dot{\mathbf{u}} = \dot{\bar{\mathbf{u}}} \qquad \text{on} \qquad A_u \tag{6.10}$$

Finally, we can remark that the previous developments prove the *duality* of Markov's and Hill's principles in the sense of convex analysis.

6.2.2. *Theorems of limit analysis.*

According to Mandel [6.55], the theorems providing lower and upper bound of the limit load are direct consequences of the minimum principles of Hill and Markov, applied to the particular case of proportional loading. The following demonstration was proposed by Nguyen Dang Hung [6.11] and [6.12].

So, we suppose that :

1. The body has well fixed supports, which means that :

$$\dot{\bar{\mathbf{u}}} = 0 \qquad \text{on} \qquad A_u$$

2. The load increases in proportion with the loading parameter α :

$$\mathbf{X} = \alpha\, \mathbf{X}^o \qquad \text{and} \qquad \overline{\mathbf{X}} = \alpha\, \overline{\mathbf{X}}^o .$$

Now, let us prove the lower bound theorem as a direct consequence of the proof of the minimum of Hill's functional. Let $\boldsymbol{\sigma}_-$ a licit stress field corresponding to the loading $\alpha_- \mathbf{X}^o$ and $\alpha_- \overline{\mathbf{X}}^o$. Let $\boldsymbol{\sigma}$ and $\dot{\boldsymbol{\varepsilon}}$ the collapse stress and strain rate fields. We write the equation of virtual power for the stress field $(\boldsymbol{\sigma} - \boldsymbol{\sigma}_-)$ which satisfies the equilibrium equations and $\dot{\boldsymbol{\varepsilon}}$ which is licit. Using (6.5), we obtain :

$$(\alpha_1 - \alpha_-)\left[\int_V \mathbf{X}^{oT}\, \dot{\mathbf{u}}\, dV + \int_{A_\sigma} \overline{\mathbf{X}}^{oT}\, \dot{\mathbf{u}}\, dA_\sigma\right] = \Pi\,(\sigma_-) - \Pi\,(\sigma) \geq 0\,.$$

As the bracket in the left hand member of the above relation is nonnegative, we obtain :

$$\alpha_1 \geq \alpha_- .$$

In a similar way, let us prove that the upper bound theorem is a direct consequence of the minimum of Markov's functional. Let $\dot{\mathbf{u}}_+$, $\dot{\varepsilon}_+$ be a set of licit displacement and strain rate fields. The corresponding multiplier is defined by :

$$\alpha_+\left(\int_V \mathbf{X}^{oT}\; \dot{\mathbf{u}}_+\, dV + \int_{A_\sigma} \overline{\mathbf{X}}^{oT}\, \dot{\mathbf{u}}_+\, dA_\sigma\right) = \int_V D(\dot{\boldsymbol{\varepsilon}}_+)\, dV\,. \tag{6.11}$$

Similarly, for the collapse fields, we have :

$$\alpha_1\left(\int_V \mathbf{X}^{oT}\; \dot{\mathbf{u}}_+\, dV + \int_{A_\sigma} \overline{\mathbf{X}}^{oT}\, \dot{\mathbf{u}}\, dA_\sigma\right) = \int_V D(\dot{\boldsymbol{\varepsilon}})\, dV\,. \tag{6.12}$$

Let us consider now the body at collapse. Markov's minimum principle (6.2) furnishes the relation :

$$\Phi(\dot{\varepsilon}_+) - \Phi(\dot{\boldsymbol{\varepsilon}}) = \int_V \left[D(\dot{\varepsilon}_+) - D(\dot{\boldsymbol{\varepsilon}}) \right] dV$$

$$- \alpha_1 \left[\int_V \mathbf{X}^{oT} (\dot{\mathbf{u}}_+ - \dot{\mathbf{u}})\, dV + \int_{A_\sigma} \overline{\mathbf{X}}^{oT} (\dot{\mathbf{u}}_+ - \dot{\mathbf{u}})\, dA_\sigma \right] \geq 0 .$$

The elimination of the first term of the right-hand member of the above expression, owing to (6.11) and (6.12) gives :

$$(\alpha_+ - \alpha_1) \left[\int_V \mathbf{X}^{oT} \dot{\mathbf{u}}_+ \, dV + \int_{A_\sigma} \overline{\mathbf{X}}^{oT} \dot{\mathbf{u}}_+ \, dA_\sigma \right] = \Phi(\dot{\varepsilon}_+) - \Phi(\dot{\varepsilon}) \geq 0 .$$

As the bracket in the left hand member is positive, taking account of (6.11),we obtain :

$$\alpha_+ \geq \alpha_1 .$$

Now we may state the fundamental theorems of limit analysis in the form of constrained extremum problems. First the statical theorem leads to seek the maximum of the multiplier corresponding to licit stress fields :

$$\max \alpha_-$$

$$\alpha_- , \boldsymbol{\sigma}_-$$

constraints :

$$\begin{aligned} &\Delta \sigma_- + \alpha_- \mathbf{X}^o = 0 && \text{in} \quad V \\ &\mathbf{L} \sigma_- = \alpha_- \overline{\mathbf{X}}^o && \text{on} \quad A_\sigma \\ &\sigma_R(\sigma_-) \leq \sigma_Y && \text{in} \quad V. \end{aligned} \tag{6.13}$$

For the kinematical approach, it is convenient to impose the normalization condition

$$\int_V \mathbf{X}^{oT} \dot{\mathbf{u}}_+ \, dV + \int_{A_\sigma} \overline{\mathbf{X}}^{oT} \dot{\mathbf{u}}_+ \, dA_\sigma = 1 . \tag{6.14}$$

This is allowed, because the velocity fields are always determined to within a positive scalar factor. Owing to (6.11) and (6.14), the multiplier α_+ is equal to the total dissipation. So the kinematical theorem leads to seek the minimum of the multipliers corresponding to licit strain rate fields :

$$\min_{\dot{\boldsymbol{\varepsilon}}_+,\, \dot{\mathbf{u}}_+} \int_V D(\dot{\varepsilon}_+)\, dV$$

constraints :

$$\dot{\varepsilon}_+ = \Delta^T \dot{\mathbf{u}}_+ \qquad \text{in} \qquad V$$

$$\dot{\mathbf{u}}_+ = 0 \qquad \text{on} \qquad A_\sigma$$

$$\int_V \mathbf{X}^{oT}\, \dot{\mathbf{u}}_+\, dV + \int_{A_\sigma} \overline{\mathbf{X}}^{oT}\, \dot{\mathbf{u}}_+\, dA_\sigma = 1\,. \tag{6.15}$$

Finally, let us deduce the Euler-Lagrange equations of the two above variational problems. For the statical problem, we introduce the associated *lagrangian L_-* :

$$L_- = \alpha_- - \int_V \lambda \left(\sigma_R^2 - \sigma_Y^2 \right) dV - \int_V \dot{\mathbf{u}}_+^T \left(\Delta\, \boldsymbol{\sigma}_- + \alpha_- \mathbf{X}^o \right) dV$$

$$+ \int_{A_\sigma} \dot{\mathbf{u}}_+^T \left(\mathbf{L}\, \boldsymbol{\sigma}_- - \alpha_- \overline{\mathbf{X}}^o \right) dA_\sigma.$$

It is obtained by considering the initial functional, i.e. the multiplier α_- and adding terms which involves the constraints (yield condition, equilibrium equations and corresponding boundary conditions). Each constraint is multiplied by a corresponding unknown field, denoted *Lagrange's multiplier.*

The solution of the constrained statical problem is the same that one of the saddle point problem associated to the lagrangian [6.38].

As in the previous Section, it is easy to show that the Euler-Lagrange equations of the statical theorem are the kinematical conditions (6.15). In a similar manner, let us introduce the lagrangian

$$L_+ = \int_V D\,(\dot{\varepsilon}_+)\, dV - \int_V \sigma_-^T \left[\dot{\varepsilon}_+ - \Delta^T \dot{\mathbf{u}}_+ \right] dV$$

$$- \alpha_- \left(\int_V \mathbf{X}^{oT} \dot{\mathbf{u}}_+ \, dV + \int_{A_\sigma} \overline{\mathbf{X}}^{oT} \dot{\mathbf{u}}_+ \, dA_\sigma - 1 \right).$$

So we can prove that the Euler-Lagrange equations of the kinematical problem are the statical constraints (6.13). Finally, we see that the lower and upper bound problems are *dual* in the sense of convex analysis.

6.2.3. *Two-field principle.*

The previous functionals constitute the theoretical foundation for the use of the pure finite elements of the displacement or equilibrium type. The strength of these approaches is that it provides dependable bounds, a virtue which is sometimes lost by relaxing plastic yield condition or approximating contributions to the dissipation integral. Another approach is to use two-field principles. This provides an approximation of the limit multiplier, which is shown in the literature to be a better estimate than the bounds. Such two-field principles are the theoretical foundation for the use of mixed and hybrid finite elements. It also provides simultaneous approximations of the collapse fields for stresses and velocities.

First, let us deduce the mixed variational problems. For this, consider the Markov's minimum principle. It results from the *theorem of maximum dissipation* (see (3.3)) that we have :

$$\int_V D(\dot{\varepsilon}) \, dV = \max_{\sigma} \int_V \boldsymbol{\sigma}^T \dot{\varepsilon} \, dV ,$$

$$\sigma_R(\boldsymbol{\sigma}) \leq \sigma_Y \qquad \text{in} \qquad V. \tag{6.16}$$

If, furthermore, we substitute $\dot{\mathbf{u}}$ to $\dot{\boldsymbol{\varepsilon}}$ in (6.16) by using the internal compatibility condition (6.9), Markov's minimum principle becomes the following saddle-point problem :

$$\underset{\dot{\mathbf{u}}}{\text{Min}} \ \underset{\sigma}{\text{Max}} \left[\int_V \frac{1}{2} \sigma^T \Delta^T \dot{\mathbf{u}} \, dV - \int_V \mathbf{X}^T \dot{\mathbf{u}} \, dV - \int_{A_\sigma} \overline{\mathbf{X}}^T \dot{\mathbf{u}} \, dA_\sigma \right],$$

constraints :

$$\dot{\mathbf{u}} = 0 \qquad \text{on} \qquad A_u$$

$$\sigma_R(\boldsymbol{\sigma}) \leq \sigma_Y \qquad \text{in} \qquad V. \tag{6.17}$$

It can be shown easily that the Euler-Lagrange equations of this problem are the equilibrium equations (6.6) and (6.7) and the internal compatibility condition (6.9). In this sense, it may be said that this new principle generalizes Markov's principle relaxing the internal compatibility. It constitutes an extension to rigid-perfectly-plastic body of the well-known Hellinger-Reissner principle.

A variant of this approach is to apply this reasoning to the upper bound problem of the previous Section.

Taking account of the relation (6.16) and eliminating the rate strain field by the internal compatibility condition (6.9), we change the upper bound problem into the following saddle-point problem :

$$\underset{\dot{\mathbf{u}}}{\text{Min}} \quad \underset{\sigma}{\text{Max}} \quad \int_V \frac{1}{2} \sigma^T \Delta^T \dot{\mathbf{u}} \, dV \, ,$$

constraints :

$$\dot{\mathbf{u}} = 0 \qquad \text{on} \qquad A_u$$

$$\int_V \mathbf{X}^{oT} \dot{\mathbf{u}} \, dV + \int_{A_\sigma} \overline{\mathbf{X}}^o \dot{\mathbf{u}} \, dA_\sigma = 1$$

$$\sigma_R(\boldsymbol{\sigma}) \leq \sigma_Y \qquad \text{in} \qquad V . \tag{6.18}$$

Euler-Lagrange equations of this problem are also (6.6), (6.7) and (6.9).

Anticipating on the content of Section 6.3, we suppose that the body V is subdivided into regions called finite elements. Alternatively to the last development above, we may relax the compatibility on the interelement connection during the discretization of the problem. So we generalize the hybrid displacement principle to the rigid perfectly-plastic body. This variational principle was first presented by Nguyen Dang Hung [6.13]. Let be :

- V_e some finite element,

- A any interface between finite elements,

- **n** any normal unit vector exterior to any finite element or the body,

- **T** Lagrange's multiplier field on interfaces.

Then, we may introduce the following saddle-point problem :

$$\underset{\mathbf{T}}{\text{Min}} \quad \underset{\dot{\mathbf{u}}}{\text{Max}} \left[\int_V D(\dot{\varepsilon})\, dV - \int_V \mathbf{X}^T \dot{\mathbf{u}}\, dV - \int_{A_\sigma} \overline{\mathbf{X}}^T \dot{\mathbf{u}}\, dA_\sigma - \sum_A \int_A \mathbf{T}^T \dot{\mathbf{u}}\, dA \right],$$

constraints :

$$\dot{\boldsymbol{\varepsilon}} = \frac{1}{2} \Delta^T \dot{\mathbf{u}} \qquad \text{in each} \qquad V_e$$

$$\dot{\mathbf{u}} = \dot{\bar{\mathbf{u}}} \qquad \text{on} \qquad A_u . \tag{6.19}$$

The Euler-Lagrange equations of this problem are the equilibrium equation (6.6) and (6.7) and :

$$\begin{aligned} \dot{\mathbf{u}}] &= 0 && \text{on each} \quad A \\ \mathbf{L}\,\boldsymbol{\sigma} &= \mathbf{T} && \text{on each} \quad A . \end{aligned} \tag{6.20}$$

Similarly, by relaxing the equilibrium on the interelement connection during the discretization of Hill's principle, we obtain the generalization of the hybrid stress-principle of Pian [6.14].

Calling $\boldsymbol{\mu}$ the Lagrange multiplier field on interfaces, we obtain the saddle-point principle :

$$\underset{\sigma}{\text{Max}} \quad \underset{\boldsymbol{\mu}}{\text{Min}} \left[-\int_{A_u} \sigma^T \mathbf{L}^T \dot{\bar{u}}\, dA_u + \sum_A \int_A \sigma^T \mathbf{L}^T \boldsymbol{\mu}\, dA \right],$$

constraints :

$$\Delta\,\boldsymbol{\sigma} + \mathbf{X} = 0 \qquad \text{in each} \qquad V_e$$

$$\mathbf{L}\,\boldsymbol{\sigma} = \overline{\mathbf{X}} \qquad \text{on} \qquad A_\sigma . \tag{6.21}$$

Euler-Lagrange equations of the problem are the compatibility conditions (6.9) and (6.10), and :

$$\begin{aligned} \mathbf{L}^+ \sigma^+ + \mathbf{L}^- \boldsymbol{\sigma}^- &= 0 && \text{on each} \quad A \\ \dot{\mathbf{u}} &= \boldsymbol{\mu} && \text{on each} \quad A . \end{aligned} \tag{6.22}$$

6.3. The finite element method (F.E.M.).

6.3.1. *The finite element concept.*

The basic concept of the method, when applied to problems of structural analysis, is that a continuum (the total structure) can be modeled analytically by its subdivision into regions (the finite elements) considered interconnected at joints called nodes or nodal joints. In each of the elements, the behaviour is described by a separate set of assumed functions representing the displacements or stresses in that region ; these sets of functions should be chosen in a form that ensures continuity of the described behaviour throughout the complete continuum. The unknown magnitudes of these functions depend on parameters associated with the nodal points. By means of a variational principle, a set of equations is obtained for each element to be assembled to represent the equilibrium or compatibility of the entire body. The special advantages of the method reside in its suitability for automation of the equation formation process and in the ability to represent highly irregular and complex structures and loading situations.

Papers by Turner [6.56] and Argyris [6.57] in the late fifties are generally considered as the starting point of the method as a practical tool for engineers. In the past twenty years, finite elements have been the object of numerous publications ; hundreds of them are referenced in the now classical monograph by Zienkiewicz [6.58]. Within the frame the present book, we must restrict ourselves to a short presentation of the method.

Readers interested by more details can consult references such as Oden [6.59], Strang and Fix [6.60] or Ciarlet [6.61] for the mathematical foundation of the method, Gallagher for a conventional textbook [6.62], Robinson [6.63] for a parallel development of the displacement or stress models, Desai and Abel [6.64] for applications, Pilkey [6.65] for computer codes.

Because they were the first to be used and most intuitively underscore the analogy between real discrete elements and finite portions of a continuum domain, finite elements of displacement types deriving from the minimization of the total potential energy will be first considered. They will be followed by equilibrium models minimizing the total complementary energy . Other elements including hybrid or mixed models based upon a two-field variational principle will finally be covered.

6.3.2. *Analysis procedure.*

The following six steps summarize the finite elements analysis procedure.

1. Idealization of the continuum. This step, common to every analysis, be it by finite elements or analytical procedure, is never too much emphasized. Take a bridge for instance ; should it be idealized as part of a continuous beam, as a grid of girders, an orthotropic plate or a three-dimensional continuum? How deep should the soil be taken into account in a foundation analysis? Engineering judgment and the accuracy of desired results should decide.

2. Discretization of the continuum. Having fundamentally assumed the behaviour of the structure by idealizing it as a plate, shell solid or plane problem, an appropriate choice must be made in the library of elements available for this type of structure or part of structure. Not only should the element types be chosen but a finite element mesh must be decided. This again requires engineering experience and some prior knowledge about the general flow of efforts through the structure. It must always be remembered that the best elements and the most powerful computer program can never compensate for a poor idealization or discretization. For repetitive well defined structures, automatic meshing program do now exist but it is still left to human intervention to decide the level of refinement to be used in different areas.

3. Derivation of the connection matrices for the elements corresponding to the variational principle. These matrices link different quantities. For the equilibrium elements, the matrix relates the parameters of the stress field to the forces at the nodal points (nodal forces). For the displacement elements, matrix connects the parameters of the strain field to the displacements at the nodal points (nodal displacements). Complementaries matrices associated to the yield condition must be also derived. Corresponding instructions are written in finite element programs, which automatically repeat the numerical computations for all elements involved in the discretization.

4. Assembly of the equations and connection matrices for the overall discretized continuum. This operation is automatically performed by the computer when the topology of the structure and its division into finite elements have been input as data.

5. Solution of the discretized variational problem, taking into account the appropriate boundary conditions, defined in terms of nodal forces or displacements. In linear elastic problems, the solution is relatively straightforward because obtained by resolving a linear system of simultaneous equation. In limit and shakedown analysis, the solution is relative to a discreet unconstrained or constrained optimization problem and can be obtained by applying *mathematical programming techniques*.

6. Computation of the dual quantities in mathematical programming sense. These quantities are called “weak” or “generalized”, i. e. known only by means of averages. Because they are only approximate and their accuracy depends on the discretization, great care must be exercised in the interpretation of these results.

6.4. Pure equilibrium finite elements.

The application of the equilibrium finite elements to limit analysis was carried out by Belytschko and Hodge [6.15] and [6.16], Zavelani-Rossi and his co-workers [6.17] and [6.30], Lysmer [6.18], Cascaro and Di Carlo [6.19], Bottero and his co-workers [6.20], Anderheggen [6.27], Biron [6.28] and [6.66], Hutula [6.29], Robinson [6.31], Peano [6.32], Nguyen Dang Hung and his co-workers [6.12, 6.24 and 6.33], Faccioli and Vitiello [6.69].

In the equilibrium finite element formulation, it is assumed that the stress field σ within each finite element is a polynomial field depending on a system of arbitrary parameters h_k, components of a vector **h**, as follows :

$$\boldsymbol{\sigma} = \mathbf{S}(\mathbf{x})\,\mathbf{h} + \boldsymbol{\sigma}_*(\mathbf{x})\,. \tag{6.23}$$

In this expression, $\boldsymbol{\sigma}_*$ is particular solution of the non-homogeneous equations of internal equilibrium (6.6). Obviously, if there are zero body forces, this term vanishes. **S** must be chosen so that the stress field **S h** is a solution of the corresponding homogeneous equations of internal equilibrium, whatever the stress parameters h_k may be.

Then, the stress field (6.23) is a priori in internal equilibrium. Besides, we must satisfy a priori the boundary equilibrium conditions (6.7) and also the equilibrium at interfaces between continuous finite elements. So, we obtain obviously linear conditions on the stress parameters, that we may write in a general way as follows :

$$\mathbf{C}\,\mathbf{h} = \bar{\mathbf{g}} \tag{6.24}$$

where **C** is the so-called statical *connection matrix.*

Let us note that admissible discontinuities, such as (5.91) and (5.96) for plates, are allowed only on the element interfaces. By the way, let us focus our attention on the strain field. For sake of convenience, let us suppose momentarily that σ_* vanishes. Putting (6.23) in the total internal dissipation, we have :

$$\int_V \boldsymbol{\sigma}^T \dot{\boldsymbol{\varepsilon}}\, dV = \mathbf{h}^T <\dot{\mathbf{v}}>\,.$$

So, the strain rate field is defined by :

$$<\dot{\mathbf{v}}> = \int_V \mathbf{S}^T \dot{\boldsymbol{\varepsilon}}\, dV\,,$$

which is a vector of strain rates weighted by the shape functions of the discretized stress field. We say that the strain rate field is a *weak quantity* and that the averages, components of the vector $<\dot{\mathbf{v}}>$, are the generalized rate strains.

The vectors associated to weak quantities are denoted between brackets. Conversely, the stress field is called a *strong quantity.*

6.4.1. *Lower bound problems.*

Now, we make the hypothesis of proportional loading. Consequently, the relations (6.23) and (6.24) become obviously :

$$\boldsymbol{\sigma}_- = \mathbf{S}\,\mathbf{h} + \alpha_-\,\sigma_*^{o} \tag{6.25}$$

$$\mathbf{C}\,\mathbf{h} = \alpha_-\,\bar{\mathbf{g}}^{o}\,. \tag{6.26}$$

Besides, we must satisfy also the yield *anywhere* in the body, that is, owing to (6.25) :

$$\sigma_R\,(\mathbf{S}\,\mathbf{h} + \alpha_-\,\sigma_*^{o}) \leq \sigma_Y\,. \tag{6.27}$$

To this aim, it is necessary to associate a finite number of discretized inequalities to $\mathbf{h}$ and α_-, enough to ensure that the yield condition is strictly respected throughout the body :

$$f_i(\mathbf{h}\,,\,\alpha_-) \leq 0\,. \tag{6.28}$$

Finally, the discretized form of the lower bound problem is :

$$\underset{\alpha_-,\,\mathbf{h}}{\text{Max}} \quad \alpha_-$$

constraints :

$$\mathbf{C}\,\mathbf{h} = \alpha_-\,\bar{\mathbf{g}}^{o}.$$
$$f_i(\mathbf{h}\,,\,\alpha_-) \leq 0. \tag{6.29}$$

To deduce the Euler-Lagrange equations of this problem, let us introduce the associated Lagrangian :

$$L_{h_-} = \alpha_- + \langle\dot{\mathbf{q}}\rangle^T\,(\mathbf{C}\,\mathbf{h} - \alpha_-\,\bar{\mathbf{g}}^{o}) - \sum_i \langle\lambda_i\rangle\,f_i\,.$$

So, we have the following variation equations :

$$\bar{\mathbf{g}}^{oT}\langle\dot{\mathbf{q}}\rangle + \sum_i \langle\lambda_i\rangle\,\frac{\partial f}{\partial\,\alpha_-} = 1 \tag{6.30}$$

$$\mathbf{C}^T <\dot{\mathbf{q}}> = \sum_i <\lambda_i> \frac{\partial f}{\partial \mathbf{h}} . \tag{6.31}$$

We may recognize there the kinematical conditions (6.15) but in a *weak form.* They are satisfied only in the sense of the mean. More precisely, (6.30) is the discretized form of the normalization condition (6.14) and (6.31) is the compatibility condition :

$$\mathbf{C}^T <\dot{\mathbf{q}}> = <\dot{\mathbf{v}}>$$

between the vector $<\dot{\mathbf{q}}>$ of weak displacements and the vector :

$$<\dot{\mathbf{v}}> = \sum_i <\lambda_i> \frac{\partial f}{\partial \mathbf{h}} \tag{6.33}$$

of weak generalized rate strains.

Now, we must discuss some important points.

1. Remark on the way the yield condition is enforced.

First, let us note that, for the purpose of obtaining true lower bound, it is necessary to enforce the yield criterion anywhere. A first method was introduced by Belytschko and Hodge [6.15], [6.16]. These authors use quadratic polynomial shape functions in each element. For instance, if we consider a plate element, we introduce the following quadratic field :

$$\begin{aligned} M_x &= h_1 + h_2\, x + h_3\, y + h_4\, x^2 + h_5\, xy + h_6\, y^2 \\ M_y &= h_7 + h_8\, x + h_9\, y + h_{10}\, x^2 + h_{11}\, xy + h_{12}\, y^2 \\ M_{xy} &= h_{13} + h_{14}\, x + h_{15}\, y + h_{16}\, x^2 - (h_4 + h_{12})\, xy + h_{17}\, y^2 . \end{aligned} \tag{6.34}$$

Momentarily, for the sake of clarity, we assume that the distributed load p vanishes. The moments (6.34) satisfy a priori the homogeneous equation of internal equilibrium :

$$\frac{\partial^2 M_x}{\partial x^2} + 2\,\frac{\partial^2 M_{xy}}{\partial x\, \partial y} + \frac{\partial^2 M_y}{\partial y^2} = 0 \tag{6.35}$$

obtained by eliminating the reactions V_x and V_y between (5.92), (5.93) and (5.94). Besides, in each element, this field must satisfy the Von Mises' yield criterion (5.29) for the plates which have the following form :

$$\sigma_R (h_1 ,..., h_{17} , x , y) \leq \sigma_Y. \tag{6.36}$$

For purpose of implementing a discreet mathematical programming algorithm for the lower bound problem, it is necessary to replace this functional inequality by a finite number of discrete inequalities :

$$f_i (h_1 ,..., h_{17}) = \sigma_R (h_1 ,..., h_{17} , x_i , y_i) - \sigma_Y \leq 0. \tag{6.37}$$

By choosing points (x_i , y_i) at which (6.36) is enforced.

If the set of (x_i , y_i) includes all *relative maxima* of the left hand member of (6.36) within the finite element, it follows that the conditions (6.37) will imply (6.36). However, the location of the relative maxima of σ_R is not a well-behaved function of the parameters h_k, because, as the parameters change, relative maxima may appear or disappear or more into or out of an element.

Further, since the locations of the relative maxima are functions of h_k, the relative maxima of σ_R must be calculated within each step of the mathematical programming algorithm. This process is relatively time-consuming and must be repeated a considerable number of times, placing a severe restriction on the size of problems, that can be solved within a reasonable amount of computer time. In conclusion, this technique leads to formulate the whole method as a highly non-linear programming problem.

Another method, introduced by Zavelani-Rossi [6.17], consists to choose in each element a point mesh where the yield condition is checked. So the optimal loading parameter obtained is not necessarily a lower bound. A postoptimal check is performed on the basis of the solution, with reference to a very dense regular mesh within each element. On the basis of the final check, the solution multiplier is reduced so as to ensure the respect of the yield condition for the denser mesh.

To enforce the yield criterion anywhere in an easier way, an alternative method was used by Zavelani-Rossi [6.17], Lysmer [6.18], Casciaro and Di Carlo [6.19], Botero and his co-workers [6.20]. Linear variations of shape functions are assumed within each element. Then, if a condition such as (5.29) is verified at two distinct points, A and B, within the same element, it turns out to be verified at every point of segment AB. Consequently, the simple checking of yield condition (5.29) at the nodal points ensures that it is nowhere violated in the element.

2. Remark on mathematical programming algorithms.

The role of the numerical technique is also very important and plays a key-role in the efficiency of the method. The most attractive form of the lower bound problem arises when the yield condition is linearized, as in references [6.17], [6.18], [6.20], [6.27], [6.30], [6.32] and [6.69]. Indeed, the problem becomes a *linear programming problem* and may be solved efficiently for very large scale systems by existing algorithms based on the well-known *Simplex method* [6.21]. Now, the set of inequalities (6.28) may be written as a system of linear inequalities and the lower bound problem becomes as follows :

$$\underset{\alpha\_,\mathbf{h}}{\text{Max}} \quad \alpha\_$$

constraints :

$$\mathbf{C}\,\mathbf{h} = \alpha\_\,\bar{\mathbf{g}}^{o}$$

$$\mathbf{N}\ \mathbf{h} + \alpha\_\,\bar{\mathbf{k}}_{*}^{o} \leq \mathbf{k}_{Y}\,. \tag{6.38}$$

Let us note that the dual quantities $<\dot{\mathbf{q}}>$ and $<\boldsymbol{\lambda}>$ may be immediately obtained after computation of optimal solution in the Simplex method by means of *marginal costs* [6.17], [6.21].

Then, the collapse mechanism is known, but only in weak sense.

It is also important to hear in mind that, in this method, the time comsumption is toughly proportional to the number of variables and to the cube of the constraints. So, it is sometimes more interesting as in [6.30] to resolve directly the *dual program* [6.21] which has the form of an upper bound problem, but whose the variables are weak quantities :

$$\underset{\boldsymbol{\lambda},<\dot{\mathbf{q}}>}{\text{Min}} \quad \mathbf{k}_{Y}^{T}<\boldsymbol{\lambda}>$$

constraints :

$$\mathbf{C}^{T}<\dot{\mathbf{q}}> = \mathbf{N}^{T}<\boldsymbol{\lambda}>$$

$$\bar{\mathbf{g}}^{o^{T}}<\dot{\mathbf{q}}> + \bar{\mathbf{k}}_{*}^{o^{T}}<\boldsymbol{\lambda}> = 1\,. \tag{6.39}$$

The constraints (6.39) are identical to the Euler-Lagrange equations (6.30) and (6.31) of the primal problem.

Another important problem is the choice of an *initial feasible vector*. Obviously, the zero vector is feasible for the lower bound problem. Nevertheless, experience carried out by Zavelani-Rossi, showed than the Simplex algorithm did not succeed sometimes in finding a basic vector when starting from the origin. A general method is proposed in [6.17] to get a non-zero feasible initial vector.

Now, when a non-linear yield criterion is considered, as in [6.12], [6.15], [6.16], [6.19], [6.28], [6.29], [6.33] and [6.66], a nonlinear mathematical programming technique is required. The method used in the previous references is the Sequential Unconstrained Minimization Technique (SUMT), developed by Fiacco and Mc Cormick [6.22], [6.23], [6.41], [6.42] and [6.43]. First of all, it is necessary to eliminate the equality constraints of equilibrium (6.26). For this, let us split up the vector **h** (resp. the connection matrix **C**) in two parts $\mathbf{h}_1$ and $\mathbf{h}_2$ (resp. $\mathbf{C}_1$ and $\mathbf{C}_2$) :

$$\mathbf{C}_1\,\mathbf{h}_1 + \mathbf{C}_2\,\mathbf{h}_2 = \alpha\_\,\bar{\mathbf{g}}^{o} \tag{6.40}$$

such that $\mathbf{C}_1$ is a regular square matrix.

Then, all the stress parameters depend on the loading multiplier $\alpha\_$ and hyperstatic unknowns, components of $\mathbf{h}_2$:

$$\mathbf{h}_1 = \alpha\_\,\mathbf{C}_1^{-1}\,\bar{\mathbf{g}}^{o} - \mathbf{C}_1^{-1}\,\mathbf{C}_2\,\mathbf{h}_2 \tag{6.41}$$

Practically, the solution of (6.40) is performed by means of a Gauss elimination. Next, the lower bound problem is put in the following reduced form :

$$\underset{\alpha\_,\mathbf{h}_2}{\mathrm{Max}} \quad \alpha\_$$

constraints :

$$f_i\,(\mathbf{h}_2\,,\,\alpha\_) \leq 0\,. \tag{6.42}$$

The technique embodies the inequality constraints in the objective function in the form

$$\pi = \alpha\_ + \varepsilon \sum_i \frac{1}{f_i}\,. \tag{6.43}$$

Starting from a feasible point, π is maximized for a sequence of decreasing values of ε. As ε approaches zero, the sequence of solutions π approaches of $\alpha\_$.

6.4.2. *Relaxed problems.*

Another solution, based on Hill's principle, was been proposed by Nguyen Dang Hung [6.12 and 6.24]. Compared to the previous ones, it possess the following new features :

1- Introduction of a criterion of yield "in the mean", leading to a relaxed principle.

2- emphasis on the similarity between rigid-plastic analysis and traditional elastic analysis.

First, let us introduce the basic concept of the criterion of the mean. The key idea consists to remark that von Mises yield criterion (1.34) may be written in matrix form as follows :

$$\sigma_R^2 = \frac{1}{2}\sigma^T \mathbf{D}'\boldsymbol{\sigma} \leq \sigma_Y^2. \tag{6.44}$$

Hencky has shown [6.25] that the density of distortion elastic energy , U_d, is proportional to σ_R :

$$U_d = \frac{3\,\sigma_R^2}{2\,G}. \tag{6.45}$$

From this, it can be argued that yielding occurs when the distortion energy density reaches the ultimate value $3\,\sigma_Y^2 / 2\,G$ which is Hencky's energetic interpretation of von Mises criterion. Accordingly, we adopt the criterion that yielding occurs when the mean density in the volume V reaches the limit value $3\,\sigma_Y^2 / 2\,G$. Owing to (6.44) and (6.45), et may be easily shown that the criterion of the mean becomes in matrix form :

$$\frac{1}{2V}\int_V \sigma^T \mathbf{D}'\boldsymbol{\sigma}\,dV - \sigma_Y^2 \leq 0. \tag{6.46}$$

The concept of the yield condition in "the mean" presented here has the following meaning : the yielding condition is relaxed and we must expect to obtain only an approximation of the lower bound. Numerical comparisons with other methods show that the results obtained are very good even for a moderately refined mesh [6.24].

For sake of clarity, we assume zero body forces, so that the discretized field is given by (6.23) with σ_* vanishing. The equilibrium equation (6.24), in turn, generate a parallel system of generalized displacements q

$$\Pi = -\int_{A_u} \sigma^T \mathbf{L} \dot{\bar{\mathbf{u}}}\, dA = -\mathbf{g}^T \dot{\mathbf{q}} = -\mathbf{h}^T \mathbf{C}^T \dot{\mathbf{q}} \tag{6.47}$$

where Π represents Hill's functional (6.2). Introducing (6.23) into (6.46), we can write the criterion of the mean

$$\frac{1}{2}\mathbf{h}^T \mathbf{F}\,\mathbf{h} - \sigma_Y^2 \leq 0 \tag{6.48}$$

where

$$\mathbf{F} = \frac{1}{V}\int_V \mathbf{S}^T \mathbf{D'}\,\mathbf{S}\, dV\,. \tag{6.49}$$

Now, we can remark that the criterion of the mean may be associated to Hill's principle, by applying it to the finite element V and by considering a Lagrange parameter field λ constant on the element, to relax the yield criterion (6.44). So, the lagrangian (6.8) becomes :

$$L = \lambda\left(\frac{1}{2}\mathbf{h}^T \mathbf{F}\,\mathbf{h} - \sigma_Y^2\right) - \mathbf{h}^T \mathbf{C}^T \langle\dot{\mathbf{q}}\rangle. \tag{6.50}$$

The Euler-Lagrange equation corresponding to the variation of the stress parameter is the compatibility condition, but in a weak form :

$$\mathbf{C}^T \langle\dot{\mathbf{q}}\rangle = \lambda\, \mathbf{F}\,\mathbf{h}\ . \tag{6.51}$$

Eliminating **h** between this last condition and the equilibrium equation (6.24), we obtain :

$$\mathbf{K}\langle\dot{\mathbf{q}}\rangle = \lambda\,\mathbf{g} \tag{6.52}$$

where

$$\mathbf{K} = \mathbf{C}\,\mathbf{F}^{-1}\,\mathbf{C}^T. \tag{6.53}$$

The value of the parameter λ corresponding to the element may be determined from (6.48) to be

$$\lambda = \left(\frac{2}{2\,\sigma_Y^2} <\dot{q}>^T \mathbf{K} <\dot{q}> \right)^{1/2} \tag{6.54}$$

and is the mean strain intensity rate of the considered element.

It is interesting to remark that the constraint matrix **K** given in (6.53) can be deduced directly from the equivalent stiffness matrix for elastic analysis. Indeed, the matrix **D** occuring in the yield criterion (6.44) is identical to an elastic stiffness matrix $\mathbf{D}_e$ of the Hooke's law with Poisson's ratio equal to a half because of the incompressibility of von Mises' rigid-plastic material.

Consequently, all equilibrium finite elements formulated for elastic analysis may be used with slight changes for rigid-plastic analysis.

For sake of convenience, the yield criterion of the mean (6.48) is now rewritten in a dimensionless form for element number e :

$$\varphi_e = \frac{1}{2}\,\mathbf{h}_e^T\,\mathbf{F}_e\,\mathbf{h}_e \le 1\,. \tag{6.55}$$

Hill's functional (6.47) for the whole structure is the sum of these functionals defined in each element :

$$\Pi\,(\mathbf{h}) = -\sum_e \mathbf{h}_e^T\,\mathbf{C}_e^T <\dot{\mathbf{q}}_e>\,. \tag{6.56}$$

According to Hill's principle (6.2), the limit analysis problem is reduced to minimizing (6.56), subject to inequalities (6.55), with respect to arbitrary stress parameters **h**. This is a particular case of the general nonlinear programming problem. In [6.12], it is transformed into a sequence of unconstrained optimization problems by introducing an appropriate penalty function [6.22 and 6.26]. The penalty function chosen by Nguyen Dang Hung is the same as that one proposed by Casciaro and Di Carlo [6.19], because of its remarkable stability during the interactive process.

6.5. Pure displacement finite elements.

With a view to estimate the size of the error, it is interesting to compute both the lower and upper bounds. For this, several previous authors, Hodge and Belytschko [6.15], Bottero and his co-workers [6.20], Anderheggen [6.27], Hutula [6.29], Biron [6.28, 6.66], Nguyen Dang Hung and his co-workers [6.12, 6.24 and 6.33], have developed dual solutions, based on displacement finite elements.

The kinematical solution alone was also developed by Hayes and Marcal [6.67].

To realize a displacement formulation, we must make some assumptions on the velocity field :

$$\dot{\mathbf{u}} = \mathbf{M}\,(\mathbf{x})\,\dot{\mathbf{q}} \tag{6.57}$$

where $\mathbf{M}\,(\mathbf{x})$ is a matrix of shape functions and $\dot{\mathbf{q}}$ the vector of *nodal displacement rates*. To assure a priori the compatibility, we must enforce the boundary condition, the continuity of the displacement rate field (6.57) on the interelement connection and the internal compatibility condition (6.9). The strain rate components are then also, in each element, polynomial fields, depending on a system of arbitrary parameters $\dot{v}_k$, components of a vector $\dot{\mathbf{v}}$:

$$\dot{\boldsymbol{\varepsilon}} = \mathbf{R}(\mathbf{x})\dot{\mathbf{v}}. \tag{6.58}$$

The strain parameters are connected to the nodal displacement rates by linear compatibility conditions :

$$\dot{\mathbf{v}} = \mathbf{C}^T\,\dot{\mathbf{q}}. \tag{6.59}$$

Naturally, we can make the same remark as in the previous Section concerning the dual field.

Introducing (6.58) in the total internal dissipation, we have :

$$\int_V \boldsymbol{\sigma}^T\,\dot{\boldsymbol{\varepsilon}}\,dV = \langle\mathbf{h}\rangle^T\,\dot{\mathbf{v}}$$

where the vector of generalized stresses

$$\langle\mathbf{h}\rangle = \int_V \mathbf{R}^T\,\boldsymbol{\sigma}\,dV$$

represents stresses weighted by the shape functions of the discretized strain field. This time, the stress field is known only by the weak quantities $\langle\mathbf{h}\rangle$.

6.5.1. *Upper bound problem.*

Because of (6.51), it is obvious that the total internal dissipation may be expressed in terms of the strain rate parameter vector :

$$\Omega(\dot{\mathbf{v}}) = \int_V D\ (\mathbf{R}(x)\,\dot{\mathbf{v}}\,)\ dV. \tag{6.60}$$

The discretized form of the normalization condition is obtained by putting (6.57) in (6.14) :

$$<\bar{\mathbf{g}}^{o}>^{T}\ \dot{\mathbf{q}} = 1 \tag{6.61}$$

with

$$<\bar{\mathbf{g}}^{o}> = \int_V \mathbf{M}^T\,\mathbf{X}^o\,dV + \int_{A_\sigma} \mathbf{M}^T\,\bar{\mathbf{X}}^o\,dA_\sigma\ . \tag{6.62}$$

Finally, taking account of the compatibility condition (6.52), the upper bound problem becomes, after discretization :

$$\underset{\dot{\mathbf{q}}}{\text{Min}} \qquad \Omega\left(\mathbf{C}^T\,\dot{\mathbf{q}}\right) \tag{6.63.)}$$

Constraint :

$$<\bar{\mathbf{g}}^{\,o}>^{T}\,\dot{\mathbf{q}} = 1. \tag{6.64}$$

Now, to deduce the Euler-Lagrange equations of this problem, let us introduce the associated lagrangian :

$$L_{h_+} = \Omega\left(\mathbf{C}^T\,\dot{\mathbf{q}}\right) - \alpha\left(\,<\bar{\mathbf{g}}^{\,o}>^{T}\,\dot{\mathbf{q}} - 1\right).$$

So, we obtain the following variation equation :

$$\mathbf{C}\,\frac{\partial\,\Omega}{\partial\,\dot{\mathbf{v}}} = \alpha<\bar{\mathbf{g}}^{\,o}>, \tag{6.64}$$

We may recognize the equilibrium equation (see (6.13)), but in a weak form. More especially, (6.64) is the equilibrium relation between the generalized forces $\alpha<\bar{\mathbf{g}}^{\,o}>$ and the weak stress vector :

$$<\mathbf{h}> = \frac{\partial\,\Omega}{\partial\,\dot{\mathbf{v}}}, \tag{6.65}$$

formally identical to the first constraint (6.29).

Finally, we shall say some words on the mathematical programming algorithm. Of course, the most interesting form of the upper bound problem is obtained when the dissipation function Ω is linear [6.20 and 6.27]. This occurs when the plastic yielding criterion is linearized. For convenience, the strain rate parameter vector is often expressed with respect to positive or zero new parameters, components of a vector $\boldsymbol{\lambda}$:

$$\dot{\mathbf{v}} = \mathbf{N}^T \boldsymbol{\lambda}$$

so that the total dissipation (6.60) becomes linear :

$$\Omega = \mathbf{k}_Y^T \boldsymbol{\lambda} .$$

Finally, the upper bound problem (6.63) becomes :

$$\min_{\boldsymbol{\lambda},\dot{\mathbf{q}}} \quad \mathbf{k}_Y^T \boldsymbol{\lambda}$$

constraints :

$$\begin{aligned} &\mathbf{C}^T \dot{\mathbf{q}} = \mathbf{N}^T \boldsymbol{\lambda} \\ &<\bar{\mathbf{g}}^o>^T \dot{\mathbf{q}} = 1 . \end{aligned} \qquad (6.66)$$

Let us remark that this problem is formally similar to problem (6.39) but, this time, the unknowns $\boldsymbol{\lambda}$ and $\dot{\mathbf{q}}$ are strong variables.

Now, for the general case of a non linear criterion, we must use a non-linear programming algorithm. In the parts of the body which remain rigid, the stresses are undetermined.

Thus, when the dissipation vanishes, the concept of derivative (6.65) disappears (*). So, a method not requiring derivative is necessary. In reference [6.15], the minimization is performed by the (non linear) Simplex of Nelder and Mead [6.37], a technique that does not require the derivatives of the objective function. In [6.33], Best's algorithm of feasible conjugated directions [6.39], with the special method of Goldstein-Ritter-Armijo-Best

(*) In convex analysis [6.38] the generalized stress vector (6.56) occuring in (6.55), belongs to the subderivative to Ω, notion which generalizes the classic derivative.

[6.40] seems to be well adapted to the kinematical problem (6.63).

6.5.2. *Relaxed problems.*

To obtain an approximate value for the limit load, Markov's principle (6.1) may also be employed as in the method of Nguyen Dang Hung [6.12 and 6.24]. Let us remark that for von Mises yield criterion (1.34), the dissipation may be written as follows :

$$D(\dot{\boldsymbol{\varepsilon}}) = \left(\frac{1}{2} \dot{\boldsymbol{\varepsilon}}^T \mathbf{D}'^{-1} \dot{\boldsymbol{\varepsilon}} \right)^{1/2} \tag{6.67}$$

In matrix form, the total dissipation becomes :

$$\Omega = \int_V \frac{\frac{1}{2} \dot{\boldsymbol{\varepsilon}}^T \mathbf{D}'^{-1} \dot{\boldsymbol{\varepsilon}}}{\left(\frac{1}{2} \dot{\boldsymbol{\varepsilon}}^T \mathbf{D}' \dot{\boldsymbol{\varepsilon}} \right)^{1/2}} \, dV \, .$$

An important assumption is now made in order to cancel the nonlinear term in the functional (6.60) : there exists in the finite element a mean value

$$\lambda = \left(\frac{1}{V} \int_V \frac{1}{2} \dot{\varepsilon} \, \mathbf{D}'^{-1} \dot{\varepsilon} \, dV \right)^{1/2} = \left(\frac{1}{2} \dot{\mathbf{v}}^T \, \mathbf{H} \, \dot{\mathbf{v}} \right)^{1/2} \tag{6.68}$$

where the strain rate vector has been replaced by its value defined in (6.58) and

$$\mathbf{H} = \frac{1}{V} \int_V \mathbf{R}^T \mathbf{D}'^{-1} \mathbf{R} \, dV \, .$$

Taking account of (6.58) and (6.68), the functional Ω may now be written :

$$\Omega(\dot{\mathbf{v}}) = \frac{1}{2\lambda} \dot{\mathbf{v}}^T \mathbf{H} \dot{\mathbf{v}} \, .$$

Besides, owing to (6.59) and introducing (6.57) in both last terms of Markov's functional (6.1), we obtain :

$$\Phi = \frac{1}{2\lambda} \dot{\mathbf{q}}^T \mathbf{K} \dot{\mathbf{q}} - \langle \bar{\mathbf{g}} \rangle^T \dot{\mathbf{q}}$$

where

$$\mathbf{K} = \mathbf{C}^{\mathrm{T}} \mathbf{H} \, \mathbf{C}$$

and

$$<\overline{g}> = \int_V \mathbf{N}^{\mathrm{T}} \mathbf{X} \, dV + \int_{A_\sigma} \mathbf{N}^{\mathrm{T}} \overline{\mathbf{X}} \, dA_\sigma . \tag{6.69}$$

Expressing the minimum property of Φ, the Euler-Lagrange equation

$$\mathbf{K} \dot{\mathbf{q}} = \lambda <\overline{g}>$$

is obtained, which is identical to the relation (6.52) found in the static approach. This demonstrates the very important fact that it is possible to carry out the numerical computations for the kinematic approach using the same algorithms as for the static approach (see 6.4.2). As in 6.4.2, the constraint matrix **K** can be deduced from the equivalent stiffness matrix for elastic analysis by setting the Poisson's ratio equal to one half because of the incompressibility of the rigid-plastic material.

6.6. Mixed finite elements.

A disadvantage of the pure finite elements is that a field is strongly known and the dual only weakly. With a view to determine simultaneously the stresses and displacements in a strong manner, it is attractive to introduce mixed finite elements as in the approach of Christiansen [6.34] or Casciaro and Di Carlo [6.35 and 6.36].

To realize such a formulation, we must discretize the stress field :

$$\boldsymbol{\sigma} = \mathbf{S} \, (\mathbf{x}) \, \mathbf{h} . \tag{6.70}$$

and the displacement field by means of (6.57). So, the strain rate field is given by (6.58), owing to the compatibility conditions (6.59). Putting (6.70) and (6.58) in the expression of the total dissipation, and taking account of (6.59), we obtain :

$$\int_V \frac{1}{2} \, \sigma^{\mathrm{T}} \Delta^{\mathrm{T}} \, \dot{\mathbf{u}} \, dV = \dot{\mathbf{q}}^{\mathrm{T}} \mathbf{G} \, \mathbf{h} \tag{6.71}$$

with

$$\mathbf{G} = \mathbf{C} \int_V \mathbf{R}^{\mathrm{T}} \mathbf{S} \, dV . \tag{6.72}$$

As in the Section 6.4.1, we add a finite number of discretized inequalities concerning **h**, enough to ensure that the yield condition is strictly respected anywhere :

$$f_i(\mathbf{h}) \leq 0 . \tag{6.73}$$

6.6.1. *Mixed problem.*

This is the discretized form of the saddle-point problem (6.18).

The discretized form of the normalization is again (6.61), with the definition (6.62) of the weak nodal force vector. Then, the problem (6.18) is written as follows :

$$\underset{\dot{\mathbf{q}}}{\text{Min}} \quad \underset{\mathbf{h}}{\text{Max}} \quad \dot{\mathbf{q}}^T \mathbf{G}\,\mathbf{h}$$

constraints :

$$\begin{gathered} <\bar{\mathbf{g}}^{o}>^T \dot{\mathbf{q}} = 1 \\ f_i(\mathbf{h}) \leq 0 . \end{gathered} \tag{6.74}$$

The associated Lagrangian is obviously :

$$L_{hm} = \dot{\mathbf{q}}^T \mathbf{G}\,\mathbf{h} - \alpha\left(<\bar{\mathbf{g}}^{o}>^T \dot{\mathbf{q}} - 1\right) - \sum_i \lambda_i f_i(\mathbf{h}) . \tag{6.75}$$

We can deduce from it the Euler-Lagrange equations, respectively the equilibrium equations :

$$\mathbf{G}\,\mathbf{h} = \alpha <\bar{\mathbf{g}}^{o}> \tag{6.76}$$

and the compatibility conditions :

$$\mathbf{G}^T \dot{\mathbf{q}} = \sum_i \lambda_i \frac{\partial f_i}{\partial \mathbf{h}} . \tag{6.77}$$

The comparison of (6.76) with (6.24) shows that matrix **G** may be interpreted as a statical connection matrix.

Let us remark that value of the objective function at the saddle-point provides only an approximation of the limit load, but which can be expected to be a better estimate than the bounds.

Let us note also that it is sometimes more attractive as in [6.34] to solve the associated maximum problem

$$\max_{\alpha,\mathbf{h}} \quad \alpha$$

constraints :

$$\mathbf{G}\,\mathbf{h} = \alpha <\bar{\mathbf{g}}^{\,o}>$$
$$f_i\,(\mathbf{h}) \le 0 \tag{6.78}$$

which has the form of a lower bound problem, with the new connection matrix **G**, defined by (6.72). The model presented above was introduced by Christiansen in [6.34].

6.6.2. *Mixed principle.*

Another variational problem, leading to a mixed finite element, is the saddle-point problem (6.17) which was first used by Casciaro and Di Carlo [6.35 and 6.36]. The power of external forces in the functional (6.17) is discretized as the same term in Markov's functional. So, taking account of (6.69), the discretized form of the saddle-point problem (6.17) is :

$$\min_{\dot{\mathbf{q}}} \quad \max_{\mathbf{h}} \quad \left(\dot{\mathbf{q}}^T\, \mathbf{G}\,\mathbf{h} - <\bar{\mathbf{g}}>^T\, \dot{\mathbf{q}} \right)$$

constraints :

$$f_i\,(\mathbf{h}) \le 0\,. \tag{6.79}$$

It is easy to show that the Euler-Lagrange equations are the following equilibrium and compatibility equations

$$\mathbf{G}\,\mathbf{h} = <\bar{\mathbf{g}}> \tag{6.80}$$

$$\mathbf{G}^T\,\dot{\mathbf{q}} = \sum_i \lambda_i \,\frac{\partial f_i}{\partial\,\mathbf{h}}\,. \tag{6.81}$$

The algorithm adopted in [6.35] computes the solution of the equilibrium equations (6.80), by means of a Gauss elimination. By comparison with (6.40), we have :

$$\mathbf{h}_1 = \mathbf{G}_1^{-1} <\bar{\mathbf{g}}> - \mathbf{G}_1^{-1}\, \mathbf{G}_2\, \mathbf{h}_2\,.$$

Then, the minimization is performed, using the Davidon-Fletcher-Powell method.

6.7. Hybrid finite elements.

Another way to obtain a strong approximation on stress and displacements is provided by the hybrid approach. For instance, let us present the hybrid model used by Nguyen Dang Hung in [6.12] and [6.24].

For sake of clearness, let us assume that the body forces vanish. The equilibrated stress field is a polynomial depending on the system of parameters h_k by (6.70). Introducing (6.70) into the criterion of the mean, we find the constraints (6.48). As in the Section 6.4.2, the matrix **F** defined in (6.49) may be generated by the flexibility matrix of the elastic analysis. An approximated system of nodal velocities $\dot{\mathbf{q}}$, which completely determines the Lagrange's multipliers fields $\boldsymbol{\mu}$ on the border of each element, may be defined as follows :

$$\boldsymbol{\mu} = \mathbf{M}(\mathbf{x})\,\dot{\mathbf{q}}\,. \tag{6.82}$$

Then, taking account of (6.70), (6.82) one has

$$\sum_{A} \int_{A} \sigma^T L^T \mu \, dA = \mathbf{h}^T \mathbf{G}^T \dot{\mathbf{q}} \tag{6.83}$$

with

$$\mathbf{G} = \sum_{A} \int_{A} \mathbf{S}^T \mathbf{L}^T \mathbf{M} \, dA. \tag{6.84}$$

To relax the criterion of the mean (6.48), as in Section 6.3.2, we introduce the following lagrangian, associated with the discretization of the principle (6.21) :

$$L_{hh} = \lambda\left(\frac{1}{2}\,\mathbf{h}^T \mathbf{F}\,\mathbf{h} - \sigma_V^2\right) - \langle\bar{\mathbf{g}}\rangle^T \dot{\mathbf{q}} + \mathbf{h}^T \mathbf{G}^T \dot{\mathbf{q}}\,. \tag{6.85}$$

So, the Euler-Lagrange equation are respectively the equilibrium and compatibility equations :

$$\mathbf{G}\,\mathbf{h} = \langle\bar{\mathbf{g}}\rangle \tag{6.86}$$

$$\mathbf{G}^T \dot{\mathbf{q}} = \lambda\,\mathbf{F}\,\mathbf{h} \tag{6.87}$$

(6.87) permits **h** to be eliminated from (6.86), to finally arrive at the classical results :

$$\mathbf{K}\dot{\mathbf{q}} = \lambda <\bar{\mathbf{g}}> \tag{6.88}$$

with :

$$\mathbf{K} = \mathbf{G}^T \mathbf{F}^{-1} \mathbf{G} . \tag{6.89}$$

(6.88) is formally the same that (6.52), but the nodal velocities are now strongly known.

As in Section 6.5.2, we conclude that it is possible to carry out the numerical computations for the hybrid approach using the same algorithm as for the static approach (see 6.4.2).

6.8. Variational principles for shakedown problems.

We have seen in Chapter 4 that the fundamental theorems of shakedown theory were similar to previous ones of limit analysis. The parallel can be drawn also between the variational principles of both these theories. Nevertheless, owing to the small amount of works published on this subject, we shall limit ourselves to the principles only used in literature for finite elements. The reader can find a good synthesis about this topic in recent publications such as [6.87, 6.88 and 6.89].

6.8.1. *Basic principles.*

So, we introduce again Markov's and Hill's principles, but we must modify them in the following way :

- we consider the history of the structure over the whole collapse cycle ;
- the unknowns are now the plastic strain rate and residual stress fields ;
- they must be admissible in the shakedown sense ;
- the external actions are represented by the stress field in the corresponding fictitious e-

lastic body, denoted σ^*.

These preliminary remarks allow to introduce the following variational principles :

Theorem 11 : *Markov's principle over a cycle : among the admissible plastic strain rate fields, the true one makes the functional*

$$\Phi_s = \oint \int_V D(\dot{\varepsilon}^p)\, dVdt - \oint \int_V \sigma^{*T}\, \dot{\varepsilon}^p\, dVdt \tag{6.90}$$

an absolute minimum.

Theorem 12 : *Hill's principle over a cycle : among the admissible residual stress fields, the true one makes the functional*

$$\pi_s = -\int_{A_u} \bar{\boldsymbol{\rho}}^T L^T \boldsymbol{\Delta}\bar{\mathbf{u}} \, dA_u \tag{6.91}$$

an absolute minimum.

First, let us prove that the true plastic strain rate field $\dot{\boldsymbol{\varepsilon}}^p$ minimizes (6.90) among all admissible field $\dot{\boldsymbol{\varepsilon}}^{p'}$. Indeed, we have

$$\Phi_s(\dot{\boldsymbol{\varepsilon}}^{p'}) - \Phi_s(\dot{\boldsymbol{\varepsilon}}^p) = \oint \int_V \left[D(\dot{\boldsymbol{\varepsilon}}^{p'}) - D(\dot{\boldsymbol{\varepsilon}}^p) - \sigma^{*T} (\dot{\boldsymbol{\varepsilon}}^{p'} - \dot{\boldsymbol{\varepsilon}}^p) \right] dVdt$$

or, using the convexity inequality (6.3) :

$$\Phi_s(\dot{\boldsymbol{\varepsilon}}^{p'}) - \Phi_s(\dot{\boldsymbol{\varepsilon}}^p) \geq \oint \int_V \left(\boldsymbol{\sigma} - \sigma^* \right)^T (\dot{\boldsymbol{\varepsilon}}^{p'} - \dot{\boldsymbol{\varepsilon}}^p) \, dVdt \, .$$

Now, to the true field $\boldsymbol{\sigma}$ corresponds a time independent field of residual stresses $\bar{\boldsymbol{\rho}}$. As $\dot{\boldsymbol{\varepsilon}}^{p'}$ and $\dot{\boldsymbol{\varepsilon}}^p$ are both admissible fields, and taking account of the *theorem of virtual powers*, we have :

$$\Phi_s(\dot{\boldsymbol{\varepsilon}}^{p'}) - \Phi_s(\dot{\boldsymbol{\varepsilon}}^p) \geq P\left(\bar{\boldsymbol{\rho}}, \oint (\dot{\boldsymbol{\varepsilon}}^{p'} - \dot{\boldsymbol{\varepsilon}}^p) \, dt \right) = 0 \tag{6.92}$$

which proves theorem 11.

In a similar way, let us prove that the true stress field $\bar{\boldsymbol{\rho}}$ minimizes (6.91) among all admissible residual stress fields $\bar{\rho}'$. Indeed, we have :

$$\Pi_s(\bar{\rho}') - \Pi_s(\bar{\rho}) = \int_{A_u} (\bar{\boldsymbol{\rho}} - \bar{\boldsymbol{\rho}}')^T \mathbf{L}^T \boldsymbol{\Delta}\bar{\mathbf{u}} \, dA_u \, .$$

Now, to the true field $\boldsymbol{\rho}$ corresponds the plastic strain rate field $\dot{\boldsymbol{\varepsilon}}^p$ which must be admissible :

$$\oint \dot{\boldsymbol{\varepsilon}}^p \, dt = \Delta^T \boldsymbol{\Delta}\mathbf{u} \qquad \text{in} \qquad V \tag{6.93}$$

$$\Delta \mathbf{u} = \Delta \bar{\mathbf{u}} \qquad \text{on} \qquad A_u \tag{6.94}$$

So, using again the *theorem of virtual powers,* we have :

$$\int_{A_u} (\bar{\boldsymbol{\rho}} - \bar{\boldsymbol{\rho}}')^T \mathbf{L}^T \Delta \bar{\mathbf{u}} \, dA_u = P\ (\bar{\boldsymbol{\rho}} - \bar{\boldsymbol{\rho}}',\ \oint \dot{\boldsymbol{\varepsilon}}^p \, dt) = \oint P\ (\bar{\boldsymbol{\rho}} - \bar{\boldsymbol{\rho}}',\ \dot{\boldsymbol{\varepsilon}}^p)\, dt\,.$$

Finally, the *theorem of maximum dissipation* gives

$$\Pi_s (\bar{\boldsymbol{\rho}}') - \Pi_s (\bar{\boldsymbol{\rho}}) = \oint [\int_V D(\dot{\boldsymbol{\varepsilon}}^p)\, dV - P (\bar{\boldsymbol{\rho}}',\ \dot{\boldsymbol{\varepsilon}}^p)]\, dt \geq 0. \tag{6.95}$$

Next, we may deduce the Euler-Lagrange equations of these basic principles. Then let us take the variation of the functional of Markov's principle over a cycle with respect to the definitions (2.1) of the specific rate and (4.15) of the residual stresses :

$$\delta\, \Phi_s = \oint P\ (\bar{\boldsymbol{\rho}},\ \delta\, \dot{\boldsymbol{\varepsilon}}^p)\, dt\,.$$

Besides, let us assume that $\bar{\boldsymbol{\rho}}$ is time-independent :

$$\delta\, \Phi_s = P\ (\boldsymbol{\rho}, \oint\ \delta\, \dot{\boldsymbol{\varepsilon}}^p\, dt)\,. \tag{6.96}$$

Now, as the plastic strain rate fields are admissible, their variation verifies the homogeneous compatibility conditions :

$$\oint\ \delta\ \dot{\boldsymbol{\varepsilon}}^p\, dt = \Delta^T\, \delta\, \Delta \mathbf{u} \qquad \text{in} \qquad V \tag{6.97}$$

$$\delta\ \Delta \mathbf{u} = 0 \qquad \text{on} \qquad A_u\,. \tag{6.98}$$

Finally, integrating by parts in (6.96) and taking account of the above relations (6.97) and (6.98), we have :

$$\delta\, \Phi_s = \oint \left[\int_{A_\sigma} \bar{\rho}^T\, \mathbf{L}^T\, \delta\ \Delta \mathbf{u}\, dA_\sigma - \int_V \bar{\rho}^T\, \Delta^T\, \delta\, \Delta \mathbf{u}\, dV \right] dt = 0\,.$$

So, the Euler-Lagrange equations are auto-equilibrium equations :

$$\Delta\, \bar{\boldsymbol{\rho}} = 0 \tag{6.99}$$

$$\mathbf{L}\,\bar{\boldsymbol{\rho}} = 0. \tag{6.100}$$

For the variation equations of the Hill's principle over a cycle, let us introduce as in (6.8) the following lagrangian :

$$L_s = \oint \int \lambda \left[\left(\sigma_R\,(\bar{\rho} + \sigma^*) \right)^2 - \sigma_Y^2 \right] dV\,dt - \int_{A_u} \bar{\rho}^T\,\mathbf{L}^T\,\Delta\bar{\mathbf{u}}\,dA_u. \tag{6.101}$$

Applying the Green-Ostrogradsky formula and taking account of equations (6.99) and (6.100), we obtain easily :

$$\delta L_s = \int_V \boldsymbol{\delta}\,\bar{\rho}^T \left[\oint \dot{\boldsymbol{\varepsilon}}^p\,dt - \Delta^T\,\Delta u \right] dV$$

$$- \int_{A_u} \delta\,\bar{\rho}^T\,\mathbf{L}^T\,(\,\boldsymbol{\Delta}\mathbf{u} - \boldsymbol{\Delta}\bar{\mathbf{u}})\,dA_u\ ]$$

consequently, the variation equations are the compatibility conditions (6.93) and (6.94).

The above approach, similar to Mandel's one in limit analysis (see Section 6.2.2), is due to the junior author. More extensive developments are available in his Ph. D. Thesis [4.46].

6.8.2. *Theorems of the plastic shakedown.*

Now, we can provide lower and upper bounds of the shakedown load as direct consequence of the previous basic principles. Naturally, we suppose in the following developments that :

1. the body has fixed supports, in the following sense :

$$\boldsymbol{\Delta}\bar{\mathbf{u}} = 0 \qquad \text{on} \qquad A_u \tag{6.102}$$

2. the load domain is varying in a radial way (see (4.28)).

First let $\bar{\rho}_-$ given by (4.29) be an admissible residual stress field for which the structure shakes down.

Let $\bar{\boldsymbol{\rho}}$, given by (4.30) and $\dot{\boldsymbol{\varepsilon}}^p$, be the collapse residual stress and plastic strain rate fields.

Owing to (6.95), we have :

$$\Pi_s(\bar{\boldsymbol{\rho}}_-) - \Pi_s(\bar{\boldsymbol{\rho}}) = \oint P \;(\boldsymbol{\sigma} - \boldsymbol{\sigma}_-,\; \dot{\boldsymbol{\varepsilon}}^p)\, dt \geq 0. \tag{6.103}$$

As the selfstress fields are time independent and taking account of (4.26), (4.29) and (4.30), we obtain :

$$P\;(\bar{\boldsymbol{\rho}} - \bar{\boldsymbol{\rho}}_-,\; \Delta\varepsilon^p) + (\alpha_s - \alpha_-)\oint P\;(\sigma^{*o},\; \dot{\boldsymbol{\varepsilon}}^p)\, dt \geq 0.$$

Let us remark that $(\bar{\boldsymbol{\rho}} - \bar{\boldsymbol{\rho}}_-)$ is self-equilibrated. On the other hand, the collapse field $\dot{\boldsymbol{\varepsilon}}^p$ is admissible.

Then, it satisfies the compatibility conditions with the homogeneous boundary kinematical condition (6.102). So, because of the *theorem of virtual powers*, the first term of the left hand, member of above relation vanishes. Besides, the integral of the second term is positive. Finally, we obtain :

$$\alpha_s \geq \alpha_-. \tag{6.104}$$

In a similar way, let us prove the upper bound theorem. Let be $\dot{\boldsymbol{\varepsilon}}^p_+$ an admissible plastic strain rate field. The corresponding loading parameter α_+ is defined by (4.36). Similarly, for the collapse field $\dot{\varepsilon}^p$, we have the shakedown load given by (4.37). Now, let us consider the body at collapse.

Applying Markov's principle over a cycle (see (6.90)), furnishes the relation :

$$\Phi_s(\dot{\boldsymbol{\varepsilon}}^p_+) - \Phi_s(\dot{\boldsymbol{\varepsilon}}^p) = \oint \int_V \left[D\;(\dot{\boldsymbol{\varepsilon}}^p_+) - D\;(\dot{\boldsymbol{\varepsilon}}^p) \right] dV\, dt$$

$$- \alpha_s \oint P\;(\sigma^{*o},\; \dot{\boldsymbol{\varepsilon}}^p_+ - \dot{\boldsymbol{\varepsilon}}^p)\, dt \geq 0\,.$$

The elimination of the first term of the right hand member of above expression, owing to (4.36) and (4.37), gives

$$(\alpha_+ - \alpha_s)\oint P\,(\sigma^{*o},\; \dot{\boldsymbol{\varepsilon}}^p_+)\, dt = \Phi_s(\dot{\boldsymbol{\varepsilon}}^p_+) - \Phi_s(\dot{\boldsymbol{\varepsilon}}^p) \geq 0\,.$$

As the integral in the left hand member is positive, we obtain :

$$\alpha_+ \geq \alpha_s. \tag{6.105}$$

Now, we may state the fundamentals theorems of plastic shakedown in form of constrained extremum problems.

In the statical theorem, we seek the maximum of multipliers corresponding to admissible residual stress fields :

$$\max_{\alpha_-, \boldsymbol{\rho}_-} \quad \alpha_-$$

constraints :

$$\Delta \bar{\boldsymbol{\rho}}_- = 0 \qquad \text{in} \qquad V$$

$$\mathbf{L}\, \bar{\boldsymbol{\rho}}_- = 0 \qquad \text{on} \qquad A_u$$

$$\sigma_R\,(\alpha_-\, \sigma^{*o} + \bar{\boldsymbol{\rho}}_-) \leq \sigma_Y \qquad \text{in} \qquad V. \tag{6.106}$$

For all points, σ^{*o} belonging to the domain of reference.

For the kinematical approach, it is convenient to impose the following condition to the admissible plastic strain rate fields :

$$\oint \int_V \sigma^{*oT}\, \dot{\boldsymbol{\varepsilon}}^p_+ \, dV\, dt = 1\,. \tag{6.107}$$

Owing to (4.37) and (6.107), the multiplier α_+ is equal to the total dissipation over a cycle. So, the kinematical theorem leads to seek the minimum of multipliers corresponding to admissible plastic strain rate fields :

$$\min_{\dot{\boldsymbol{\varepsilon}}^p_+, \Delta u_+} \quad \oint \int_V D\ (\dot{\boldsymbol{\varepsilon}}^p_+)\, dV\, dt$$

constraints :

$$\oint \dot{\boldsymbol{\varepsilon}}^p_+\, dt = \Delta^T\, \boldsymbol{\Delta u}_+ \qquad \text{in} \qquad V$$

$$\boldsymbol{\Delta u}_+ = 0 \qquad \text{on} \qquad A_u$$

$$\oint \int_V \sigma^{*oT}\, \dot{\boldsymbol{\varepsilon}}^p_+ \, dV\, dt = 1\,. \tag{6.108}$$

Finally, let us deduce the Euler-Lagrange equations of the two variational problems above. For the statical problem, we introduce the associated lagrangian :

$$L_{s-} = \alpha_- - \oint \int_V \lambda \left[\sigma_R^2(\alpha_- \sigma^{*o} + \bar{\rho}_-) - \sigma_Y^2 \right] dV\, dt - \int_V \Delta\mathbf{u}^{+T} \Delta \bar{\rho}_- \, dV$$

$$+ \int_{A_\sigma} \Delta\mathbf{u}_+^T \mathbf{L} \bar{\rho}_- \, dA_\sigma .$$

It is easy to show that the Euler-Lagrange equations of the statical theorem are the kinematical conditions (6.108). In a similar way, let us introduce the lagrangian :

$$L_{s+} = \oint \int_V D\ (\dot{\varepsilon}_+^p)\, dV\, dt - \int_V \bar{\rho}_-^T \left[\oint \dot{\varepsilon}_+^p \, dt - \Delta^T \Delta\mathbf{u}_+ \right] dV$$

$$- \alpha_- \left(\oint \int_V \sigma^{*oT} \dot{\varepsilon}_+^p \, dV\, dt - 1 \right)$$

from which we can prove that the Euler-Lagrange equations of the kinematical problem are the statical relations (6.106).

Finally, we see that the lower and upper bounds problems of the plastic shakedown are *dual* in the sense of the mathematical programming. The first rigourous prove of this duality is due to Debordes [6.68].

6.8.3. *Two-fields principles.*

In this domain, the literature is presently very poor. Therefore, we limit ourselves to expound the hybrid stress principle used by Morelle in [6.44]. We want to relax the equilibrium on the interelement connection during the discretization of the lower bound problem. For this, let us introduce the Lagrange's multiplier field $\Delta\mu$ on interfaces. With the same notations as in Section 6.3.3, we may introduce the following saddle-point problem :

$$\underset{\Delta\mu}{\text{Max}} \quad \underset{\alpha,\, \bar{\rho}_-}{\text{Min}} \quad (-\alpha + \sum_A \int_A \bar{\rho}^T \mathbf{L}\, \Delta\mu\, dA)\}$$

constraints :

$$\Delta \bar{\rho} = 0 \qquad \text{in each finite element } V_e$$

$$\mathbf{L}\bar{\boldsymbol{\rho}} = 0 \qquad \text{on} \qquad A_\sigma$$

$$\sigma_R\,(\alpha\,\sigma^{*o} + \bar{\boldsymbol{\rho}}) \le \sigma_Y \qquad \text{in each} \qquad V_e. \tag{6.109}$$

The Euler-Lagrange equations of the problem are the compatibility conditions (6.93-94), (6.102) and :

$$\mathbf{L}^+\,\bar{\rho}^+ + \mathbf{L}^-\;\bar{\rho}^- = 0 \qquad \text{on each A}$$

$$\oint \dot{\mathbf{u}}_+\,dt = \boldsymbol{\Delta}\,\boldsymbol{\mu} \qquad \text{on each A.} \tag{6.110}$$

6.9. Pure equilibrium finite elements for plastic shakedown.

The application of the equilibrium finite elements in plastic shakedown has followed that obtained in limit analysis in the recent years. To the author's knowledge, these studies were performed by Belytschko [6.45], Corradi and Zavelani [6.46], Nguyen Dang Hung and Palgen [6.47], Morelle and Nguyen Dang Hung [6.48, 6.50, 6.51and 6.88], Weichert and Gross-Weege [6.84], Franchi and Genna [6.85 and 6.86].

For simplicity, we shall assume that the body forces vanish.

Within each element, the same polynomial interpolation is performed for the stress field in the fictitious elastic body and the residual stress field :

$$\sigma^{*o} = \mathbf{S}\,(\mathbf{x})\,\mathbf{h}^{*o} \tag{6.111}$$

$$\bar{\boldsymbol{\rho}}_- = \mathbf{S}\,(\mathbf{x})\,\mathbf{h}\,. \tag{6.112}$$

By the way, the shape functions must be chosen so that the stress fields are solutions of the homogeneous equations of internal equilibrium, however the stress parameters h_k and h_k^{*o} may be.

Besides, $\bar{\boldsymbol{\rho}}_-$ must satisfy a priori the boundary equilibrium equations (6.100) and also the equilibrium at the interfaces between continuous finite elements.

So, we obtain the following homogeneous linear system (*) :

$$\mathbf{C}\,\mathbf{h} = 0. \tag{6.113}$$

As in the Section 6.3, the strains are of course known only in a weak form. Indeed, putting (6.112) in the work of the residual stresses on the plastic strain increment (4.26), we obtain :

$$\int_V \bar{\rho}_-^T \, \Delta\varepsilon^p \, dV = \mathbf{h}^T <\Delta v^p>$$

where

$$<\Delta v^p> = \int_V \mathbf{S}^T \, \Delta\varepsilon^p \, dV$$

is a vector of generalized plastic strain increments.

Finally, we must satisfy also the yield criterion *anywhere* in the body and for *all the points* σ^{*o} of the load domain of reference, that is, owing to (6.111) and (6.112) :

$$\sigma_R\left(\mathbf{S}\,(\alpha_- \, \mathbf{h}^{*o} + \mathbf{h}) \right) \leq \sigma_Y \, . \tag{6.114}$$

With regard to the way in which the yield condition is enforced anywhere, the remark of the Section 6.3.1 is still valid. Naturally, we can relax this condition, as in the method of the yield criterion in the mean [6.12 and 6.24] proposed for shakedown problems in [6.53], and first used in [6.47].

Now, we define the *convex enveloppe* of a set K as the smallest convex set which contains K. Then, to satisfy inequality (6.114) for all the points of the domain of reference, following classical theorem can easily be demonstrated, under the condition that the plastic condition is convex [6.49].

(*) At this point of view, we must note the work of Corradi and Zavelani [6.46] is a special application of the statical theorem because, as the stresses are derived from a displacement model, they are autoequilibrated only in a weak sense. So, in the general case, we must expect to obtain only an approximation of the lower bound.

Theorem 13 : *Shakedown will occur for the domain of reference if and only if it occurs for its convex envelope.*

For a domain such as the one defined by (4.27 bis), from the previous theorem the inequality (6.114) needs only be verified for the vertices of the load domain of reference. If there are k independent loads, we have then 2^k vertices to consider. To each vertex σ_m^{*o}, is associated a vector $\mathbf{h}_m^{*o}$ which can be computed by a classical elastic finite element algorithm.

So, to enforce the yield condition anywhere in the body and for all the points of the load domain of reference, we need add a *finite number* of discretized inequalities :

$$f_i\left(\alpha\_\ \mathbf{h}_m^{*o} + \mathbf{h}\right) \le 0. \tag{6.115}$$

Finally, the discretized form of the lower bound problem (6.106) is :

$$\max_{\alpha\_,\ \mathbf{h}} \quad \alpha\_$$

constraints :

$$\mathbf{C}\,\mathbf{h} = 0$$

$$f_i\left(\alpha\_\ \mathbf{h}_m^{*o} + \mathbf{h}\right) \le 0. \tag{6.116}$$

To deduce the Euler-Lagrange equations of the previous problem, let us introduce the associated lagrangian :

$$L_{sh-} = \alpha\_ + \langle\mathbf{\Delta q}\rangle^T\,\mathbf{C}\,\mathbf{h}\_ - \sum_i \sum_m \langle\lambda_{im}\rangle\, f_i\,(\alpha\_\ \mathbf{h}_m^{*o} + \mathbf{h})\ .$$

So, we have the following variation equations :

$$\sum_i \sum_m \langle\lambda_{im}\rangle \mathbf{h}_m^{*o} \frac{\partial f_i}{\partial \mathbf{h}} = 1 \tag{6.117}$$

$$\mathbf{C}^T \langle\mathbf{\Delta q}\rangle = \sum_i \sum_m \langle\lambda_{im}\rangle \frac{\partial f_i}{\partial \mathbf{h}}. \tag{6.118}$$

It is the weak form of the kinematical conditions (6.108). More precisely, (6.117) is the discretized form of the normalization condition (6.107), and (6.118) is the compatibility condition :

$$\mathbf{C}^T <\mathbf{\Delta q}> = <\mathbf{\Delta v}^p>$$

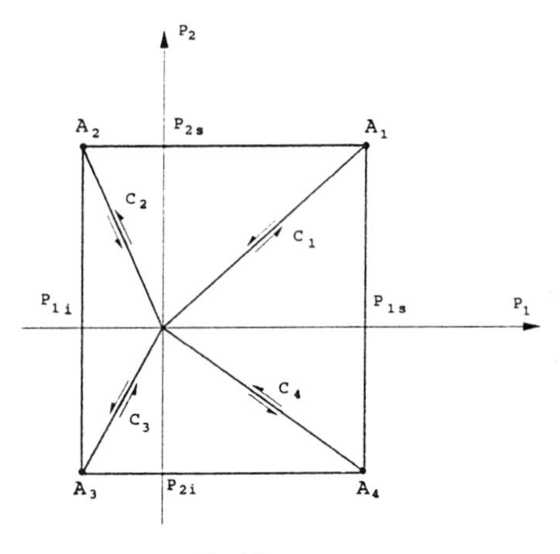

Fig. 6.2.

between the vector $<\mathbf{\Delta q}>$ of weak displacements and the vector

$$<\mathbf{\Delta v}>^p = \sum_i \sum_m <\lambda_{im}> \frac{\partial f_i}{\partial \mathbf{h}}$$

of the weak generalized plastic strain increments.

The previous interpretation of the Lagrange multipliers was first introduced by Nguyen Dang Hung and Palgen in [6.47].

Before examining the other kinds of finite elements, we need discuss the following points.

Remark on step-by-step algorithms.

Let us consider k independent loads $P_1, \ldots, P_k$ and the corresponding load domain S of which the vertices are $A_1, \ldots, A_{2^k}$.

Now, for any vertex A_m, we perform a classical step-by-step analysis for a repeated load cycle Cm varying between zero and A_m. Then, a shakedown occurs for all previous cycles, theorem 13 ensures the structure shakes down for any loading history contained in the load domain S which is obviously the convex envelope of the family $A_1, \dots, A_{2^k}$.

So, theorem 13 appears as a powerful tool to do shakedown analysis even if only step-by-step computer packages are available. Nevertheless, it may be noted the step-by-step computations up to the limit state can be very time-consuming by comparison with the mathematical programming methods.

Remark on mathematical programming algorithms.

In the shakedown analysis by finite elements, the problem of the size of the optimization problems solvable within a reasonable computer time is still more critical than in limit analysis. Indeed, as seen before, the number of plastic constraints must be multiplied by the number of the possible combinations of the independent loads.

This problem arises especially when the plastic yielding condition is nonlinear. In [6.45], Belytschko uses the classical Sequential Unconstrained Minimization Technique (SUMT) of Fiacco and Mc Cormick [6.22 and 6.23] (see Section 6.4.1.). In [6.47], Nguyen Dang Hung and Palgen prefer the ACDPAC nonlinear programming algorithm, developed by Best [6.52].

Nevertheless, the formulations with linearized plastic yielding condition seem presently the more promising for numerical purposes, because they allow different reductions of the size of the problem and can be solved efficiently by existing algorithms based on the Simplex Method [6.21]. Now, the lower bound problem (6.116) becomes :

$$\max_{\alpha_-, \mathbf{h}} \quad \alpha_-$$

constraints :

$$\mathbf{C}\,\mathbf{h} = 0$$

$$\mathbf{N}\,\mathbf{h} + \alpha_- \mathbf{k}_m^{*o} \leq \mathbf{k}_Y \tag{6.119}$$

with

$$\mathbf{k}_m^{*o} = \mathbf{N}\,\mathbf{h}_m^{*o} \; .$$

It is well known that the computational effort required by solving linear programming problems increases very rapidly with the number of constraints, while the dimension of the unknown vector is not so crucial. Therefore, the resolution of the dual program

$$\operatorname*{Min}_{<\lambda_m>,<\Delta q>} \sum_m \mathbf{k}_Y^T <\lambda_m>$$

constraints :

$$\mathbf{C}^T <\Delta \mathbf{q}> = \sum_m \mathbf{N}^T <\lambda_m>$$

$$\sum_m \mathbf{k}_m^{*oT} <\lambda_m> = 1 \qquad (6.120)$$

which has the form of an upper bound problem with weak variables, appears computationally preferable [6.46]. However, for two- or three-dimensional problems, even this formulation implies the numerical solution of a large problem.

An alternative formulation is proposed by Corradi and Zavelani in [6.46]. The admissible domain at a check point of the body, for a given vertex of the load domain, is defined by a polyhedron in the space of variables $\boldsymbol{\rho}$, $\alpha\_$. All points of this domain may be represented by a *convex combination* of the vectors which define the vertices of the polyhedron. So, the positive coefficients of the convex combination, components of the vector $\boldsymbol{\alpha}$, can be taken as the new variables. The plastic inequality constraints, which define the polyhedron, can then be replaced by a single constraint

$$\mathbf{u}^T \boldsymbol{\alpha} \leq 1 ,$$

u being a vector with unit components. The evaluation of the coordinates of the polyhedron vertices is explained in [6.46]. It was already mentioned that the critical factor in solving linear programming problems is the number of constraints. So, for plane stress or plane strain problems, the new formulation and the previous one (6.120) have roughly the same number of constraints. For general three-dimensional problems however, the new formulation permits a substantial gain.

Besides, let us observe that, by cancelling some of the vertices at a check point, the vector (ρ, $\alpha\_$) is constrained in a polyhedron contained in the initial one. By virtue of the lower bound theorem, the loading parameter so obtained cannot be greater than the true value, and therefore this procedure conserves the lower bound character. Computational experience, although small at the present time, seems to suggest that even drastic reductions do not affect the meaningful figures of the obtained value of the multiplier.

Another solution for reducing the size of the problem, which has the advantage to be rigourous, was proposed by De Donato [4.47] and Maier [6.54] and used first by Nguyen

Dang Hung and Morelle in [6.50]. Indeed, let us remark that all the inequalities in (6.119) are satisfied if we enforce :

$$\mathbf{N}\,\mathbf{h} + \alpha\_\,\mathbf{k}^{*o} \le \mathbf{k}_Y$$

with the components of the vector $\mathbf{k}^{*o}$ defined by :

$$(\mathbf{k}^{*o})_i = \sup_{1 \le m \le 2^k} (\mathbf{k}_m^{*o})_i\,.$$

So, the lower bound problem (6.116) is equivalent to the following one :

$$\operatorname*{Max}_{\alpha\_,\,\mathbf{h}} \quad \alpha\_$$

constraints :

$$\mathbf{C}\,\mathbf{h} = 0$$

$$\mathbf{N}\,\mathbf{h} + \alpha\_\,\mathbf{k}_m^{*o} \le \mathbf{k}_Y. \tag{6.121}$$

6.10. Pure displacement finite elements for plastic shakedown.

To carry out a displacement formulation, we must make some assumption on the displacement increment :

$$\Delta\mathbf{u} = \mathbf{M}(\mathbf{x})\,\Delta\,q\,,$$

where the elements of the matrix $M(\mathbf{x})$ have a polynomial form within each finite element and Δq is the vector of the *nodal displacement increments*. The plastic strain increment components are then in each element a polynominal field :

$$\Delta\varepsilon^p = \mathbf{R}(\mathbf{x})\ \Delta\mathbf{v}^p.$$

We know that the plastic strain rate field must not satisfy the internal compatibility, but only (6.93).

Nevertheless, it is convenient to choose the same polynomial dependence of the plastic strain rate field as one of the plastic strain increment field :

connection. The plastic strain increment parameters are connected to the nodal displacements increments by the linear compatibility conditions :

$$\oint \dot{\mathbf{v}}^p \, dt = \Delta \mathbf{v}^p = \mathbf{C}^T \, \Delta \mathbf{q}. \tag{6.123}$$

Let us observe that the compatibility of the plastic strain rate field is in this way satisfied only over a whole cycle.

Now, the stresses are known only in a weak form. Indeed, introducing (6.122) in the work of the residual stresses on the plastic strain increment, we obtain :

$$\int_V \bar{\boldsymbol{\rho}}^T \, \Delta\varepsilon^p \, dV = \langle \mathbf{h} \rangle^T \, \Delta \mathbf{v}^p$$

where

$$\langle \mathbf{h} \rangle = \int_V \mathbf{R}^T \bar{\boldsymbol{\rho}} \, dV$$

is a vector of generalized residual stresses.

Because of (6.122), it is obvious that the total internal dissipation may be expressed with respect to the plastic strain rate parameters :

$$\Omega_s \, (\dot{\mathbf{v}}^p) = \oint \int_V D \, (\mathbf{R}(\mathbf{x}) \dot{\mathbf{v}}^p) \, dV. \tag{6.124}$$

The discretized form of the normalization condition is obtained by putting (6.122) in (6.107) :

$$\oint \langle \mathbf{h}^{*o} \rangle^T \, \dot{\mathbf{v}}^p \, dt = 1 \tag{6.125}$$

with

$$\langle \mathbf{h}^{*o} \rangle = \int_V \mathbf{R}^T \, \boldsymbol{\sigma}^{*o} \, dV.$$

Now, we must perform the time discretization of the plastic strain rate field (6.122). Theorem 13 suggests that we should consider a cycle containing all the vertices $\langle \mathbf{h}_m^{*o} \rangle$ of the load domain of reference in the space of the generalized elastic stresses of reference.

Finally, when the vector $<\mathbf{h}^{*o}>$ reaches the vertex $<\mathbf{h}_m^{*o}>$, it shall be agreed that the value of the generalized plastic strain rates vector $\dot{\mathbf{v}}^p$ (t) is $\dot{\mathbf{v}}_m^p$. Owing to (6.123), (6.124) and (6.125), the upper bound problem (6.108) becomes :

$$\operatorname*{Min}_{\mathbf{\Delta q},\dot{\mathbf{v}}_m^p} \quad \sum_m \int_V D(\mathbf{R}(x)\,\dot{\mathbf{v}}_m^p)\, dV$$

constraints :

$$\mathbf{C}^T \mathbf{\Delta q} = \sum_m \dot{\mathbf{v}}_m^p$$

$$\sum_m <\mathbf{h}_m^{*o}>^T \dot{\mathbf{v}}_m^p = 1. \tag{6.126}$$

Now, to deduce the Euler-Lagrange equations of this problem, let us introduce the associate lagrangian :

$$L_{sh+} = \Omega_s(\dot{\mathbf{v}}_m^p) - \alpha\left(\sum_m <\mathbf{h}_m^{*o}>^T \dot{\mathbf{v}}_m^p - 1\right) + <\mathbf{h}>^T\left(\mathbf{C}^T\, \mathbf{\Delta q} - \sum_m \dot{\mathbf{v}}_m^p\right).$$

So, we obtain the following variational equations :

$$\mathbf{C}<\mathbf{h}> = 0\,, \tag{6.127}$$

$$<\mathbf{h}> + \alpha <\mathbf{h}_m^{*o}> = \frac{\partial \Omega_s}{\partial \dot{\mathbf{v}}_m^p}\,. \tag{6.128}$$

We may recognize in (6.127) the auto-equilibrium equation in a weak form and in (6.128), the yield criterion.

Both above relation, we can easily deduce the following equilibrium equations :

$$\mathbf{C}\frac{\partial \Omega_s}{\partial \dot{\mathbf{v}}_m^p} = \alpha <\mathbf{g}_m^{*o}> \tag{6.129}$$

$$\text{with} \quad \langle \mathbf{g}_m^{*o} \rangle = \mathbf{C} \langle \mathbf{h}_m^{*o} \rangle . \tag{6.130}$$

Let us note that (6.129) is formally similar to the relation (6.64), arising in limit analysis by displacement finite elements.

Remark on mathematical programming algorithm.

The only existing finite element programs based on the kinematical formulation of the shakedown analysis were performed by Nguyen Dang Hung and Morelle [6.50 and 6.51], Ponter and Karadeniz [6.80 to 6.83]. As in the previous authors use a piecewise linearized plastic yield condition, we consider only the linear programming algorithm. As in Section 6.5.1, for sake of easiness, the plastic strain rate parameters are expressed with respect to positive or zero new parameters :

$$\dot{\mathbf{v}}_m^p = \mathbf{N}^T \boldsymbol{\lambda}_m ,$$

so that the total dissipation occurring in the problem (6.126) is a linear form :

$$\Omega_s = \sum_m \mathbf{k}_Y^T \boldsymbol{\lambda}_m .$$

Finally, the upper bound problem (6.126) becomes :

$$\underset{\boldsymbol{\lambda}_m, \Delta \mathbf{q}}{\text{Min}} \quad \sum_m \mathbf{k}_Y^T \boldsymbol{\lambda}_m$$

constraints :

$$\mathbf{C}^T \Delta \mathrm{q} = \sum_m \mathbf{N}^T \lambda_m$$

$$\sum_m \langle \mathbf{k}_m^{*o} \rangle^T \boldsymbol{\lambda}_m = 1 \tag{6.131}$$

with

$$\langle \mathbf{k}_m^{*o} \rangle^T = \mathbf{N} \langle \mathbf{h}_m^{*o} \rangle .$$

Now, let us remark that it is equivalent to minimize the ratio :

$$\frac{\sum_m \mathbf{k}_Y^T \boldsymbol{\lambda}_m}{\sum_m \langle \mathbf{k}_m^{*o} \rangle^T \lambda_m}$$

for any set λ_m which defines the compatibility condition (6.131) without inforcing the normalization condition. As the components of the vector λ_m are positive quantities, the numerator of above ratio is bounded in the following way :

$$\sum_m \langle \mathbf{k}_m^{*o} \rangle^T \lambda_m \leq \langle \mathbf{k}^{*o} \rangle^T \sum \lambda_m \tag{6.132}$$

with

$$(\langle \mathbf{k}^{*o} \rangle)_i = \sup_{1 \leq m \leq 2^k} (\langle \mathbf{k}_m^{*o} \rangle)_i \tag{6.133}$$

Besides, the bound is reached in (6.132), when we choose :

$$(\lambda_{M_i})_i > 0$$

if the upper bound is realized in (6.133) for the vertex number M_i of the load domain and

$$(\lambda_m)_i = 0$$

for the i subscripts corresponding to the other vertices. So, we define the new vector $\boldsymbol{\lambda}$ such that $(\boldsymbol{\lambda})_i = (\lambda_{M_i})_i$.

Finally, the minimum of above ratio is :

$$\frac{\mathbf{k}_Y^T \boldsymbol{\lambda}}{\langle \mathbf{k}^{*o} \rangle^T \boldsymbol{\lambda}}$$

and the lower bound problem (6.131) is equivalent to the following one.

$$\underset{\Delta \mathbf{q},\boldsymbol{\lambda}}{\text{Min}} \quad \mathbf{k}_Y^T \boldsymbol{\lambda}$$

constraints :

$$\mathbf{C}^T \Delta \mathbf{q} = \mathbf{N}^T \boldsymbol{\lambda}$$

$$<\mathbf{k}^{*o}>^T \boldsymbol{\lambda} = 1. \tag{6.134}$$

In the latter formulation, used in [6.50 and 6.51], the size of the numerical problem is considerably reduced.

6.11. Hybrid finite elements for shakedown.

To obtain strong approximation on stresses and displacements, Morelle proposes in [6.44] an hybrid finite element approach deduced from the saddle-point problem (6.109). The residual stress field is a polynomial depending on the system of parameters **h** by (6.112), with the matrix **S** (**x**) such that $\bar{\boldsymbol{\rho}}$ satisfies the internal self-equilibrium equation within each finite element. An appropriate system of nodal displacement increments $\Delta \mathbf{q}$ determines completely the Lagrange's multipliers field $\Delta \mu$ on the border of each element :

$$\Delta \boldsymbol{\mu} = \mathbf{M}(\mathbf{x}) \, \Delta \mathbf{q}.$$

With the notation (6.83), we obtain as before

$$\sum_A \int_A \bar{\boldsymbol{\rho}}^T \mathbf{L}^T \Delta \boldsymbol{\mu} \, dA = \mathbf{h}^T \mathbf{G}^T \Delta \mathbf{q}$$

with **G** defined by (6.84). Finally, the saddle-point problem (6.109) becomes :

$$\underset{\Delta q}{\text{Max}} \quad \underset{\alpha\_ , \mathbf{h}}{\text{Min}} \quad (-\alpha\_ + \mathbf{h}^T \mathbf{G}^T \Delta q)$$

constraints :

$$f_i (\alpha\_ \, \mathbf{h}_m^{*o} + \mathbf{h}) \leq 0 \tag{6.135}$$

In the solution, the equilibrium at the interelement connection is satisfied only in the following weak sense :

$$\mathbf{G}\,\mathbf{h} = 0.$$

Of course, as in [6.52], we can solve the following equivalent problem which has the form of a lower bound problem :

$$\underset{\alpha\_,\mathbf{h}}{\text{Max}} \qquad \alpha\_$$

constraints :

$$\mathbf{G}\,\mathbf{h} = 0$$

$$f_i\,(\alpha\,\mathbf{h}_m^{*o} + \mathbf{h}) \leq 0\,. \tag{6.136}$$

References.

[6.1] D.C.A. KOOPMAN and R.H. LANCE, "On linear programming and plastic limit analysis", Jl. Phys. Solids, **13**, pp. 7-87, 1965.

[6.2] G. CERADINI and C. GAVARINI, "Calcolo a rottura e programmazione lineaire", Giornale del Genio Civile, **103**, pp. 48-64, 1965.

[6.3] G. SACCHI, "Contribution à l'analyse limite des plaques minces en béton armé à l'aide des solutions statiquement admissibles", Instituto Lombardo,Academia de Geologiche (a), **100**, pp. 529-554, 1966.

[6.4] J.H. ARGYRIS, "Elasto-plastic matrix displacement analysis of three-dimensional stress systems by finite element method", Int. J. Mech. Sci., **9**, pp. 143-155, 1967.

[6.5] G.G. POPE, "a discrete element method for analysis of plate elasto-plastic strain problems", (R.A.P. Farnborough T.R. , 65028), 1965.

[6.6] P.V. MARCAL and I.P. KING, "Elasto-plastic analysis of two-dimensional stress systems by the element method", Int. J. Mech. Sci., **9**, pp. 143-155, 1967.

[6.7] Y. YAMADA, N. YOSHIMURA and T. SAKURAI, "Plastic stress-strain matrix and its application for the solution of elastic-plastic problems by the finite element method", Int. J. Mech. Sci., **9**, pp. 343-351, 1967.

[6.8] O.C. ZIENKIEWICZ, S. VALLIAPAN and I.P. KING, "Elasto-plastic solutions of engineering problems - "initial stress" , finite element approach", Int. J. Num. Meth. Enq., **1**, pp. 75-100, 1969.

[6.9] A.A. MARKOV, "On variational principles on theory of plasticity", Prik. Mat. Mekh., **11**, pp. 339-350, 1947.

[6.10] R. HILL, "Mathematicla theory of plasticity", Oxford Univ. Press, London, 1950.

[6.11] NGUYEN DANG HUNG, "sur les principes variationnels en plasticité", Collection des Publications de la Faculté des Sciences Appliquées, Université de Liège, n° 84, pp. 61-76, 1980.

[6.12] NGUYEN DANG HUNG, "Sur la plasticité et le calcul des états limites par éléments finis", Thèse de doctorat spécial, Université de Liège, 1984.

[6.13] NGUYEN DANG HUNG, "Principes variationnels à deux champs pour le matériau de Saint-Venant-Von Mises", Bulletin de l'Académie Polonaise des Sciences, Série des sciences techniques, **14**, n 7-8, pp. 325-331, 1976.

[6.14] T.H.H. PIAN, "Formulation of finite element methods for solid continua", US Japan Seminar on Structural Analysis, Tokyo, 1969.

[6.15] P.G.J. HODGE and T. BELYTSCHKO, "Numerical methods for the limit analysis of plates", J. Appl. Mech. ASME, **35**, pp. 769-802, 1967.

[6.16] T. BELYTSCHKO and P.G.J. HODGE, "Plane stress limit analysis by finite elements", J. Appl. Mech. ASME, **96**, pp. 931-944, 1970.

[6.17] A. ZAVELANI-ROSSI, "Finite element techniques in plane limit problems", Meccanica, 1974.

[6.18] J. LYSMER, "Limit analysis of plane problems in soil mechanics", J. Soil Mech. Found. div. ASCE, July 1970.

[6.19] R. CASCIARO and A. DI CARLO, "Formulatione delle analisi limite delle piastre come problema di minimax mediante una rapprentazione agli elementi finite del campo delle tensioni", Gironale del Genio Civile, (2), pp. 87-108, 1970.

[6.20] A. BOTTERO, R. NEGRE, J. PASTOR and S. TURGEMAN, "Finite element method and limit analysis theory for soil mechanics problems", Comp. Meth. Appl. Mech. Eng., **22**, pp. 131-149, 1980.

[6.21] G.B. DANTZIG, "Linear programming and extensions", Princeton University Press, 1963.

[6.22] A.V. FIACCO and G.P. CORMICK, "Nonlinear programming : sequential unconstrained minimization techniques", Wiley, New-York, 1968.

[6.23] A.V. FIACCO and G.P. CORMICK, "Programming under nonlinear constraints by unconstrained minimization : a primal-dual method", research analysis corp., Mc Lean, Va, RAC-TP-96, 1963.

[6.24] NGUYEN DANG HUNG, Direct limit analysis via rigid-plastic finite elements, Comp. Meth. aPPL. Mech. Eng., **8**, pp. 81-116, 1976.

[6.25] H. HENCKY, Zur theorie plastischer deformationen und der hierdurch in material herforgerufenen nachspannung, Z. Angew. math. Mech., **4**, pp. 323-334, 1924.

[6.26] S.L.S. JACOBY, J.S. KOWALIK and J.T. PIZZO, "Interactive methods for nonlinear optimization problems", Prentice-Hall, 1972.

[6.27] E. ANDERHEGGEN, "Finite element analysis assuming rigid-ideal-plastic material behaviour", Limit analysis using finite elements, ASME, pp. 1-18, 1976.

[6.28] A. BIRON, "On results and limitations of lower bound limit analysis through nonlinear programming", Limit analysis using finite elements, ASME, pp. 19-34, 1976.

[6.29] D.N. HUTULA, "Finite element analysis of two-dimensional plane structures", Limit analysis using finite elements, ASME, pp. 35-52, 1976.

[6.30] A. ZAVELANI-ROSSI, A. PEANO and L. BINDA, "Lower bounds to collapse pressure of axisymmetric vessels", Limit analysis using finite elements, pp. 53-66, 1976.

[6.31] M. ROBINSON, "Lower bound limit pressures for the cylinder-cylinder intersection : a parametric survey", Limit analysis using finite elements, ASME, pp. 87-101, 1976.

[6.32] A. PEANO, "Limit analysis via stress functions", Limit analysis using finite elements, ASME, pp. 67-86, 1976.

[6.33] NGUYEN DANG HUNG, M. TRAPLETTI and D. RANSART, "Bornes quasi-inférieures et bornes supérieures de la pression de ruine des coques de révolution par la méthode des éléments finis et par la programmation non linéaire", Int. J. Nonlinear Mech., **13**, pp., 79-102.

[6.34] E. CHRISTIANSEN, "Computation of limit loads", Int. J. Num. Meth. Eng., **17**, pp. 1547-1570, 1981.

[6.35] R. CASCIARO, A. DI CARLO and G. VALENTE, "Plane stress limit analysis by finite elements", Proc. of the Int. Symp. on Discrete Meth. in Eng., 19-20 Sept. 1974.

[6.36] R. CASCIARO and A. DI CARLO, "Mixed finite element models in limit analysis ", in Comp. Meth. in Nonlinear Mech. (ed. J.T. ODEN), Austin, Texas, pp. 171-181, 1974.

[6.37] J.A. NELDER and R. MEAD, "A simplex method for function minimization", The Computer Journal, 7, pp. 308-315, 1965.

[6.38] I. EKELAND and R. TEMAM, "convex analysis and variational problems", New-york, North-Holland, 1975.

[6.39] J.M. BEST, "FCDPAK, A Fortran IV subroutine to solve differentiable mathematical programmes", Dep. of Combinatories and optimization, Univ. of Waterloo, Canada, 1972.

[6.40] J.M. BEST and K. RITTER, "An accelerated conjugate direction method to solve linearly constrained minimization problems", Research report CORR 73-16, Univ. of Waterloo, Canada, 1973.

[6.41] C.W. DAVIDON, "variable metric method for minimization", AEC Research and Development Reprot, ANL-5990 (rev.), 1959.

[6.42] R. FLETCHER and C.N. REEVES, "Function minimization by conjugate gradients", Comp. J., 7, 1964.

[6.43] A.V. FIACCO and G.P. Mc CORMICK, "Nonlinear programming : sequential unconstrained minimization techniques", RAC Research Series, New-York, John Wiley and Sons, Inc., 1968.

[6.44] P. MORELLE, "Etude d'un élément fini hybride de contrainte de solide axisymétrique et de son application au calcul d'une borne inférieure du multiplicateur d'adaptation plastique", Internal report LMMSC, n° 146, Univ. of Liège, 1984.

[6.45] T. BELYTSCHKO,Plane stress shakedown analysis by finite elements", Int. J. Mech. Sci., **14**, pp. 619-625, 1972.

[6.46] L. CORRADI and A. ZAVELANI, "A linear programming approach to shakedown analysis of structures", Comp. Meth. Appl. Mech., **3**, pp. 37-53, 1974.

[6.47] NGUYEN DANG HUNG and L. PALGEN, "Shakedown analysis by displacement method and equilibrium finite elements", Trans. of the CSME, **6**, n°1, pp. 34-40, 1980-1981.

[6.48] P. MORELLE and NGUYEN DANG HUNG, "Etude numérique de l'adaptation plastique des plaques et des coques de révolution par les éléments finis équilibrés", J. de Mécanique Théorique et Appliquée, **2**, n° 4, pp. 567-599, 1983.

[6.49] J.A. KONIG, "Engineering applications of shakedown theory", Lecture note for the session organized in Udine, Italy, October 1977.

[6.50] NGUYEN DANG HUNG and P. MORELLE, "Accuracy problems in shakedown shell analysis using dual finite elements", Proceedings Congress ARFEC, Lisbonne, pp. 285-299, June 1984.

[6.51] P. MORELLE, "Numerical shakedown analysis of axisymmetric sandwich shells : an upper bound formulation", Int. J. Num. Meth. Eng., **23**, pp. 2071-2088, 1986.

[6.52] J.M. BEST, "ACDPAC : A Fortrain IV subroutine to solve differentiable mathematical programs", Dept. Comb. Opt., Univ. of Waterloo Canada, 1976.

[6.53] NGUYEN DANG HUNG and J.A. KONIG, "A finite element formulation for shakedown problems using a yield criterion of the mean", Comp. Meth. Appl. Mech. Eng., n°8, pp. 179-192, 1976.

[6.54] G. MAIER, "A shakedown matrix theory allowing for work hardening and second order geometric effects", Fondation of plasticity, **1**, ed. A. SAWCZUK, Leyden, Nordhoff, pp. 417-433, 1973.

[6.55] J. MANDEL, "Cours de mécanique des milieux continus", Tome II, Mécanique des solides, Gauthiers-Villars, Paris, 1966.

[6.56] J.M. TURNER, R.W. CLOUGH, H.C. MARTIN and L.J. TOPP, "Stiffness and deflection analysis of complex structures", J. Aero. Sci., **23**, pp. 805-823, 1956.

[6.57] J.H. ARGYRIS and S. KELSEY, "Energy theorems and structural analysis", Butterworth, 1960 (reprinted from Aircraft Eng., 1954-1955).

[6.58] O.C. ZIENKIEWICZ, "The finite method", Mc Graw-Hill, 1977.

[6.59] J.T. ODEN, "Finite elements of nonlinear continua", Mc Graw-Hill, 1972.

[6.60] G. STRANG and G.J. FIX, "An analysis of the finite element method", Prentice-Hall, 1973.

[6.61] P. CIARLET, "The finite element method for elliptic problems", North-Holland, 1978.

[6.62] R.H. GALLAGHER, "Finite element analysis fundamentals", Prentice-Hall, 1975.

[6.63] J. ROBINSON, "Integrated theory of finite element methods", John Wiley, 1973.

[6.64] C.S. DESAI and J.F. ABEL, "Introduction to the finite element method", Van Nostrand, 1972.

[6.65] W. PILKEY and N. PERRONE, "Structural mechanics software series", University Press of Virginia, 1977.

[6.66] A. BIRON and G. CHARLEUX, "Limit analysis of axisymmetric pressure vessel intersections of arbitrary shape", Int. J. of Mech. Sc., **14**, pp. 25-41, 1972.

[6.67] D.J. HAYES and P.V. MARCAL, "Determination of upper bounds for problems in plane stress using finite element techniques", Int. J. of Mech. Sc., **9**, pp. 245-251, 1967.

[6.68] O. DEBORDES, "Contribution to the theory and calculation of the asymptotic elastoplasticity" (in French), Dr. Sc. thesis, Univ. of Provence, Aix-Marseille, 1977.

[6.69] E. FACCIOLI and E. VITIELLO, "A finite element, Linear programming method for the limit analysis of thin plates", Int. J. Num. meth. Eng., **5**, pp. 311-325, 1973.

[6.70] G. MAIER, "A quadratic programming approach for certain classes of nonlinear structural problems", Meccanica, **3**, n°2, 1968.

[6.71] G. MAIER and O. DE DONATO, "Historical deformation analysis of elastoplastic structures as a parametric linear complementary problem", Meccanica, **11**, pp. 166-171, 1976.

[6.72] M.S. COHN and G. MAIER, "Engineering plasticity by mathematical programming", Pergamon, 1979.

[6.73] G. MAIER, J. MUNRO, "Mathematical programming application to engineering plastic analysis", Appl. Mech. Rev., **35**, pp. 1631-1643, 1982.

[6.74] R. CONTRO, L. CORRADI, R. NOVA, "Large displacement analysis of elasto-plastic structures - a nonlinear programming approach", Sol. Mech. Arch., **2**, n° 4, pp. 433-476, 1977.

[6.75] D.E. GRIERSON, A. FRANCHI, O. DE DONATO, L. CORRADI, "Mathematical programming and nonlinear finite element analysis", Comp. Meth. Appl. Mech. Eng., 17/18, pp. 497-518, 1979.

[6.76] C. POLIZZOTTO, "A quadratic programming approach to dynamic elastoplasticity", Proc. SMIRT-8 Conf., Brussels, **B**, paper 5/4, 1985.

[6.77] I. KAMEKO, "Complete solutions for a class of elastic-plastic structures", Comp. Meth. Appl. Mech. Eng., **21**, pp. 193-209, 1980.

[6.78] J.A.T. FREITAS, D.L. SMITH, "Elastoplastic analysis of planar stuctures for large displacements", J. Struct. Mech., **12**, 1984.

[6.79] A. BORKOWSKI, "Statics of elastic and elastoplastic skeletal structures", Elsevier-PWN, Warsaw, 1986.

[6.80] S. KARADENIZ, A.R.S. PONTER, "A linear programming upper bound approach to the shakedown limit of thin shells subjected to variable thermal loading", J. Strain Analysis, **19**, n°4, pp. 221-230, 1984.

[6.81] S. KARADENIZ, "The development of upper bound and associated finite element techniques for the plastic shadedown of thermally loaded structures, Ph.D. Thesis, University of Leicester, 1983.

[6.82] A.R.S. PONTER, S. KARADENIZ, "An extended shakedown theory for structures that suffer cyclic thermal loading, Part 1 : Theory", J. Appl. Mech., **52**, pp. 877-882, 1985.

[6.83] A.R.S. PONTER, S. KARADENIZ, "An extended shakedown theory for structures that suffer cyclic thermal loading, Part 2 : applications", J. Appl. Mech., **52**, pp. 883-889, 1985.

[6.84] J. GROSS-WEEGE, D. WEICHERT, "Zur numerischen Untersuchung des einspielverhaltens elasto-plastischer scheiben unter mechanischen und thermischen wechsellasten", Ing. Archiv., **57**, pp.297-306, 1987.

[6.85] A. FRANCHI, F. GENNA, "Incremental elastic-plastic and shakedown analysis by finite elements", Proc. Euromech Colloquium "Inelastic structures under variable loads", C. Polizzotto, A. Sawzuk Ed., Palermo, pp. 269-284, 1983.

[6.86] A. FRANCHI, F. GENNA, "A numerical scheme for integrating the rate plasticity equations with an "a priori" error control", Comp. Meth. Appl. Mech. Eng., **60**, pp. 317-342, 1987.

[6.87] M. SAVE, G. DE SAXCE, A. BORKOWSKI, "Computation of shakedown loads, feasibility study", Commission of the European Communities, Nuclear Science and Technology, Report EUR 13618 EN, 1991.

[6.88] H. NGUYEN DANG, P. MORELLE, "Plastic shakedown analysis", in "Mathematical programming methods in structural plasticity", Ed. by D.L. SMITH, Springer-verlag , 1990.

[6.89] H. NGUYEN DANG, "Variational and computational plastic limit and shakedown analysis", Birkhauser-verlag , 1995.

[6.90] P. MORELLE, "Analyse duale de l'adaptation plastique des structures par la méthode des éléments finis et la programmation mathémathique",Doctor thesis, University of Liège , 1989.

Part two : Applications to Plates, Shells and Disks.

7. Metal plates.

7.1. Introduction.

Consider a plane horizontal region A, which is bounded by one or several curves. At a generic point P of this region, erect a straight vertical segment of length t that has P as its midpoint.

Imagine that the point P occupies all possible positions in A and assume :

(1) that t is a continuous and "sufficiently smooth" function of the position of P ; and (2) that t always remains small with respect to the dimensions of A. The body generated in this manner is a *plate* or a *disk* (fig. 7.1).

It is called a plate when loaded transversally to its median plane. It then carries the applied loads essentially by bending stresses ; its median plane will bend but not stretch or contract (See Sections 7.2 and 7.7 for further details.)

In many important cases, the region A will be a circle, a circular ring, or a polygon, and the thickness will be constant. When assumptions 1 and 2 are satisfied, material normals to the median plane will transform into material normals of the curved surface into which the median plane deforms, both in the plastic and the elastic ranges. When the thickness t does not vary smoothly enough in some regions, elastic stress concentrations will occur there, which later on may be leveled out by plastic flow.

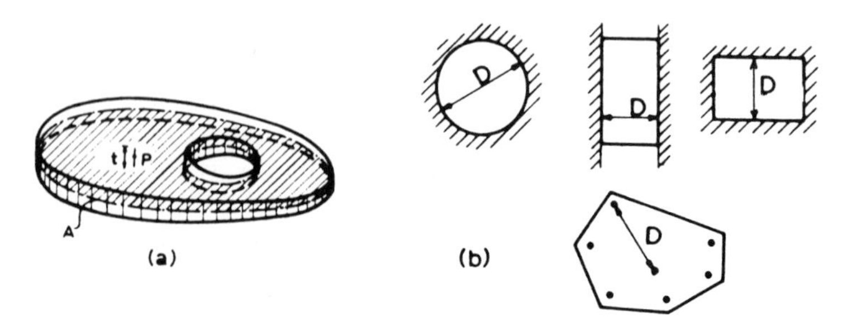

Fig. 7.1. (a) and (b)

Alternatively, very abrupt thickness variation may result in a local elastic state of stress with three principal tensile stresses and this is likely to cause brittle failure (see Com. V., Section 5.2). Hence, condition 1 must be imposed when the emphasis is on plastic deformation. Condition 2 is the basis for material normals to the median plane to remain normal to the median surface in the deformed state. To make this condition more precise, we may introduce the "slenderness" of the plate, that is a geometrical nondimensional parameter defined by

$$\mu = \frac{D}{t}, \tag{7.1}$$

where D is some "span" of the plate with thickness t. This span will be taken to be [see fig. 7.1 (b)] :

1. for circular or annular plates : the outside diameter ;

2. for rectangular plates : (a) simply supported of built-in along two parallel edges : the span ; (b) simply supported or built-in along four edges : the smallest span ;

3. for plates supported on columns : the largest span between two neighbouring supports.

The parameter μ is the extension to plates of the span-to-height ratio used in beam theory. For a plate of a given material to carry transverse loads essentially by bending stresses, μ must be bounded both from below and from above. Indeed, if μ is unduly small, we no longer have a thin plate but a body with comparable horizontal and vertical dimensions.

On the other hand, if μ is unduly large, we have a membrane that carries transverse loads by direct stresses after undergoing deflections that are comparable to its thickness. This situation is more often encountered with metallic plates that tend to be very slender, than with reinforced concrete plates.

The limit load obtained from simple plastic theory based on the rigid perfectly plastic scheme will thus prove to have a real physical meaning only for a limited domain of values of μ, that we shall define as precisely as possible for the various problems treated. Even in this domain of μ the limit load will not correspond to large plastic deformations under constant load, such as usually occur in frame structures. In most cases, favourable geometry changes due to unrestricted plastic flow will cause membrane action that eventually enables the plate to carry a load in excess of the limit load (see Sections 3.5 and 7.2).

Since the treatment of reinforced concrete plates justifies a complete chapter, in this chapter we will only consider metal plates, either isotropic or (structurally) orthotropic.

7.2. Experimental information on metal plates.

Tests on metal plates up to plastic collapse are unfortunately in limited number, and all of them [7.1-7.7] deal with circular plates except two works on rectangular and square plates [7.8] [7.9].

The essential results of the experiments on circular plates are given by the load versus deflection diagrams of figs. 7.2-7.8. The theoretical limit loads have been computed as indicated in the following sections, assuming a rigid perfectly plastic material. The theoretical load versus deflection diagram coincides with the load axis up to the limit value of the load and, for this value, becomes a parallel to the deflection axis when changes in geometry caused by unrestricted plastic flow are neglected.

On the other hand, when these changes are taken into account, the parallel to the deflection axis at the level of the limit load must be replaced by a raising curve of parabolic shape, as drawn in figs. 7.3 to 7.5.

Simple inspection of these figures show that the limit load cannot be regarded as a physical failure load or carrying capacity.

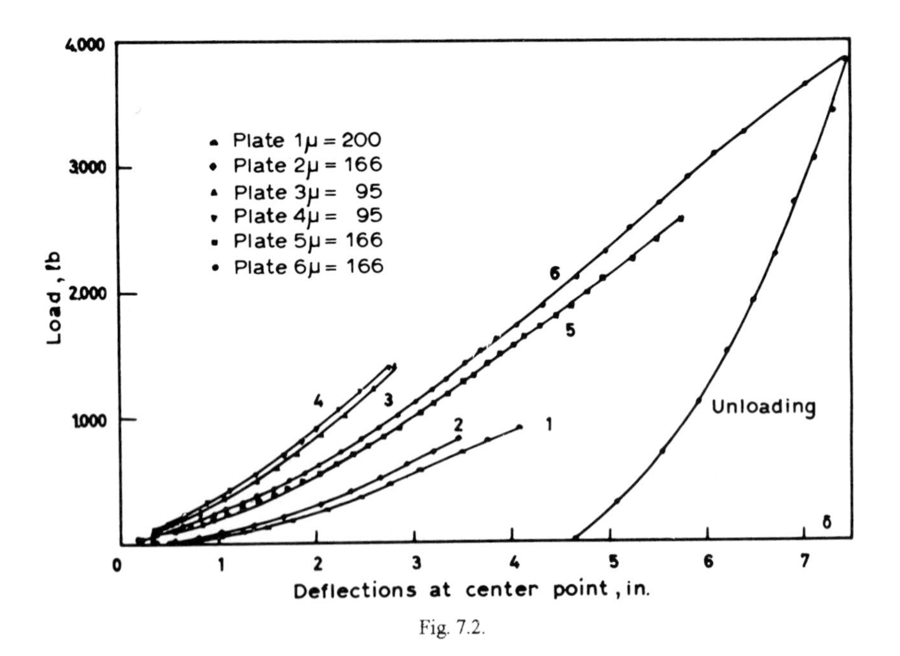

Fig. 7.2.

Indeed, the main difficulty is to define what will be considered as failure of the plate. In the case of beams and frames, experimental diagrams exhibit a sharp bend near the limit load P_1, preceded by very small deflections and followed by very large permanent deflections for very small further load increments. In experiments on plates, the sharp bent in the load versus deflection diagram tends to disappear (see fig. 7.6, D/t = 40). Both the slenderness D/t of the plate and the boundary conditions are responsible for this smoothing of the curve.

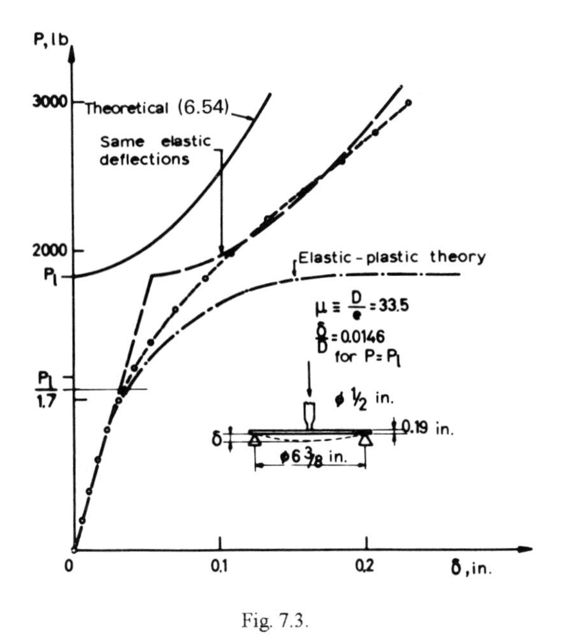

Fig. 7.3.

For extremely slender plates as those studied by Cooper and Shifrin [7.6] (see fig. 7.2), with slenderness μ varying from 95 to 200, membrane effects are prominent from the very beginning of loading and bending analysis is meaningless, even in the elastic range. Plates of this kind must be treated as membranes, as will be done in Section 7.7. The load versus deflection curves then obtained are rays (represented by dashed lines in figs. 7.4 and 7.5). From the experiments summarized in figs. 7.3 to 7.7, we see that bending is the relevant phenomenon for isotropic plate with $\mu \equiv D/t \leq 40$.

When fatigue or buckling failures are not to be considered, failure of the plate may reasonably be defined as exceeding a given limit set to the maximum deflection. To choose this limit in the light of currently available experiments, we note that :

1. For $P = P_1$, the total deflection δ remains a small fraction of the diameter, always smaller than 2% (except for the simply supported plate with $\mu = 40$, fig. 7.6, where δ/D = 2.36%);

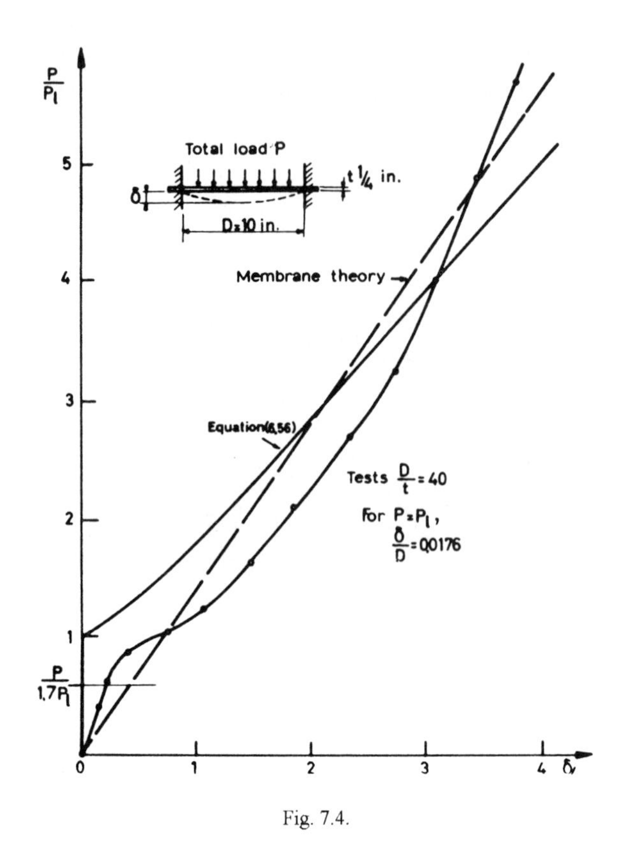

Fig. 7.4.

2. For $P = P_l/1.7$ the plate is in the elastic range and its maximum deflection does not exceed 1% of the diameter;

3. For the built-in plates of fig. 6.7, the average ratio δ/P increases rapidly as P increases from $0.8P_l$ to P_l.

The two-segment approximations to the diagrams give the following results :

for D/t = 10, δ/P becomes 11 times larger ;

D/t = 20, δ/P becomes 15 times larger ;

D/t = 40, δ/P becomes 5 times larger .

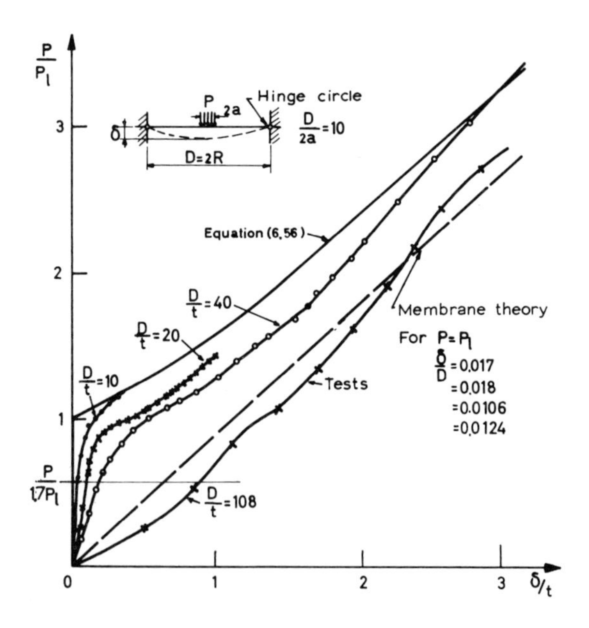

P/P_l
Hinge circle
D/2a = 10
D=2R
Equation (6.56)
D/t = 40
D/t = 20
D/t = 10
Membrane theory
For P=P_l
δ/D = 0.017
= 0.018
= 0.0106
= 0.0124
Tests
P/1.7P_l
D/t = 108
δ/t

Fig. 7.5.

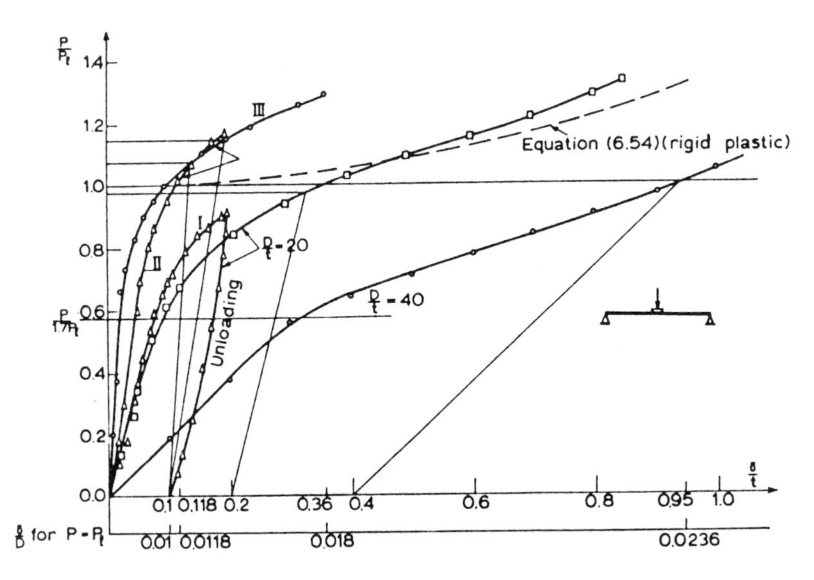

P/P_l
Equation (6.54)(rigid plastic)
III
II
I
D/t = 20
Unloading
D/t = 40
P/1.7P_l
δ/t
0.1 0.118 0.2 0.36 0.4 0.6 0.8 0.95 1.0
δ/D for P = P_l
0.01 0.0118 0.018 0.0236

Fig. 7.6.

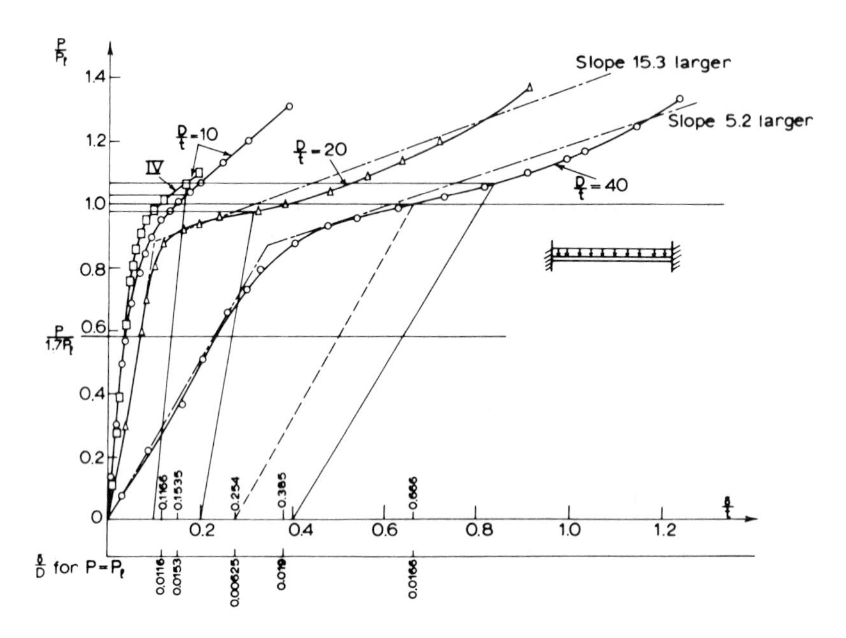

Fig. 7.7.

Hence, the built-in plate with D/t = 20 is such that its limit load may be regarded as the failure load. In this particular case we have, for $P = P_l$, $\delta/D = 1.9\%$. Consequently, *we choose the failure deflection as 2% of the diameter*(*).

It is readily verified that the corresponding experimental load is always larger than P_l, except for the simply supported plate with D/t = 40, where it is approximately 0.9 P_l. It is interesting to remark that a *permanent* defection of 1% of the span corresponds to loads between 0.975 P_l and 1.15 P_l for the simply supported plate, between 0.975 P_l and 1.055 P_l for the built-in plate when

$$10 \leq D/t \leq 40. \tag{7.2}$$

(*) This rule is also accepted in the Belgian Code on Plastic design of Beams and Frames.

Tests by Ohashi and Murakami [7.10, 7.11] and by Ohashi, Murakami and Endo [7.12] give further support to the preceding conclusions. Though their main purpose was to

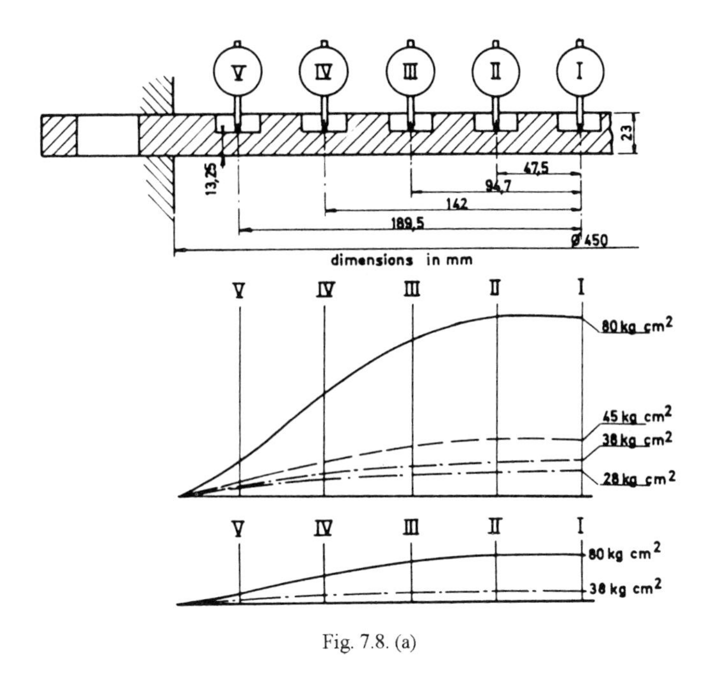

Fig. 7.8. (a)

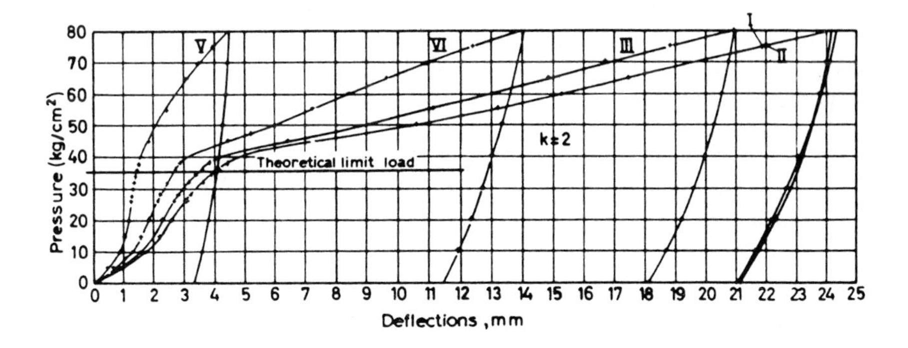

Fig. 7.8. (b)

verify the validity of an elastic plastic analysis, they show that, for a circular plate with D/t = 25, subjected to uniformly distributed load, and built-in or simply supported [7.10, 7.11], relative deflections δ/D at limit load are 1.4% and 2% respectively, with a sharp bent in the load versus deflection diagram near the limit load. Tests on annular plates [7.12] with D/t = 40 show that the bent in the experimental load versus deflection diagram is less and less noticeable with increasing central hole diameter. From the preceding discussion it may be concluded that, in the slenderness range defined by eq. (7.2), the limit load may be regarded as the actual failure load and its determination is therefore useful. Moreover, not only has the limit load a physical significance, but the predicted collapse mechanism is fairly well supported by the experiments [7.1]. *Design for safety with respect to the theoretical limit load will result in safety with respect to failure deflection, without necessitating the more difficult analysis that would have furnished the load versus deflection curve.*

Experiments on built-in, circular structurally orthotropic, uniformly loaded plates [7.7] give very similar results. Typical diagrams (as well as dial locations) are shown in fig. 7.8. When the breadth of the reinforcing rings is close to the plate thickness, the limit load can be given the same physical interpretation as above provided that

$$22.9 \leq \frac{D}{t_{av}} \leq 38 , \tag{7.3}$$

when $t_{av} = \dfrac{t_{max} + t_{min}}{2}$.

When the load can be reversed, as for the simply supported plate with the central concentrated load of fig. 7.9, the theoretical load versus deflection diagram is shown in fig. 7.10, based on the rigid perfectly plastic scheme [7.2]. The experimental curves differ from the theoretical ones in that, upon reversal of the load, unrestrained plastic flow occurs in the neighbourhood of the (negative) limit load (range AB, fig. 7.9). Hence, this later limit load is here a real carrying capacity. A similar situation is seen on the diagrams of fig. 7.11, dealing with reversed loading of six rectangular built-in beams of 10-in. span, 1 in. thick, subjected to a concentrated load at their central point, and with 1, 3, 5, 7, 8.88, and 18 in. breadths, respectively [7.8].

With the exception of the two first, these beams are actually rectangular plates built-in along the two short edges and free along the others. Plastic collapse can occur either by a mechanism appropriate for the built-in beam or by a local plate mechanism around the central load, or by a combination of both.

The slenderness ratio is $\mu = 38.6$ and it is seen that, as for circular plates, when $P = P_l$ the total maximum deflection is about 2% of the span and the permanent maximum deflection is about 1% of the span.

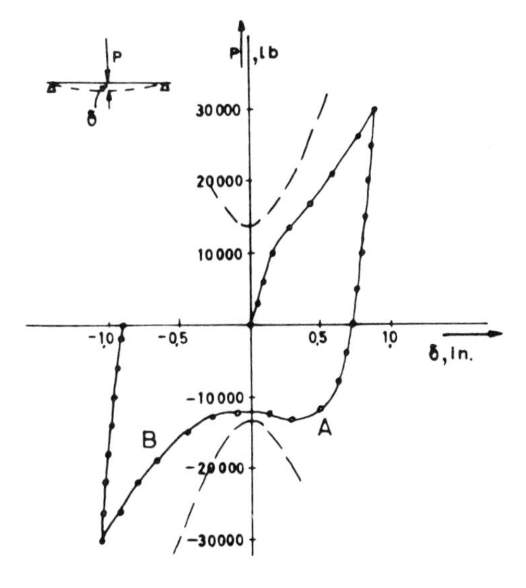

Fig. 7.9.

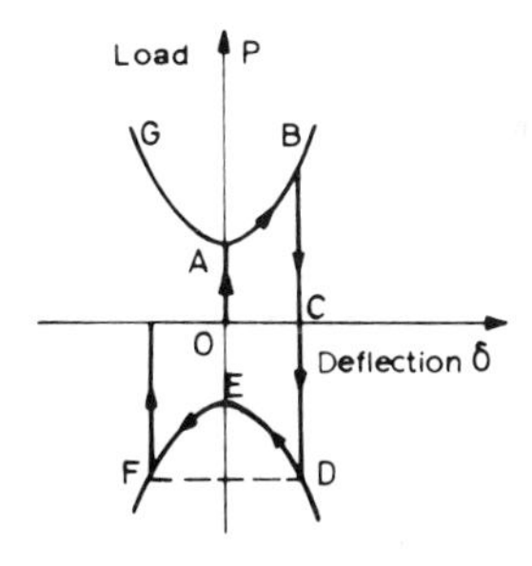

Fig. 7.10.

Finally it should be emphasized that much more experimental information is needed, especially concerning noncircular plates and plates made of other metals than steel, before the limit load can be attributed a physical meaning of wider applicability.

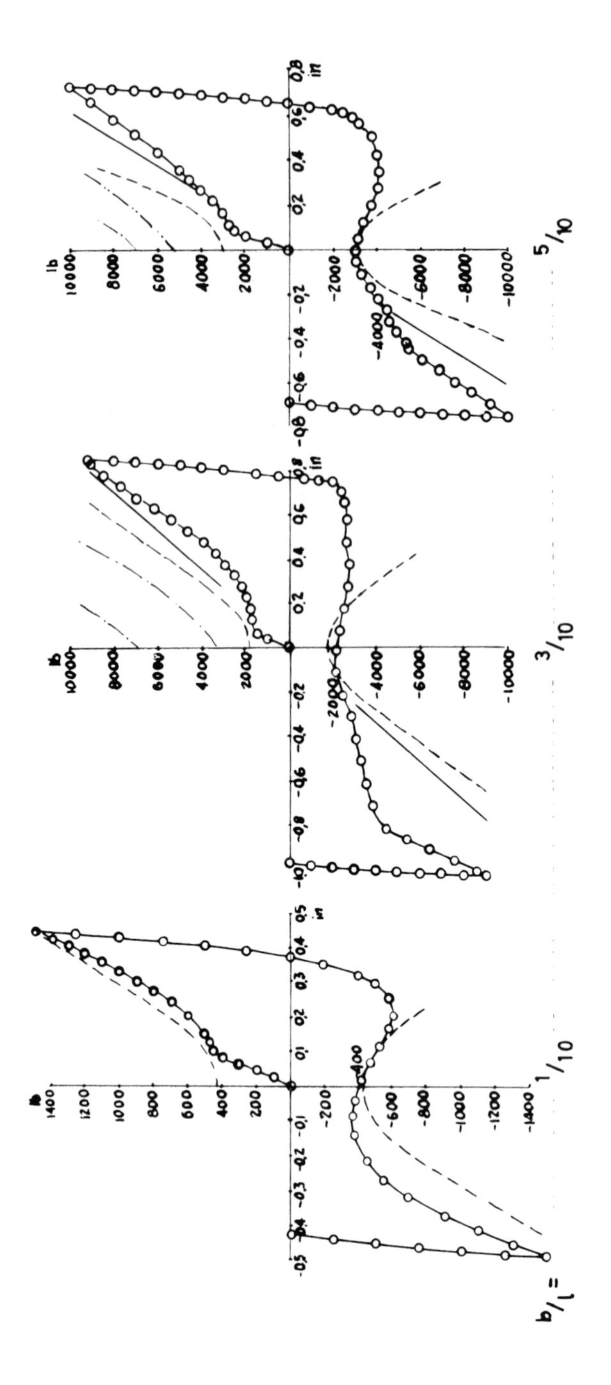

Fig. 7.11.

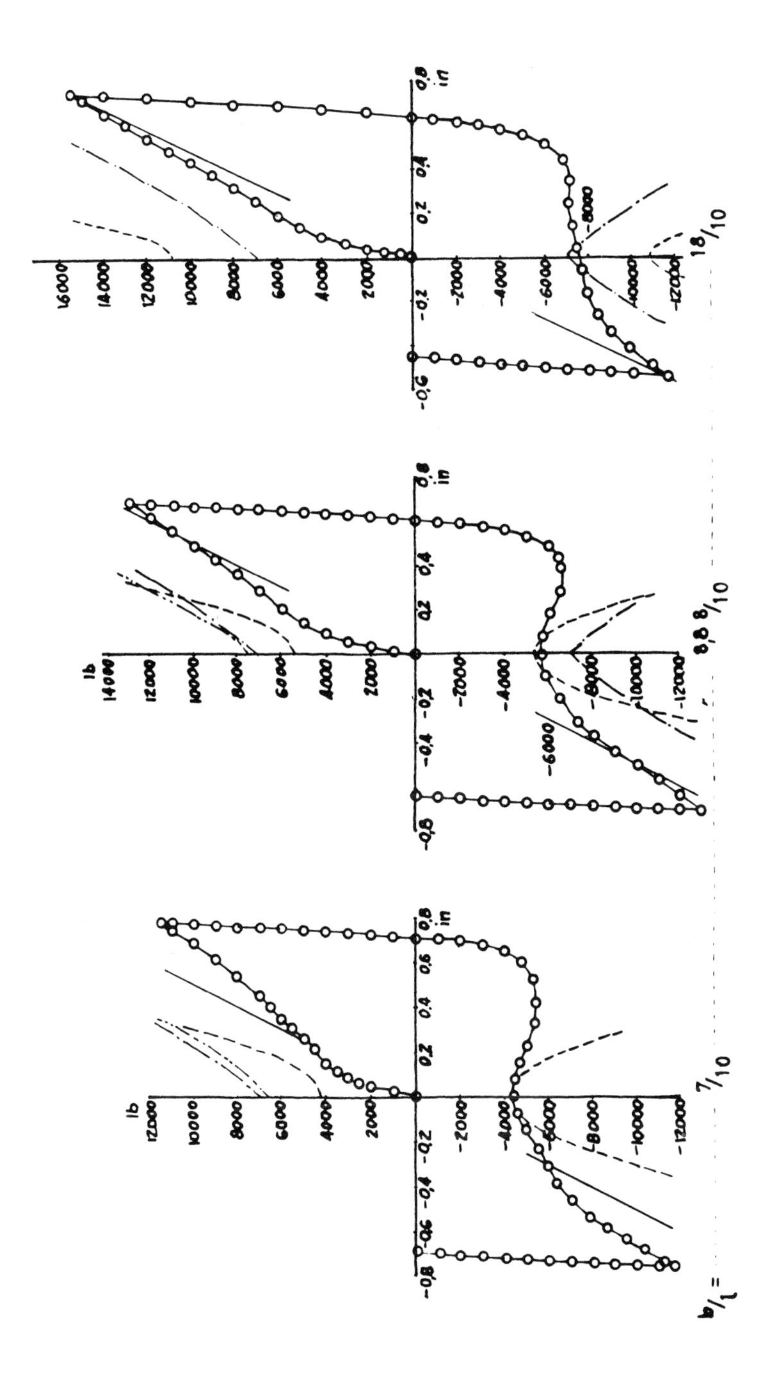

Fig. 7.11. (continued)

7.3. Circular isotropic plates.

7.3.1. *General relations.*

We refer to fig. 7.12, for notations. The distributed load p is assumed to depend on r exclusively and, hence, the problem is axially symmetric.

Equilibrium of the transverse forces acting on a circular central part with radius r requires that

$$2\pi r V = -\int_0^r p\, 2\pi r\, dr\,, \tag{7.4}$$

where p is the load per unit surface. Moment equilibrium of an annular plate element (fig.7.12) requires that

$$\frac{d}{dr}(r M_r) = M_\theta + rV\,. \tag{7.5}$$

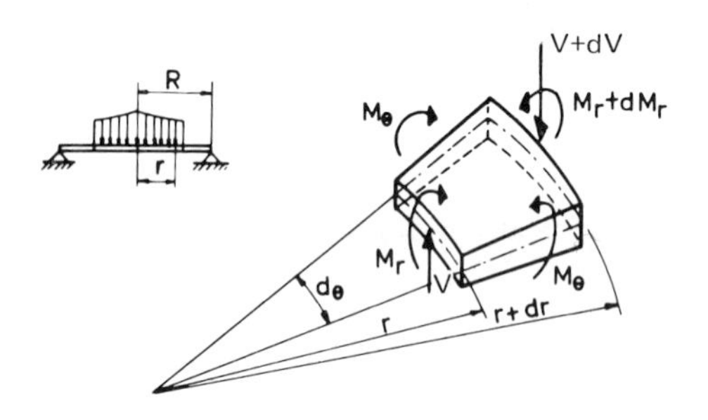

Fig. 7.12.

Elimination of V from eqs. (7.4) and (7.5) furnishes the following fundamental equation :

$$\frac{d}{dr}(r M_r) = M_\theta - \int_0^r pr\, dr. \tag{7.6}$$

A velocity field will be described by a function $\dot{w}$ of r only. The corresponding curvature field is

$$\dot{\kappa}_r = -\frac{d^2\dot{w}}{dr^2}, \qquad \dot{\kappa}_\theta = -\frac{1}{r}\frac{d\dot{w}}{dr}. \tag{7.7}$$

7.3.2. *Simply supported plate, circular loading.*

The load is uniformly distributed over a central circular area of radius a, and the plate is simply supported along its outer edge, which has radius R [7.13] (fig. 7.13).

We first assume Tresca's yield condition to hold. The state of stress at the various points of the plate will be represented by stress points located on or inside the yield curve. The locus of these stress points will be called the "stress profile". The stress profile must start from point A for r=0, because the axial symmetry requires that $M_r = M_\theta$ at the centre. The stress profile must end at point B for r=R where $M_r = 0$. Regime AF must be excluded because, according to normality law, eq. (7.7), and boundary conditions, it is not compatible with nonvanishing $\dot{w}$.

Hence, assuming the plate to be entirely plastic at collapse, we have

$$M_\theta = M_p \qquad \text{for} \qquad 0 \le r \le R. \tag{7.8}$$

Substituting M_p for M_θ in eq. (7.6) and integrating we obtain :

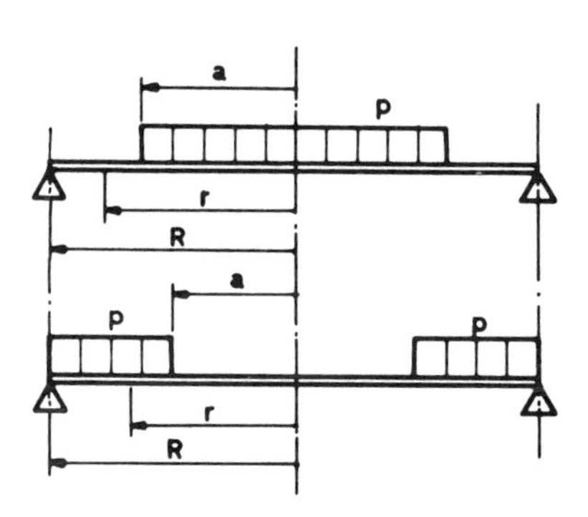

Fig. 7.13. (a)

for $0 \le r \le a$, where $\int_0^r pr\,dr = p\,\frac{r^2}{2}$,

$$M_r = M_p - \frac{pr^2}{6} + \frac{C_1}{r}. \tag{7.9}$$

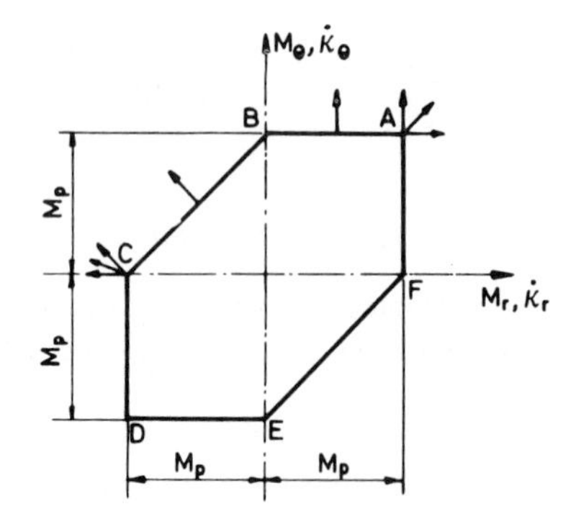

Fig. 7.13. (b)

for $a \leq r \leq R$, where $\int_0^r pr\,dr = p\,\frac{a^2}{2}$,

$$M_r = M_p - \frac{pa^2}{2} + \frac{C_2}{r}. \tag{7.10}$$

Because M_r remains finite (and equal to M_p) at r=0, we have $C_1 = 0$.

Continuity of M_r for r=a then yields $C_2 = \frac{pa^3}{3}$. Finally, $M_r = 0$ at r=R yields the load

$$p = \frac{6M_pR}{a^2\,(3R - 2a)}. \tag{7.11}$$

The corresponding moment field is obtained upon substitution of expression (7.11) for p in eqs. (7.9) and (7.10), with the values of C_1 and C_2 found above.

We obtain

$$M_r = M_p\left[\,1 - \frac{r^2}{a^2\,(3 - 2a/R)}\right]$$

$$M_\theta = M_p \qquad\qquad 0 \leq r \leq a \tag{7.12}$$

$$M_r = M_p\left[1 - \frac{3 - 2a/r}{3 - 2a/R}\right]$$

$$M_\theta = M_p \qquad\qquad a \le r \le R. \qquad (7.13)$$

To verify that eq. (7.11) actually gives the exact limit load, we must find a kinematically admissible velocity field corresponding to the moment field (7.12) and (7.13). For the stress profile AB, the normality law requires that

$$\dot{\kappa}_r = 0, \qquad \dot{\kappa}_\theta \ge 0.$$

Using the first of these conditions, eq. (7.7), and the boundary condition $\dot{w}(R) = 0$, we obtain the conical collapse mechanism

$$\dot{w} = \dot{w}_o\left(1 - \frac{r}{R}\right) \qquad (7.14)$$

where $\dot{w}_o$ denotes the transversal velocity of the central point.

From eqs. (7.7) and (7.14) it is readily verified that $\dot{\kappa}_\theta > 0$. We thus have found a complete solution and eq. (7.11) furnishes the exact limit load p_l. Let $P_l = \pi a^2 p_l$ denote the total load. Eq. (7.11) may also be written as

$$P_l = \frac{6\pi M_p}{3 - (2a/R)}. \qquad (7.15)$$

When a tends to zero or to R, we obtain the *limiting value of a concentrated central load,*

$$P_l = 2\pi M_p, \qquad (7.16)$$

and *the limit value of a uniformly distributed load,*

$$P_l = 6\pi M_p, \qquad (7.17)$$

respectively. Note that, for a concentrated load, the moment field is $M_r = 0$, $M_\theta = M_p$ (given by eq.(7.13) with a=0) with a singularity for r=0. It should however be kept in mind that bending solutions neglect the influence of shear forces and, consequently, cannot be

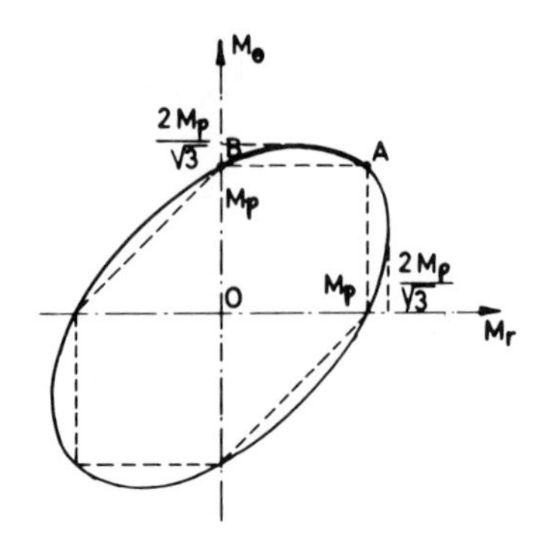

Fig. 7.14.

expected to be valid for concentrated loads. For $D/t \geq 20$, it has been found [7.14] that bending theory neglecting shear forces effects applies when $a/t > 1.5$.

The preceding problem has been treated by Hopkins and Wang using the von Mises

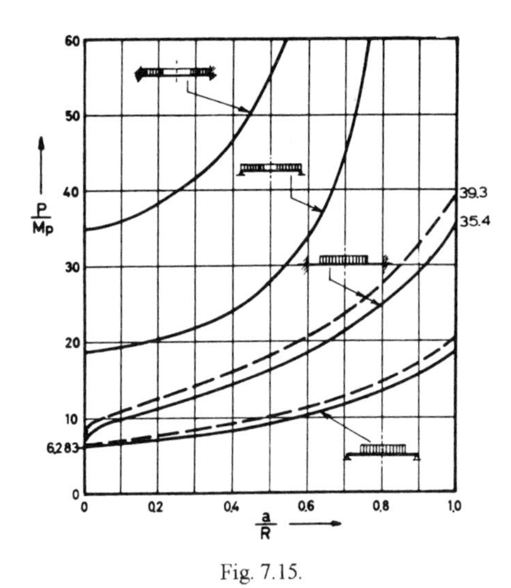

Fig. 7.15.

yield condition [7.15]. The curved stress profile AB on fig. 7.14 is used. From the very nature of von Mises' condition, the resulting differential equilibrium equation is nonlinear and must be integrated numerically. The limit load versus the ratio a/R is given in fig. 7.15.

Velocity fields that correspond to the moment fields obtained in this manner can be found. The limit load is therefore exact. The results of Hopkins and Wang practically coincide with those found by Sokolovsky [7.16] by the use of a deformation theory. This is in accordance with our remark at the end of Section 3.6. For a/R =1, that is for a *uniformly loaded plate*, we have

$$p_l = \frac{6.51\, M_p}{R^2} \qquad \text{from [7.15]},$$

$$p_l = \frac{6.46\, M_p}{R^2} \qquad \text{from [7.16]}, \tag{7.18}$$

the discrepancy being smaller that 1%.

For a concentrated load P, the limit value is found to be

$$P_l = 2\,\pi\, M_p, \tag{7.19}$$

as for the Tresca criterion. This is easily understood. Indeed, in both cases the stress profile reduces to two points, A (at the centre) and B (everywhere else) that are on both yield curves.

7.3.3. *Simply supported circular plate, annular loading.*

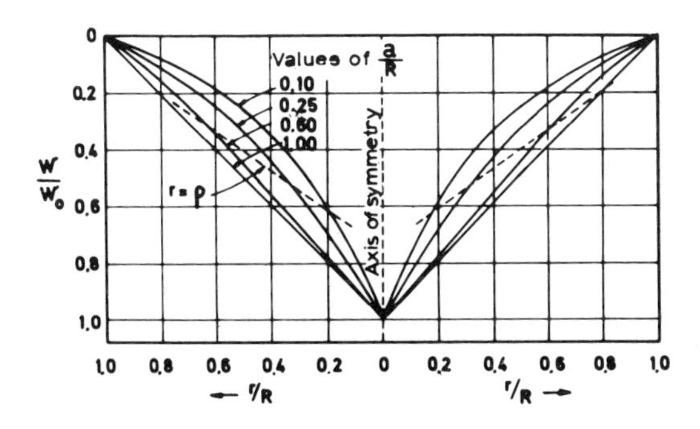

Fig. 7.16.

The load is uniformly distributed over an external annulus [7.13]. The analysis is similar to that of Section 6.3.2 and the limit load is shown in fig. 7.15. Corresponding solution for the von Mises condition can be found in the Atlas of Limit Loads for Metal Plates, Shells and Disks [7.26].

7.3.4. *Built-in circular plate, circular loading.*

For Tresca's yield condition, the stress profile is ABC, fig. 7.13. The side CD is excluded because the requirements of the normality law cannot be reconciled with eqs. (7.7) for nonvanishing $\dot{w}$. Complete solutions are obtained as in Section 7.3.2, but the limit load depends on a/R through the radius ρ at which the stress regime changes from AB to BC, and this radius is given by a transcendental equation that must be solved numerically. Limit loads are given in fig. 7.15 and velocity distributions in fig. 7.16. For a/R =1, the limit load is $P_l \approx 35.4\ M_p$, approximately 88% higher than for the simply supported plate. Limit loads with the von Mises condition [7.15] are also given in fig. 7.15.

With the von Mises condition the limiting intensity of a uniform load acting over the entire plate is (a/R =1)

$$p_l \approx \frac{12.5\ M_p}{R^2}, \tag{7.20}$$

or,

$$P_l \equiv p_l\, \pi\, R^2 = 39.3\ M_p . \tag{7.21}$$

The limit value of a concentrated central load is

$$P_l = \frac{2}{\sqrt{3}}\, 2\, \pi\, M_p = 7.26\ M_p . \tag{7.22}$$

This compares with $2\, \pi\, M_p$ for the condition of Tresca.

7.3.5. *Built-in plate, annular loading.*

The limit load is given in fig. 7.15, for the yield condition of Tresca [7.13].

7.3.6. *Circular line load with radius a and total magnitude Q.*

From Sawczuk and Jaeger [7.17] we have :

for the simply supported plate

$$Q_l = 2\, \pi\, M_p\, \frac{1}{1 - a/R}, \tag{7.23}$$

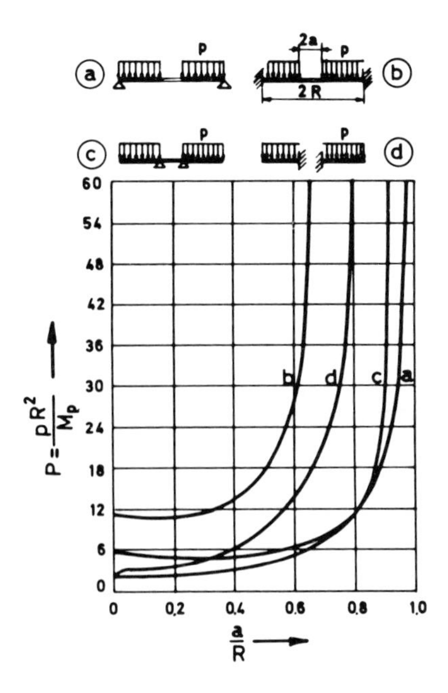

Fig. 7.17.

for the built-in plate

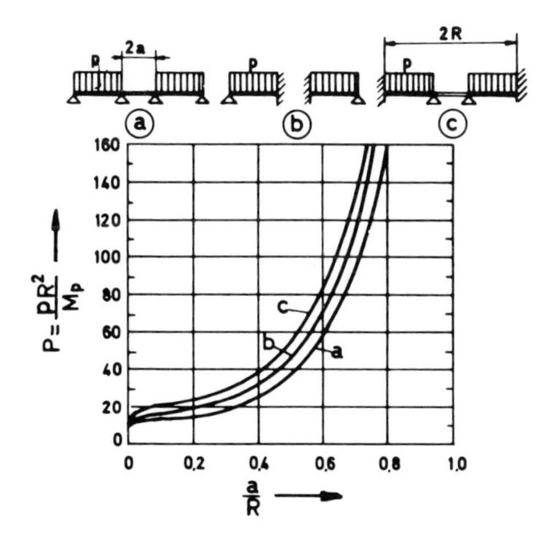

Fig. 7.18.

$$Q_1 = 2\,\pi\, M_p \; 1 - \frac{1}{\ln\rho}\,, \tag{7.24}$$

where ρ is given by

$$\rho - \frac{a}{R}(1 - \ln\rho) = 0\,. \tag{7.25}$$

7.3.7. *Annular plates subjected to uniform load.*

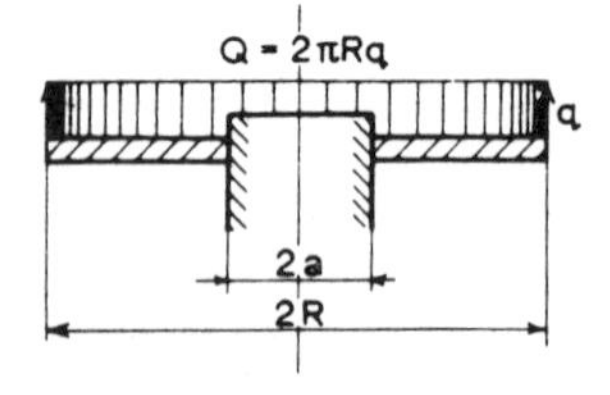

Fig. 7.19 (a) $\frac{Q}{M_p} = 2\,\pi\; 1 + \frac{1}{\ln a/R}$

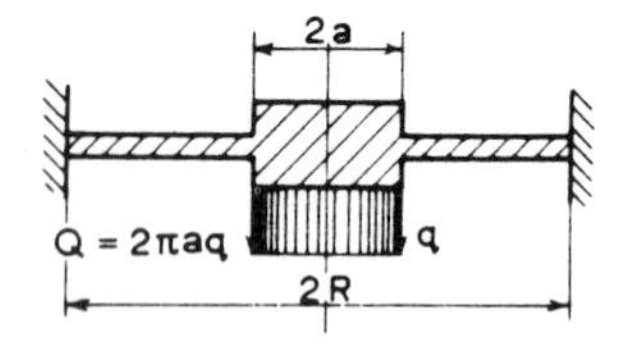

Fig. 7.19 (b) $\frac{Q}{M_p} = 2\,\pi$

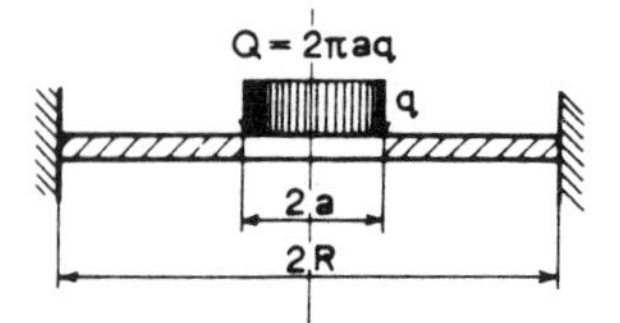

Fig. 7.19 (c) $\frac{Q}{M_p} = \frac{1}{1 - a/R}$

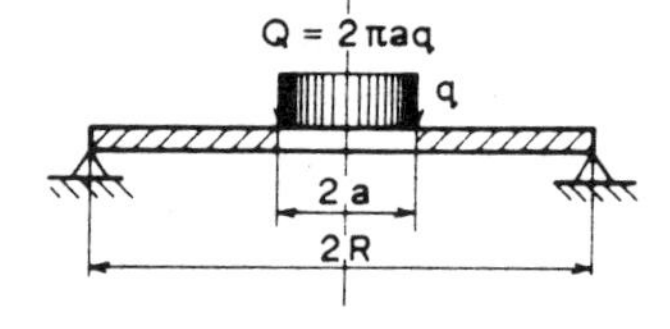

Fig. 7.19 (d) $\frac{Q}{M_p} = 2\,\pi\,\kappa$

with $\frac{\kappa}{\kappa-1} e^{\,1/\kappa - 1} = \frac{R}{a}$

Limit loads for various boundary conditions are given in figs. 7.17 and 7.18 from Sawczuk and Jaeger [7.17].

7.3.8. *Various plates with line loads.*

Various plates with line loads are shown in fig. 7.19. The relations in this figure are taken from Iliouchine ([1.2] pp. 237-244)(). They were obtained by a purely statical approach with the Tresca yield condition. Hence they are lower bounds Q_-, and are represented in fig. 7.20. Relation (7.25) was found independently by Hodge [3.9] as an exact limit load.

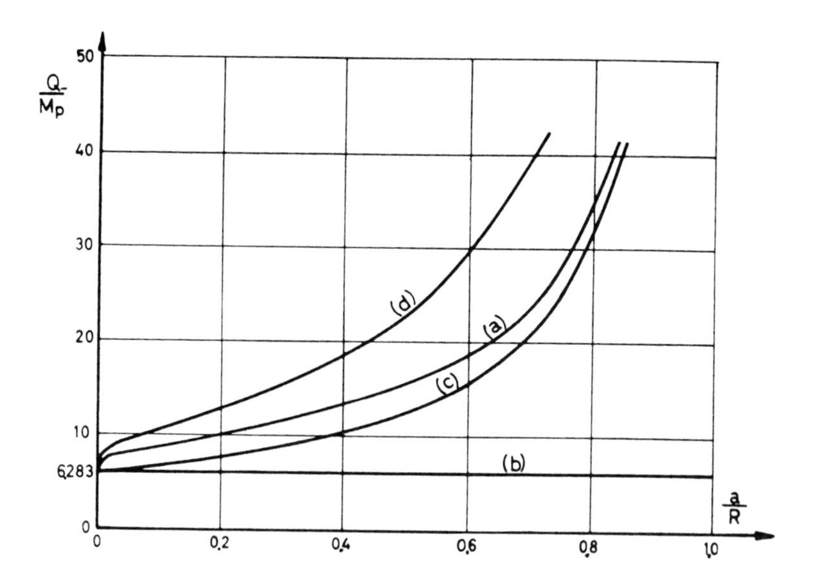

Fig. 7.20.

7.3.9. *Plate with overhang.*

Drucker and Hopkins [7.18] have discussed a plate of radius αR ($\alpha \geq 1$), simply supported on a circle of radius R, and subjected simultaneously to a uniform pressure p over the area interior to the circle of support and to a central concentrated load P (fig. 7.21).

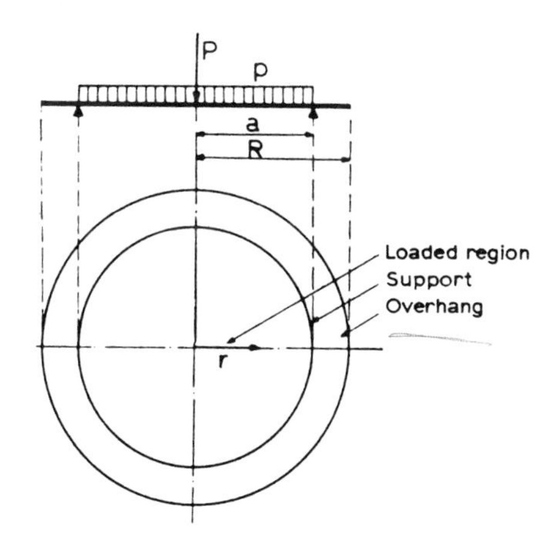

Fig. 7.21.

This problem is of particular interest because it contains the simply supported plate ($\alpha = 1$) and the built-in plate (α sufficiently large) as limiting cases, and is likely to represent more accurately real support conditions intended to provide simple support. The yield condition of Tresca is used. The results of the integration of eq. (7.6) in the various plastic regimes are given on fig. 7.22. In the present case, the relevant regimes are AB and BC.

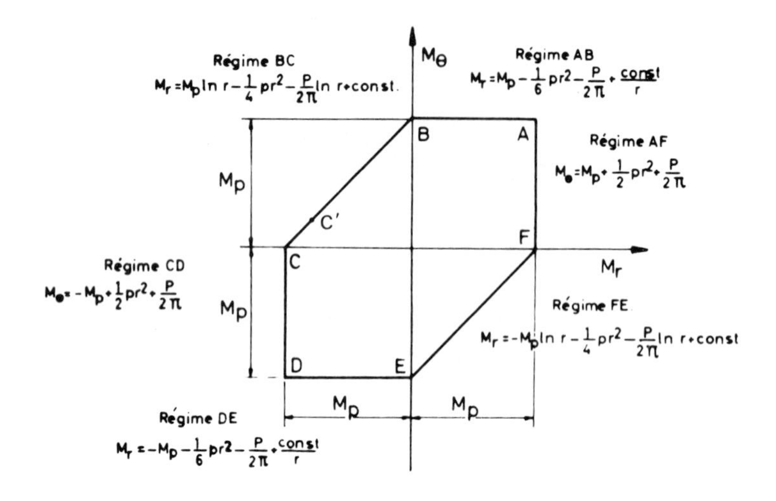

Fig. 7.22.

For small values of r, we first have

$$M_r = M_p - \frac{P}{2\pi} - p\,\frac{r^2}{6}, \qquad 0 \le r \le r_1 \tag{7.26}$$

where r_1 corresponds to $M_r = 0$, that is,

$$r_1 = \frac{6}{p}\left(M_p - \frac{P}{2\pi}\right) \tag{7.27}$$

Note that there is a singularity at r=0 because M jumps from M_p to $M_p - P/2\pi$. This is due to the use of a bending theory in the presence of the concentrated load P.

If $\alpha = 1$ (simply supported plate), $r_1 = R$ and we get from eq. (7.27)

$$\frac{p R^2}{6 M_p} + \frac{P}{2 \pi M_p} = 1 . \tag{7.28}$$

Hence, we see that the uniformly distributed load p and the central concentrated load P for collapse are linearly related. Relation (7.28) reduces to eq. (7.16) or eq. (7.17) when p or P vanish, respectively. If $\alpha > 1$, higher loads are possible. A detailed analysis [7.18] shows that

1. If $0 < \ln \alpha < 1$, that is, if $1 < \alpha < 2.718$, the plastic regime is ABC'B, points C' and B corresponding to $r=R$ and $r=\alpha R$, respectively. The abscissa of point C' is $M_r(R) = -M_p \ln \alpha > -M_p$. Eq. (7.28) is replaced by

$$p \pi R^2 \left(\frac{1 - \frac{r_1^2}{R^2}}{4} \right) + \frac{P}{4 \pi} \ln \frac{R^2}{r_1^2} = M_p \left(\ln \frac{R}{r_1} + \ln \alpha \right) \tag{7.29}$$

where r_1 is given by eq. (7.27).

2. If $\ln \alpha=1$, that is if $\alpha = 2.718$, point C' coincides with point C. We have $M_r(R) = -M_p$ and relation (7.29) holds, with $\ln \alpha = 1$.

3. If $\ln \alpha > 1$, that is , if $\alpha > 2.718$, the overhang remains rigid and horizontal, and eq. (7.29) must be used with $\ln \alpha = 1$ as in case 2.

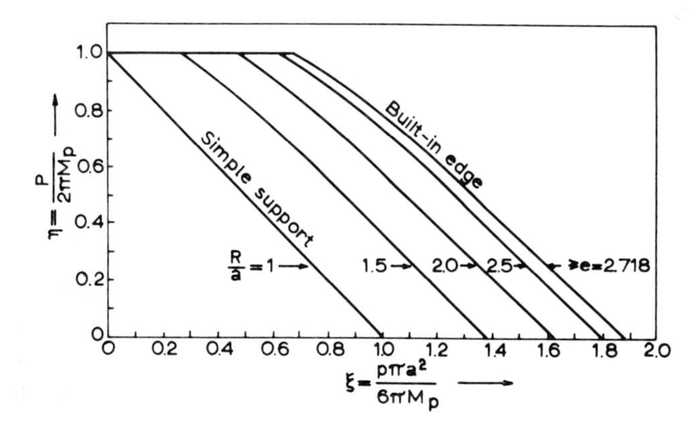

Fig. 7.23.

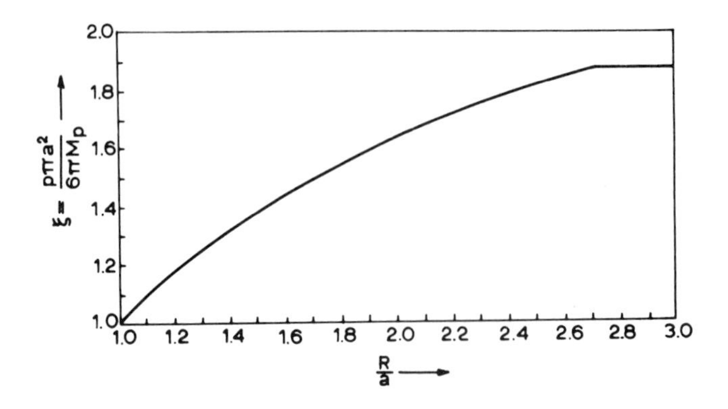

Fig. 7.24.

Results are summarized in figs. 7.23 and 7.24, where dimensionless variables have been used.

7.4. Circular orthotropic plates.

7.4.1. *Introduction.*

Many plates used in practice are strengthened by stiffeners, to achieve high strength with small structural weight. Stiffeners are most often placed along the lines of an orthogonal net and the plate so constructed exhibits structural plane orthotropy, that is, structural plane anisotropy with two orthogonal axes of symmetry.

Material orthotropy can also arise from the cold forming process : yield stresses in various directions are then different. Whatever the source, the orthotropy will be described by a proper yield condition, which must ultimately be justified by experimental evidence. In the following, two types of orthotropy will be considered, generalizing the yield condition of Tresca. The normality law is assumed to apply to the yield conditions used, and the principal directions for the stresses and the orthotropy are supposed to be the radial and circumferential directions.

Material orthotropy can also arise from the cold forming process : yield stresses in various directions are then different. Whatever the source, the orthotropy will be described by a proper yield condition, which must ultimately be justified by experimental evidence. In the following, two types of orthotropy will be considered, generalizing the yield condition of Tresca. The normality law is assumed to apply to the yield conditions used, and the principal directions for the stresses and the orthotropy are supposed to be the radial and circumferential directions.

7.4.2. *Orthotropy of the first kind.*

The assumed yield condition is shown in fig. 7.25 [7.19, 7.20]. The full plastic moments $M_{\theta p}$ and M_{rp} in the circumferential and radial directions, respectively, differ by a given factor $\kappa = \frac{M_{\theta p}}{M_{rp}}$, called the orthotropy coefficient.

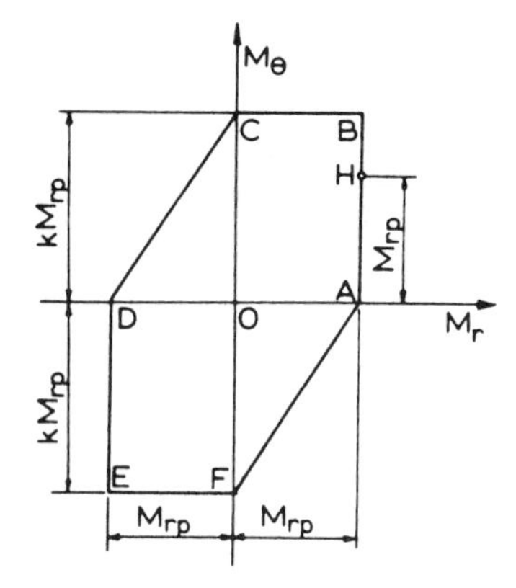

Fig. 7.25. (a)

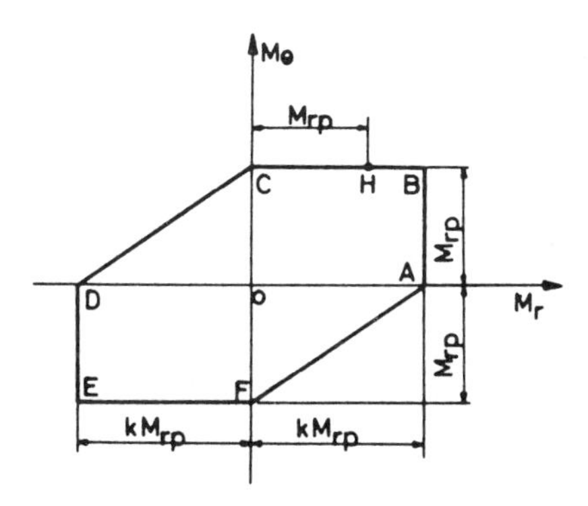

Fig. 7.25. (b)

Restricting our considerations to axisymmetric loading, we immediately note that the central point of the plate is special. Indeed, any two orthogonal directions at this point are principal directions for stresses and orthotropy. Hence, the plate is locally isotropic at the centre. Any stress profile must start, at r=0, from point H in fig. 7.25, corresponding to $M_r = M_\theta$.

Suppose first that $k \geq 1$, $M_{\theta p} \geq M_{rp}$, and consider a simply supported uniformly loaded plate (fig. 7.26). The stress profile is HBC in fig. 7.25 (a).

Integration of eq. (7.6) with this stress profile and adequate continuity conditions at the radius r_o corresponding to point B gives the moment field shown in fig. 7.26.

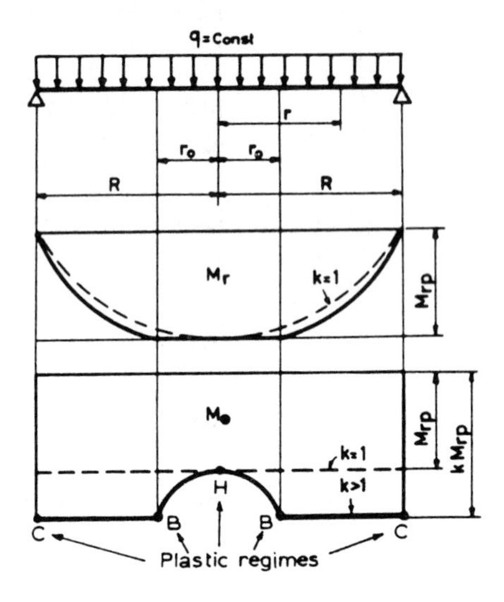

Fig. 7.26.

The value of r_o is given by

$$2\left(\frac{r_o}{R}\right)^3 (k-1) - 3\,\frac{r_o k}{R} + k - 1 = 0\,. \tag{7.30}$$

The limit load is

$$P_l \equiv p_l\,\pi R^2 = 6\,\pi\,\frac{k-1}{\left(\frac{r_o}{R}\right)^2} M_{rp}, \tag{7.31}$$

and the velocity field is represented in fig. 7.27. The limit load (7.31) is shown graphically in fig. 7.28, with the limit load for the built-in plate obtained in a similar manner. An approximate formula for this latter case is

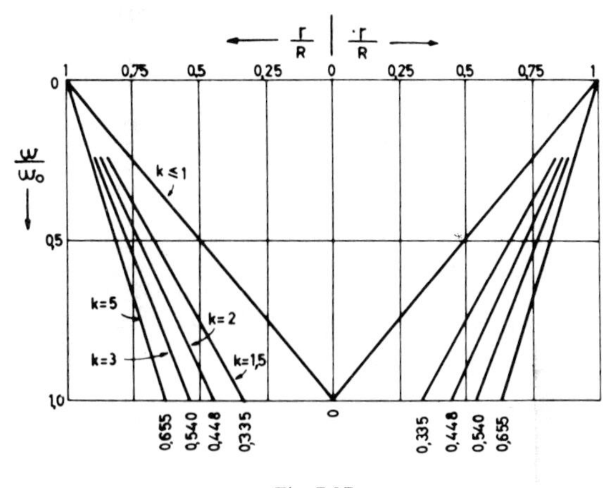

Fig. 7.27.

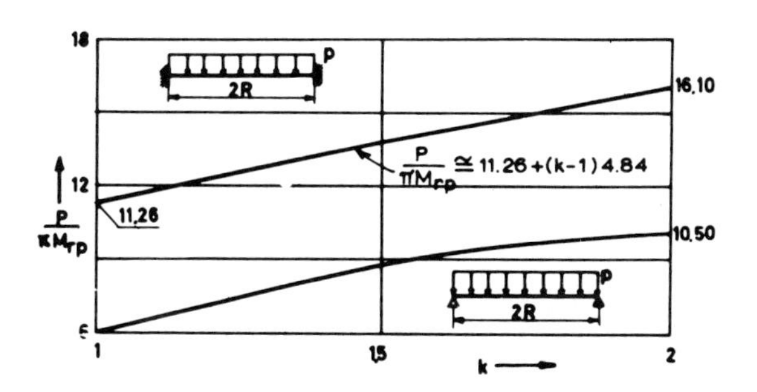

Fig. 7.28.

$$P_l = \pi M_{rp}[\ 11.26 + 4.84(k-1)]. \qquad (7.32)$$

When $k \leq 1$, the stress profile for simple supports is BC in fig. 7.25 (b), as for an isotropic plate with plastic moment M_p equal to $M_{\theta p}$. Hence, the limit load is

$$P_l = 6\,\pi\, M_{\theta\, p} < 6\,\pi\, M_{rp}\,,$$

and it is concluded that purely radial strengthening is completely useless.

The preceding results, taken form Olszak and Sawczuk [7.19, 7.20, 7.21] have been extended to various loading cases by Markowitz and Hu [7.22]. The reader is referred to the Atlas of Limit Loads [7.26] for results and to the original publication for details.

We finally remark that, for large central bosses or large central holes, shear action may become important and the limit load from bending theory might lose its physical significance (see Section 7.7.2, fig. 7.41). As already noted in Section 7.2, experiments [7.7] support formula (7.32) for a structural orthotropy with numerous reinforcing rings.

7.4.3. *Orthotropy of the second kind.*

An alternative yield condition has been proposed by Sawczuk [7.21]. It is shown in fig. 7.29.

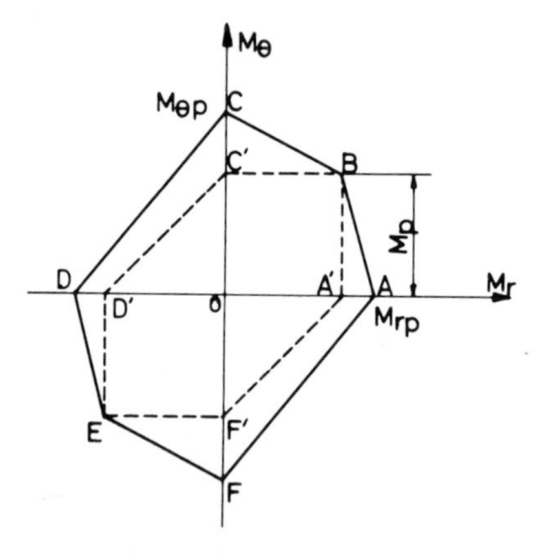

Fig. 7.29.

For a simply supported, uniformly loaded plate, the stress profile is BC as r varies from 0 to R ; it furnishes the statically admissible moment field (fig. 7.30).

$$M_r = \left[1 - \left(\frac{r}{R} \right)^2 \right] M_p ,$$

$$M_\theta = \left[1 + \left(\frac{r}{R} \right)^2 (k - 1) \right] M_p , \qquad (7.33)$$

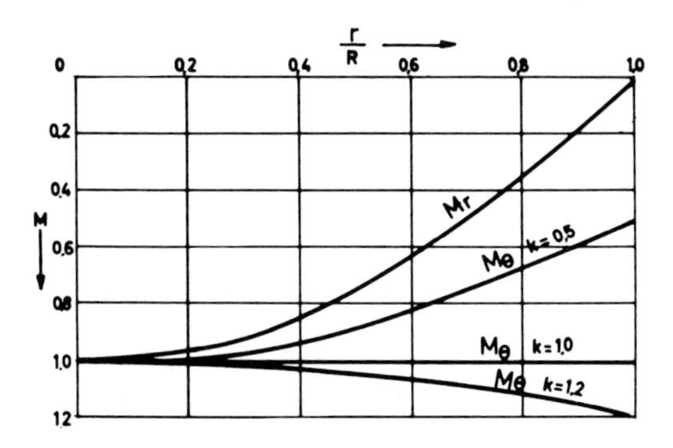

Fig. 7.30

where $k = \frac{M_{\theta p}}{M_p}$.

Associated kinematically admissible velocity fields are shown in fig. 7.31, and the exact limit load is

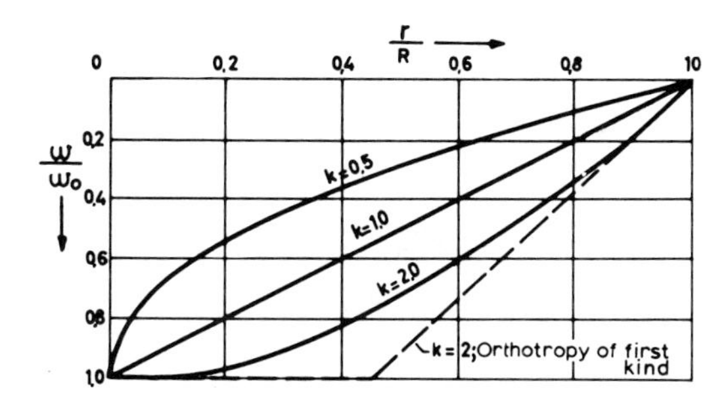

Fig. 7.31.

$$P_1 \equiv p_1 \pi R^2 = 6 \pi M_p \frac{2+k}{3}. \tag{7.34}$$

We note that the limit load does not depend on $\frac{M_{rp}}{M_p}$ as it was to be expected from the stress profile used. Hence, radial strengthening is useless as far as the limit load is concerned.

For the yield curve of fig. 7.29 to remain convex, the condition

$$k \geq \frac{M_{rp}}{M_p + M_{rp}}$$

must be satisfied [7.21].

7.5. Isotropic rectangular plates.

7.5.1. *Introduction.*

Limit analysis of rectangular metal plates is still in its infancy. Contrary to the case of circular plates, the principal directions of bending moment and curvature are not known beforehand. The mathematical difficulties therefore are considerably greater. On the other hand, experiments are completely lacking, despite the obvious practical interest of the problem. Hence, we restrict ourselves to presently available theoretical solutions.

7.5.2. *Tresca yield condition.*

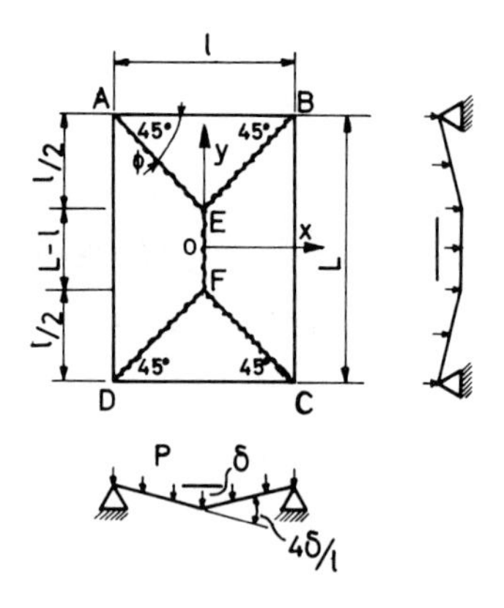

Fig. 7.32.

Consider a simply supported rectangular plate, with side lengths l and L, subjected to uniformly distributed load. In an elementary approach, Drucker [7.23] imagines a mechanism formed of straight yield lines (hinge lines) (fig. 7.32), as is customary in reinforced concrete (see Chapter 9). The hinge lines must be regarded as the limiting case of narrow strips exhibiting cylindrical curvature. Hence, in the considered mechanism, the curvature rates vanish everywhere except in the hinge lines, where the principal directions of curvature rate are that of the hinge line and the normal to it. The curvature rate along the hinge line always vanishes. Because of isotropy and normality law, principal directions of the moment tensor coincide with those of the curvature rate tensor.

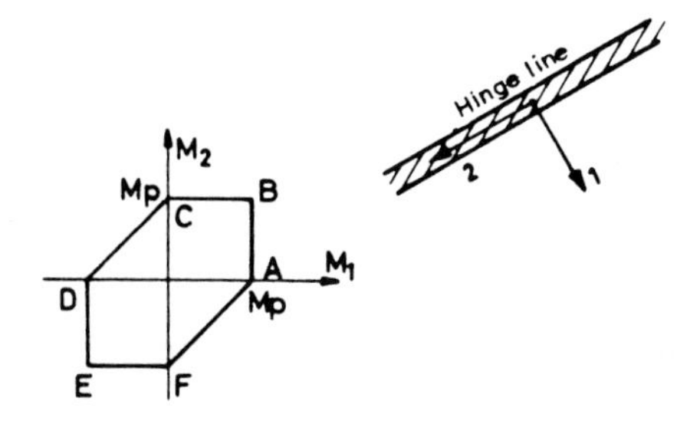

Fig. 7.33.

From these considerations, we see that the plastic regime in a (straight) hinge line must be either AB or DE (fig. 7.33) and the dissipation per unit length of hinge line is simply the product $M_p \, |\dot{\theta}|$, where $\dot{\theta}$ denotes the relative rotation rate of adjacent parts.

Inspection of fig. 7.32 shows that the relative rotation rates are $\frac{4\,\delta}{l}$ for segment EF ; and $2\left(\frac{2\,\delta}{l}\right)^{1/2}$ for segments AE, EB, FD, and FC, where δ is the transversal velocity of segment EF. The power of the applied load is p times the volume bounded by the initial midplane of the plate and its deformed situation at collapse.

From the kinematic theorem, we then obtain

$$p_+ \left[\frac{1}{3}\, l^2\, \delta + l\,(L - l)\,\delta \right] = M_p \left[4\, \frac{\delta}{l}\,(L - l) + 4\,(2\sqrt{2}\,)\, \frac{\delta}{l} \frac{l}{\sqrt{2}} \right],$$

or

$$p\, \frac{l^2}{M_p} = 24\, \frac{1 + \frac{L}{l}}{3\, \frac{L}{l} - 1}. \tag{7.35}$$

To obtain a lower bound p_, let us imagine a moment field with no twist, inspired from the analogous beam problem :

$$M_x = C_1\left(\frac{l^2}{4} - x^2\right),$$

$$M_y = C_2\left(\frac{L^2}{4} - y^2\right),$$

$$M_{xy} = 0 . \tag{7.36}$$

To satisfy the equilibrium equation (Shear forces are given by $V_x = \frac{\partial M_x}{\partial x} + \frac{\partial M_{xy}}{\partial y}$, $V_y = \frac{\partial M_y}{\partial y} + \frac{\partial M_{xy}}{\partial x}$).

$$\frac{\partial^2 M_x}{\partial x^2} + \frac{\partial^2 M_y}{\partial y^2} + 2\frac{\partial^2 M_{xy}}{\partial x \partial y} = -p , \tag{7.37}$$

we must have

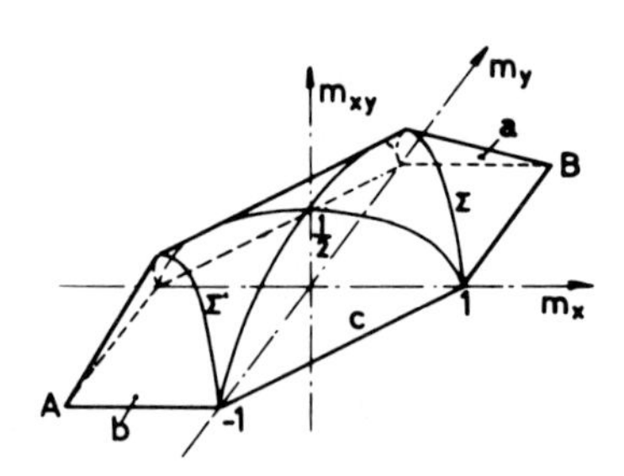

Fig. 7.34.

$$C_1 + C_2 = \frac{p}{2} . \tag{7.38}$$

Maximum moments occurs for x=0 and y=0. Highest load is obtained with regime B (fig. 7.33) at central point, because both maximum moments attain the plastic value simultaneously. substitution of x=y=0 and $M_x = M_y = M_p$ in eqs. (7.36) results in

$$C_1 = 4\frac{M_p}{l^2} ,$$

$$C_2 = 4\frac{M_p}{L^2} . \tag{7.39}$$

From eqs. (7.38) and (7.39) we obtain the lower bound

$$p\_ = 8 M_p\left(\frac{1}{l^2} + \frac{1}{L^2}\right). \tag{740}$$

For a square plate with side l, eqs. (7.35) and (7.40) are combined into

$$16 \leq \frac{p_l \, l^2}{M_p} \leq 24 . \tag{7.41}$$

The bounds given by eq. (7.41) are far apart because the lower bound is much too crude.

A refined approach to the problem of the rectangular plate has been given by Shull and Hu [7.24]. An improvement of the moment field is obtained by setting

$$M_x = M_y = C\left(1 - \frac{8x^3}{l^2}\right)\left(1 - \frac{8y^3}{L^2}\right). \tag{7.42}$$

Substitution of expressions (7.42) for M_x and M_y in eq. (7.37) furnishes M_{xy} as a function of p, C, x, and y. Now, the yield condition of Tresca in the moment space M_x, M_{yx}, M_{xy} is needed. We know that relations (1.12) and (1.13) are also valid for the (plane) moment tensor, by simply replacing σ_1, σ_2, σ_x, σ_y, τ_{xy} by M_1, M_2, M_x, M_y, M_{xy}, respectively. Hence, condition (7.30) can be written

$$\text{(a)} \qquad \frac{m_x + m_y}{2} + \left[\left(\frac{m_x - m_y}{2} \right)^2 + m_{xy}^2 \right]^{1/2} = \pm 1,$$

$$\text{(b)} \qquad \frac{m_x + m_y}{2} - \left[\left(\frac{m_x - m_y}{2} \right)^2 + m_{xy}^2 \right]^{1/2} = \pm 1,$$

$$\text{(c)} \qquad 2\left[\left(\frac{m_x - m_y}{2} \right)^2 + m_{xy}^2 \right]^{1/2} = \pm 1, \qquad (7.43)$$

using dimensionless moments as defined by eqs. (5.25).

Conditions (7.43) is represented in fig. 7.34. The surface consists of a cylindrical central part c with eq. (7.43c) and two conical caps a and b with eqs. (7.43a) and (7.43b) respectively. The intersections $\sum$ and $\sum$ of the surfaces above satisfy the condition

$$m_x m_y = m_{xy}^2 . \qquad (7.44)$$

Because we have chosen a particular moment field (7.42) in which $M_x = M_y$, the yield surface reduces to its intersection polygon with the plane $m_x = m_y$. The equations of the sides of this polygon are

$$m_x + |m_{xy}| = 1 ,$$
$$m_x - |m_{xy}| = -1 ,$$
$$2\, m_{xy} = \pm 1 . \qquad (7.45)$$

For the stress point to be on or within the polygon with eq. (7.45) for $0 \le x \le 1$ and $0 \le y \le L$, p must not be larger than some values that depend on C and $\eta = \frac{1}{L}$. Best lower bounds were obtained for each η, adjusting C to maximize the limiting values of p, with the aid of an IBM-650 computer. The upper bound [eq. (7.35)] is improved by considering the length λ of segment EF as a parameter in the mechanism (or equivalently, the angle φ, which was previously taken as 45°). The value of λ is chosen to minimize the load, according to the kinematic theorem. As will be shown in details in Section 9.4.8, the resulting load is

$$p_{+} = \frac{24\,M_p}{l^2\left[\left(3 + (l/L)^2\right)^{1/2} - l/L\right]^2}. \qquad (7.46)$$

Results are summarized in table 7.1 and fig. 7.35.

Table 7.1. Load-carrying capacities of rectangular plates.

Ratio of sides $\eta = l/L$	Lower bound p_*^-	Upper bound p_*^+	p_*	Max. possible percent error
1.00	0.826	1.000	0.913 ± 0.087	9.6
0.95	0.870	1.055	0.963 ± 0.092	9.6
0.90	0.921	1.117	1.019 ± 0.098	9.6
0.85	0.979	1.186	1.083 ± 0.103	9.6
0.80	1.049	1.274	1.162 ± 0.112	9.6
0.75	1.108	1.372	1.240 ± 0.132	10.6
0.70	1.193	1.494	1.344 ± 0.150	11.2
0.65	1.293	1.639	1.466 ± 0.173	11.8
0.60	-	1.828	-	-
0.55	1.561	2.053	1.087 ± 0.246	13.6
0.50	1.744	2.358	2.050 ± 0.308	15.0
0.45	1.979	2.747	2.363 ± 0.384	16.3
0.40	2.289	3.289	2.789 ± 0.500	17.9
0.35	2.708	4.049	3.379 ± 0.670	19.8
0.30	3.309	5.236	4.273 ± 0.963	22.5
0.25	4.222	7.042	5.632 ± 1.410	25.0
0.20	5.471	10.417	8.079 ± 2.338	28.9
0.15	8.615	17.241	12.928 ± 4.303	33.3
$p_* = pL^2/24M_p$				

Developing the idea of Shull and Hu, Del Rio Cabrera [7.9] used analytical functions with several unknown parameters in his Master thesis to build up statically and plastically admissible moment fields for uniformly loaded rectangular Tresca and Mises plates. The heuristic procedure used to obtain lower bound p_- was the following

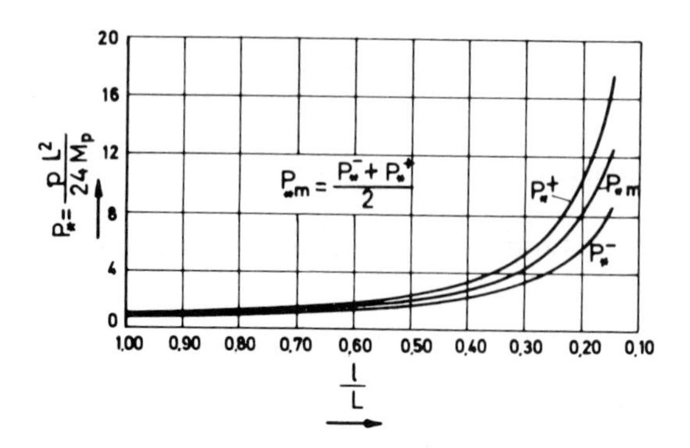

Fig. 7.35. Upper and lower bounds on load-carrying capacities of supported Tresca rectangular plate

$$p_* = pL^2/24M_p.$$

Table 7.2. Rectangular supported Tresca plates.

Ratio of sides, ε	Static approach ($\bar{p}_*$)			Kinematic approach (p_*^+)		
	Shull and Hu (1963)	Ranaweera and Leckie (1970)	Del Rio Cabrera	Shull and Hu (1963)	Ranaweera and Leckie (1970)	Del Rio Cabrera *
0.10	-	-	28.926	-	-	37.008
0.15	8.615	-	14.347	17.241	-	17.241
0.20	5.741	9.500	8.810	10.417	10.510	10.168
0.25	4.222	-	6.063	7.042	-	6.814
0.30	3.309	4.640	4.504	5.236	5.166	4.962
0.35	2.708	-	3.524	4.049	-	3.826
0.40	2.289	2.945	2.857	3.289	3.230	3.079
0.45	1.979	-	2.384	2.747	-	2.563
0.50	1.744	2.102	2.037	2.358	2.308	2.188
0.55	1.561	-	1.775	2.053	-	1.904
0.60	-	1.635	1.572	1.828	1.788	1.686
0.65	1.293	-	1.412	1.639	-	1.514

0.70	1.193	1.337	1.293	1.494	1.460	1.376
0.75	1.108	-	1.179	1.372	-	1.262
0.80	1.049	1.146	1.104	1.274	1.240	1.168
0.85	0.979	-	1.018	1.186	-	1.090
0.90	0.921	1.002	0.970	1.117	1.084	1.023
0.95	0.870	-	0.902	1.005	-	0.966
1.00	0.826	902	0.879	1.000	0.973	0.916

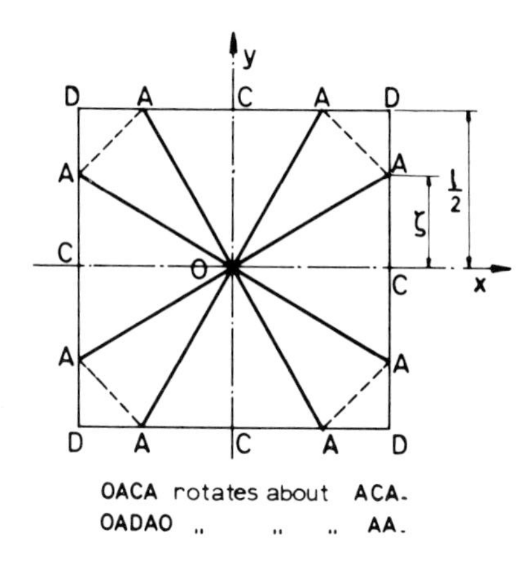

Fig. 7.36. OACA rotates about ACA, OADAO rotates about AA.

*Corners are allowed to lift up.

(a) starting from an assumed moment field that contains parameters, determine those parameters that make the plate fully plastic and that maximize the minimum value of p (x, y) ;

(b) load the plate uniformly with this minimum value, keep the bending moments obtained at the end of the preceding step, and modify the twisting moments to restore equilibrium ;

(c) check the resulting field for plastic admissibility and possibly diminish p_ accordingly.

Lower bounds for the limit load obtained in this manner are compared in table 7.2 with other contemporary solutions and with upper bounds given by collapse mechanisms similar to those used by Shull and Hu. Results for built-in plates, and with the Mises yield condition can be found in the original work [7.9] and in refs. [7.25 and 7.26].

Side dimensions are L and εL, $p_* = \frac{pL^2}{6M_p}$, where p is uniformly distributed pressure.

The yield condition of von Mises has been applied to the problem of a square, uniformly loaded, simply supported plate, with the corners prevented to lift. Hodge [3.9] has found

$$p_- = 20.6 \frac{M_p}{l^2} \leq p_l \leq 27.7 \frac{M_p}{l^2} = p_+ , \tag{7.47}$$

and Iliouchine [1.2] has obtained

$$p_- \leq p_+ = 26.4 \frac{M_p}{l^2} . \tag{7.48}$$

When corners are permitted to lift, Hodge uses the mechanism of fig. 7.36 to find

$$p_+ = 25.65 \frac{M_p}{l^2} .$$

An extension of Hodge's results to some class of orthotropic plates was given by Kao, Mura, and Lee [7.27].

7.6 . Review of other work.

Complete limit analysis solutions for plates resting on an ideally plastic foundation have been given by Sawczuk and Kaliszky ([7.28] ; see also [7.17], p. 186). The plate satisfies the Tresca yield condition and the subgrade is considered as yielding in uniaxial compression, with a yield stress that may vary with position (nonhomogeneous foundation). An infinite plate uniformly loaded over a finite circular area, and a circular plate centrally loaded by a concentrated force have been studied.

The influence of shear forces on the limit loads of circular plates has been theoretically investigated by Sawczuk and Duszek [7.29].

A first step towards accounting precisely for structural orthotropy due to stiffening has been made by Nemirovski [7.30], but did not furnish many practical results.

Circular plates with either continuous or discontinuous thickness variations (plastically nonhomogeneous) have been widely studied, mainly by Polish authors. [7.19, 7.20, 7.21, 7.31].

With a suitably linearized yield condition, limit analysis (and minimum-weight design) of plates may be formulated as problems of linear programming, as was advocated by Koopman and Lance [7.32]. Indeed, this approach is quite general [7.33, 7.34, 7.35], and some applications to reinforced concrete plates will be considered in Section 9.5.9. In their paper, Koopman and Lance [7.32] give the lower bound.

$$p_- \frac{L^2}{24\, M_p} = 0.964\,,$$

for the simply supported square Tresca plate uniformly loaded, to compare with

$$p_- \frac{L^2}{24\, M_p} = 0.826\,,$$

found by Shull and Hu [7.24]. Caution should be exercised, however, in using the former value because the mesh used in the finite difference approach was rather coarse and, hence, the yield condition might be violated at intermediate points. Indeed, bounds by Del Rio [7.9] are 0.879 and 0.916, showing that the value 0.964 is erroneously high. Finite elements have been used in refs. [6.15], [7.36], [7.37] and many others, the results of which are collected in ref. [7.26].

According to Hodge and Belytschko [6.15], the only exact two-dimensional solution available for the Mises yield criterion is the built-in square plate subjected to a concentrated load. This solution can be constructed from the corresponding solution to the circular plate obtained by Hopkins and Wang [7.15] by inscribing a circle in the square, using the circular plate moment and velocity distributions within the circle and then extending a constant moment field and zero velocity field into the remainder of the square. The corresponding limit load is $P_l = 7.257\, M_p$.

The concentrated load problem was solved for the simply supported square plate by Hopkins and Wang. They found the following bounds :

$$P_- = 6.238\, M_p < P_l < 6.283\, M_p = P_+\,.$$

Numerical bounds for the same problem were obtained by Hodge and Belytschko [6.15], by using equilibrium finite elements and mathematical programming algorithm (see

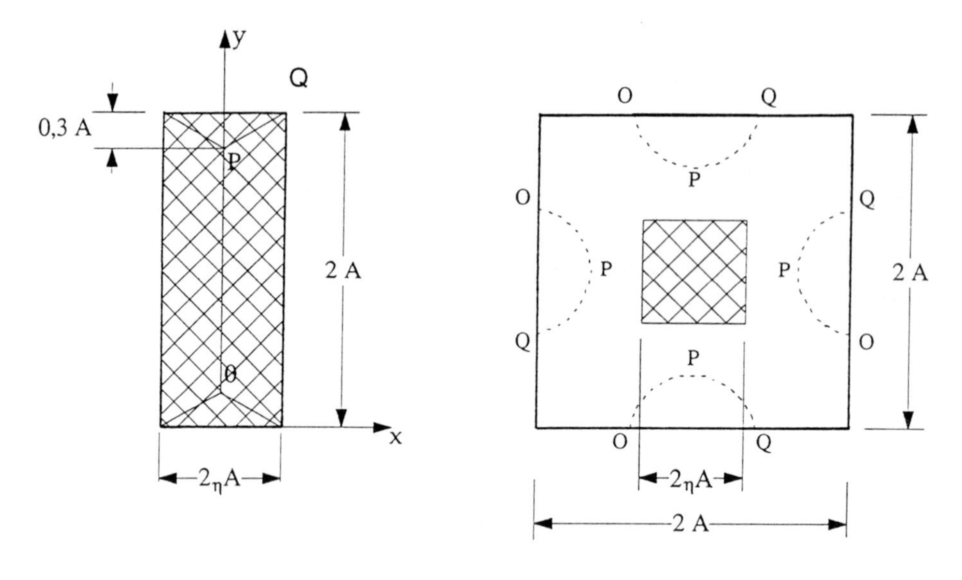

Fig. 7.37 (a)

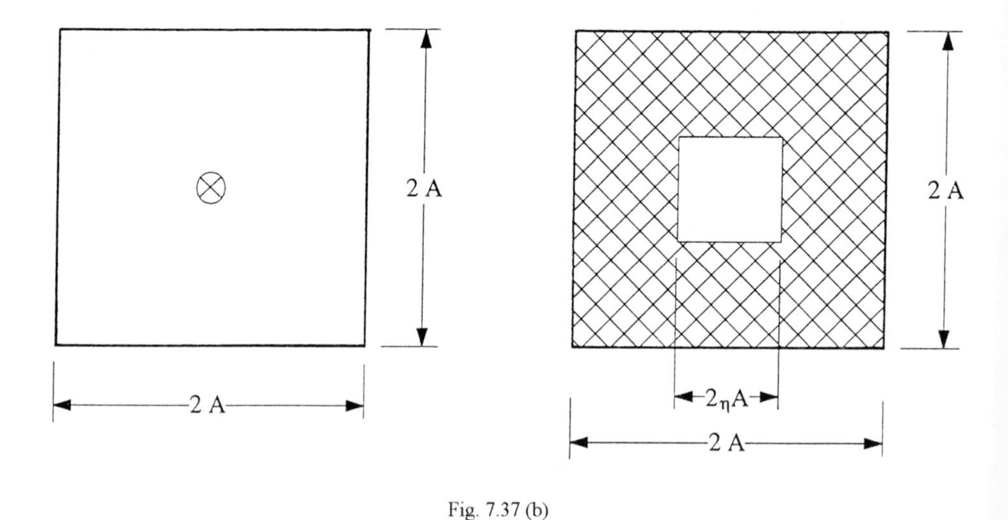

Fig. 7.37 (b)

Section 6.31). The results of all problems solved are listed in table 7.3 and the geometries of the problems illustrated in fig. 7.37.

For the built-in square plate with concentrated load, the lower and upper bounds obtained differ from the exact solution by 2%. For the simply supported square plate with

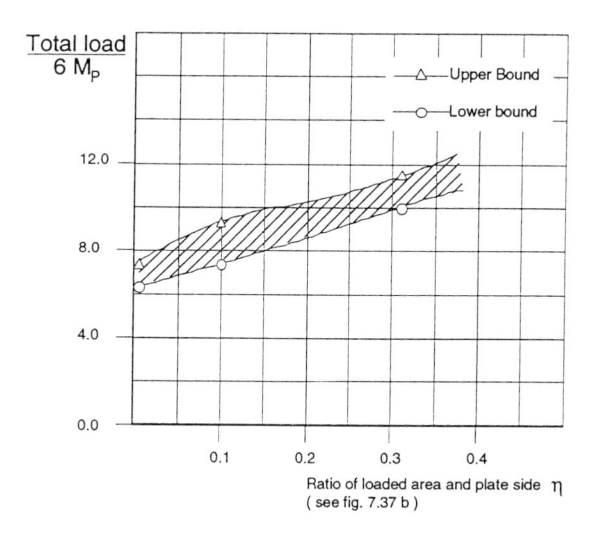

Fig. 7.38.

concentrated load, the lower bound solution is 6.40 M_p, which is greater that the exact solution for the corresponding circular plate problem [7.15]. This indicates that unlike the results for the Tresca criterion [7.40], [7.41], [7.42] for which the concentrated load solution is independent of the supports or shape, the results for the Mises criterion depend on the supports, and in the simply supported case, on the shape of the plate.

Fig. 7.38 shows results for simply supported square plates loaded uniformly over a square area at the centre of the plate. These results indicate that the total load carrying capacity increases markedly as the area of load application is enlarged. It was also noted that the moment and velocity distributions change significantly as the loaded area is increased. For the cases where the load is applied over a small area or concentrated at a point, most of the plate yields at the yield-point load. As the loaded area is increased, regions which approximately coincide with OPQ in fig. 7.37(b) become rigid and the yielding regions are confined to the areas about the diagonals of the square. At the same time the velocity distribution changes from a distorted cone to a pyramid with rounded corners.

Results were also obtained for simply supported, uniformly loaded rectangular plates with slenderness ratios of 1:1 to 5:1. These results are shown in fig. 7.39 along with the results of Shull and Hu [7.24] which were obtained with the Tresca criterion. As can be seen from fig. 7.39, their results diverge greatly as the slenderness ratio increases. The results obtained in [6.15] behave qualitatively like Shull's upper bound results. Upper bound calculated from Shull's velocity fields using the Mises criterion (given in table 7.3) are only 5 to 10 percent higher than the yield-point loads obtained in [6.15]. The results for the square plate found analytically by Hodge [3.9] are indicated in fig. 7.39 by squares. The results obtained in this investigation are between the previous bounds and, compared to the maximum possible error of 10% for the previous results, the present value for the yield-point load has a maximum possible error of 3.3%.

The result for the rectangular plate with slenderness ratio of 5:1 was obtained with an accuracy of 2.26 percent. Examination of the resulting moment and velocity fields indicates that the two fields are consistent with the flow law in most of the yielding regions. Along the line OP (fig. 7.37(a)), the moments have the following constant values :

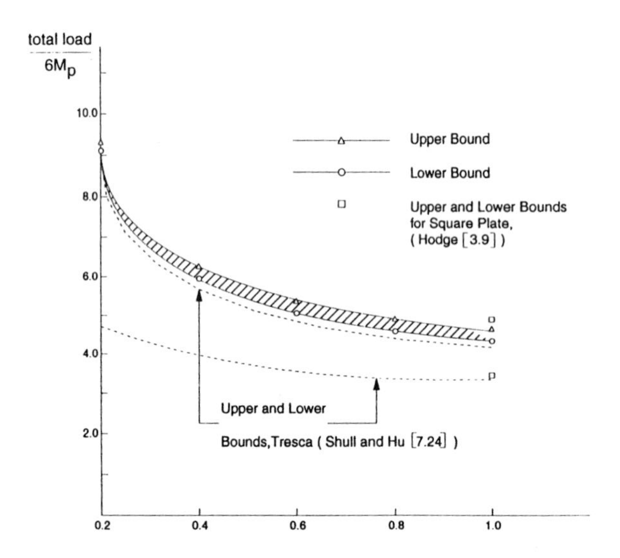

Fig. 7.39. Yield-point load of uniformly loaded, simply supported rectangular plates [6.15].

$m_x = 1.154$; $m_y = 0.577$; $m_{xy} = 0.0$. The velocity field has a hinge line along OP. According to the flow law, the only nonzero strain rate along OP should be $\dot{\kappa}_x$ which is indeed the case for that hinge line. The behaviour along the line PQ is more complex. The velocity field consists of a combination of hinge lines and curved regions, which is evidently an attempt by the method to approximate a velocity field of higher than second order. The predominant moments and strain rates are m_{xy} and $\dot{\kappa}_{xy}$ respectively, which is consistent with the flow law. However the strain rates $\dot{\kappa}_x$ and $\dot{\kappa}_y$, though small, are not consistent with the moment fields. Since the regions of the plate where this type of yielding occurs are much smaller than the OP hinge line, it explains why the results for this problem are however quite accurate. Results are presented in table 7.3 for the following additional problems : a simply supported square plate with a square hole ; a square plate with one edge free, three edges simply supported ; a built-in square plate. All of these plates were uniformly loaded.

Table 7.3. ρ = (total load) / M_p

	Solutions from [6.15]			Max. possible percent error	other solutions		
	ρ^-	ρ^+	$(\rho^+ + \rho^-)/2$		lower bound	upper bound	Approximation
Built in plate, conc. loads, fig.7.37 c	7.120	7.831	7.745	4.76	7.257 [7.19]	7.257 [7.15]	7.439 [6.36]
Simply sup. plate, conc. loads, fig. 7.37 c	6.401				6.238 [7.15]	6.283 [7.15]	6.528 [6.36]
Simply sup. plate, sq. ld $\eta = 0.333$, fig. 7.37 b	3.727	4.243	3.985	6.47			
Simply sup. plate, sq. ld $\eta = 0.100$, fig. 7.37 b	30.67	39.31	34.99	12.35			
Simply sup. plate, unif. ld $\eta = 1.0$, fig. 7.37 a	1.036	1.106	1.070	3.32	0.859 [3.9]	1.155 [3.9]	
Simply sup. plate, unif. ld $\eta = 0.3$, fig. 7.37 a	1.307	1.407	1.357	3.68	1.049 [7.24]	1.471 [7.24]	
Simply sup. plate, unif. ld $\eta = 0.6$, fig. 7.37 a	1.888	2.024	1.956	3.48		2.111 [7.24]	
Simply sup. plate, unif. ld $\eta = 0.4$, fig. 7.37 a	3.448	3.670	3.559	3.12	2.989 [7.24]	3.799 [7.24]	
Simply sup. plate, unif. ld $\eta = 0.2$, fig. 7.37 a	11.321	11.844	11.582	2.26	5.471 [7.24]	12.032 [7.24]	
Simply sup. plate, with hole, unif. ld. $\eta = 0.333$, fig. 7.37 d	0.915	0.977	0.946	3.28			
Built-in plate, unif. ld $\eta = 1.0$, fig. 7.37 a	1.786	2.052	1.919	6.93		2.310 [7.17]	
3 edges simply sup. , 1 edge free unif. ld., $\eta = 1.0$, fig. 7.37 a	0.395	0.654	0.624	4.72		0.680 [7.17]	

Previously available results [7.17] gave upper bounds for the built-in plate which were from 12 to 24 percent above the yield-point load.

Finally, note that solution have also been obtained by Casciaro and Di Carlo [6.36] in a numerical way (last column of table 7.3). As they use mixed finite elements (see Section 6.5), the values obtained are only approximations but no true bounds.

Simply supported plates of arbitrary shape have been considered by Villagio [7.38], using a complex variable technique and the yield condition of von Mises. Presently available numerical applications of the method compare poorly with previous solutions.

Complete solution for metal plates with polygonal (or more complicated) boundaries seem otherwise to be absent. Schumann has shown [7.39] that the limit load p of a uniformly loaded plate with arbitrary simply supported edge satisfies

$$p_l \geq \frac{6 \pi M_p}{S} \tag{7.49}$$

where S is the plate area. In the case of a regular polygonal plate we have $S = \pi R^2 \tan(\pi/n)$ where n is the number of sides and R the radius of the inscribed circle. An upper bound can be obtained with a pyramidal mechanism, as will be seen in Section 7.4. It gives

$$p_l \geq \frac{6 \pi M_p}{R^2} .$$

Hence,

$$\frac{6 \pi r^2}{n \tan (^n\!/\!_\pi)} \leq p_l \leq \frac{6 \pi M_p}{R^2} .$$

We finally note once again that the limit value of a concentrated load acting on a plate obeying the Tresca condition is

$$P_l = 2 \pi M_p \tag{7.50}$$

for any boundary condition (buit-in, simply supported, partially restrained) and any load location [7.40, 7.41, 7.42].

More detailed information on plate problems can be found in the book by Sawczuk and Jaeger [7.17]. Limit analysis of circular metal plates with piecewise constant thickness is extensively treated (together with optimal design) in the thesis by Lamblin [7.43].

7.7. Deformations of metal plates.

7.7.1. Introduction.

In the case of beams and frames the (theoretical) limit load P_l can most often be regarded as a real failure load because changes in geometry, either in elastic-plastic range or due to the collapse mechanism, very seldom have an stabilizing effect (except, for example, for a built-in beam [3.7]). Hence, the limit load can as a rule be used as a good approximation to the carrying capacity of the structure. Calculation of the limit load must be supplemented by a sufficiently accurate estimate of maximum deflections at impending collapse, to ensure that the assumption of negligible changes in geometry is satisfied up to the onset of unrestricted plastic flow. Fortunately, there exist general methods and numerous computer codes for evaluating these deflections.

For metal plates, the situation is quite different. Deformations, either prior to attaining the limit load or subsequent, always have a stabilizing effect and the load can be considerably increased beyond the (theoretical) limit value. Hence, *the failure load will be determined by a limit assigned to the deflection.* As was seen in Section 7.2, for circular plates with a slenderness μ ranging from 10 to 40, a maximum deflection of 2% of the

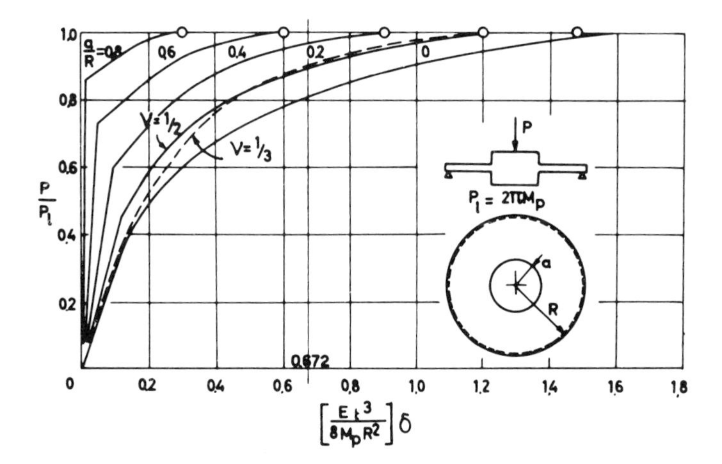

Fig. 7.40.

diameter gives a failure load very near to the limit load P_l. Unfortunately, we do not know of a general method to obtain the load versus deflection diagram of a plate up to large deflections, except numerical step-by-step analyses performed by big computer codes able to include material and geometrical non-linearities. The use of these codes may however turn out to be exceedingly expensive. In the following, we discuss currently available particular solutions that are useful in estimating whether, in every particular situation, the

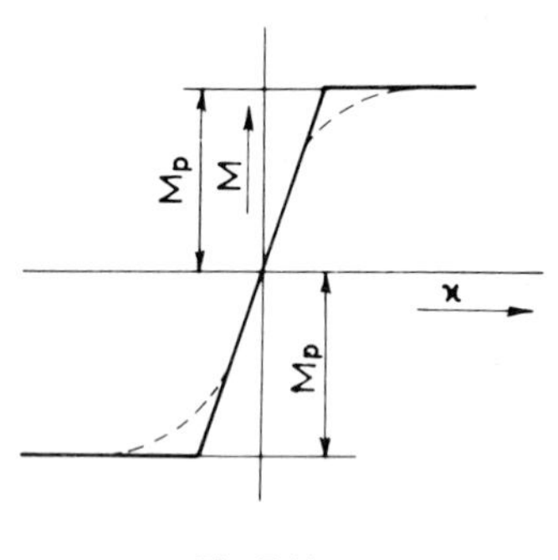

Fig. 7.41.

deflections under the limit load are acceptable. It must be emphasized, however, that both theory and experiments on the subject of noncircular plates remain rather rate.

7.7.2. Elastic-plastic range.

In the elastic-plastic range, displacements remain of the order of magnitude of elastic displacements. Accordingly, the analysis refers to the undeformed state, as usual in elasticity.

Haythornthwaite [7.3] considers a circular plate, simply supported and loaded on a rigid central boss of radius a by a load P (fig. 7.40). He uses Tresca's yield condition, the associated flow law, and the sandwich-type moment versus curvature diagram of fig. 7.41. He obtains the load versus deflection diagrams shown in fig. 7.40. Deflections corresponding to the limit load P_l are given with good accuracy by the relation

$$\delta_{pl} = \frac{12\,(1 - a/R)\,R^2}{Et^3} M_p , \qquad (7.51)$$

where E is Young's modulus. They correspond to the small circles in fig. 7.40. We remark that for a/R=0, we obtain the central concentrated load.

On the same basis, Tekinalp [7.44] has studied the built-in plate, uniformly loaded.

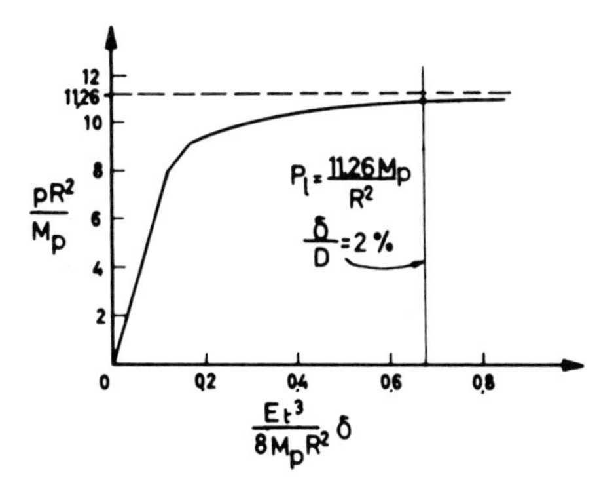

Fig. 7.42.

The load versus deflection diagram is given in fig. 7.42. In both cases, we have indicated points corresponding to $\delta/D = 2\%$ with $\mu = D/t = 50$, $E = 210000\ N/mm^2$ mm^2 $\sigma_Y = 250\ N/mm^2$, that is to

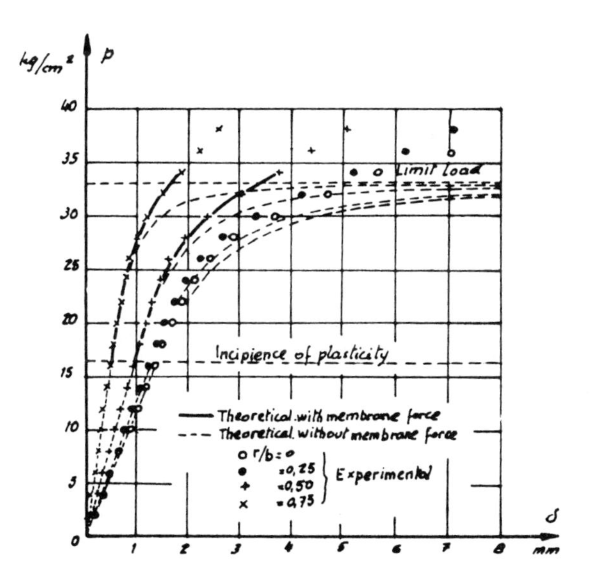

Fig. 7.43. (a) Built-in plate, uniformly loaded ; dials at indicated r/R.

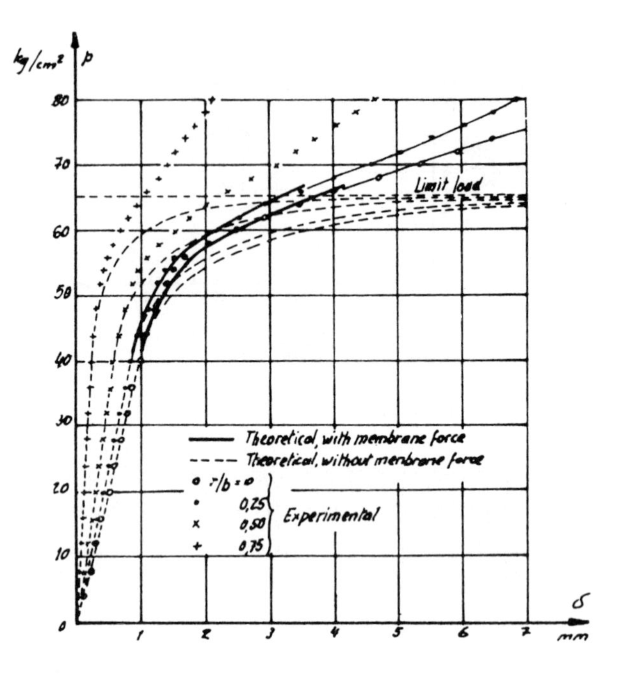

Fig. 7.43. (b) Simply supported plate, uniformly loaded ; dials at indicated r/R.

$$\frac{E\,t^3\,\delta}{8\,M_p\,R^2} = 0.672.$$

It is seen that the corresponding load is very near to P_1 for the built-in plate and remains larger than 0.8 P_1 for the simply supported plate.

The actual situation is better, because favourable membrane action begins already in the elastic-plastic range (*) as found by Oshaki and Murakami [7.10, 7.11], from whose paper we take fig. 7.43. For plates with large holes [7.12] the limit load has less practical meaning (fig. 7.44).

(*) Word hardening may also contribute to some strengthening of the plate.

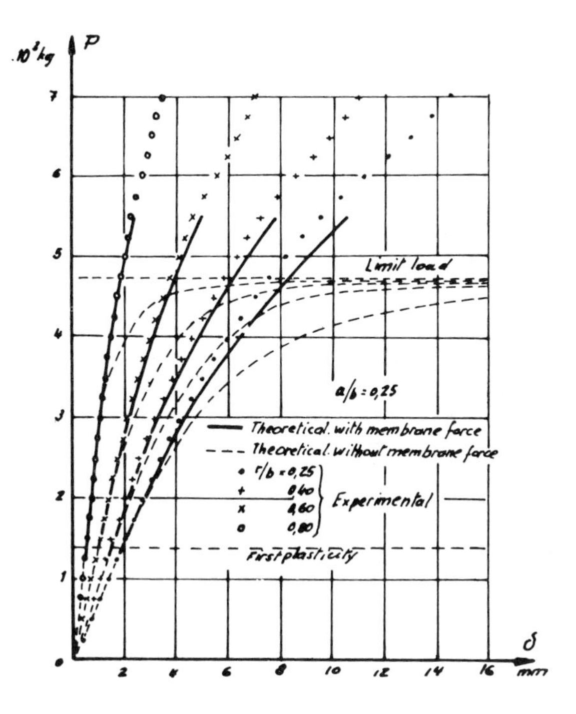

Fig. 7.44. (a)

7.7.3. *Post-limit behaviour.*

The behaviour of the plate at loads larger than the limit load P_l will be called "post-limit" behaviour. It is essentially dependent on the changes of geometry of the structure during the plastic flow. Hence, the analysis can be made assuming rigid perfectly plastic material. Approximation will be poor in the neighbourhood of P_l but will become very good for large deflections. The basic difficulty of the problem is that the collapse mechanism may not remain the same as deflections become larger (see [7.45], for example). Fortunately, for very flat simply supported conical shells considered by Onat [3.10], fig. 7.45, the collapse mechanism is also conical and the problem is greatly simplified. Using the Tresca yield condition, and with the notations $\alpha = a/R$ and $\beta = \delta/t$, the following relation is obtained :

$$P = 2\pi M_p \left[\frac{1}{1-\alpha} + \frac{\beta^2}{3}\left(1 + \alpha - 5\alpha^2 + 3\alpha^3 - \frac{3t^2}{R^2(1-\alpha)} \right) \right]. \tag{7.52}$$

If we let $\beta = 0$ in eq. (7.52) we first obtain the limit load

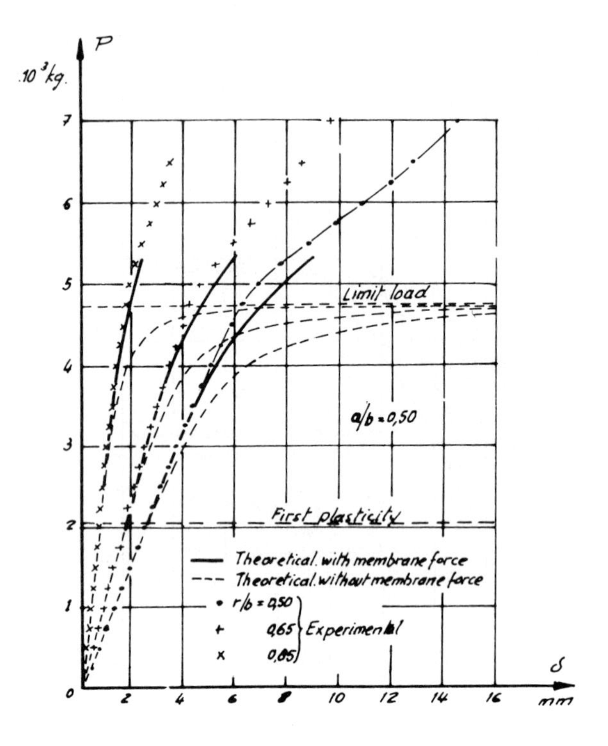

Fig. 7.44. (b)

$$P_l = \frac{2\pi M_p}{1-\alpha} \tag{7.53}$$

of the flat plate. Next, assuming small boss sizes so that $\alpha << 1$, and taking account that $\frac{t}{R(1-\alpha)} << 1$ we have at moderate δ

$$\frac{P}{P_l} = 1 + \frac{1}{3}\left(\frac{\delta}{t}\right)^2, \tag{7.54}$$

where P_l is given by eq. (7.53).

Eq. (7.54) is theoretically valid only for

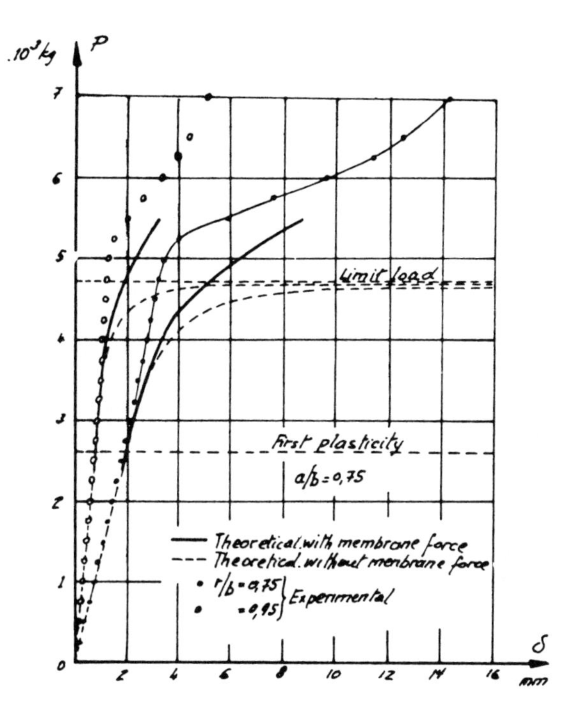

Fig. 7.44 (c). Simply supported plates, central hole with radius a. Total line load P along hole P.

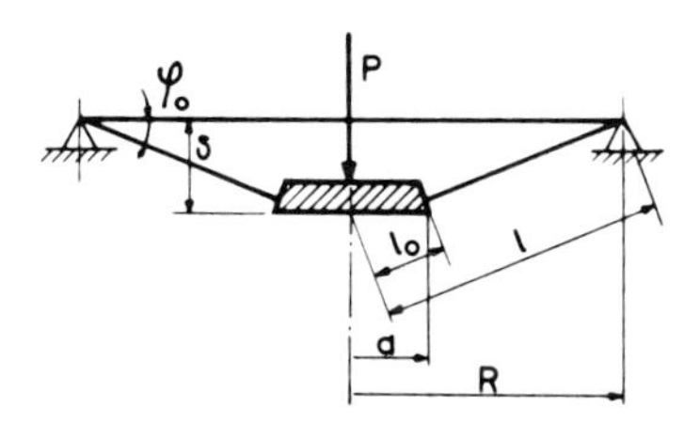

Fig. 7.45.

$$\beta \leq \frac{3}{2} \quad \frac{\alpha}{1-\alpha} \quad \frac{1}{1-\frac{\alpha}{2}(1+\alpha)+\frac{3}{2}(\alpha^4+\alpha^3)} \tag{7.55}$$

but comparison with experimental data in figs. 3.6 and 7.46 shows that it is usable with good accuracy even for larger β.

The built-in circular plate under "circular loading" (see Section 6.3.2) had been considered earlier by Onat and Haythornthwaite [7.5], who made additional assumptions concerning the deformation. They obtained

$$\frac{P}{P_1} = \begin{cases} 1 + \alpha_1 \left(\frac{\delta}{t} \right) + \alpha_2 \left(\frac{\delta}{t} \right)^2 & \text{for} \quad \frac{\delta}{t} \le \frac{1}{2} + \frac{1}{2} \ln \frac{R}{\rho} \\ \beta_1 + \beta_2 \left(\frac{\delta}{t} \right) + \beta_3 \left(\frac{t}{\delta} \right) & \text{for} \quad \frac{\delta}{t} \ge \frac{1}{2} + \frac{1}{2} \ln \frac{R}{\rho} \end{cases} \tag{7.56}$$

where

$$\alpha_1 = \frac{1 + 2 \ln \frac{R}{\rho}}{\left(2 + \ln \frac{R}{\rho}\right)\left(1 + \ln \frac{R}{\rho}\right)},$$

$$\alpha_2 = \frac{2\left(1 + 3 \ln \frac{R}{\rho}\right)}{3\left(2 + \ln \frac{R}{\rho}\right)\left(1 + \ln \frac{R}{\rho}\right)^2},$$

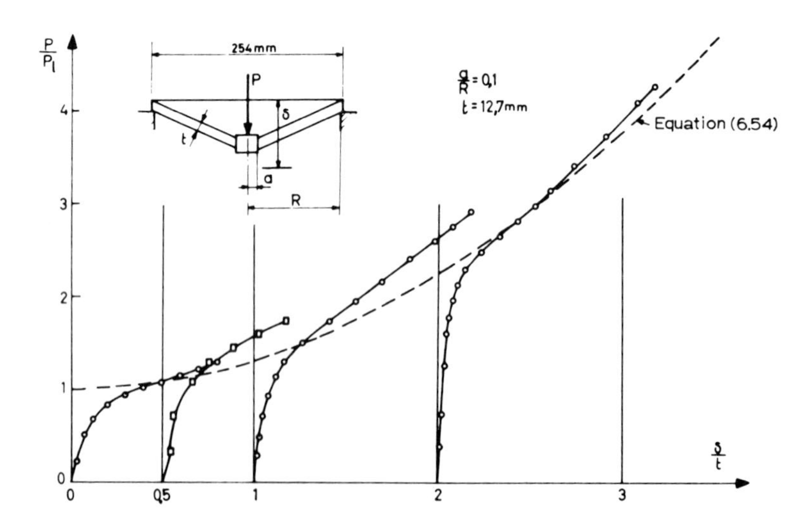

Fig. 7.46.

$$\beta_1 = \frac{3 + \ln \frac{R}{\rho}}{2\left(2 + \ln \frac{R}{\rho}\right)},$$

$$\beta_2 = \frac{2\left(1 + 2\ln \frac{R}{\rho}\right)}{\left(2 + \ln \frac{R}{\rho}\right)\left(1 + \ln \frac{R}{\rho}\right)},$$

$$\beta_3 = \frac{1 + \ln \frac{R}{\rho}}{12\left(2 + \ln \frac{R}{\rho}\right)}.$$

Here, ρ is given by the following equation :

$$1 - \frac{2}{3}\frac{a}{\rho}\left(1 + \ln \frac{R}{\rho}\right) = 0 \qquad \text{if} \qquad \frac{a}{R} \leq 0.606,$$

$$1 - \frac{a^2}{\rho^2}\left(1 + 2\ln \frac{R}{\rho}\right) + \frac{2}{3}\left(1 + \ln \frac{R}{\rho}\right) = 0 \qquad \text{if} \qquad \frac{a}{R} \leq 0.606.$$

In the formulas above, we have,

$$P_1 = 2\,\pi\, M_p \frac{A}{B},$$

where

$$A = 2 + \ln \frac{R}{\rho},$$

$$B = \begin{cases} 1 + \ln \dfrac{R}{\rho} - \dfrac{2a}{3\rho} & \text{if} \qquad \rho \geq a, \\[2ex] \dfrac{1}{2} + \ln \dfrac{R}{a} - \dfrac{\rho^2}{6\,a^2} & \text{if} \qquad \rho \leq a. \end{cases}$$

Curves of $\frac{P}{P_1}$ versus $\frac{\delta}{t}$ for a/R=1 and a/R=1/10 are given in fig. 7.4 and 7.5, respectively, together with experimental results.

7.7.4. *Membrane analysis.*

For large D/t ratios (for example, D/t = 108 in fig. 7.5) there is no difference in the plate behaviour for $P < P_1$ and for $P > P_1$. Membrane forces are of primary importance from the very beginning. Hence, we must consider plastification by stretching without bending of a very thin circular plate hinged at the boundary and loaded on a central circular area with radius a. The plate deforms into a surface of revolution. We assume that the slope φ of the meridian curve of that surface remains small. Consequently, the meridian curvature is given by d^2y/dr^2 and $\sin\varphi \approx \tan\varphi \approx \frac{dy}{dr}$. Equilibrium equations (see, for example, [7.46]) are accordingly simplified. They contain the unknown functions y (r), dy/dr, d^2y/dr^2, the membrane forces N_φ and N_θ, and the load p. It turns out that the plastic regime, $N_\varphi = N_\theta = N_p$, that must exist at the central point by symmetry, is valid in the whole structure. Integration of equilibrium equations then furnishes the deflected shape y (r) and the corresponding total load

$$P = 2\pi N_p \frac{\delta}{\frac{1}{2} + \ln\frac{R}{a}}, \tag{7.57}$$

where $P = \pi a^2 p$ and $\delta = y(0)$. We see that δ is directly proportional to P. Eq. (7.57) was used in the cases D/t = 40 and D/t = 108 in figs. 7.4 and 7.5, respectively.

7.7.5. *Summary of practical results for circular plates.*

For circular plates made of mild steel, the following three cases will be considered :

1. For D/t < 10, the theory does not apply because of the influence of work hardening (and shear forces). The theoretical limit load P_1 remains useful, however, as a conventional failure load corresponding to a relative deflection δ/D smaller than 2%.

2. For D/t ranging from 10 to approximately 30 for simply supported plates and approximately 50 for built-in plates, the load versus deflection relations are eqs. (7.54) and (7.56), respectively, applicable for $P > P_1$. *In these relations, δ is to be regarded as purely*

plastic. For every value of P, the corresponding elastic deflection δ_E should be added to δ. It is given by

$$\delta_E = \frac{P}{16 \pi F} \left[\frac{3+\nu}{1+\nu} R^2 + a^2 \ln \frac{a}{R} - \frac{7+3\nu}{4(1+\nu)} a^2 \right] \qquad (7.58)$$

for the simply supported plate (see [7.46]), and by [7.47]

$$\delta_E = \frac{P}{4} \frac{1}{16 \pi F} \left[4 R^2 + 4 a^2 \ln \frac{R}{a} - 3 a^2 \right], \quad a \neq 0, \qquad (7.59)$$

for the built-in plate. In the two preceding formulas, E is Young's modulus, ν is Poisson's ratio, $F = \frac{Et^2}{12(1-\nu^2)}$, and $P = \pi a^2 p$.

This procedure is not valid in the vicinity of P_l (approximately, for $0.9 P_l < P < 1.1 P_l$) where there is a sharp bent in the P , δ curve due to elastic-plastic behaviour. In the range $0.9 P_l < P < 1.1 P_l$, we shall estimate the deflection to be approximately $2 \delta_E$;

3. For D/t > 30, plates hinged at the boundary, D/t > 50, built-in plates, eq. (7.57) will be used.

Simple supports with no lateral restraints at all will not be considered.

7.7.6. *Strip subjected to concentrated load.*

We consider, after Haythornthwaite and Boyce [7.8], a rectangular plate built-in along two parallel supports, free along the two others and loaded at the centre by a uniformly distributed load over a small circular area of radius a.

Hence, the total load is $P = \pi a^2 p$.

1. *The breadth b of the plate between the free edges is small* [3.7]. We actually deal with a built-in beam with rectangular cross section, loaded at mid span. The limit load is known to be

$$P_l = \frac{8 M_p b}{l}, \qquad (7.60)$$

where l is the span and M_p the plastic moment per unit of breadth. The beam has three sections of plastic hinges A, B, C (fig. 7.47) where the rotation axes must be at the common level of the centroids of the sections in the undeformed state under P_l (impending collapse).

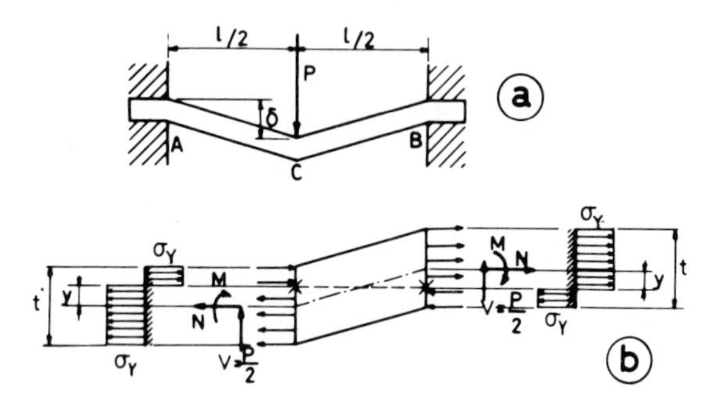

Fig. 7.47.

When the deflection at C is δ, an axial force N appears, which displaces the rotation axes of sections A, B, C. Because of the symmetry, these axes remain at a common level in sections A and B. The beam being rigid except in the sections of hinges, the deflection at C is possible only if the rotation axis in section C is at the same level as in A and B. In the conditions above, an infinitely small increase of deflection Δ δ is possible for every position of the level of the rotation axes. This level is located at the same distance y beneath the centroids of sections A and B and above the centroid of section C, for equilibrium of half a beam in the longitudinal direction (fig. 7.47) to be satisfied.

Therefore,

$$y = \frac{\delta}{2}. \qquad (7.61)$$

On the other hand, the stress distributions shown in fig. 7.47 indicate that

$$\frac{N}{N_p} = \frac{2y}{t} \qquad (7.62)$$

(N and N_p being axial forces per unit breadth). From rotational equilibrium of one half of the beam [fig. 7.47 (b)], we have

$$P = (2M + \delta N)\,\frac{4b}{l}. \qquad (7.63)$$

Dividing the two sides of eq. (7.63) by the corresponding sides of eq. (7.60), we obtain

$$\frac{P}{P_l} = \frac{M}{M_p} + \delta \frac{N}{2 M_p}. \tag{7.64}$$

We set $\frac{M}{M_p} = m$ and $\frac{N}{N_p} = n$. Because $\frac{N_p}{M_p} = \frac{t}{\frac{t^2}{4}} = \frac{4}{t}$, eq. (7.64) becomes

$$\frac{P}{P_l} = m + 2\,\delta\,\frac{n}{t}. \tag{7.65}$$

Taking account of eq. (7.61), eq. (7.62) can be rewritten as

$$n = \frac{\delta}{t}. \tag{7.66}$$

We also know that in a rectangular section with a plastic hinge,

$$|m| + n^2 = 1. \tag{7.67}$$

Using eqs. (7.66) and (7.67) in eq. (7.65) to eliminate m and n, we obtain

$$\frac{P}{P_l} = 1 + \left(\frac{\delta}{t}\right)^2,$$

valid for $\delta \le t$. When $\delta \ge t$, the neutral axes fall outside the cross sections according to eq. (7.61). Bending moments vanish and condition (7.67) is replaced by n=1, m=0. From relation (7.65) we then find $P/P_l = 2\,(\delta/t)$. Hence, the "post-limit" behaviour of a rigid-plastic built-in beam with rectangular cross section, loaded at midspan by a concentrated force, is described by

$$\frac{P}{P_l} = \begin{cases} 1 + \left(\dfrac{\delta}{t}\right)^2 & \text{for } \dfrac{\delta}{t} \le 1, \\ 2\,\dfrac{\delta}{t} & \text{for } \dfrac{\delta}{t} \ge 1 \end{cases}. \tag{7.68}$$

The preceding analysis was refined by Haythornthwaite to account for elastic deformability of the beam [7.48]. The load versus deflection relation cannot be obtained in closed form anymore, but is determined numerically.

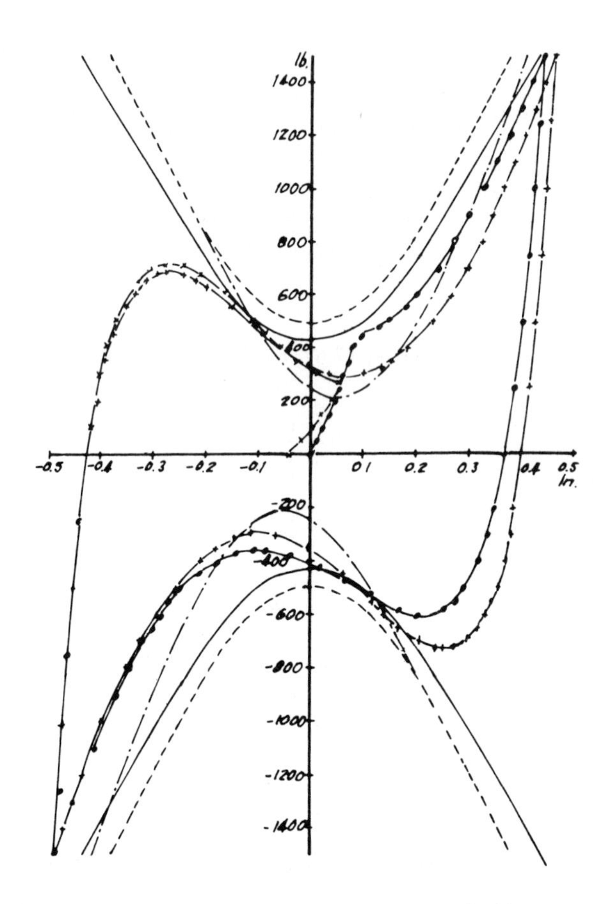

Fig. 7.48. Cyclic loading test of beam (6mm depth).

Theoretical and experimental curves are given in figs. 3.4 and 7.48.

2. *The breadth b of the plate between free edges is very large* (strip problem).The plate will exhibit a local collapse mechanism with axial symmetry about the line of action of the concentrated load. As noted in Section 7.6, the limit value is $P = 2\pi M_p$.

3. *The breadth b of the plate between free edges is "intermediate"* (0.5l, for example). The mechanism is most likely a combination of the preceding mechanisms. It

has been found [7.8] that the range of that combined mechanism is confined to b/l ≤ 1. The purely local mechanism already is a good approximation for b=l.

Experiments on six mild steel plates with b/l ratios of 0.1, 0.3, 0.5, 0.888, and 1.8 have resulted in the diagrams in fig. 7.11. The conclusions are :

1. For the first loading sequence, the simple beam mechanism is most adequate, up to b/l = 0.888, as regards both the limit load ($P_l = 8\ M_p\ b/l$) and the post-limit load versus deflection curve, provided that elastic deflections are added. For b/l = 1.8, the local mechanism should be considered ;

2. At the first reversal in loading sign, the theoretical limit load recovers its significance of real collapse load, at which large deformations occur for very small load increments.

7.8. Shakedown problems.

7.8.1. *Introduction.*

When loading is variable, the analytical evaluation of the collapse multiplier is much more difficult because of the extensive number of yield conditions (in statical approach) or mechanisms (in kinematical approach) to consider. For more information on analytical solutions, the reader is referred to the paper [7.49] and the survey course [4.24] of Köning.

For plates and shells, it is important to remark that the way to obtain the yield condition is generalized variables from one in local stresses is not as straightforward as in limit analysis and requires the introduction of the new following concepts :

1. *elastic locus* ; by elastic locus for a given section, we mean a domain in the space of generalized variables where the response at any point of the section remains perfectly elastic ;

2. *pseudo-residual stresses* ; by such stresses, we mean stresses for which **Q** vanishes at a considered section.

In such a formulation, shakedown analysis divides into two stages : (1) evaluation of an appropriate residual stress field ; (2) construction of the necessary elastic loci.

General properties of the elastic loci where established by Köning in [4.26]. Concerning the application of these concepts to shakedown theory of plates, the reader may consult also [7.49].

7.8.2. *Finite element method.*

As said before, the set of problems analytically solvable is significantly reduced by the number of equations to consider.

So, numerical methods seem to be promising formulations for shakedown problems. The purposes of this Section are to show feasibility of the finite element method and to turn reader'attention to the reduction of the collapse multiplier in variable loading with respect to proportional loading.

Some problems have been investigated by Morelle and Nguyen Dang Hung [6.48], [6.12]. These authors use equilibrium finite elements (for theoretical foundation of the method, see Section 6.8).

First, we consider a simply supported circular sandwich plate of radius a and loaded by a uniform pressure q(t) varying with time. The plate material obeys Tresca's yield condition. The load q(t) varies between the values $-p_\alpha$ and α. The theoretical solution is given in [7.49] (see also problems 4.7.2 and 4.7.3.). If

$$p \leq \frac{7 - 3\,\nu}{3\,(3 + \nu)},$$

the solution is that of limit analysis (ratchet mechanism)

$$\alpha_s = \frac{6\,M_p}{a^2}.$$

Otherwise, collapse occurs by plastic fatigue at the centre of the plate, and we have :

$$\alpha_s = \frac{1}{1 + p}\,\frac{32}{3 + \nu}\,\frac{M_p}{a^2}.$$

Of course, proportional loading is a more optimistic assumption. The comparison between the numerical results and the theoretical values is excellent. For example, with 15 finite elements, the shakedown multiplier is found equal to $\alpha = 0.80834$ instead of the theoretical value $\alpha_{th} = 0.80808$.

A similar behaviour may be observed for the plate with hole, numerically computed by Morelle and Nguyen Dang Hung [6.48] (see fig. 7.49).

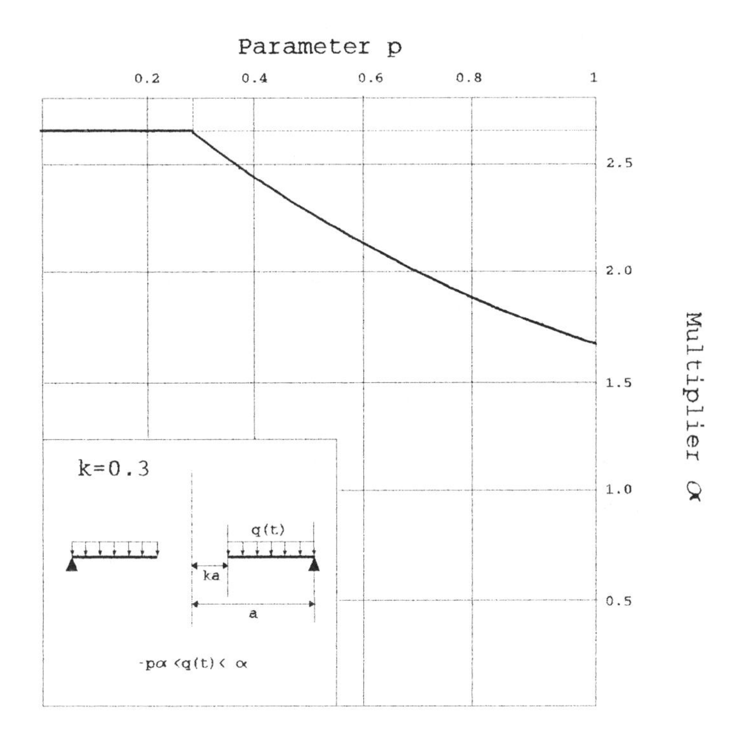

Fig. 7.49. Supported circular plate with circular hole subjected to variable loading. Variation of γ in function of loading. [6.48]

A more accurate numerical approach based on a multi-layer equilibrium finite element related to a special discretization of the pseudo-residual stress field is presented in [7.50] and [6.90].

7.9. Problems.

7.9.1. A circular isotropic plate is loaded as shown in fig. 7.50. Find the formula relating the total loads P_1 and P_2 for collapse, on the basis of a statically admissible moment field and a corresponding mechanism. The yield condition of Tresca should be used. Verify that the theorem of Section 7.4.7 applies.

Answer :

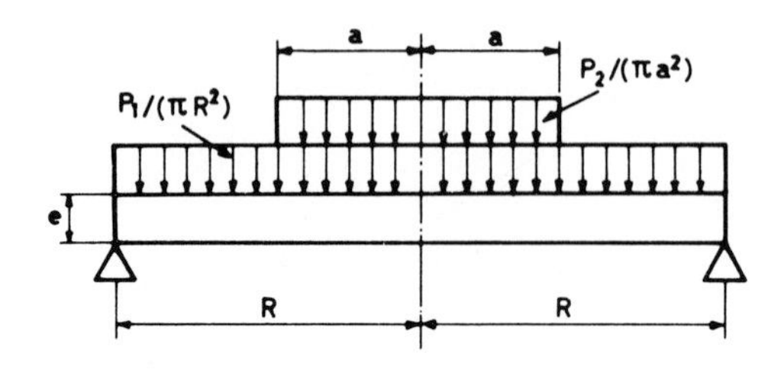

Fig. 7.50.

$$P_1 + P_2\left(3 - \frac{2a}{R}\right) = 6\,\pi\,M_p.$$

7.9.2. . Find the analytic expressions of the moment field at collapse of a circular isotropic, built-in plate with a central concentrated load P (Tresca condition). Show that a conical yield mechanism with a hinge circle at the edge corresponds to this moment field. By extending the moment field in a statically admissible manner, show that $P = 2\,\pi\,M_p$ is the limit load for a built-in edge of arbitrary shape.

7.9.3. Find bounds for the limit value of the uniform pressure acting on a circular isotropic plate, simply supported and obeying the yield condition of von Mises. *Hint :*

Lower bound - use $M_r = cp\,(R^2 - r^2)$, $M_\theta = cp\,(R^2 - \gamma r^2)$, and adjust the parameters c and γ.

Upper bound - use the result of Problem 3.8.1 to obtain the dissipation as a function of $\dot{w}$ and its derivatives. Use the field

$$\dot{w} = R - \frac{\rho}{2} - \frac{r^2}{2\rho} \qquad \text{for} \qquad 0 \le r \le \rho,$$

$$\dot{w} = R - r \qquad \text{for} \qquad \rho \le r < R,$$

and adjust ρ.

7.9.4. Prove relation 7.28, fig. 7.21.

7.9.5. Find the lower bound $p_{-} = \dfrac{20.6 M_p}{l^2}$ in relation (7.47). *Hint :* use the field

$$m_x = C\,(1 - x^2),$$
$$m_y = C\,(1 - y^2),$$
$$m_{xy} = (2\,C - 3\,p^*)\,xy,$$

with $p^* = \dfrac{pl^2}{24\,M_p}$, and adjust the parameter C.

7.9.6. Determine the upper bound $p_{+} = \dfrac{25.65\,M_p}{l^2}$ for the problem of fig. 7.36.

7.9.7. Show that relation 7.28 gives the exact limit load. *Hint* : use a yield mechanism of the form of a truncated cone. To compute the rate of dissipation, see the fan mechanisms of Section 9.4.5.

References.

[7.1] R.H. LANCE and E.T. ONAT, "A Comparison of Experiments and Theory in the Plastic Bending of Plates", J. Mech. Phys. Solids, **10**, 301, 1962.

[7.2] R.M. HAYTHORNTHWAITE and E.T. ONAT, "The Load-carrying Capacity of Initially Flat Circular Steel Plates under Reversed Loading", J. Aer. Sci., **22** : 12, December 1955.

[7.3] R.M. HAYTHORNTHWAITE, "The Deflection of Plates in the Elastic-Plastic Range", Proc. 2nd U.S. Nat. Cong. Appl. Mech., A.S.M.E., Ann Arbor, Mich., 1954.

[7.4] J. FOULKES and E.T. ONAT, "Tests on the Behaviour of Circular Plates under Transverse Load", Brown Univ. Tech. Rep. 00R-3172/3, May 1955.

[7.5] R.M. HAYTHORNTHWAITE and E.T. ONAT, "The Load-carrying Capacity of Circular Plates at Large Deflections", J. Appl. Mech. , **23** : 49, 1956.

[7.6] R.M. COOPER and G.A. SHIFRIN, "An Experiment on Circular Plates in the Plastic Range", Proc. 2nd U.S. Nat. Cong. Appl. Mech., A.S.M.E., Ann Arbor, Mich., 1954.

[7.7] M.A. SAVE, "Vérification expérimentale de l'analyse plastique des plaques et des coques en acier doux", (Experimental verification of plastic limit analysis of mild steel plates and shells), C.R.I.F. Report, M.T.21, Fabrimetal, 21, rue des Drapiers, Brussels, February 1966.

[7.8] R.M. HAYTHORNTHWAITE and W.C. BOYCE, "The Load-carrying Capacity of Wide Beams at Finite Deflection", Proc. 3rd U.S. Nat. Cong. Appl. Mech., 541-550, June 1958.

[7.9] L. DEL RIO CABRERA, "Analyse limite des plaques rectangulaires", Thèse de maîtrise en sciences appliquées, Faculté Polytechnique de Mons, 1970.

[7.10] Y. OHASHI and S. MURAKAMI, "The Elasto-plastic Bending of a Clamped Thin Circular Plate", Proc. 11th Int. Cong. Appl. Mech., Munich 1964, 212-221, H. Görtler, ed., Springer, Berlin, 1966.

[7.11] Y. OHASHI and S. MURAKAMI, "Large Deflection in Elasto-plastic Bending of a Simply Supported Circular Plate under a Uniform Load", Trans. A.S.M.E., ser. E, J. Appl. Mech., **33** : 4, 866, 1966.

[7.12] Y. OHASHI and S. MURAKAMI, "Elasto-platic Bending of an Annular Plate at Large Deflection", Ing. Archiv., **35** : 5, 340, 1967.

[7.13] H.G. HOPKINS and W. PRAGER, "The Load-carrying Capacity of Circular Plates", J. Mech. Phys. Solids, **2** : 1, 1953.

[7.14] C.A. ANDERSON and R.T. SHIELD, "On the Validity of the Plastic Theory of Structures for Collapse under Highly Localized Loading", Trans. A.S.M.E., J. Appl. Mech., **23** : 629, September 1966.

[7.15] H.G. HOPKINS and A.J. WANG, "Load-carrying Capacities for Circular Plates of Perfectly-plastic Material with Arbitraty Yield Conditions", J. Mech. Phys.

Solids, **3** : 117, 1954.

[7.16] V.V. SOKOLOVSKI, "Elasto-plastic Bending of Circular and Annular Plates", Brown Univ. Tech. Rep., 3, November 1955.

[7.17] A. SAWCZUK and T. JAEGER, "Grenztragfähigkeits-Theorie der Platten", 522, Springer, 1963.

[7.18] D.C. DRUCKER and H.G. HOPKINS, "Combined Concentrated and Distributed Load on Ideally Plastic Circular Plates", Proc. 2nd U.S. Nat. Cong. Appl. Mech., A.S.M.E., Ann Arbor, Mich., 1954.

[7.19] A. SAWCZUK, "Some Problems of Load-carrying Capacities of Orthotropic and Nonhomogeneous Plates", Zakblad, Mech. Osrod. Ciagl. Polsk. Akad. Nauk, **VIII** : 4, Warsaw, 1956.

[7.20] W. OLSZAK and A. SAWCZUK, "Théorie de la capacité portante des constructions non-homogènes et orthotropes", Suppl. to annales de l'Inst. Tech. du Bat. et des Trav. Pub., 149, May 1960.

[7.21] A. SAWCZUK, "Linear Theory of Plasticity of Anisotropic Bodies and its Applications to Problems of Limit Analysis", Zadblad, Mech. Osrod. Ciagl. Polsk. Akad. Nauk, **XI** : 5, Warsaw, 1959

[7.22] J. MARKOWITZ and L.W. HU, "Plastic Analysis of Orthotropic Circular Plates", Proc. A.S.C.E., J. Eng. Mech. Div., **90** : EM5, 251, October 1965.

[7.23] D.C. DRUCKER, "Plastic Design Methods, Advantages and Limitations", Trans. Soc. Nav. Arch. Eng., **65** : 172, 1958.

[7.24] H.E. SHULL and L.W. HU, "Load-carrying Capacities of Simply Supported Rectangular Plates", Trans. A.S.M.E., J. Appl. Mech., **30** : 617, December 1963.

[7.25] M.A. SAVE, "Limit Analysis of Plates and Shells : Research over two Decades", J. Struct. Mech., **13** : 343-370, 1985.

[7.26] M.A. SAVE, "Atlas of Limit Loads for Metal Plates, Shells and Disks", Elsevier Science Pub. 1995.

[7.27] J.S. KAO, T. MURA and S.L. LEE, "Limit Analysis of Orthotropic Plates", J. Mech. Phys. Solids, **11** : 429, November 1963.

[7.28] A. SAWCZUK and S. KALISZKY, "On the Limit Analysis of Plates Supported by a Nonhomogeneous Plastic Subgrade under Rotational Symmetry Conditions", Acta Technica Ac. Sc. Hung, Tomus 48, Fasc. 1-2, 1964.

[7.29] A. SAWCZUK and M. DUSZEK, "A Note on the Interaction of Shear and Bending in Plastic Plates", Arch. Mech. Stosowanej, **15** : 411, 1963.

[7.30] U.V. NEMIROVSKI, "Carrying Capacity of Ribbed Reinforced Circular Plates" (in Russian), Izv. Nauk. U.S.S.R., Mekh. Mach., 2, 163, 1962.

[7.31] W. OLSZAK and W. URBANOWSKI, "Plastic Nonhomogeneity. A Survey of Theoretical and Experimental Research", Nonhomogeneity in Elasticity and Plasticity, Proc. I.U.T.A.M. Symp. Warsaw, 1958, W. Clszak ed., 259-298, Pergamon Press, London, 1959.

[7.32] D.C.A. KOOPMAN and R.H. LANCE, "On Linear Programming and Plastic Analysis", J. Mech. Phys. Solids, **13** : 2, April 1965.

[7.33] W. PRAGER, "Programmation linéaire en théorie des construction" (Linear programming in structural analysis) Mémoires du C.E.R.E.S., **3** : 33, Liège 1962.

[7.34] C. GAVARINI, "I teoremi fondamentali del calcolo a rottura e la dualità in programmazione linear" (Fundamental theorems of limit analysis and duality in linear programming), Ingegneria Civile, **18** , 1966.

[7.35] M.Z. COHN and G. MAIER, editors, "Engineering Plasticity by Mathematical Programming", Pergamon Press, 1979.

[7.36] M.P. RANAWERA and F.A. LECKIE, "Finite Element Techniques in Solid Mechanics : Bound Methods in Limit Analysis", University of Southhampton, Department of Civil Engineering, 15-17, April 1970.

[7.37] E. CHRISTIANSEN and S. LARSER, "Computation in Limit Analysis for Plastic Plates", S.I.A.M., Int. J. for Numerical Meth. in Engineering, n°19 : 169-184, 1983.

[7.38] P. VILLAGGIO, "Analisi limite di piastre sottili appoggiate plastico rigide" (Limit Analysis of Thin Simply Supported Rigid-Plastic Plates), Giorn. Genio Civile, **103** : 3, 133, March 1965.

[7.39] W. SCHUMANN, "On Isoperimetric Inequalities in Plasticity", Quart. Appl. Math., **16** : 3, 300, 1958.

[7.40] W. SCHUMANN, "On Limit Analysis of Plates", Quart. Appl. Math., **16** : 1, 61, 1958.

[7.41] R.M. HAYTHORNTHWAITE and R.T. SHIELD, "A Note on the Deformable Region in a Rigid-Plastic Structure", J. Mech. Phys. Solids, **6** : 127, 1958.

[7.42] M. ZAID, "On the Carrying Capacity of Plates of Arbitrary Shape and Variable Fixity under a Concentrated Load", J. Appl. Mech., **25** : 4, 598, 1958.

[7.43] D.O. LAMBLIN, "Analyse et dimensionnement plastique de coût minimum de plaques circulaires", Thèse de doctorat en Sciences appliquées, Faculté Polytechnique de Mons, Belgium, 1975.

[7.44] B. TEKINALP, "Elastic-plastic Bending of a Built-in Circular Plate under a Uniformly Distributed Load", J. Mech. Phys. Solids, **5** : 135, 1957.

[7.45] R.M. HAYTHORNTHWAITE, "Mode Change During the Plastic Collapse of Beams and Plates", Developments in Mechanics, Plenum Press, New York, 1961.

[7.46] S. TIMOSHENKO and WOINOWSKY-KRIEGER, Theory of Plates and Shells, Mc Graw-Hill, New York, 1959.

[7.47] R.J. ROARK, Formulas for Stress and Strain, 3rd ed., Mc Graw-Hill, New York, 1954.

[7.48] R.M. HAYTHORNTHWAITE, "Plastic Behaviour of Beams with Elastic End Constraints", Proc. 9th Int. Cong. Appl. Mech., **VIII** : 59, Brussels, 1956.

[7.49] J.A. KÖNING, "Shakedown Theory of Plates", Archivum Mechaniki Stosowanej, **5** : n°21, 623-636, 1969.

[7.50] P. MORELLE and G. FONDER, “Shakedown and Limit Analysis of Shells - A Variational and Numerical Approach”, Proc. Int. Symp. IASS on “Practical Aspects in the Computation of Shell and Spatial Structures”, Leuven, Belgium, 1986.

[7.51] G. CERADINI, C. GAVARINI and M. PETRANGELI, “Steel Orthotropic Plates under Alternate Loads”, J. Struct. Div., Proc. ASCE, **101** : 2015-2026, 1975.

8. Metal shells.

8.1. Introduction.

Because the median surface of a shell exhibits at least one nonvanishing principal curvature, applied forces can be balanced exclusively, from the very beginning of the loading process, by forces acting at every point of the median surface in the corresponding tangent plane. The shell is then said to act as a membrane. This situation occurs when the shell has a very small flexural rigidity, supports with reactions in the tangent planes to the shell median surface at the boundary, and discontinuities in, neither geometry, (curvature, thickness) nor loading (concentrated loads). When the preceding conditions are not satisfied, flexural stresses arise in addition to membrane stresses.

Because shells frequently have a large slenderness ratio (span-to-thickness), the conservation of material normals to the median surface is well verified and transverse shear forces may therefore be classified as "reactions". On the other hand, large slenderness ratios may result in appreciable elastic and elastic-plastic deflections and particular attention must therefore be given to the danger of buckling. This problem, however, is beyond the scope of this book. The reader is referred to both theoretical ([8.1, 8.2, 8.3]) and experimental ([8.4 to 8.8]) papers on the subject. The dangers of fatigue and brittle fracture ([8.9, 8.10, 8.11]), are also very important, especially in welded shells with holes and nozzles. If, however, ductility is preserved by suitable precautions, like stress-relieving or annealing, plastic flow will level out stress concentrations due to various discontinuities. These stress concentrations in the elastic range can be calculated rather easily in many cases [8.12, 8.13, 8.14] but are not taken into account by building codes that implicitly rely on the ductile behaviour of the material. Further development of limit analysis should provide a stronger basis for the specifications of the building codes [8.15].

8.2. Experiments on metal shells.

8.2.1. *Conical shells.*

Figs. 3.6 and 7.46 show typical load versus deflections diagrams of the conical shells with slenderness ratios of $\mu = 10/0.5 = 20$ tested by Onat [3.9]. We recall that P_{1o} is the limit load for the plate, regarded as a cone with vanishing height δ, and that the theoretical limit load for any cone corresponds to the intersection of the dashed curve (representing eq. (7.54)) with the parallel to the load axis at the distance δ/t from the origin. By inspection of these figures we conclude that :

1. For shells subjected to tension (fig. 7.46) the load versus deflection curve tends to exhibit a sharp bent in the neighbourhood of the limit load, especially for the cone with

the largest δ/t ($\delta/t = 2$). The limit state si nevertheless stable and changes of geometry require the load for continued plastic flow to increase with the deflection.

2. For shells in compression (fig. 3.6) at the limit load (or even at a lower load as for shell with $\delta/t = 3$), large deflections occur at constant (or decreasing) load. The limit load then is a true failure load.

8.2.2. *Cylindrical shells.*

Cylindrical shells of mild steel with external annular stiffeners were subjected to external hydrostatic pressure by M.E. Lunchick [8.16]. Without referring to load versus deflection diagrams, he gives experimental values of a "collapse pressure". Because the annular stiffeners did not participate in the collapse, the shell can be regarded as built-in at the successive annuli, and subjected to external pressure and axial force due to the pressure on the ends. In table 8.1 the theoretical limit pressure given by eq. (8.27) is compared with the experimental collapse pressure.

We see that simple plastic theory, which neglects geometry changes, is quite satisfactory.

An experiment similar to test in table 8.1, but with a cylinder machined from a solid piece rather than welded, gave an experimental collapse load of 0.97 times the limit load [8.17]. It is worth noting that the slenderness $\mu = \frac{2R}{t}$ of the test cylinders varied from 113 (test 1) to 210 (test 6).

Table 8.1.

Test number	shell parameter ω	Spacing of stiffeners $l'/2R$	Experimental collapse pressure / Theoretical limit pressure
1	0.932	0.114	0.914
2	0.857	0.095	0.905
3	0.857	0.095	0.975
4	1.388	0.110	1.029
5	1.951	0.202	0.971
6	1.290	0.109	1.026
7	2.077	0.168	1.063

Note : the notations used in this table are as follows : l', distance between median planes of adjacent stiffeners ; R, mean radius of shell ; l, clear distance between stiffeners ; t, shell thickness, $\omega^2 = l^2/2tR$.

Aluminium cylinders with internal annular stiffeners were tested by Dehart and Basdekas [8.18] also under hydrostatic pressure. Though the main purpose of the experiments was to determine spacing and cross section of the stiffeners to avoid buckling of the shell and collapse of the stiffeners, the collapse pressure can be used for comparison with the limit pressure [eq. (8.27)]. We obtain table 8.2. The slenderness was $\mu = 17.1$, except for test 3G where $\mu = 18.3$.

Table 8.2.

Test number	shell parameter ω	Spacing of stiffeners $l'/2R$	Experimental collapse pressure / Theoretical limit pressure
3F	4.36	1.060	0.905
3H	1.64	0.398	0.985
3G	3.40	0.790	0.87
3J	3.275	0.795	0.92

Very careful experiments by Demir and Drucker [8.19] on steel and aluminium cylinders subjected to a ring of force with total magnitude F show a sharp bent in the load versus deflection diagram, followed by a flat part. Hence, changes of geometry do not interfere with the collapse mechanism by an appreciable amount. Experimental limit loads,

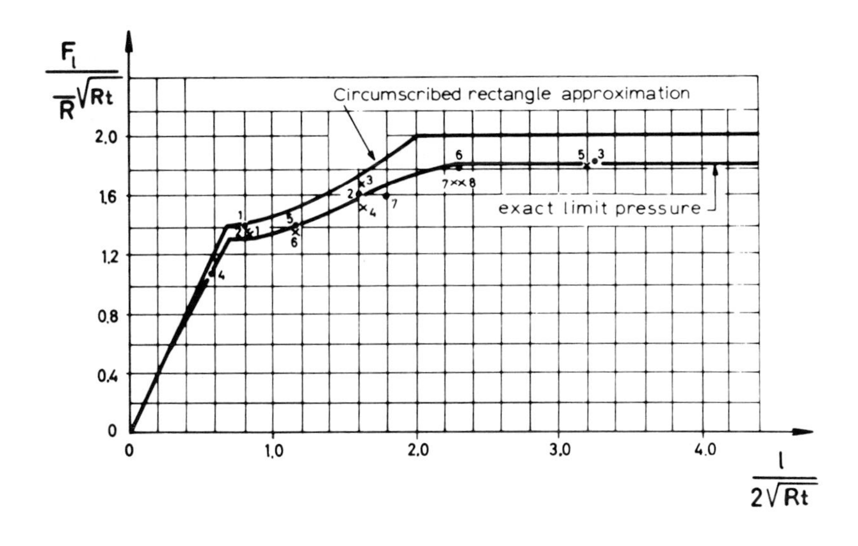

Fig. 8.2.

corresponding approximately to the end of the sharp bent in the load versus deflection diagram, agree very well with theoretical prediction, as can be seen in fig. 8.1.

Twelve mild-steel cylinders, built-in at the bottom end and closed by a welded flat plate at the upper end (fig. 8.2) were subjected by Save [8.20] to internal pressure up to very large plastic deformations. Typical load versus deflection diagrams and deformed shapes are given in fig. 8.3 and 8.4. The predicted collapse mechanisms did actually occur and experimental limit pressures (corresponding to the end of the elastic-plastic bent in the load versus deflection diagrams, points marked ^) differ from the theoretical values by less than 7% (but are all lower that the theoretical values).

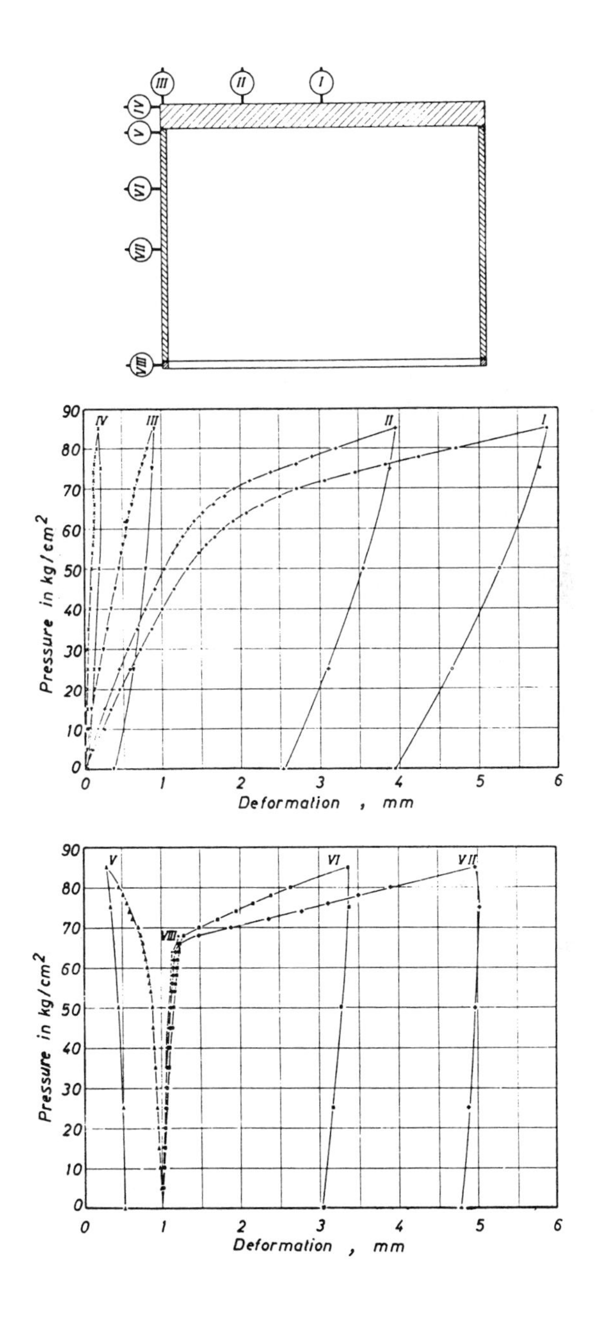

Fig. 8.3. Background plate (flat bottom).

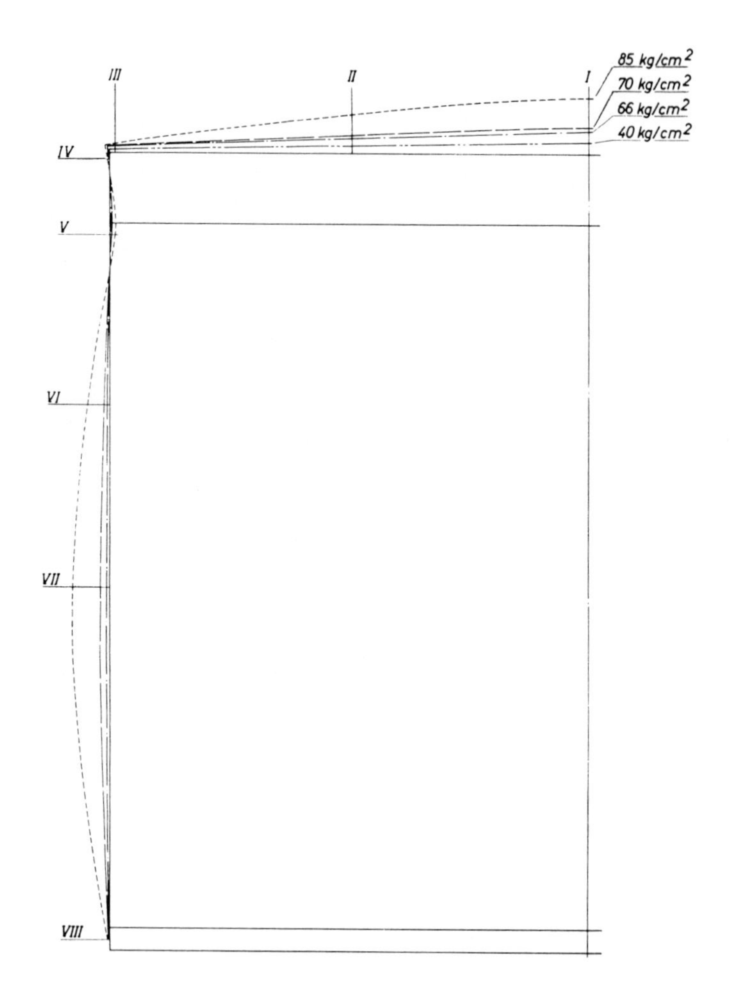

Fig.. 8.4. Plastic deformation pattern : cylindrical vessel with flat head.

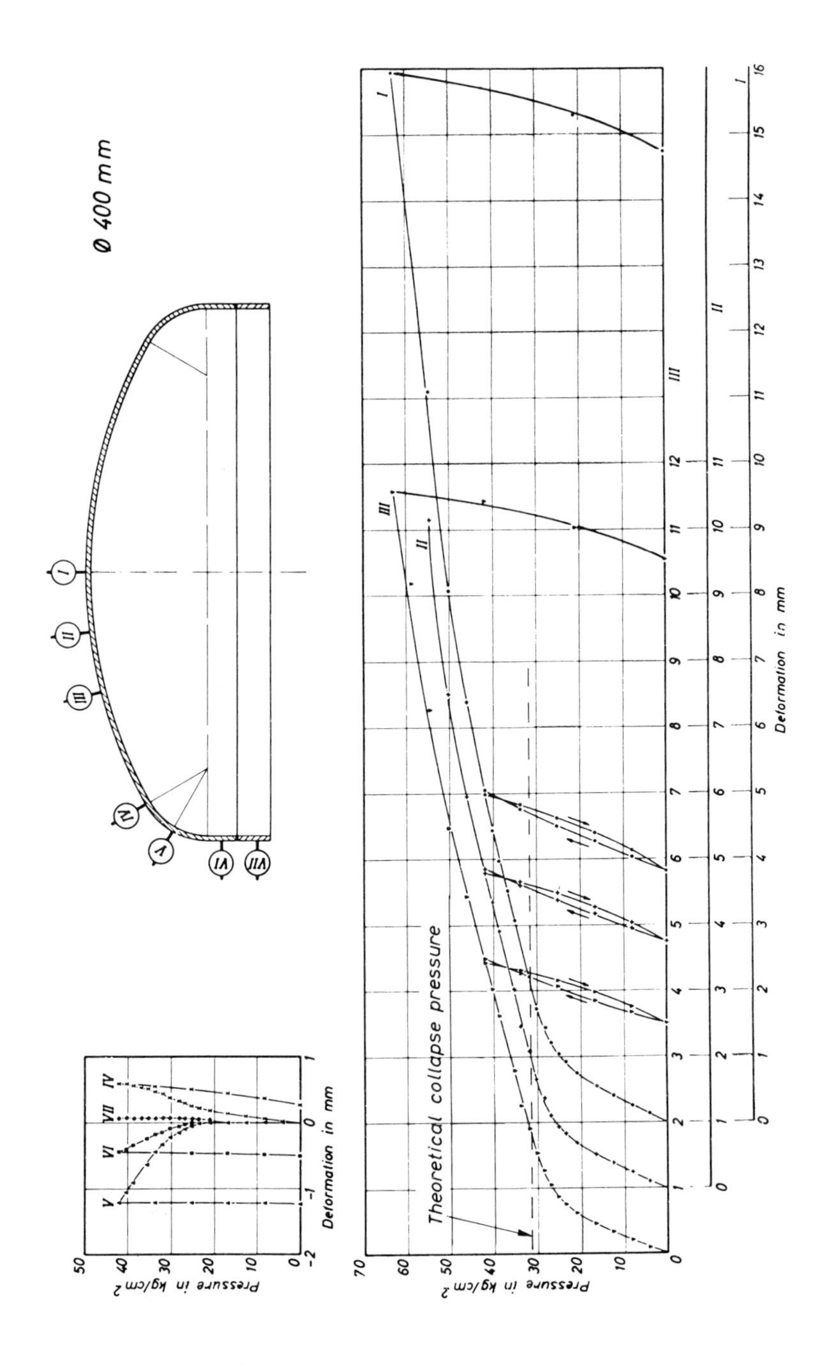
Ø 400 mm
Theoretical collapse pressure
Pressure in kg/cm²
Deformation in mm

Fig. 8.5.

8.2.3. *Torispherical and toriconical pressure vessel heads.*

Two toriconical and nine torispherical heads were each welded to a rigid annulus and subjected to internal pressure by Save [8.20].

These test shells were annealed mild-steel industrial heads made by deep-drawing. Typical load versus deflection diagrams and deformed shapes are shown in fig. 8.5 and 8.6. The predicted collapse mechanisms actually occur without being appreciably influenced by changes of geometry. Deflection at the theoretical limit pressure remains less than 1% of the diameter. Experimental values of the limit pressure are larger than theoretical values by 0 to approximately 10%. The expected slip lines were clearly visible (fig. 8.7).

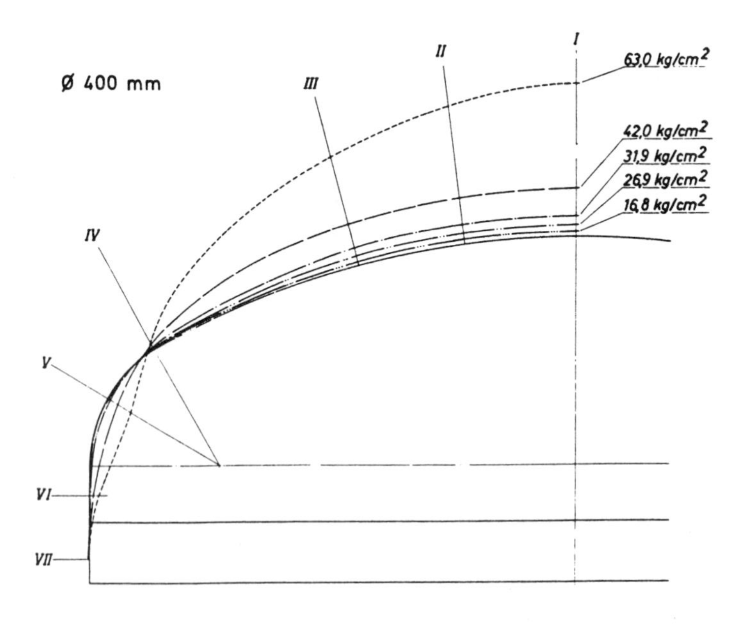

Fig. 8.6. Torispherical head deformations.

Fig. 8.7.

8.2.4. *Conclusions.*

From the preceding review of experimental results, it can be concluded that shells are far less "degenerate" problems (that is, exhibiting a lesser influence of geometry change) than plates, and that limit loads actually correspond to strong modifications of the shell behaviour, resulting in very large permanent deflections. Hence, limit loads can be regarded as real failure loads when unrestrained plastic flow is to be avoided.

8.3. Circular cylindrical shells axisymmetrically loaded.

8.3.1. Introduction.

We refer to the cylindrical coordinate system x, θ, r of fig. 8.8, with the origin at one end of the shell with length l. Because of the symmetry of revolution, circumferential displacements v vanish, whereas longitudinal displacements u and radial displacement w (positive inwards) are functions of x only. As noted in Section 5.14, the curvature rate $\dot{\kappa}_\theta$ vanishes and, consequently, the generalized stresses are M_x, N_x and N_θ (fig. 8.8). The corresponding generalized strain rates are

$$\dot{\kappa}_x = -\frac{d^2 \dot{w}}{d x^2},$$

$$\dot{\varepsilon}_x = \frac{d \dot{u}}{d x},$$

$$\dot{\varepsilon}_\theta = -\frac{\dot{w}}{R}, \tag{8.1}$$

The rate of dissipation per unit median surface is

$$D = -M_x \frac{d^2 \dot{w}}{d x^2} + N_x \frac{d \dot{u}}{d x} - N_\theta \frac{\dot{w}}{R}. \tag{8.2}$$

Using "the reduced stresses"

$$n_x = \frac{N_x}{N_p},$$

$$m_x = \frac{M_x}{M_p},$$

$$n_\theta = \frac{N_\theta}{N_p}, \tag{8.3}$$

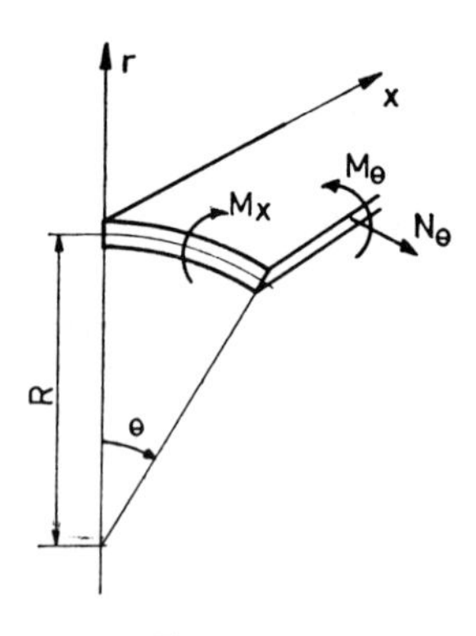

Fig. 8.8.

where $N_p = \sigma_Y t$ and $M_p = \sigma_Y t^2/4$, relation (8.2) becomes

$$D = \sigma_Y \left[m_x \left(-\frac{t^2}{4} \frac{d^2 \dot{w}}{d x^2} \right) + n_x \, t \, \frac{d \dot{u}}{d x} + n_\theta \left(-t \frac{\dot{w}}{R} \right) \right]. \tag{8.4}$$

Hence, the generalized strain rates corresponding to the reduced stresses are

$$\begin{aligned} \dot{\Phi}_x &= -\frac{t^2}{4} \frac{d^2 \dot{w}}{d x^2}, \\ \dot{\lambda}_x &= t \frac{d \dot{u}}{d x}, \\ \dot{\lambda}_\theta &= -t \frac{\dot{w}}{R} \end{aligned} \tag{8.5}$$

and relation (8.4) is written as

$$D = \sigma_Y \left(m_x \, \dot{\Phi}_x + n_x \dot{\lambda}_x + n_\theta \, \dot{\lambda}_\theta \right). \tag{8.6}$$

When the shell is subjected to an external radial pressure p (that depends on x only), the equilibrium equations of a shell element are (fig. 8.8)

$$\begin{aligned} &\text{(a)} \qquad R \frac{dV}{dx} + N_\theta + Rp = 0, \\ &\text{(b)} \qquad \frac{dM_x}{dx} = V. \end{aligned} \tag{8.7}$$

Elimination of V from these two equations yields

$$\frac{d^2 M_x}{dx^2} + \frac{M_\theta}{R} + p = 0 . \tag{8.8}$$

Defining a "reduced pressure"

$$p^* \equiv \frac{pR}{\sigma_Y \, t}, \tag{8.9}$$

Eq. (8.8) can be rewritten

$$\frac{tR}{4}\frac{d^2 m_x}{dx^2} + n_\theta + p^* = 0. \tag{8.10}$$

Except where otherwise stated, we shall use in the following the linearized Tresca yield surface represented in fig. 5.30, with eqs. (5.89) for the various planes.

8.3.2. *Built-in cylindrical shell under external uniform pressure p.*

As noted in Section 8.2, the present case represents fairly accurately the behaviour of a part of a shell between rigid stiffening rings. The part of the shell is built-in at the ring to the extent that strains must vanish there, but the shell is subjected to an axial compressive force.

$$2\,\pi\,R\,N_x = p\,\pi\,R^2. \tag{8.11}$$

Hence, there is a constant longitudinal normal stress

$$n_x = \frac{-p\,\pi\,R^2}{2\,\pi\,R\,\sigma_Y\,t} = -\frac{p^*}{2}. \tag{8.12}$$

Considering first very short shells, we can reasonably assume that they will collapse by axial compression, with $n_x = -1$. For sufficiently short shells, we thus may expect $-1 \le n_x \le -\frac{1}{2}$. Because the shell will certainly contract circumferentially, we have $n_\theta < 0$ and $\dot{\lambda}_\theta < 0$. Hence, the stress point is likely to lie on a "stress profile" such as the segment ab in fig. 5.30, but located on face I' symmetrical of face I with respect to the origin. The equation of face I' is

$$n_\theta = -1. \tag{8.13}$$

The normality law gives :

$$\dot{\lambda}_\theta \equiv -t\,\frac{\dot{w}}{R} = -1\,,$$

$$\dot{\lambda}_x \equiv t\,\frac{d\dot{u}}{dx} = 0\,,$$

$$\dot{\Phi}_x \equiv -\,\frac{t^2}{4}\,\frac{d^2\dot{w}}{dx^2} = 0\,,$$

wherewith we obtain

$$\dot{w} = C_1\,x + C_2\,,$$

$$\dot{u} = C_3\,. \tag{8.14}$$

The shell tends to deform into two conical parts (fig. 8.9) with "hinge circles" at x=0, x=l/2 and x=l. At these hinge circles, which are similar to the plastic hinges in beam theory, the rate of slope, $\frac{d\dot{w}}{dx}$, in the meridional plane varies discontinuously. This situation can occur only if the stress points are on edges of the yield surface. For x=0, the stress point is on the intersection of faces I' and III'. Equation of face III' is $-n_x - m_x = 1$, obtained by substituting $-m_x$ for m_x in eq. (5.89) of face III.

On this face, from the normality law and eqs. (8.5) we have :

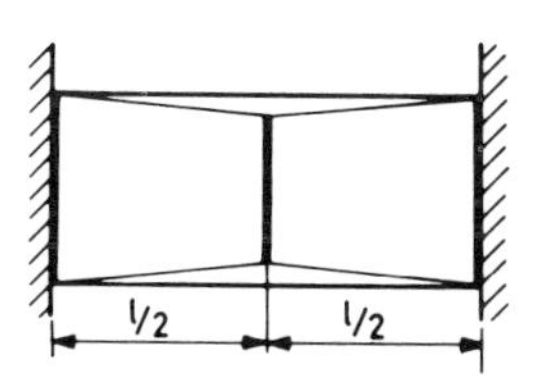

Fig. 8.9.

$$\text{(a)}\qquad \frac{t^2}{4}\,\frac{d^2\dot{w}}{dx^2} = 1\,,$$

$$\text{(b)}\qquad \dot{w} = 0\,,$$

$$\text{(c)}\qquad t\,\frac{d\dot{u}}{dx} = -\,1\,. \tag{8.15}$$

From eqs. (8.15 a) and (8.15 c) we obtain

$$\frac{d^2\dot{w}}{dx^2} = -\frac{4}{t}\frac{d\dot{u}}{dx}. \tag{8.16}$$

Relation (8.16) shows that the slope $\frac{d\dot{w}}{dx}$ is discontinuous by an amount equal to $-\frac{4}{t}\dot{u}$.

For $x=\frac{1}{2}$, the stress point is on the intersection of faces I' and III. The equation of face III is $m_x - n_x = 1$.

For $0 < x < \frac{1}{2}$, the stress point is in the face I' with the equation $n_\theta = -1$. Therefore, taking into account eq. (8.12), we have, for x=0,

$$m_x = -1 - n_x = -1 + \frac{p^*}{2}, \tag{8.17}$$

and for $x = \frac{1}{2}$,

$$m_x = 1 + n_x = 1 - \frac{p^*}{2}, \qquad \frac{dm_x}{dx} = 0. \tag{8.18}$$

Integration of eq. (8.10) with $n_\theta = -1$, and use of boundary conditions (8.17) and (8.18) yields

$$\text{(a)} \qquad m_x = \frac{2l^2}{tR}\left(1 - p^*\right)\left(\frac{x^2}{l^2} - \frac{x}{l}\right) - 1 + \frac{p^*}{2},$$

$$\text{(b)} \qquad n_x = -\frac{p^*}{2},$$

$$\text{(c)} \qquad n_\theta = -1, \tag{8.19}$$

and the limit value

$$p^* = 1 + \frac{1}{1 + \frac{l^2}{2tR}}, \tag{8.20}$$

obtained by substituting $\frac{l}{2}$ for x and $1 - \frac{p^*}{2}$ for m_x in eq. (8.19 a). The collapse mechanism corresponding to the stress field (8.19) is given by eqs. (8.14) and the discontinuity conditions at x=0 and $x=\frac{l}{2}$. Assuming vanishing rates of displacement at the left built-in end (x=0), we have,

$$\dot{w} = \dot{w}_0 x \qquad \text{for} \qquad 0 < x < \frac{l}{2}, \tag{8.21}$$

where $\dot{w}_0$ is a positive parameter which is the product of l by half the rate of deflection in the median cross section. From eq. (8.21) we see that $\frac{d\dot{w}}{dx} = \dot{w}_0$. Hence, the jump of $\dot{u}$ at x=0 is $\dot{u}] = -\left(\frac{t}{4}\right)\dot{w}_0$ and, from eq. (8.14),

$$\dot{u} = -\frac{t}{4}\dot{w}_0 \qquad \text{for} \qquad 0 < x < \frac{l}{2}. \tag{8.22}$$

In the interval $\frac{l}{2} < x < l$, we have

$$\dot{w} = \dot{w}_0(l - x), \tag{8.23}$$

$$\frac{d\dot{w}}{dx} = -\dot{w}_0, \tag{8.24}$$

$$\dot{u} = \frac{t}{4}\dot{w}_0. \tag{8.25}$$

The discontinuity in slope at $x = \frac{l}{2}$ is $2\dot{w}_0$. Having found a statically admissible stress field, eqs. (8.19), and a corresponding kinematically admissible mechanism, eqs. (8.21) to (8.25), the limiting value (8.20) of the pressure will be exact if the considered stress field and mechanism remain admissible for any possible value of the shell parameter $\frac{l^2}{2tR}$. Inspection of eqs. (8.20) and (8.19 b) immediately shows that n_x is always bounded by $-\frac{1}{2}$ and - 1 regardless of the value of $\frac{l^2}{2tR}$, and that m_x is monotonically increasing as x varies from 0 to $\frac{1}{2}$. Hence, the stress profile is always adequate and, if we let

$$\omega^2 = \frac{l^2}{2tR}, \tag{8.26}$$

the limit pressure can be written, using eqs. (8.20) and (8.9), as

$$p_l = \frac{\sigma_Y t}{R}\left(1 + \frac{1}{1+\omega^2}\right). \tag{8.27}$$

B. Paul [8.21] has obtained the exact limit pressure for the shell considered above when it is strengthened by a *ring stiffener at midspan.* His results are summarized in fig. 8.10. The yield surface for the shell is that of fig. 5.30. The yield condition of the annulus was given by B. Paul [8.22]. It is greatly simplified in the present case where, on account of symmetry, the ring can collapse exclusively from compression. If the ring remains rigid, the reduced limit pressure is,

$$p_l^* = 1 + \frac{1}{1+\frac{\omega^2}{4}} \qquad \text{if} \qquad \frac{4\,\omega^2}{4+\omega^2} \leq \gamma, \tag{8.28}$$

where $\gamma = (l\ /\ 2\ M_p\ R)\ \sigma_{Ya}\ A_a$, and σ_{Ya} is the yield stress of the annulus, A_a the cross-sectional area of the annulus, l the total span of the cylinder, and R the mean radius of the cylinder. When $\gamma \leq 4\ \omega^2\ /\ (4+\omega^2)$, the ring collapses with the shell.

8.3.3. *Cylindrical shell with axial load. Various other cases.*

8.3.3.1. *Cylindrical shell simply supported at both ends and subjected to uniform pressure.*

The solution of this problem is due to Hodge and Paul [8.23] who found the following exact limit pressure :

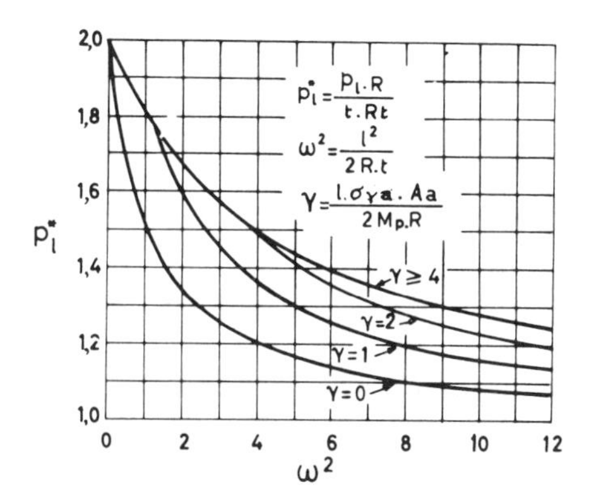

Fig. 8.10.

$$p_l = \frac{\sigma_Y t}{R}\left(1 + \frac{1}{1 + 2\,\omega^2}\right). \tag{8.29}$$

8.3.3.2. *Cylindrical shell built-in at one end, free at the other end, and subjected to uniform radial pressure and independent axial load.*

Let p be the pressure and 2 π RN the total axial load. If we replace the section of the yield surface of fig. 5.12 with the plane $n_1 = -|N| / N_p$ by the circumscribed rectangle, the following interaction formula is obtained :

$$p = \frac{N_p}{R}\left(1 + \frac{N}{N_p}\right) + \frac{2}{l^2} M_p \left[1 - \left(\frac{N}{N_p}\right)^2\right], \tag{8.30}$$

where l is the length of the shell (Onat [8.24]).

The same problem was treated in a slightly different manner by Sankaranarayanan [8.25].

8.3.3.3. *Cylindrical shell joined at both ends to rigid plates, and subjected to radial pressure and independent axial load.*

A detail solution can be found in a paper by Hodge and Panarelli [8.26] or in the book by Hodge [8.27].

Close bounds on the interaction curve p versus t are given in figs. 8.11 and 8.12 [8.26] for a Tresca and a von Mises material, respectively. The dimensionless pressure p^* and axial load n^* are defined by

$$p^* = \frac{pR}{\sigma_Y t}, \qquad n^* = \frac{N}{2\pi R \sigma_Y t},$$

where N is the total axial load, the other symbols having the same meaning as in Section 8.3.2.

8.3.3.4. *Closed cylindrical vessel subjected to internal pressure.*

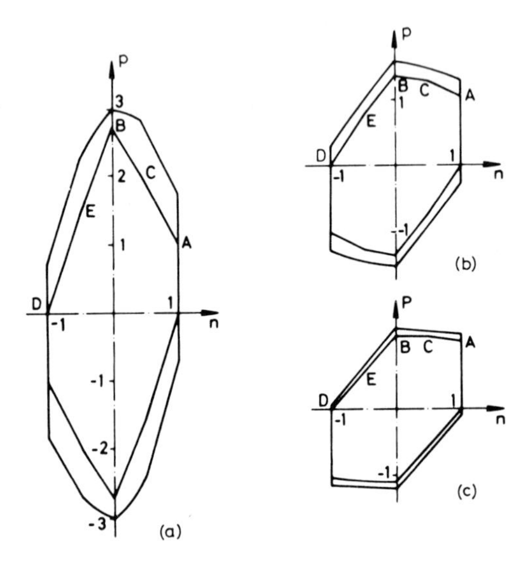

Fig. 8.11. Best bounds on interaction curves for Tresca shell. (a) ω=1. (b) ω=2. (c) ω=4.

This problem of important practical interest has been treated by Hodge [8.28] on the basis of his limited-interaction yield condition consisting of two independent Tresca hexagons for bending moments and axial forces. The solution is very simple and yet, when compared with more refined and laborious analysis [8.29, 8.30], retains all essential results with sufficient accuracy for all vessels with current proportions (that is, with $\frac{L}{R} > \left\{\frac{0.5P}{\sigma_Y}\right\}^{1/2}$, t' >t, (fig. 8.13). It can be further simplified if we note that, for a simply supported circular plate obeying the yield condition of Tresca and subjected to transversal uniformly distributed load p and edge moments M_o, the limiting value of p is given by (8.31)

$$p_l \approx \sigma_Y \left[\frac{3}{2} \left\{ \frac{t'}{R} \right\}^2 - 5.64 \frac{M_o}{\sigma_Y R^2} \right], \tag{8.31}$$

as long as M_o does not exceed half the plastic moment of the plate. The limit pressure of the cylinder is related to its thickness by

$$\frac{t}{R} = \frac{p_l}{\sigma_Y} \frac{2}{1 + \left\{ 1 + 4 p_l \, R^2 / L^2 \, \sigma_Y \right\}^{1/2}}. \tag{8.32}$$

Eq. (8.32) is a modified version of Hodge's formula, more suitable for accurate numerical computations.

Note that, because the end plates never collapse by in-plane deformation and because the yield conditions for axial forces and bending moments are independent, the solution is valid for both types of vessel shown in fig. 8.13, either with two end plates or with one end plate and one built-in end.

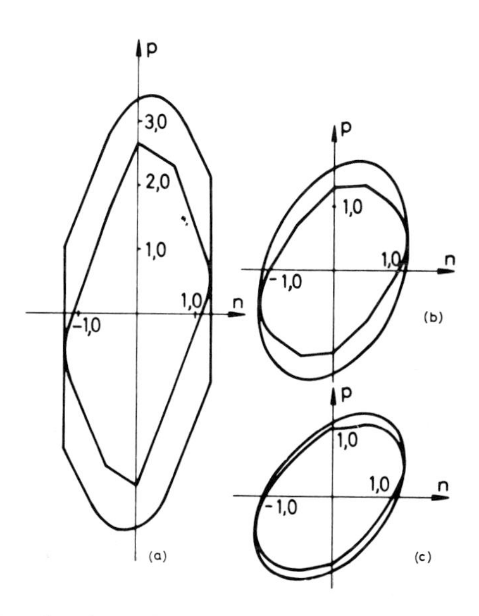

Fig. 8.12. Best bounds on interaction curves for Mises shell. (a) ω=1. (b) ω=2. (c) ω=4.

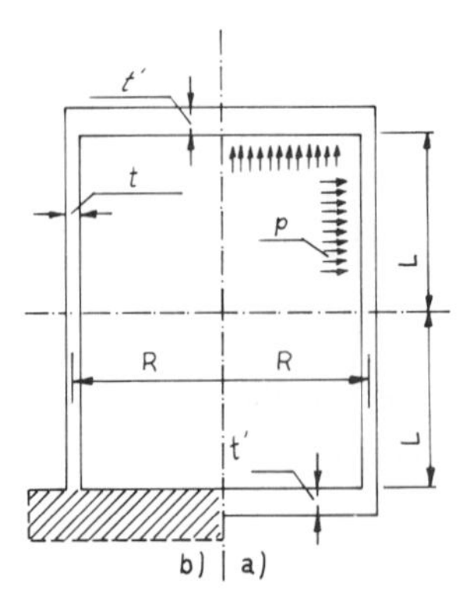

Fig. 8.13.

8.3.3.5. *Rib-reinforced cylindrical shell.*

Yield loci for rib-reinforced shells consisting of ribs and symmetric sheeting have been derived by Nemirovsky and Rabotnov [8.32]. Cylindrical shells without axial load, with purely longitudinal ribs on one side, have been treated by Biron and Sawczuk [8.33] who derived the yield curves and gave some illustrative examples. Capurso and Gandolfi [8.34] have given the yield locus of an axisymmetrically loaded shell of revolution with I reinforced ribs placed along meridians and parallels. Ribs were considered as ideal sandwich beams. These authors have treated examples of cylindrical shells with various end conditions [8.35].

The curves of figs. 8.14 and 8.15 give the reduced limit pressure p^* that depends on

$$\omega_x = \frac{N_{px}}{2\,\sigma_Y\,B_x\,t}, \qquad \omega_\theta = \frac{N_{p\theta}}{2\,\sigma_Y\,B_\theta\,t},$$

and on $\xi = \dfrac{L^2}{2RH_x^*}$ for the particularly interesting case of a closed vessel subjected to uniform pressure.

In the preceding relations, R is the radius of the sell sheet, N_p the plastic axial force of the sheet, ω_{px} and $\omega_{p\theta}$ the plastic axial forces of a longitudinal and a circumferential rib, respectively, B_x and B_θ the spacings of these ribs, t the sheet thickness, and $2L$ the length of the cylinder. H_x^* is the ratio of the plastic moment of a longitudinal rib (with a part of the cover sheet included) to its plastic axial force ω_{px}.

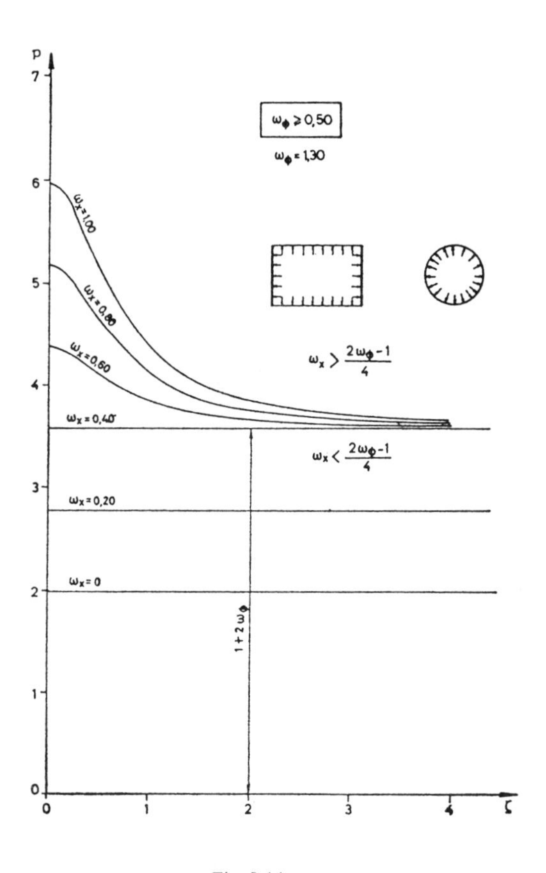

Fig. 8.14.

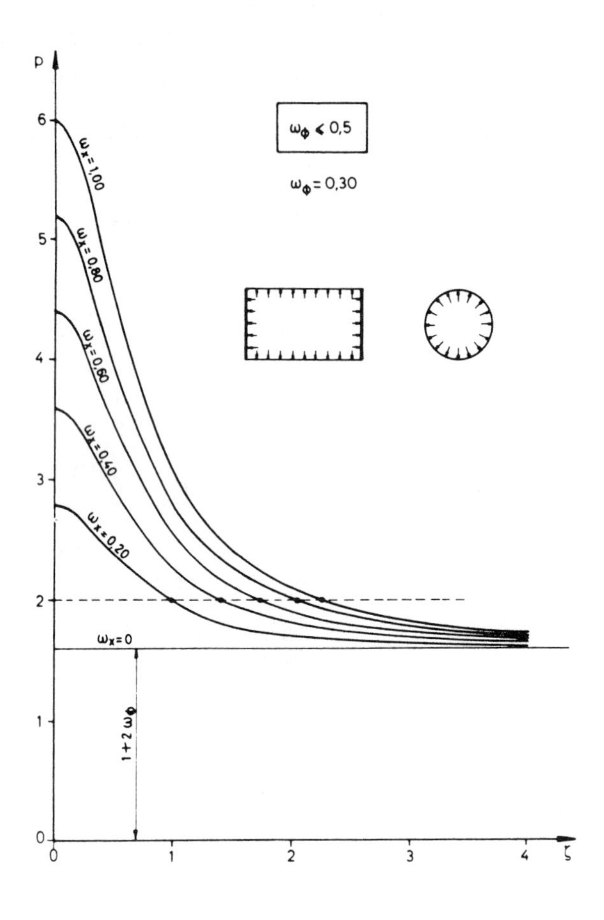

Fig. 8.15.

8.3.4. *Cylinder without axial load.*

8.3.4.1. *Cylindrical shell built-in at both ends and subjected solely to radial uniform pressure.*

In the absence of axial loads, and with the symmetry of revolution, the yield surface reduces to the M_x versus N_θ interaction curve and the problem is accordingly simplified. Using the linearized Tresca yield condition of fig. 5.24, Hodge [3.8] has obtained the diagrams of fig. 8.16 where the reduced limit pressure $p^* = \frac{p_l R}{\sigma_Y t}$ and the extent $\eta = \frac{x}{(l/2)}$ of the shell in regime ED are given as functions of the shell parameter ω. We note that there are two types of solution, depending on the value of ω with respect to 1.65.

The limit pressure given by eq. (8.20) is also represented in fig. 8.16 for comparisons.

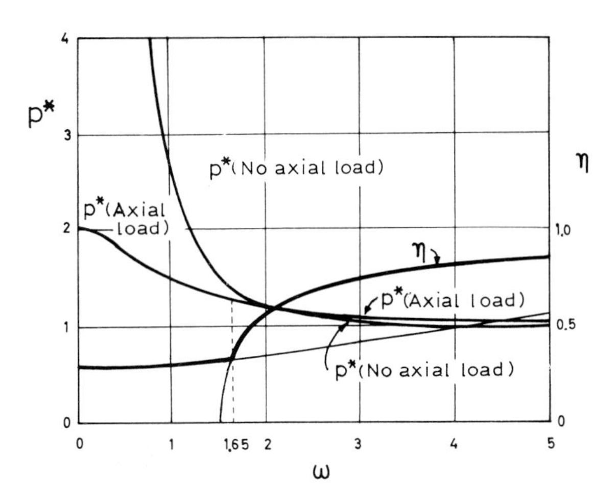

Fig. 8.16. (From Plastic Analysis of Structures by P.G.Hodge Jr. Copyright 1959, McGraw-Hill Book company. Used by permission of McGraw-Hill Book company, Inc.)

8.3.4.2. *Cylindrical shell subjected to a ring of pressure.*

When the shell is infinitely long and subjected to a ring of pressure with magnitude 2F, we have seen in Section 5.5.4 that the reduced load

$$f_{ol} = \frac{2F_{ol}}{\left(M_p \frac{N_p}{R}\right)^{1/2}}$$

depends on the yield condition as shown in table 5.4. Shells with finite length were treated by Eason and Shield [8.36] with a yield rectangle circumscribed to the Tresca yield curve of fig. 5.24, and by Eason [8.37] and Demir [8.38] with the exact Tresca curve of the same figure. When the ring of load is located in the transversal plane of symmetry of the shell, limit loads are given in fig. 8.1. Loads not located in the plane of symmetry and comparison with elastic analysis are treated in refs. [8.37] and [8.38], respectively,.

8.3.4.3. *Cylindrical tank subjected to hydrostatic pressure.*

The following yield condition is used :

$$n_\theta = \pm 1 ,$$

$$m_x = \pm 1 , \tag{8.33}$$

with

$$n_\theta = \frac{N_\theta}{N_{p\theta}} ,$$

$$m_x = \frac{M_x}{M_{px}} , \tag{8.34}$$

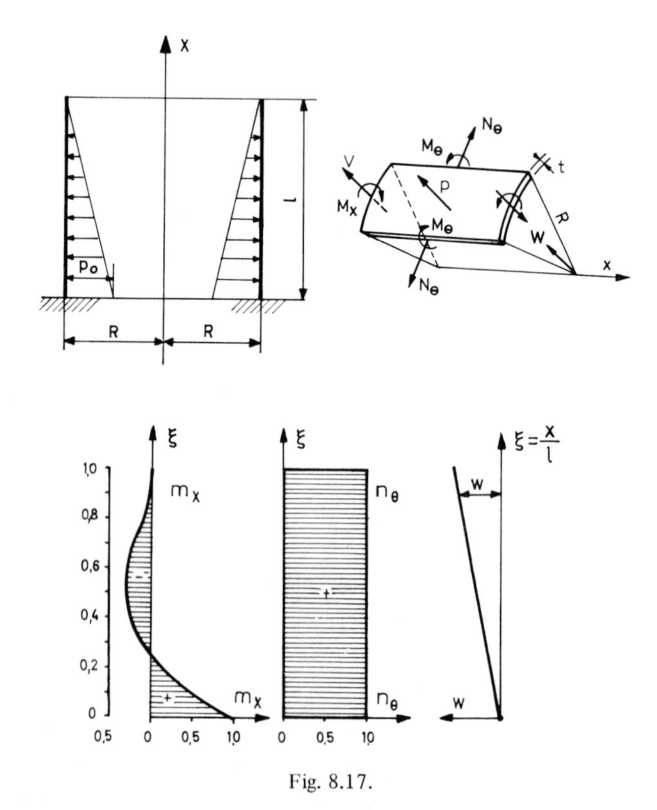

Fig. 8.17.

Condition (8.33) represents a rectangle circumscribed to the Tresca hexagon of fig. 5.24. Possible orthotropy of the shell can be taken into account regarding $N_{p\theta}$ and M_{px} as independent. *For a liquid-filled tank shell that is vertical, free at the upper edge, and built-in at the bottom edge*, the stress distributions and the collapse mechanism are shown in fig. 8.17.

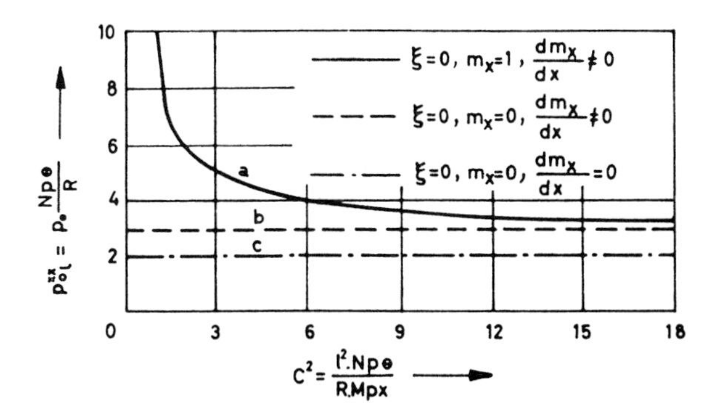

Fig. 8.18.

We have [8.39]

$$m_x = \frac{C^2}{2}\left(p_o^{**} - 1\right)\xi^2 - \frac{C^2}{6}p_o^{**}\,\xi^3 + \left(3\,C^2 - 3 - p_o^{**}\,C^2\right)\xi + 1\,,$$

where $\xi = x/l$, $C = l\left(\dfrac{N_{p\theta}}{R\,M_{px}}\right)$, $p_o^{**} = \dfrac{p_o\,N_{p\theta}}{R}$, and $p = p_o(1 - \xi)$. The limit pressure is

$$p_{ol}^{**} = 3\left(1 + \frac{2}{C^2}\right) \tag{8.35}$$

It is given by the curve a in fig. 8.18. The solution (8.35) is valid only for "short" shells, for which $C^2 < 17.1$, because the maximum absolute value of the bending moment for $x > 0$ must remain smaller than (or most equal to) M_{px}.

When the bottom edge is simply supported or free, curve a must be replaced by curve b or c, respectively.

For shells that are simply supported at both edges, the various possible mechanisms, moment distributions, limit loads, and hinge locations are given in figs. 8.19 to 8.21 [8.39]. For all cases, $n_x = \pm 1$.

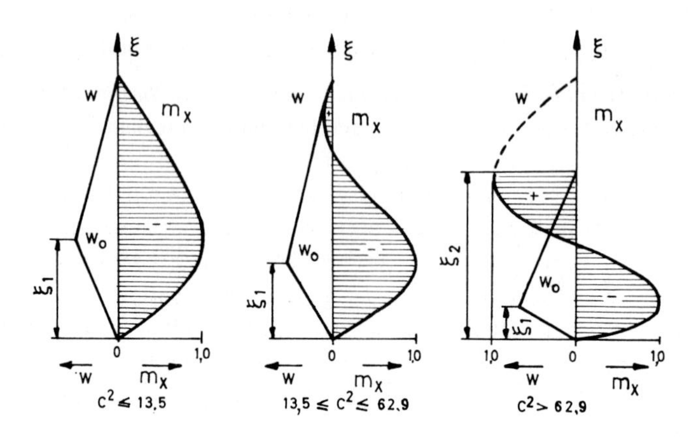

Fig. 8.19.

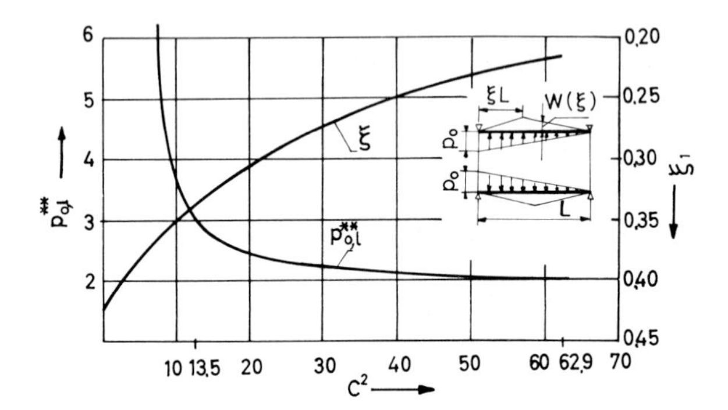

Fig. 8.20.

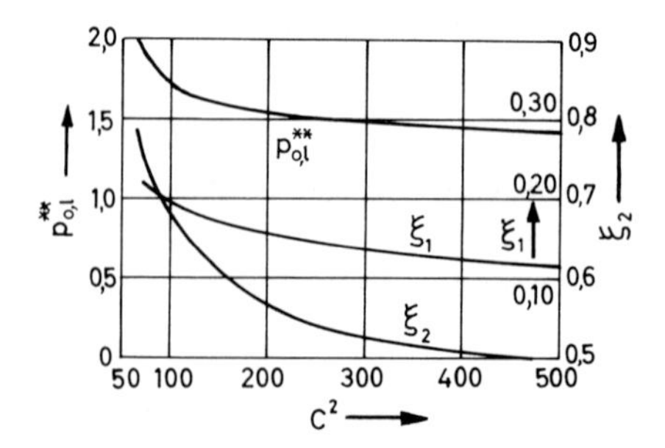

Fig. 8.21.

8.3.5. *Second-order effects in cylindrical shells.*

Limit load obtained in Section 8.3.1 to 8.3.4 are based on the rigid perfectly plastic idealization, but real shells are neither rigid nor perfectly plastic but elastic work hardening. When cylindrical shells are subjected to both transversal and axial loads, inward radial deflections provide lever arms for the compressive axial forces, introducing supplementary bending moments. The situation is similar to the divergence of equilibrium state in beams. The considered second order effects occur in elastic-plastic ranges and are likely to decrease the load corresponding to unrestrained plastic flow, if we neglect the counteracting influence of work hardening. We hereafter follow B. Paul [8.40] in studying a closed cylindrical shell subjected to uniform external pressure. Equilibrium must now be formulated in the deflected situation (fig. 8.22). Let w(x) be the radial displacements. Moment equilibrium about the centroid, in the undeformed state, of an elementary cross section with central angle $d\,\theta$ is

$$d\,M_x + N_x\,dw + d\,N_x\,w = V\,dx.$$

Because N_x is a constant, the preceding equation can be rewritten as

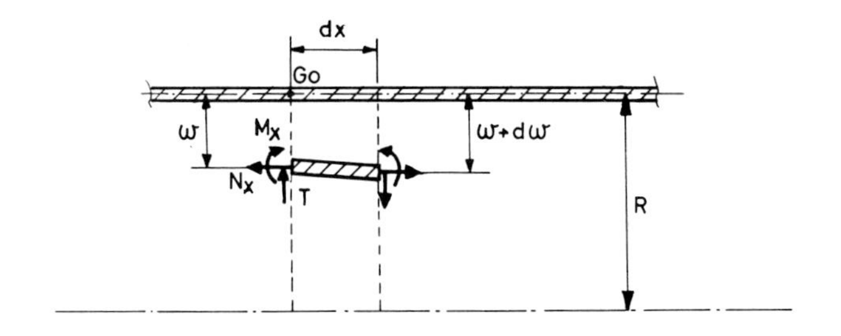

Fig. 8.22.

$$\frac{d\,M_x}{dx} + N_x\,\frac{dw}{dx} = V. \tag{8.36}$$

Eq. (8.7a) remains unaltered. Eliminating V from eqs. (8.7a) and (8.36), and using the already defined reduced variables, we obtain

$$\frac{tR}{4}\,\frac{d^2 m_x}{dx^2} + R\,n_x\,\frac{d^2\,w}{dx^2} + n_\theta + p^* = 0\,, \tag{8.37}$$

to couple with

$$n_x = -\frac{p^*}{2} \tag{8.38}$$

We try a stress profile of the type used in Section 8.3.2, a profile on face I' and its intersection with faces III and III', fig. 5.30. We have

$$\begin{gathered} -1 \leq n_x \leq -\frac{1}{2} \\ -(1 + n_x) \leq m_x \leq 1 + n_x \end{gathered} \tag{8.39}$$

Except for the end points of the stress profile, the plastic curvature κ_x^p vanishes so that the curvature is purely elastic and hence given by Hooke's law. Using reduced variables and taking account of $\kappa_\theta = 0$, we may write

$$\Phi_x \equiv -\frac{t^2}{4} \frac{d^2 w}{dx^2} = \frac{3}{4} m_x (1 - \nu^2) \frac{t \sigma_Y}{E}, \tag{8.40}$$

where E is Young's modulus and ν is Poisson's ratio for the shell material. With the origin of the abscissa x at midspan, we have

$$\text{(a)} \qquad \left(\frac{dm_x}{dx} \right)_{x=0} = 0 .$$

Substituting the expression (8.40) for $\frac{d^2 m_x}{dx^2}$ and - 1 for n_θ (stress profile on face I') in eq. (8.37), we obtain a second other differential equation with constant coefficients in the unknown m_x. Integration of this equation, and use of relation (a), yields

$$m_x = C \cos \frac{2kx}{l} + 2 \frac{\omega^2}{k^2} (1 - p^*), \tag{8.41}$$

where

$$k^2 = 3 (1 - \nu^2) \beta^2 \omega^4 p^*$$

and

$$\beta^2 = \frac{2\,\sigma_Y\,R^2}{E l^2} .$$

The integration constant C and the limit value p^* are obtained from the conditions that the extreme points of the stress profile correspond to the central and the end cross sections (sections of hinge circles). We thus obtain [8.40] :

1. For the shell built-in at both ends :

$$C = 1 - \frac{p^*}{2} - 2\,\frac{\omega^2}{k^2}\,(1 - p^*) ,$$

$$p^* = \frac{1 - \cos k + \dfrac{k^2}{2\,\omega^2}\,(1 + \cos k)}{1 - \cos k + \dfrac{k^2}{4\,\omega^2}\,(1 + \cos k)} . \tag{8.42}$$

Note that eq. (8.42) is transcendental because κ depends on p^*. If E tends to infinity, k^2 tends to zero and eq. (8.42) reduces to eq. (8.27), valid for the rigid-plastic shell. Let p_o^* be the corresponding limit value : $p_o^* = 1 + (1\ /\ 1{+}\omega^2)$. The value (8.42) of p^* decreases monotonically from p_o^* to unity when k varies from zero to π. For $k > \pi$, relation (8.39) ceases to be satisfied and the stress profile must be modified (except for $k = \pi + 2n$ with $n = 1, 2, 3,\ldots$, corresponding to $p^* = 1$). It is shown in ref. [8.40] that, for all $k > \pi$, several statically admissible stress fields can be found that all give $p^* = 1$. Moreover, corresponding mechanisms can be obtained for all these stress fields (either for $k < \pi$ or $k > \pi$). We define a new shell parameter γ by the following relation

$$\gamma^2 \equiv \frac{k^2}{p^*} = 3\,(1 - \nu^2)\,\beta^2\,\omega^4 = \frac{3}{2}\,(1 - \nu^2)\,\frac{\sigma_Y\,l^2}{E t^2} . \tag{8.43}$$

It is easily seen that,

$$\gamma^2 < \pi^2 \qquad \text{when} \qquad k^2 < \pi^2 \text{ because} \qquad p^* > 1,$$

and

$$\gamma^2 > \pi^2 \qquad \text{when} \qquad k^2 \geq \pi^2 \text{ because} \qquad p^* = 1.$$

A shell will therefore be called "short" if $\gamma^2 \leq \pi^2$, and eq. (8.42) will hold, whereas it will be called "long" when $\gamma^2 \geq \pi^2$, with $p^* = 1$.

2. Shell simply supported at both ends. It is found [8.40] that

$$p^* = \frac{\left[(k^2/2\omega^2) - 1\right]\cos k + 1}{\left[(k^2/4\omega^2) - 1\right]\cos k + 1}, \tag{8.44}$$

for short shells with $\gamma^2 \leq (\pi/2)^2$, and that $p^* = 1$ for long shells with $\gamma^2 \geq (\pi/2)^2$.

3. Shells simply supported at one end and built-in at the other.

$$p^* = (2 + 2\,\omega^2\,\psi)\,(1 + 2\,\omega^2\,\psi), \tag{8.45}$$

with $\psi = 2/k^2\ (\sec ku - 1)$ and $2\cos ku = 1 + \cos k\,(2 - u)$ for short shells with $\gamma^2 \leq (3/4\pi)^2$. For long shells with $\gamma^2 \geq (3/4\pi)^2$, $p^* = 1$.

Closed cylindrical shells subjected to external uniform pressure can thus be classified into "short shells" or "long shells" when

$$\gamma^2 = \frac{3}{2}\,(1 - \nu^2)\,\frac{\sigma_Y\, l^2}{E t^2} \tag{8.46}$$

is smaller or larger than $(n\pi/4)^2$, respectively, with n = 2, 3, or 4 for the shell simply supported at both ends, simply supported at one end and built-in at the other, or built-in at both ends, respectively. It must be emphasized that the values of the pressure obtained above should be regarded, from a rigourous point of view, exclusively as upper bounds to the actual carrying capacity, because the theorems of limit analysis do not apply when equilibrium equations are referred to the deformed state.

The "carrying capacities" p^* obtained above are given in figs. 8.23, 8.24 and 8.25. In fig. 8.23, the lower parts of the curves give the buckling pressures according to Batdorf [8.3], that have been used when these pressures are smaller than the carrying capacities. It turns out that, in the whole considered range of β, the carrying capacity obtained in Section 8.3.5 is smaller that the buckling pressure for *all* short shells. Moreover, the range of long shells in the same situation is very narrow. Accordingly, this simple rule may be accepted : long shells tend to collapse by buckling and short shells by unrestrained plastic flow. The terms *short* and *long* are defined unambiguously by means of the parameter γ. It is also worth noting that, as for all long shells $p^* = 1$, annular rigid stiffeners will influence the value of the pressure for unrestrained plastic flow only if they transform the shell into "short" subshells, that is, if their spacing is smaller that

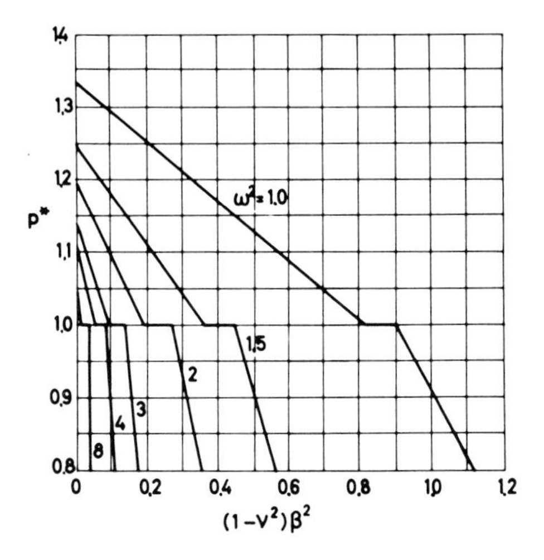

Fig. 8.23.

$$l = \pi \, \frac{t}{2} \left(\frac{8\,E\,(1-\nu^2)}{3\,\sigma_Y} \right)^{1/2} . \tag{8.47}$$

Regarding the deflections prior to unrestrained plastic flow, Hodge [8.41], and Paul and Hodge [8.23] have shown in examples that a load as high as 98% of the limit load could be attained without the maximum elastic-plastic deflection exceeding five times the maximum elastic deflection.

Work hardening seems to have a very small effect on the load-carrying capacity [8.42] ; it should in any case increase it with respect to theoretical predictions based on

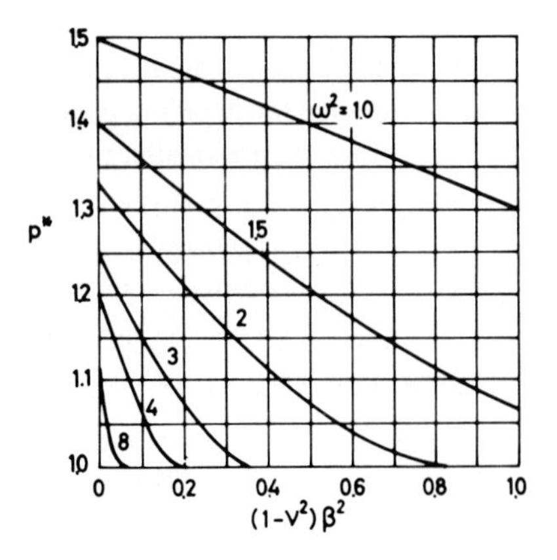

Fig. 8.24.

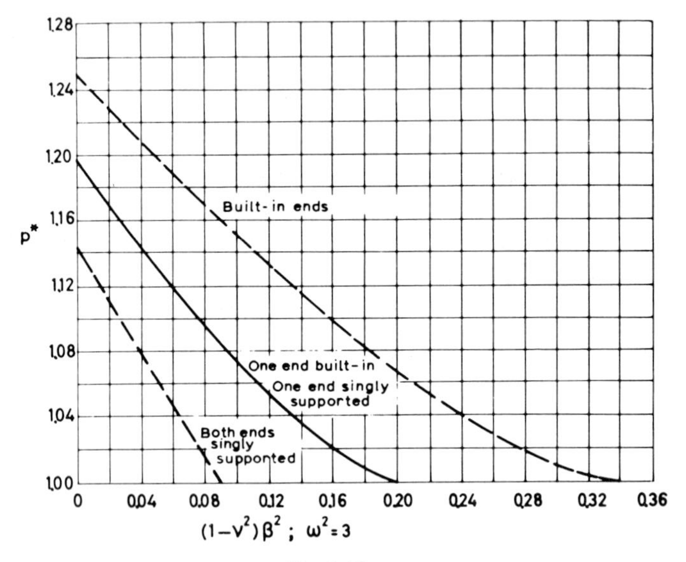

Fig. 8.25.

perfect plasticity. Also, the use of a linearized yield condition "inscribed" to the exact Tresca condition, which in turn is "inscribed" to the more realistic von Mises condition, should also result in extra safety.

8.4. Rotationnally symmetric shells.

8.4.1. *Introduction.*

Generalized stresses M_φ, M_θ, N_φ, N_θ are shown in fig. 5.8. The yield condition of von Mises for a shell that has rotational symmetry in shape and loading was derived independently by Ilyushin [1.2] and Hodge [5.14]. The Tresca condition is given in Section 5.4.2, tables 5.2 and 5.3. Both conditions are nonlinear ; thus their use is difficult. For this reason, approximations to the Tresca condition have been used to obtain better bounds or exact solutions. One possible approximation is the Tresca condition for a sandwich shell, as discussed in Section 5.5. Hodge [8.43] has suggested the use of a limited interaction yield locus, assuming that either bending moments or axial forces but not both at the same time are important. Hence, interaction of generalized stresses of the same nature is fully taken into account, whereas generalized stresses of different nature (moments and axial forces) are uncoupled. The corresponding yield hypersurface for a Tresca material is formed of two independent hexagonal hyperprisms, one for the M_φ, M_θ interaction, one the N_φ, N_θ interaction. The equations of the twelve hyperplanes, together with the components of the corresponding strain-rate vectors, are given in table 8.3, in which $m_i = \dfrac{M_i}{M_p}$, $n_i = \dfrac{N_i}{N_p}$,

$\lambda_i \equiv \dot{\varepsilon}_i$, $\Phi_i \equiv m_i \dot{\kappa}_i$, $(i = \varphi , \theta)$. Hodge [3.8] has described a way of visualizing this four-dimensional hypersurface and has shown that, except for the plate, only part of the hypersurface can furnish a collapse mechanism. Because the limited interaction yield hypersurface is completely circumscribed to the Tresca hypersurface and must be shrunk by a factor of 0.618 to become completely inscribed, an exact limit load P_o with the limited interaction condition will, according to Section 5.5.2, bound the limit load with the exact Tresca condition as follows :

$$0.618\, P_o \le P_l \le P_o . \qquad (8.48)$$

Flügge and Nakamura [8.44] have proposed the use of the six parabolic hypercylinders, the equations of which are given in table 5.3, and reproduced in table 8.4, together with the corresponding strain rates. This set of hypercylinders, when limited to their mutual intersections, coincides partly with the exact Tresca hypersurface and completely circumscribe it. An exact limit load P^* for this approximate yield hypersurface gives the following bounds

$$0.851\, P^* \le P_l \le P^* .$$

Table 8.3. Hodge's limited interaction yield locus.

Face nr	Equation	$\dot{\lambda}_\theta$	$\dot{\lambda}_\varphi$	$\dot{\Phi}_\theta$	$\dot{\Phi}_\varphi$
1	$n_\theta = 1$	1	0	0	0
2	$n_\varphi = 1$	0	1	0	0
3	$-n_\theta + n_\varphi = 1$	- 1	1	0	0
4	$-n_\theta = 1$	- 1	0	0	0
5	$-n_\varphi = 1$	0	- 1	0	0
6	$n_\theta - n_\varphi = 1$	1	- 1	0	0
7	$m_\theta = 1$	0	0	1	0
8	$m_\varphi = 1$	0	0	0	1
9	$-m_\theta + m_\varphi = 1$	0	0	- 1	1
10	$-m_\theta = 1$	0	0	- 1	0
11	$-m_\varphi = 1$	0	0	0	- 1
12	$m_\theta - m_\varphi = 1$	0	0	1	- 1

Table 8.4. Hypersurfaces of parabolic hypercylinders.

Label of hyper-surface	Yield condition	$\dot{\varepsilon}_\varphi$	$: \dot{\varepsilon}_\theta$	$: \dot{\Phi}_\theta$	$: \dot{\Phi}_\theta$
H_φ^+	$m_\varphi + n_\varphi^2 = 1$	$2n_\varphi$	:0	:1	:0
H_φ^-	$-m_\varphi + n_\varphi^2 = 1$	$2n_\varphi$	:0	:-1	:0
$H_{\varphi\theta}^+$	$m_\varphi - m_\theta + (n_\varphi - n_\theta)^2 = 1$	$2(n_\varphi - n_\theta)$	$:-2(n_\varphi - n_\theta)$	:1	:-1
$H_{\varphi\theta}^-$	$-m_\varphi + m_\theta + (n_\varphi - n_\theta)^2 = 1$	$2(n_\varphi - n_\theta)$	$:2(n_\varphi - n_\theta)$	:-1	:1
H_θ^+	$m_\theta + n_\theta^2 = 1$	0	$:2n_\theta$	:0	:1
H_φ^+	$-m_\theta + n_\theta^2 = 1$	0	$:2n_\theta$	:0	:-1

Comparison of relations (8.48) and (8.49) shows a noticeable improvement. It is shown in ref. [8.44] that the collapse mechanism obtainable with the hypercylinders H_φ^+ and H_φ^- si merely a rigid-body translation and does not represent any plastic deformation. General expressions of the velocity fields and stress fields are then obtained for the two other pairs of hypercylinders. In these expressions, only quadratures remain to be performed with the integration constants to be determined from boundary and continuity conditions. The main difficulty lies in the choice of appropriate hypercylinders for the particular problem at hand.

As will be seen in Section 8.5, the Tresca condition for a cylindrical shell can also be used successfully for some other shells of revolution, as suggested by Drucker and Shield [8.45, 8.46].

8.4.2. *Spherical cap subjected to radial uniformly distributed load.*

8.4.2.1. *Introduction.*

Let R be the radius, 2 α the cap angle and t its thickness (fig. 8.26). To begin with, assume that a certain stress profile is appropriate. With this stress profile, the equilibrium equations can often be integrated to yield a statically admissible stress field, from which a mechanism is obtained using the normality law. It then remains to be checked whether the stress field and the mechanism remain admissible for the whole range of shell parameters.

8.4.2.2. *Spherical cap hinged at the edge.*

With the limited interaction yield condition, a complete solution can be obtained [8.43]. It yields

$$p_l^* \equiv \left(\frac{p\,R}{N_p}\right)_l = 2 + 2\,k\,\sin\alpha\left[\ln\frac{1+\sin\alpha}{\cos\alpha} - \sin\alpha\right]^{-1}, \tag{8.50}$$

with the following stress field (see fig. 8.26 for definition of angle φ)

$$n_\theta = -\frac{1}{2}\left[p^* - (p^* - 2)\sec^2\varphi\right],$$

$$n_\varphi = -1,$$

$$m_\theta = 1,$$

$$m_\varphi = 1 - \frac{p^* - 2}{2\,k}\left[\frac{1}{\sin\varphi}\ln\frac{1+\sin\varphi}{\cos\varphi} - 1\right]. \tag{8.51}$$

In formulas (8.50) and (8.51), $k = \dfrac{M_p}{R\,N_p}$. The shear force is

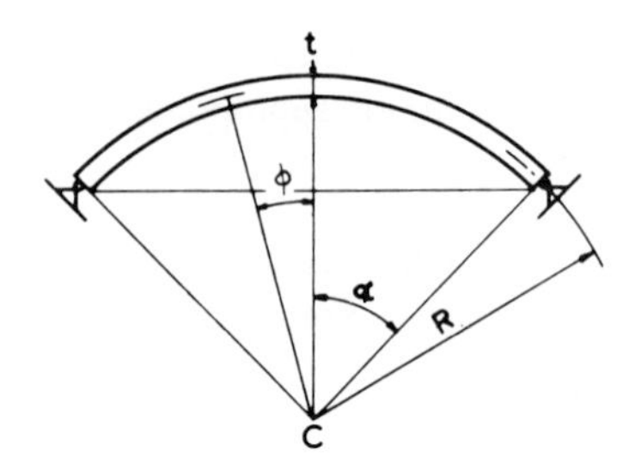

Fig. 8.26.

$$V = -N_p \frac{p^* - 2}{2} \tan \varphi. \tag{8.52}$$

For a shell with a centred cut-out, yield pressures have been obtained by Hodge and Lakshmikantham [8.47] and are given in the book by Hodge [8.27].

With the Tresca yield condition for a sandwich shell, the following good bounds are given in Hodge's book [8.27] :

$$\frac{2k}{(1+k)(1-\alpha \cot \alpha)}, \qquad \text{if} \qquad p_-^* \geq 2, \tag{8.53}$$

or

$$p_-^* = 2. \tag{8.54}$$

$$p_+^* = \frac{2k}{1 - \alpha \cot \alpha}, \qquad \text{if} \qquad \cos \alpha \geq 1 - k, \tag{8.55}$$

$$p_+^* = 2\left(\frac{\sin \alpha - \varphi_1 \cos \alpha}{\sin \alpha - \alpha \cos \alpha} - \frac{\cos \alpha}{\cos \varphi_1} \frac{\sin \alpha - \sin \varphi_1}{\sin \alpha - \alpha \cos \alpha}\right), \tag{8.56}$$

with $\cos \varphi_1 = \dfrac{\cos \alpha}{(1-k)}$ if $\cos \alpha \leq 1 - k$

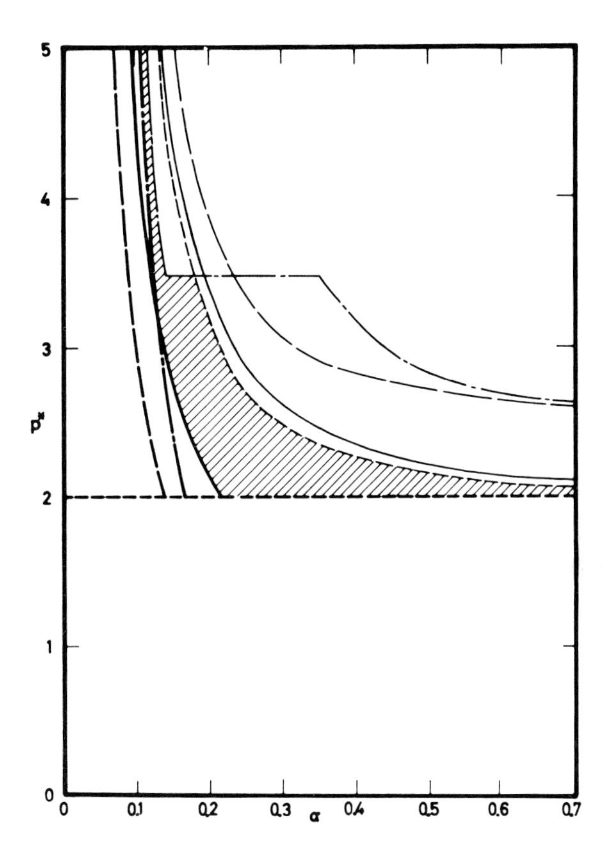

Fig. 8.27. - - - - - Bounds by Onat and Prater [5.8] ; _ _ _ _ Bounds based on a sandwich structure [3.9] ;
_ · _ · _ . Improved bounds based on a sandwich structure ;
__________ bounds based on the linited interaction yield surface,
Light lines : upper bounds, heavy lines lower bounds $k = M_p / RN_p = t/4R = 0.005$.

8.4.2.3. *Spherical cap built-in at the edge.*

With the limited interaction yield condition, the stress field (8.51) is valid for $\varphi \leq \varphi_1$, the particular value φ_1 corresponding to $m_\varphi = 0$. For $\varphi_1 \leq \varphi \leq \alpha$, n_θ and n_φ continue to be given by relation (8.51), whereas

$$m_\varphi = -1 + \frac{p^* - 2}{2\,k} \ln \frac{\cos \varphi}{\cos \alpha} - \ln \frac{\sin \alpha}{\sin \varphi},$$
$$m_\theta = m_\varphi + 1. \tag{8.57}$$

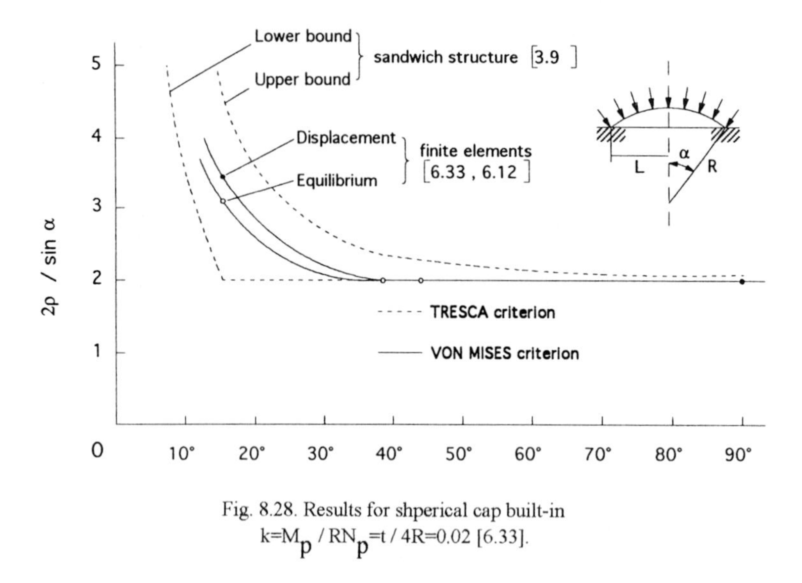

Fig. 8.28. Results for shperical cap built-in
$k=M_p / RN_p = t / 4R = 0.02$ [6.33].

The limit load p^* and the value of φ_1 are given by the following system of equations :

$$p^* = 2 + 2\,k \sin\varphi_1 \left[\ln \frac{1+\sin\varphi_1}{\cos\varphi_1} - \sin\varphi_1 \right]^{-1},$$

$$p^* = 2 + 2\,k \,\frac{1 + \ln(\sin\alpha / \sin\varphi_1)}{\ln(\cos\varphi_1 / \cos\alpha)}, \qquad (8.58)$$

Various bounds for p^* are compared in fig. 8.27. The exact values for a solid shell obeying the Tresca yield condition fall in the shaded region.

Lower and upper bounds are obtained also by Nguyen Dang Hung and coworkers [6.33], [6.12], using equilibrium and displacement finite elements (see sections 6.3.1 and 6.4.1).

These bounds are compared in fig. 8.28, with those obtained in [3.9].

8.4.3. *Conical shell loaded by a finite boss.*

When the boss in hinged to the shell (fig. 8.29) and if the *limited interaction yield condition* is used, with the following dimensionless variables

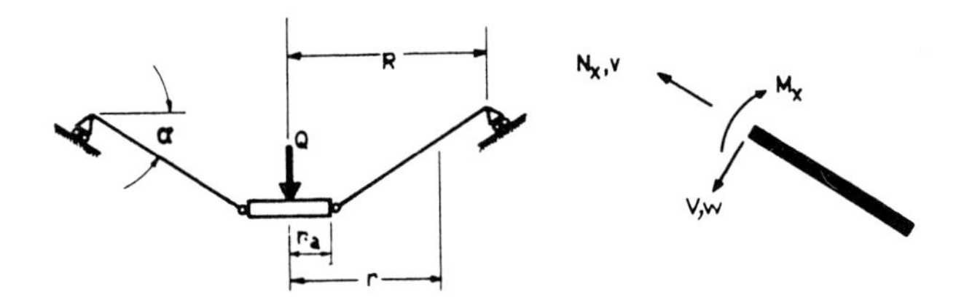

Fig. 8.29.

$$a = \frac{r_a}{R}, \tag{8.59}$$

$$q = Q \frac{1}{2 \pi R N_p \sin \alpha} - j, \tag{8.60}$$

$$j = \frac{M_p \cos^2 \alpha}{R N_p \sin \alpha}, \tag{8.61}$$

the limiting value q_l is given, together with a subsidiary unknown b, by [8.48]

$$q_l = \frac{1 - b^2 + 2 b^2 \ln b}{4 (1 - b)} = \frac{b (1 - \ln b) - (a + 1/e)}{\ln (b/a)}. \tag{8.62}$$

In relation (8.62), e is the basis of the natural logarithms. Relation (8.62) gives the exact limit load for

$$j \geq 0.134, \tag{8.63}$$

and must be regarded as an upper bound for smaller j.

Because $\frac{M_p}{N_p} = \frac{t}{4}$ (uniform shell), relation (8.61) can be rewritten as

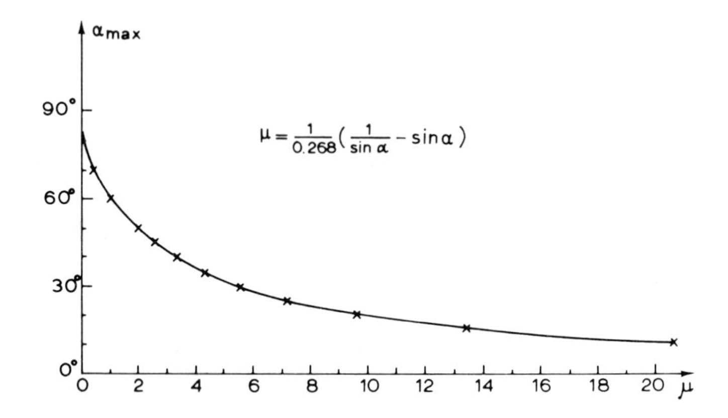

Fig. 8.30.

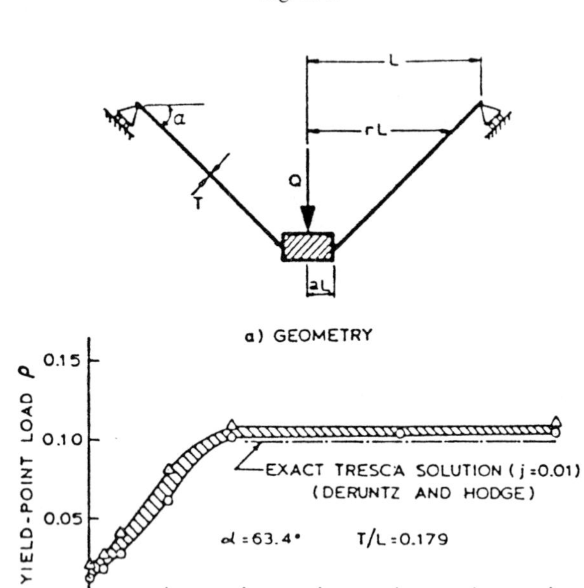

Fig. 8.31. Conical shell with rigid boss [6.28].

$$j = \frac{1}{\mu}\frac{\cos^2\alpha}{2\sin\alpha},$$

where μ is the slenderness ratio 2R/t. Condition (8.63) then becomes

$$\frac{1}{\sin\alpha} - \sin\alpha \geq 0.268\,\mu\,. \tag{8.64}$$

The curve in fig. 8.30 gives α_{max} versus μ for eq.(8.62) to give the exact limit load.

If we now consider vanishingly small boss size ($r_a \to 0$), we obtain a concentrated load at the vertex. The solution is found to be q=0, that is

$$Q = 2\,\pi\, M_p \cos^2\alpha. \tag{8.65}$$

Table 8.5.

$\beta = \frac{1}{4j} = \frac{R \sin\alpha}{t \cos^2\alpha}$	$q = \frac{Q}{2\,\pi\, M_p \cos^2\alpha}$
0	1.1111
0.1	1.1145
0.2	1.1251
0.3	1.1427
0.4	1.1673
0.418	1.1725
0.5	1.1985
0.6	1.2348
0.7	1.2760
0.8	1.3219
0.9	1.3725
1.0	1.4278

It must be emphasized that the limit load, eq. (8.65), is valid for arbitrary connection at the boss and for arbitrary edge conditions, for alls shell geometry such that $\mu \geq \frac{1}{2} \sin\alpha \cos^2\alpha$. This condition is obviously always satisfied for what can be regarded as a shell. When the edge is built-in, the limit value (8.65) is also valid with the exact Tresca yield condition and with the linearized Tresca condition, because the stress field is $n_x = n_\theta = m_x = 0,\;\; m_\theta = 1$ in the whole shell and the corresponding stress point falls on the three yield surfaces.

The *exact Tresca yield condition* has been used by Onat [3.10] and by Lance and Onat [8.49] to analyze the simply supported conical shell loaded by a central boss rigidly connected to the shell. The problem is solved numerically, in the limited range of the shell parameter $\beta = \frac{1}{4j}$ given in table 8.5.

8.4.4. *Other cases of conical shells.*

The previous problem, but with the rigid boss rigidly attached to the truncated shell, was (fig. 8.29' (a)) numerically computed by Biron [6.28] by using the finite element method (see Section 6.3.1 and 6.4.1).

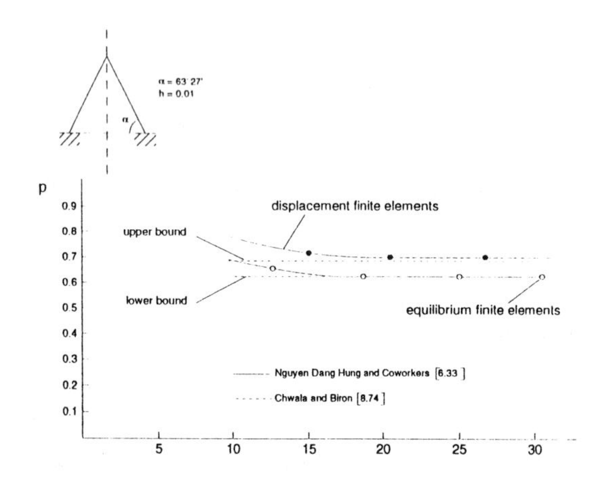

Fig. 8.32. Finite element convergence of limit pressure [6.33].

Fig. 8.31 (b) shows a typical result for a steep shell in the form of the variation of lower and upper bounds to the collapse load as a function of the dimensionless boss radius a. Also plotted is an exact solution obtained by De Runtz and Hodge [8.72] for the same problem, except that the Tresca yield condition was used instead of the Mises condition.

8.4.4. *Other cases of conical shells.*

The previous problem, but with the rigid boss rigidly attached to the truncated shell, was (fig. 8.29' (a)) numerically computed by Biron [6.28] by using the finite element method (see Section 6.3.1 and 6.4.1).

Fig. 8.31 (b) shows a typical result for a steep shell in the form of the variation of lower and upper bounds to the collapse load as a function of the dimensionless boss radius a. Also plotted is an exact solution obtained by De Runtz and Hodge [8.72] for the same problem, except that the Tresca yield condition was used instead of the Mises condition.

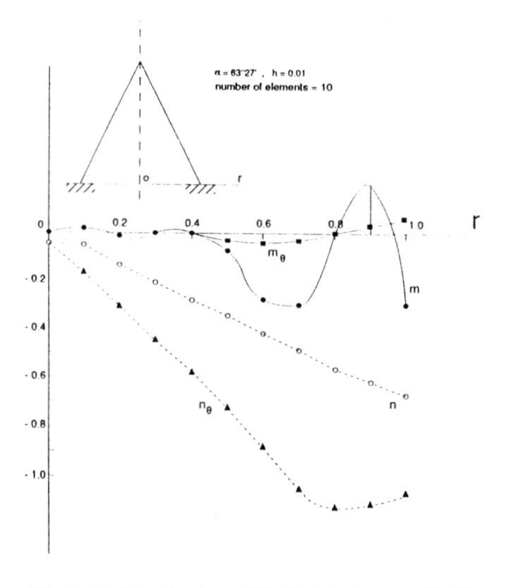

Fig. 8.33. Distribution of limit state stresses [6.33].

It is noted that for small values of the boss radius, no result could be obtained for this particular geometry from the exact solution, because of mathematical complexities. Yet the upper and lower bounds from the suggested procedure were sufficiently close together for a practical evaluation of the collapse load for the entire range of variation of a, hence for a known difficult problem.

The built-in conical shell loaded by a uniform pressure is computed by Nguyen Dang Hung and coworkers [6.33, 6.12], by using the finite element method (see Sections 6.3.1. and 6.4.1). fig. 8.32 shows the convergence of two approaches (displacement and equilibrium), and the comparison with the results of Chwala and Biron [8.74]. We remark that the statical method presents a decreasing convergence. Indeed, the yield criterion is enforced only at two points by element but not everywhere. Consequently, the computed bound is only an approximation of the lower bound, which must converge to the exact bound when the finite element mesh is sufficiently dense.

Fig. 8.33 shows the distribution of stresses, obtained by equilibrium finite elements (strong variables). Comparison of results concerning displacement rates is done in fig. 8.34.

The simply supported conical shell loaded with a uniformly distributed pressure is analyzed in the paper by Kuech and Lee [8.52] on the basis of the limited interaction yield condition.

The approximate yield hypersurface of Flügge and Nakamura [8.44] was used by them to study a conical shell with a centred cut-out along the edge of which acts a uniformly distributed shear force. The outer edge of the shell is built-in.

The reader is referred to the original papers for more information.

8.5. Torispherical and toriconical thin pressure vessels heads.

8.5.1. *Introduction.*

The correct design of metal pressure vessels and tanks, and especially of their heads, is a very important engineering problem. We therefore devote a whole section to it.

Following Drucker and Shield [8.53], consider a cylindrical vessel subjected to uniform internal pressure. The studied head is made of a torus with radii r and R (see fig. 8.35). Results will also apply when the spherical cap is replaced by the tangent cone drawn with a dashed line in fig. 8.35. The torispherical shape is used to keep the height H of the head reasonably small with the same thickness as the cylinder. An end plate would be much thicker, whereas achieving some kind of optimum (with respect to stress or weight) with a thickness not larger than that of the cylinder results in values of H ranging from 0.255D to 0.3938D [8.54, 8.55].

At present, the torispherical heads are designed according to (semi-empirical) specifications of various codes, the most widely used of which is probably the A.S.M.E.

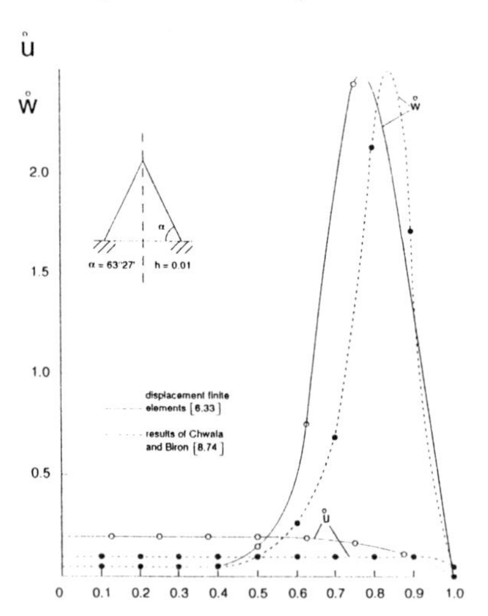

Fig. 8.34. Distribution of displacement rates [6.33].

Code for Unfired Pressure Vessels. Collapse of a large vessel with torispherical head during hydrostatic pressure tests in the United States some years ago has caused critical examination of the code formulas, which turn out to be unsafe for large $\mu = D/t$, that is, for large vessels subjected to relatively low pressure. Consider a torispherical head with a slenderness ratio $\mu = 440$. Galletly [8.56] has shown that the stresses predicted by the formulas of the A.S.M.E. Code were less than half those given by a correct elastic analysis. Compressive circumferential stresses as high as 30.000 psi under 1.5 times the service pressure, were shown to exist, though completely neglected in the code. Hence, a head of the type studied made of a steel with $\sigma_Y = 29.000$ psi, would have been regarded as "correctly designed" according to the A.S.M.E. Code, and would have exhibited, under hydrostatic testing, large plastic deformations and possibly cracks and rupture.

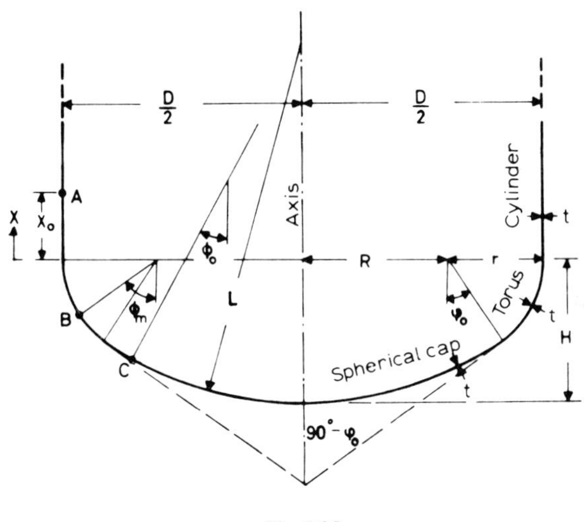

Fig. 8.35.

It is therefore of great practical interest to evaluate the limiting value of the pressure that produces a plastic collapse mechanism. As already noted in Section 8.2, the corresponding theoretical predictions are fairly well supported by experiments [8.20].

8.5.2. *Limit pressure.*

Referring to fig. 8.36, the equilibrium equations of a shell element are

$$r_0 \frac{d N_\varphi}{d \varphi} + \left(N_\varphi - N_\theta\right) r_1 \cos\varphi - V r_0 + Y r_1 r_0 = 0 , \tag{8.66}$$

$$r_0 N_\varphi + r_1 \sin \varphi N_\theta + \frac{d}{d \varphi}\left(V r_0\right) - p r_0 r_1 = 0 , \tag{8.67}$$

$$r_0 \frac{d M_\varphi}{d \varphi} + \left(M_\varphi - M_\theta\right) r_1 \cos\varphi - V r_1 r_0 = 0 . \tag{8.68}$$

These formulas may be simplified in accordance with the following arguments [8.45]. For the bending state to contribute in an appreciable manner to balancing the pressure p, the shear force V must be of the order of $N_p = \sigma_Y t$ (or at least not much smaller) as can be seen from eqs. (8.66) and (8.67). But because $M_\varphi - M_\theta$ cannot exceed

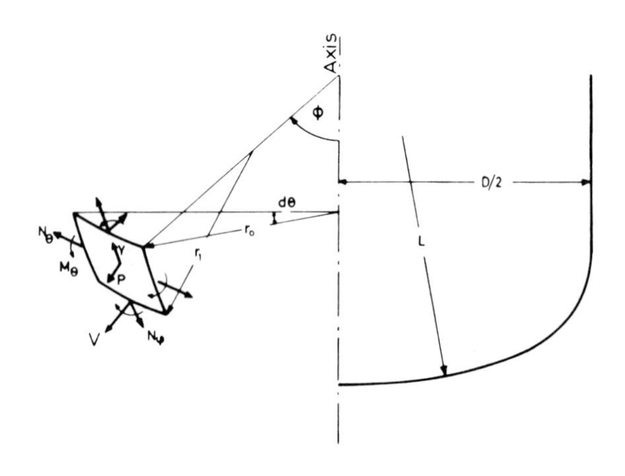

Fig. 8.36.

$2M_p = \sigma_Y t^2/2$, the term $V r_0 r_1$ in eq. (8.68) is large with respect to $\left(M_\varphi - M_\theta\right) r_1 \cos\varphi$ provided t/r_0 remains small, that is, at all points of the shell far enough from the axis. Eq. (8.68) can then be simplified into

$$\frac{d M_\varphi}{d\varphi} - V r_1 = 0 . \tag{8.69}$$

If it is now assumed that the rate of curvature $\dot{\kappa}_\theta$ remains negligible, M_θ is a reaction and the yield surface derived from the Tresca condition is shown in fig. 8.37 (a) (see Section 5.4.2). To achieve further simplification, we shall use the simpler circumscribed yield surface of fig. 8.37 (b), which consists of a parabolic cylinder and four planes.

In the momentless state of stress ($M_\theta = M_\varphi = V = 0$, pure membrane response) we only have eqs. (8.66) and (8.67). If the first is integrated with respect to φ, taking into account that $r_0/\sin\varphi$ is equal to the value r_2 of the principal radius of curvature in the plane normal to the meridian, we obtain

$$\text{(a)} \qquad N_\varphi = \frac{p\, r_2}{2}$$

$$\text{(b)} \qquad \frac{N_\varphi}{r_1} + \frac{N_\theta}{r_2} = p. \qquad (8.70)$$

A membrane reaches its limit load as soon as the most stressed point attains the yield

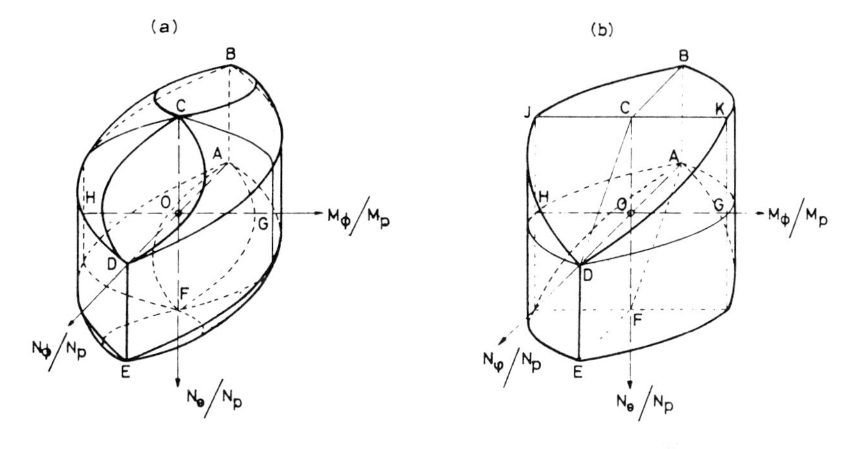

Fig. 8.37.

limit. Indeed, as the state of stresses is statically determinate, there is no possible stress redistribution and, hence, no increase in load in the absence of a change of geometry.

The dangerous point is at the junction of the torus with the spherical cap, where $N_\varphi = pL/2$. With this value of N_φ , eq. (8.70 b) gives $N_\theta = -pL\,(L/2r - 1)$ because $r_1 = r_0$. The limit pressure for the membrane is reached when $N_\varphi - N_\theta = \sigma_Y t$, that is, for

$$p^M = \frac{2\, t\, \sigma_Y\, r}{L\,(L - r)}. \qquad (8.71)$$

Table 8.6 displays the values of

$$\frac{p^M D}{2 t \sigma_Y} = \frac{r D}{L(L-r)} .$$

Actually, the torus is subjected to significant bending stresses. Under increasing pressure, the meridional axial forces N_φ applied by the cylinder and the sphere to the torus tend to depress its curvature $1/r$, that is, to push the region B in fig. 8.35 inward. If the tore thickness, though small with respect to its radius r, is large enough to avoid buckling under the compressive axial forces N_θ, a hinge circle will occur at B (fig. 8.35) enabling the central region to contract circumferentially and deflect inward. A hinge circle with the rate of rotation of the opposite sign will form in the spherical cap at C, and a similar hinge circle will occur at A, usually in the cylinder, but sometimes in the torus. The whole region ABC must be plastic to enable inward displacement. In this region, because $N_\varphi > 0$ and $N_\theta < 0$, the yield condition will be

$$N_\varphi - N_\theta = \sigma_Y t,$$

$$|M_\varphi| \le \frac{t^2}{4} \sigma_Y \left[1 - \left(\frac{N_\varphi}{t \sigma_Y} \right)^2 \right], \tag{8.72}$$

according to fig. 8.32 (b).

Assuming that M_φ exhibits analytical extremums in A, B and C, where consequently V=0, integration of the equilibrium equations, eqs. (8.66), (8.67) and (8.69) gives the stress field in the plastic region ABC and the corresponding value of the pressure. The results are :

1. in the cylinder,

$$M_\varphi = -\frac{t^2}{4} \sigma_Y \left[1 - \left(\frac{p D}{4 t \sigma_Y} \right)^2 \right] + \frac{t \sigma_Y}{2 D} \left(2 + \frac{p D}{2 t \sigma_Y} \right) \left(x_o - x \right)^2, \tag{8.73}$$

$$V = \frac{t \sigma_Y}{D} \left(2 + \frac{p D}{2 t \sigma_Y} \right) \left(x_o - x \right), \tag{8.74}$$

where x is the abscissa from the junction with the torus, and x_o the abscissa of the hinge circle A.

2. In the torus,

$$\frac{M_\varphi}{r\,t\,\sigma_Y}=\frac{t}{4\,r}\left[1-\left(\frac{p\,D}{2\,t\,\sigma_Y}\right)^2\frac{\left(R+r\sin\varphi_m\right)^2}{D^2\sin^2\varphi_m}\right]-\frac{p\,R}{2\,t\,\sigma_Y}\,\frac{\left[1-\cos(\varphi-\varphi_m)\right]}{\sin\varphi_m}$$

$$+\frac{r}{R}\cos\varphi\left[k(\varphi_m)-k(\varphi)\right]+\ln\frac{R+r\sin\varphi}{R+r\sin\varphi_m}\,, \tag{8.75}$$

$$\frac{V}{t\,\sigma_Y}=\frac{p\,R}{2\,t\,\sigma_Y}\,\frac{\sin(\varphi_m-\varphi)}{\sin\varphi_m}+\frac{r}{R}\sin\varphi\left[k(\varphi)-k(\varphi_m)\right], \tag{8.76}$$

$$k(\varphi)=\int_{\varphi_o}^{\varphi}\frac{R\,d\varphi}{R+r\sin\varphi}=\frac{2\,R}{(R^2-r^2)^{1/2}}\left[\tan^{-1}\left\{\frac{r+R\tan\varphi/2}{(R^2-r^2)^{1/2}}\right\}\right]_{\varphi_o}^{\varphi} \tag{8.77}$$

φ is the angle normal to the meridian curve with the axis of rotation ; φ_m is the angle φ corresponding to the hinge circle B. In the sphere (regarding $\varphi-\varphi_o$ as small quantity)

$$\frac{M_\varphi}{L\,t\,\sigma_Y}=\frac{t}{4\,L}\left[1-\left(\frac{p\,L}{2\,t\,\sigma_Y}\right)^2\right]+\frac{1}{2}\left(\varphi-\varphi_s\right)^2, \tag{8.78}$$

$$\frac{V}{t\,\sigma_Y}=\varphi-\varphi_s, \tag{8.79}$$

where φ_s corresponds to the hinge circle C.

The four unknowns p, φ_m, φ_s and x_o are obtained from the continuity conditions on M_φ and V at the junction of the cylinder and the tore (x=0, $\varphi=\pi/2$) and at the junction of the tore and the spherical cap ($\varphi=\varphi_o$). The four equations were solved by successive approximations for these values of the shell parameters :

$$\frac{t}{D} = 0.002\ ;\ 0.004\ ;\ 0.006\ ;\ 0.008\ ;\ 0.010\ ;\ 0.012\ ;\ 0.014\ ;$$

$$\frac{L}{D} = 1.0\ ;\ 0.9\ ;\ 0.8\ ;\ 0.6\ ;$$

$$\frac{r}{D} = 0.006\ ;\ 0.008\ ;\ 0.010\ ;\ 0.12\ ;\ 0.14\ ;\ 0.16\ .$$

The limit loads obtained are exact for the yield surface of fig. 8.37 (b) because collapse mechanisms corresponding to the stress fields are shown to exist [8.45]. They are upper bounds p_+ to exact limit load p_l for the yield surface of fig. 8.37 (a), completely inscribed to that of fig. 8.37 (b).

Lower bounds $p_- = \alpha\, p_+$ are obtained with a factor α such that the stress point $\alpha\, N_\varphi$, $\alpha\, N_\theta$, $\alpha\, M_\varphi$ is on or within the yield surface of fig. 8.37 (a), when applied to the stress field above. It turns out that

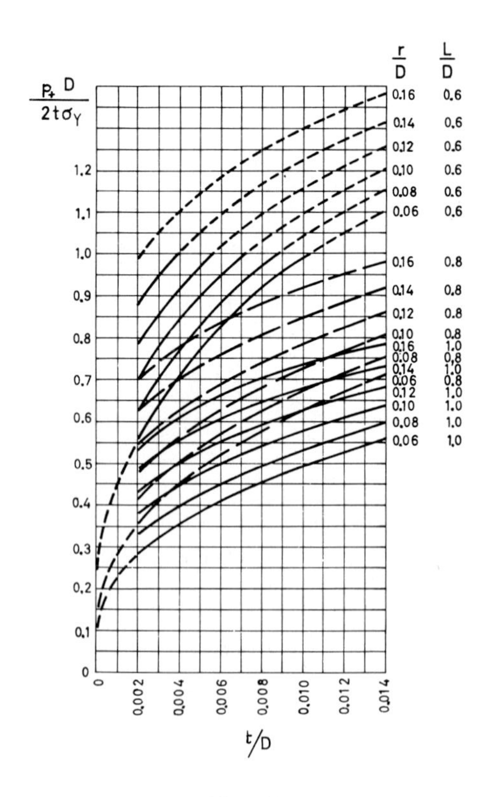

Fig. 8.38.

$$\alpha = \frac{T^2 - 4\,T + 12}{2\,(T^2 - 4\,T + 8)}, \tag{8.80}$$

with $T = \frac{(p + D)}{2\,t\,\sigma_Y}$, the critical point being A. The factor α varies from 0.82 to 0.90. The values of p_+ are given in figs. 8.38 and 8.39. The values of p_- are given in figs. 8.40 and 8.41. If p_s is the service pressure and λ the load factor (or safety factor) it will be sufficient for practical applications to use mean value.

$$p_l \equiv \lambda\,p_s = \frac{p_- + p_+}{2}. \tag{8.81}$$

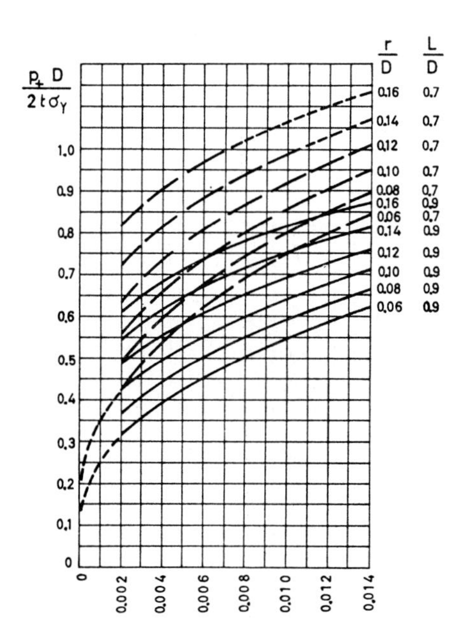

Fig. 8.39.

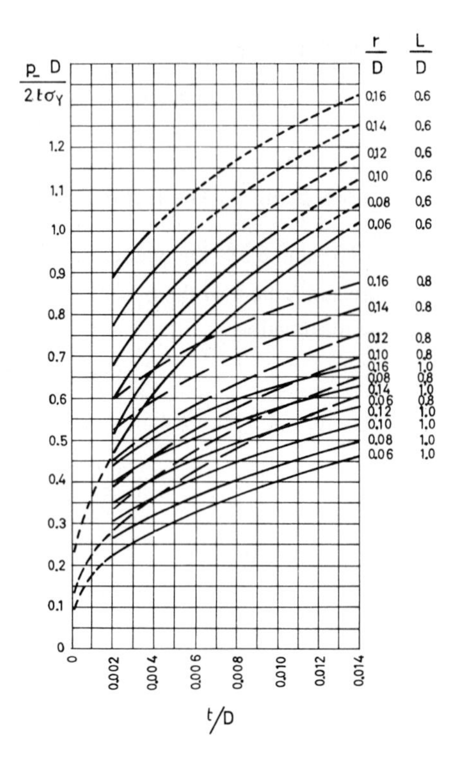

Fig. 8.40.

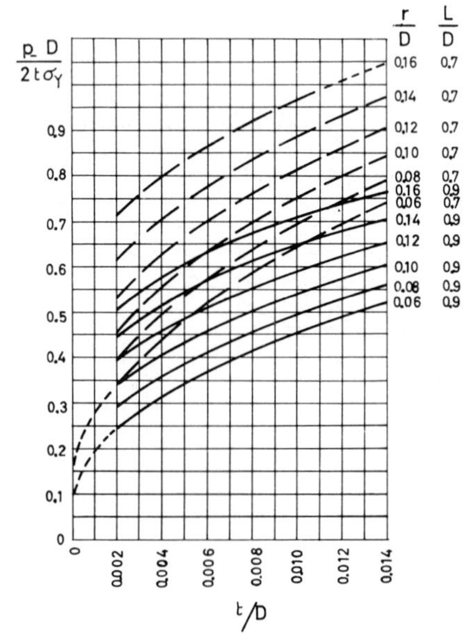

Fig. 8.41.

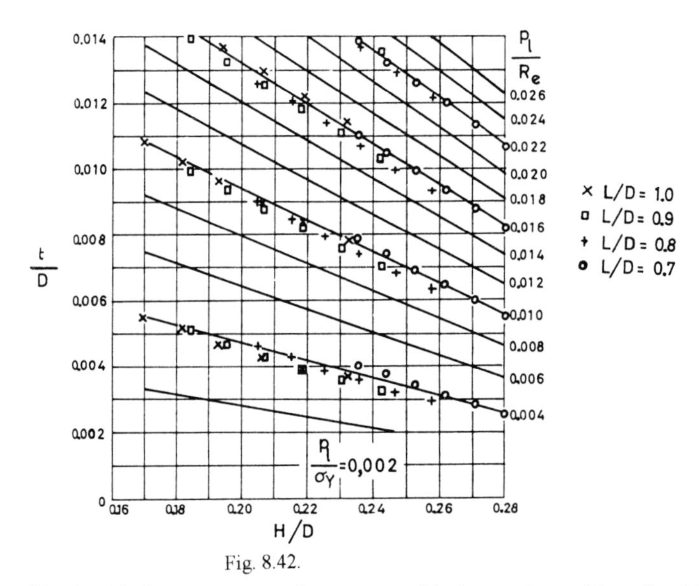

Fig. 8.42.

For a given t/D, the limit pressure p_l increases with increasing r/D and with decreasing L/D. The ratio H/D varies in a similar manner with r/D and L/D (see table 8.6).

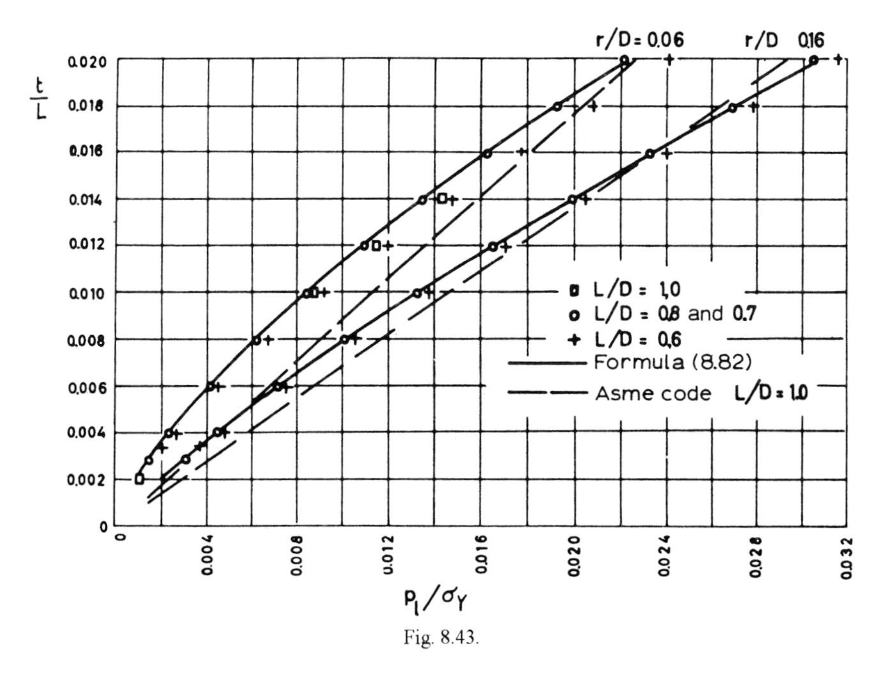

Fig. 8.43.

In fig. 8.42 the straight lines represent approximately the relations t/D versus H/D for fixed p_l/σ_Y, in the range 0.17 < H/D < 0.28 corresponding to r/D varying from 0.06 to 0.16 and

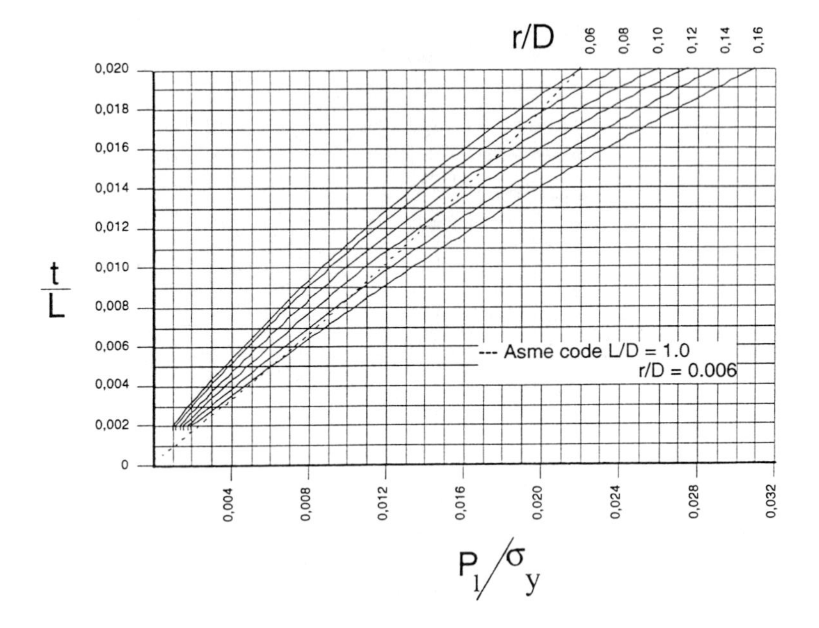

Fig. 8.44.

L/D from 0.7 to 1.0. The exact points are indicated to show the accuracy of the approximation.

It also turns out that, for given r/D, the variation of p_l/σ_Y with t/L is nearly independent of L/D. For L/D = 0.7 and 0.8, the formula

$$\frac{p_l}{\sigma_Y} = \left(0.33 + 5.5\,\frac{r}{D}\right)\frac{t}{L} + 28\left(1 - 2.2\,\frac{r}{D}\right)\left(\frac{t}{L}\right) - 0.0006 \tag{8.82}$$

gives a value of p_l/σ_Y very near of that given by eq. (8.81) (e.g., with less than 3% discrepancy for t/L = 0.01). Relation (8.82) can also be used for L/D = 1.0, 0.9, and 0.6 as shown is fig. 8.43. It is represented in fig. 8.44.

8.5.3. *Discussion.*

The analysis in Section 8.5.2 implies that the connection between the cylinder and the spherical cap is the weak part of the vessel.

This situation occurs when this torical connection is thin and of large curvature.

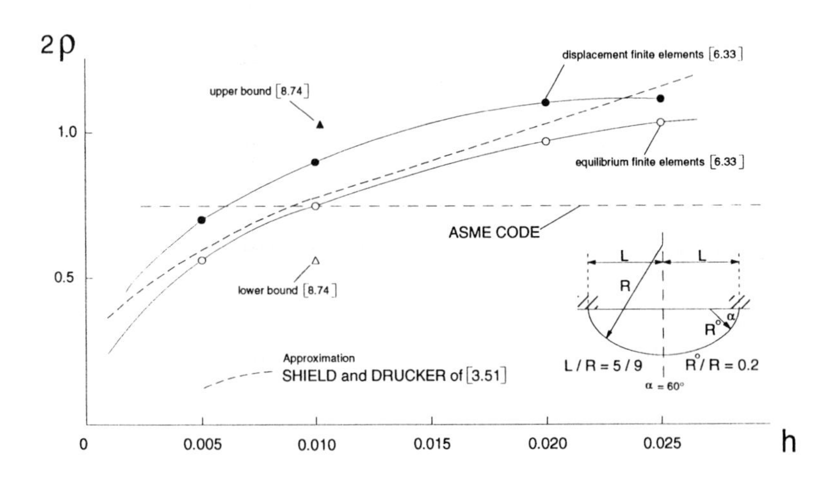

Fig. 8.45. Finite element approximation for torispherical shells [6.33].

Large storage tanks or vessels for relatively low internal pressure (1 to 5 kg/cm^2, for example), when designed according to the A.S.M.E. Code for Unfired Pressure Vessels, belong to this class. As shown in fig. 8.44 by the dashed ray, the A.S.M.E. code formulas can then be unsafe, especially in the region of small t/L.

On the other hand, for high-pressure vessels, the relative thickness of the torus, as given by the code, increases. The connection between the cylinder and the sphere becomes rigid and behaves like a reinforcing ring. Because L/D is usually greater than 0.5, the cylinder now becomes the weakest part of the vessel.

The curves giving the limit pressure are thus valid only for $pD/2t\,\sigma_Y < 1$. this is the reason why they have been drawn in dashed lines for $pD/2t\,\sigma_Y \geq 1$.

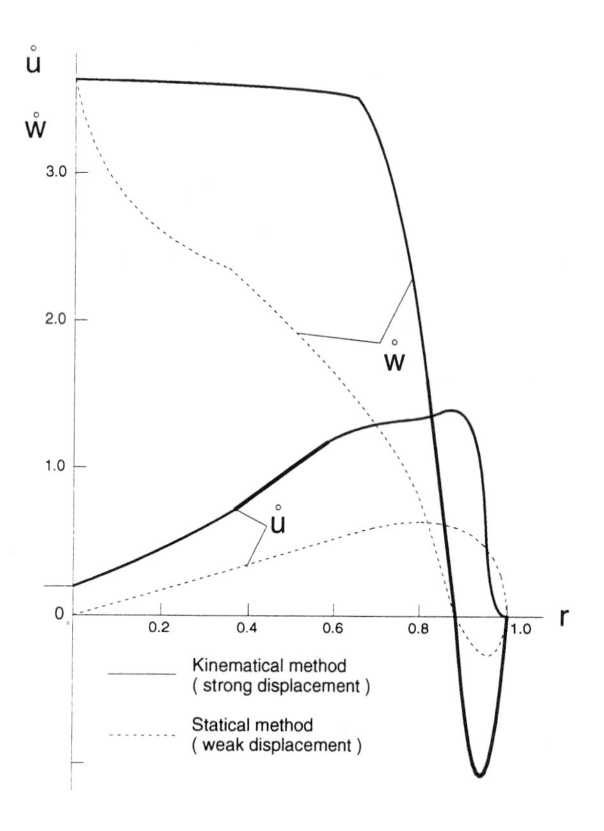

Fig. 8.46. Torispherical shell (h=0.02) - distribution of displacement rates [6.33].

8.5.4. *Finite element method.*

Nguyen Dang Hung and his coworkers have analyzed the problem of a torispherical shell by displacement and equilibrium finite elements [6.33, 6.12] (see Sections 6.3.1 and 6.4.1). The comparison is showed in fig. 8.45 with the theoretical formula proposed by Shield and Drucker [8.53] obtained by averaging the bounds given in [8.46] and [8.53] :

$$\rho_{th} = 0.165\,\frac{L}{R} + 1.375\,\frac{R_o}{R} + 56\,.\left(1 - 1.1\,\frac{R_o}{L}\right)\left(\frac{L}{R}\right)^2 h - 75/h\,.\,10^{-6}\,.$$

Finite element results are found in close agreement with those of Shield and Drucker. On the other hand, we see in fig. 8.45 that the A.S.M.E. Code is more optimist for $h > 0.01$ and more pessimist for h lesser than this value. Fig. 8.46 shows the distribution of displacement rates.

For other upper and lower bound solutions by finite elements, the reader may consult the paper of Biron and Charleux [6.66].

8.5.5. *Ellipsoidal shell heads.*

Fig. 8.47 shows the results of Nguyen Dang Hung and coworkers [6.33, 6.12] (see Sections 6.3.1 and 6.4.1), compared to those of Chwala and Biron [8.51]. The values in [6.33], obtained by finite element method are much closer. On the other hand, the statical values are greater than the kinematical ones. This anomaly is due to the fact that the yield criterion is not enforced everywhere in the equilibrium elements.

Other results of lower and upper bound by finite element method have been obtained by Biron and Charleux in [6.66].

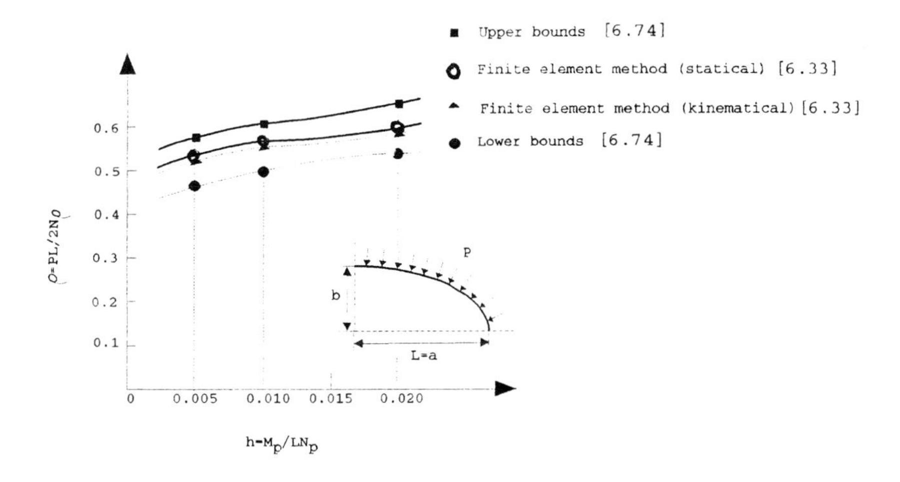

Fig. 8.47. Ellipsoidal shell [6.33].

8.6. Axisymmetric pressure vessel intersections.

8.6.1. *Introduction.*

The problem of limit analysis of pressure vessel intersections is already at a relatively advanced stage of solution, at least where rotational symmetry is concerned. Closed form solutions are obtained by analytical studies [8.59, 8.60].

Theoretical predictions have found good experimental support in six collapse tests [8.61].

Of course, these solutions require some simplifying assumptions, such as the equality of circumferential and meridional moments or the use of a limited interaction yield surface.

It is noted, however, that each specific problem became a specific research project, with few possibilities for evaluating the consequences of design modifications (such as for example other shapes of reinforcement or different head thicknesses) without starting a new investigation.

So, a considerable improvement was obtained by Biron and Charleux, by introducing numerical methods, applicable to a very wide range of geometrical configurations. While non-symmetric problems are still sufficiently difficult to warrant a special method for each problem studied, a very limited number of computer programs should be sufficient to handle in a reasonable amount of computer time the limit analysis of pressure vessel intersections of arbitrary rotationally symmetric shape subjected to pressure.

We present in the next Section, the lower and upper solutions of Biron and Charleux [6.66] by finite element method.

8.6.2. *Formulation of the problem.*

Consider the assembly of three rotationally symmetric shells under pressure, as shown in fig. 8.48. The first shell is a cylinder of radius R^c and thickness T^c, the third is a portion of a sphere of radius R^s and thickness T^s while the intermediate shell (generally corresponding to a reinforcement), of thickness T^1 and axial length L^1 may be defined by an arbitrary relation R = R (Z). Let the cylindrical coordinates be the radius R, the circumferential angle θ and the axial distance X for the cylinder and Z for the other two shells. If subscripts φ and θ denote components along the meridional and circumferential directions, respectively, the stress state is defined by five stress resultants per unit length : two direct stresses N_θ and N_φ two moments M_θ and M_φ and a transverse shear force S. Sign conventions are illustrated in fig. 8.36 and angles of intersection and radii of curvature are shown in detail in fig. 8.49.

In what follows, superscripts c, i and s are used to indicate relation to cylinder intermediate shell and sphere respectively.

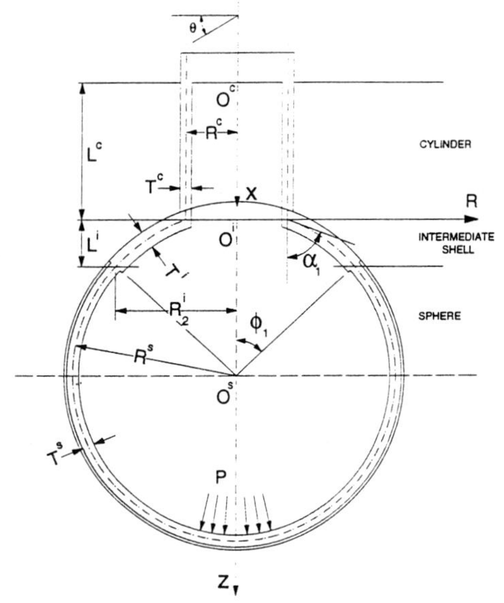

Fig. 8.48. Shell geometry [6.66].

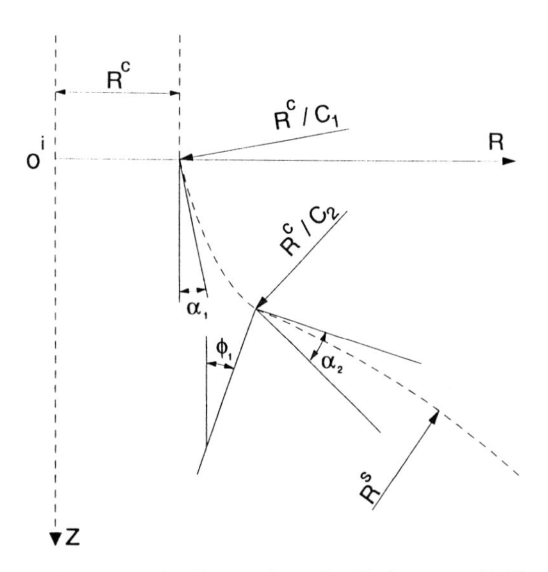

Fig. 8.49. Angle of intersection and radii of curvature [6.66].

Two different boundary conditions are considered in zone 1 (in the cylinder), namely :

(a) the end of the cylinder is fixed.

(b) There is no shear at the end of the cylinder corresponding for example to a plane of symmetry.

For the numerical procedure used, the reader may see Sections 6.3.1 and 6.4.1.

8.6.3. *Results.*

It is convenient to introduce the following dimensionless quantities :

$$\eta = \left(\frac{T^c}{R^c}\right)\left(\frac{T^s}{R^s}\right), \quad \omega = \frac{T_c}{4\,R_c}, \qquad \beta = \frac{R^c}{R^s}.$$

Fig. 8.50 and 8.51 show lower and upper bounds obtained by the present method for η =1.5 (corresponding to a weaker cylinder) and n=2.5 (corresponding to a weaker sphere). Results obtained in refs. [8.59], [8.60] with the limited interaction yield surface are also given.

The present results are of course somewhat lower than those in refs. [8.59], [8.60] because of the different yield condition, and the largest discrepancy between these two sets

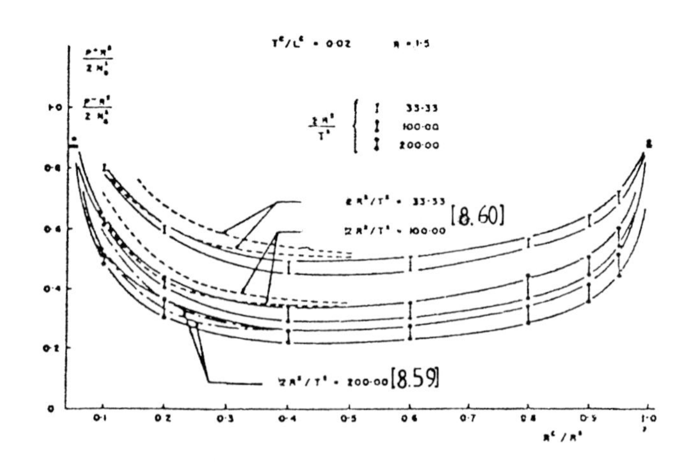

Fig. 8.50. Lower and upper bounds for sphere-cylinder intersections, n=1.5 [6.6]

of results (of therer of 20 per cent) occurs for η =2.5 with $\frac{R^c}{R^s} > 0.3$.

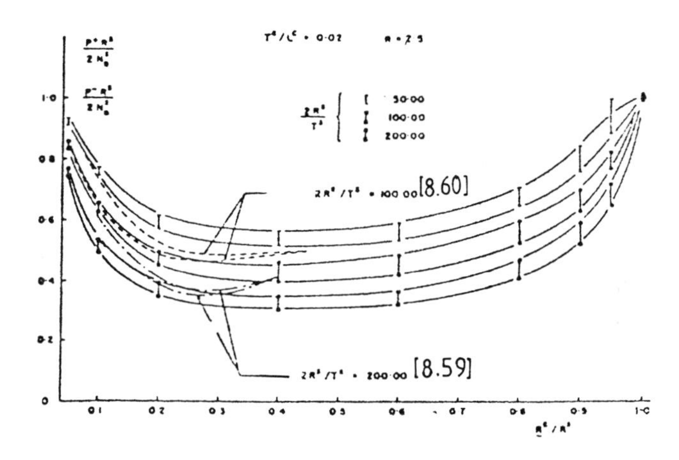

Fig. 8.51. Lower and upper bounds for sphere-cylinder intersections, n=2.5 [6.6]

Results in [8.59] and [8.60] obtained with limited interaction criterion for fig. 8.50 and 8.51.

A detailed analysis of the results of one particular problem is presented here in order to illustrate specific aspects of the program. The parameters selected were :

$$\eta = 2.5, \qquad \omega = 0.00625, \qquad \beta = 0.2$$

and the bounds obtained for that particular test were 0.348 and 0.393.

Fig. 8.52 shows, with dimensions, the geometry of the structure in the undeformed and in the deformed state, as established by the upper bound program. A total of eight zones, four in each shell, were used as indicated by circled numbers.

Fig. 8.53 shows the variation of all stress components in the cylinder and the sphere. It is noted in particular that, in the sphere, the circumferential moment m_θ^s and the meridional moment m_φ^s are different. This observation contradicts the traditional assumption made in previous analyses, [8.59], [8.60] (which is of course allowed for lower bounds) that they should be equal.

Other particular geometries are analyzed also in [8.59], particularly the case of sphere-cone-cylinder intersections.

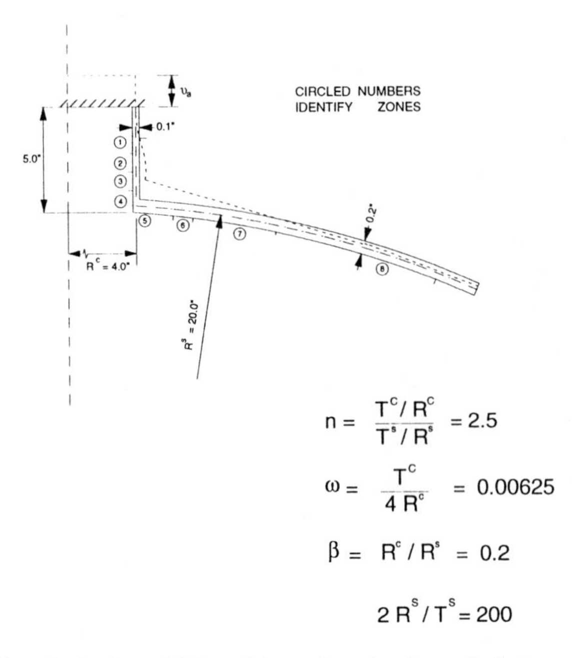

Fig. 8.52. Zone distribution and deformed shape of sample sphere-cylinder intersection.

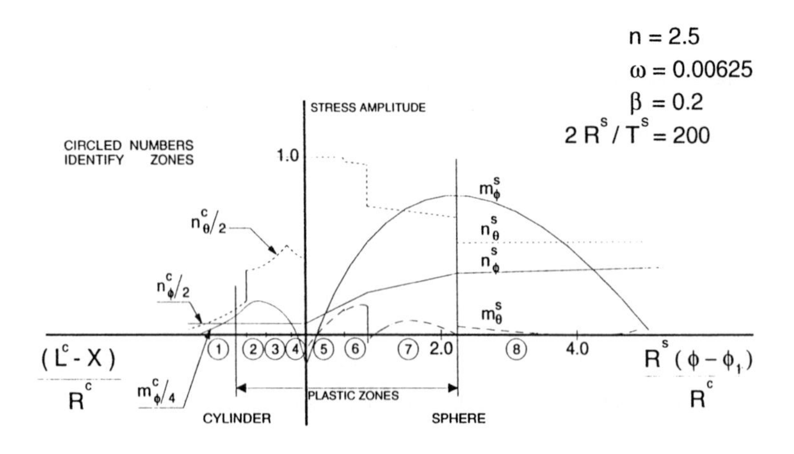

Fig. 8.53. Stress variation in sample problem as obtained from lower bound results.

8.7. Thin-wall beam with circular axis.

8.7.1. *Introduction.*

It is well known that the flanges of I beams (or channel beams) with curved axis are subjected to transversal bending because the axial resultant forces in a flange in neighbouring sections forming the angle d φ (fig. 8.54) are statically equivalent to a radial force. Hence, when the dimensions of the flanges are such that this effect becomes noticeable, the collapse mechanism does not retain the shape of the cross section but includes transversal bending of the flanges [8.62, 8.63].

8.7.2. *Wide-flange I beam.*

Consider a wide flange I beam with circular axis, subjected to a uniform bending moment diagram with magnitude M (fig. 8.54). Due to the constancy of M with φ, the deformed axis will remain circular.

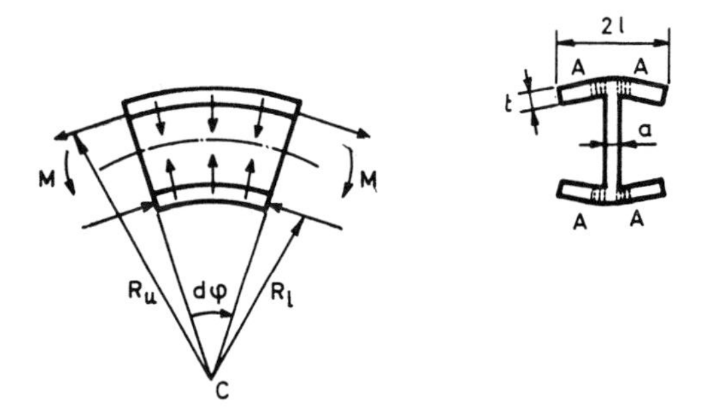

Fig. 8.54.

The flanges will behave like cylindrical, axisymmetrically loaded, shells that are built-in at the web. An element of the upper flange is shown in fig. 8.55. In the present situation, $N_x = 0$ whereas V_x and M_θ are reactions. The generalized stresses M_x and N_θ are related by the equilibrium equation

$$\frac{d^2 M_x}{d x^2} + \frac{N_\theta}{R} = 0 , \tag{8.83}$$

obtained by eliminating V_x from the equilibrium equations

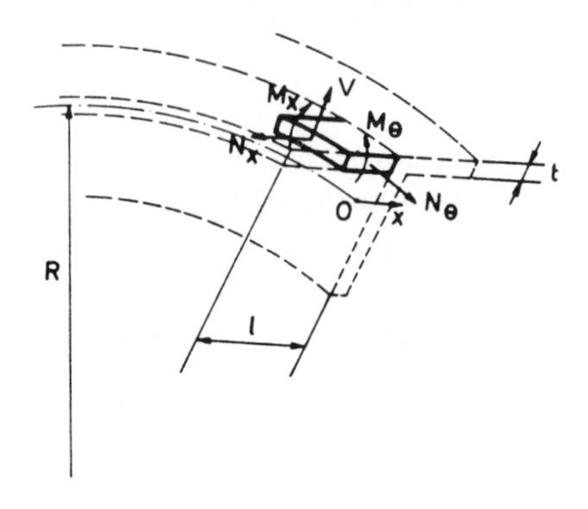

Fig. 8.55.

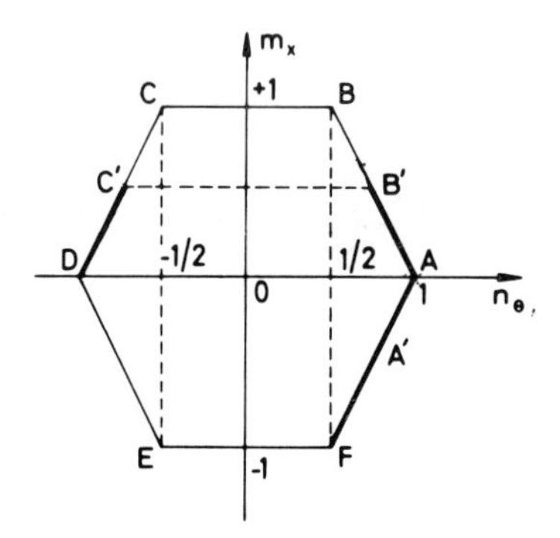

Fig. 8.56.

$$\text{(a)} \qquad \frac{d V_x}{d x} + \frac{N_\theta}{R} = 0,$$

$$\text{(b)} \qquad \frac{d M_x}{d x} = V_x . \qquad (8.84)$$

We use the hexagonal yield condition of fig. 8.56 for the flange, with $m_x = \dfrac{M_x}{M_{pf}}$ and $n_\theta = \dfrac{N_\theta}{N_{pf}}$, where $M_{pf} = \sigma_Y \dfrac{t^2}{4}$ and $N_{pf} = \sigma_Y\, t$. This is an inscribed approximation to the exact curve derived from the yield criterion of Tresca (see Section 5.4.4, fig. 5.24). Eq. (8.83) can be rewritten as

$$\frac{d^2 m_x}{d x^2} + \frac{4}{t R} n_\theta = 0 . \tag{8.85}$$

For a flange with a not-too-large span l (fig. 8.55), the bending moment M_x at the connection with the web will remain smaller that the plastic moment and the stress profile will lie entirely on side AF (fig. 8.56) with the equation

$$m_x - 2 n_\theta = -2 . \tag{8.86}$$

The boundary conditions at x=0 are

$$m_x = 0, \qquad \frac{d m_x}{d x} = 0. \tag{8.87}$$

Integration of eq. (8.85), taking account of eqs. (8.86) and (8.87), furnishes $n_\theta = \cos\left\{\left(\frac{2 x}{t R}\right)^{½}\right\}$ If we let

$$\alpha = l \left(\frac{2}{t R}\right)^{½}, \tag{8.88}$$

the expression for n_θ becomes

$$n_\theta = \cos \frac{\alpha x}{l},$$

valid for $1 \geq n_\theta \geq \frac{1}{2}$, that is, for $0 \leq x \leq \sqrt{0.548tR}$. As long as

$$l \leq \sqrt{0.548tR}, \tag{8.90}$$

we can use eq. (8.89) to calculate the *efficiency* ρ_f of the flange, defined as

$$\rho_f = \frac{\int_0^l n_\theta \, dx}{l} \equiv \frac{\int_0^l N_\theta \, dx}{N_{pf} l} . \tag{8.91}$$

When condition (8.90) is not satisfied, the stress profile must be modified. Since transversal bending deflections produce circumferential strains ε_θ with a sign opposite to those produced by the direct bending moment M on the I beam, we are led to choose the stress profile DC'B'AF. In this profile, the extremity of the flange is in regime DC', with $\dot{\varepsilon}_\theta < 0$, though the considered flange is, as an average, subjected to traction. The jump from C' to B' (discontinuity in n_θ with x) in admissible, and continuity of m_x (and $\frac{d\,m_x}{d\,x}$) is preserved. We now have the following situation :

1. Regime DC' over $0 \le x \le d_1$. The yield condition is

$$m_x = 2 + 2\, n_\theta . \tag{8.92}$$

The boundary conditions are

$$m_x = 0, \qquad \frac{d\,m_x}{d\,x} = 0, \qquad \text{at } x=0. \tag{8.93}$$

The resulting stress distribution is

$$n_\theta = -\cos\frac{\alpha\, x}{l},$$

$$m_x = 2 - 2\cos\frac{\alpha\, x}{l}. \tag{8.94}$$

2. Regime B'A over $d_1 \le x \le d_2$. The yield condition is

$$m_x + 2\, n_\theta = 2 . \tag{8.95}$$

Elimination of m_x from eqs. (8.85) and (8.95), and integration, gives

$$n_\theta = A\, e^{\alpha x/l} + B e^{-\alpha x/l} \tag{8.96}$$

3. Regime A F over $l - (d_1 + d_2)$. The yield condition is

$$m_x = -2 + 2\, n_\theta\,. \tag{8.97}$$

Substitution of eq. (8.97) into eq. (8.85), and integration yields

$$n_\theta = C \sin \frac{\alpha x}{l} + D \cos \frac{\alpha x}{l}\,. \tag{8.98}$$

The six unknowns A, B, C, D, d_1, and d_2 are obtained from the following six conditions : at

$x = d_1$	discontinuity of n_θ by mere sign change,
$x = d_1$	continuity of V_x, that is, of $\frac{dm_x}{dx}$,
$x = d_1 + d_2$,	$n_\theta = 1$,
$x = d_1 + d_2$	continuity of n_θ,
$x = d_1 + d_2$	continuity of V_x, that is, of $\frac{dm_x}{dx}$,
$x = l$,	$n_\theta = \frac{1}{2}$.

If we let $a_1 \equiv \frac{d_1}{l}$, $a_2 \equiv \frac{d_2}{l}$, the conditions just stated are

$$\cos \alpha a_1 = A e^{\alpha a_1} + B e^{-\alpha a_1}\,,$$

$$\sin \alpha a_1 = -A e^{\alpha a_1} + B e^{-\alpha a_1}\,,$$

$$1 = A e^{\alpha(a_1 + a_2)} + B e^{-\alpha(a_1 + a_2)}\,,$$

$$1 = C \sin \alpha(a_1 + a_2) + D \cos \alpha(a_1 + a_2)\,,$$

$$C \cos \alpha (a_1 + a_2) - D \sin \alpha (a_1 + a_2) = -Ae^{\alpha (a_1 + a_2)} + Be^{-\alpha (a_1 + a_2)},$$

$$C \sin \alpha + D \cos \alpha = \frac{1}{2}. \tag{8.99}$$

The system of eqs. (8.99) was solved on the Bull computer of Liège University, for α varying from $\alpha = 1.045$ (corresponding to $\cos \alpha = \frac{1}{2}$) to $\alpha = 5.0$. With the curves of A, B, C, D, a_1, and a_2 versus α, the efficiency of the flange can be calculated. It is found to be

$$\rho_f = -\frac{\sin \alpha a_1}{a_1} + \frac{A}{\alpha}\left[e^{\alpha (a_1 + a_2)} - e^{\alpha a_1}\right] - \frac{B}{\alpha}\left[e^{-\alpha (a_1 + a_2)} - e^{\alpha a_1}\right]$$
$$-\frac{C}{\alpha}\left[\cos \alpha - \cos \alpha(a_1 + a_2)\right] + \frac{D}{\alpha}\left[\sin \alpha - \sin \alpha(a_1 + a_2)\right]. \tag{8.100}$$

The curve ρ_f versus α is given in fig. 8.57.

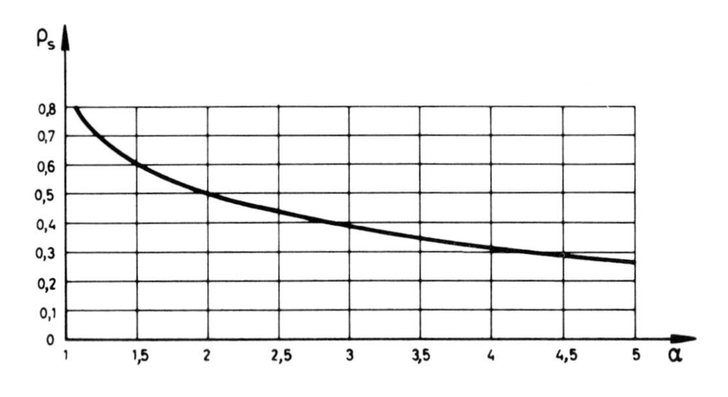

Fig. 8.57.

We now turn to the problem of the *efficiency of the web*. The total axial forces N_u and N_l in the upper and lower flange with radii R_u and R_l, respectively, can be determined from the preceding analysis. The radial stresses applied by the flanges to the web with thickness a are (see fig. 8.57 and 8.58)

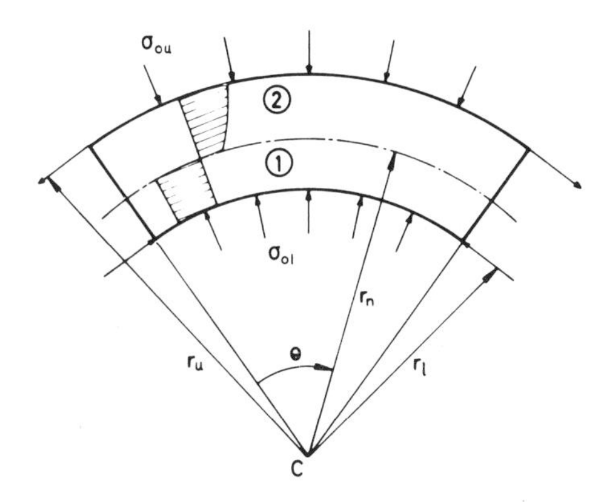

Fig. 8.58.

$$\sigma_{ou} = -\frac{N_u}{aR_u} < 0 \, ,$$

$$\sigma_{ol} = -\frac{N_l}{aR_l} < 0 \, . \qquad (8.101)$$

The web is also subjected to bending stresses σ_θ. In region 1, fig. 8.44, we have $\sigma_\theta < 0$ and $\sigma_r \leq 0$. Hence, the yield condition of Tresca is simply

$$\sigma_\theta = -\sigma_Y \qquad \text{with} \qquad -\sigma_Y \leq \sigma_r \leq \sigma_Y \, . \qquad (8.102)$$

In region 2, $\sigma_\theta > 0$ and $\sigma_r \leq 0$. The Tresca yield condition is

$$\sigma_\theta - \sigma_r = \sigma_Y \, . \qquad (8.103)$$

The equilibrium equation is (see Section 10.6, eq. (10. 8.3))

$$\sigma_\theta - \sigma_r - r\frac{d\,\sigma_r}{dr} = 0 \, . \qquad (8.104)$$

With the considered yield conditions and the boundary conditions (fig. 8.58)

$$\sigma_r = \sigma_{ou} \qquad \text{at} \qquad r = r_u ,$$
$$\sigma_r = \sigma_{ol} \qquad \text{at} \qquad r = r_l , \tag{8.105}$$

integration of eq. (8.104) yields the following relations :

$$\sigma_r = \frac{1}{r}\left[-\sigma_Y r + r_l\left(\sigma_{ol} + \sigma_Y\right)\right], \qquad \sigma_\theta = -\sigma_Y \tag{8.106}$$

in region 1, and

$$\sigma_r = \sigma_Y \ln\frac{r}{r_u} + \sigma_{ou}, \qquad \sigma_\theta = \sigma_Y\left(1 + \ln\frac{r}{r_u}\right) + \sigma_{ou} \tag{8.107}$$

in region 2. The boundary radius r_n between regions 1 and 2 is obtained from the condition that the net resultant force over the cross section vanishes. We obtain in this manner

$$r_n \ln\frac{r_u}{r_n} - (r_n - r_l) + \frac{2lt}{a}\left[\rho_u\left(1 - \frac{r_u - r_n}{R_u}\right) - \rho_l\right] = 0, \tag{8.108}$$

where ρ_u and ρ_l are the efficiencies of the upper and lower flanges, respectively.

Finally, the plastic moment of the I beam is evaluated as the total moment of the internal forces with respect to the centre of curvature.

$$M_p = \sigma_Y [2tl\left(\rho_u R_u - \rho_l R_l\right) - \frac{a}{2}\left(r_n^2 - r_l^2\right)$$
$$+ \left(\frac{a}{4} - \frac{tl\rho_u}{R_u}\right)\left(r_u^2 - r_n^2\right) + \frac{a}{2} r_n^2 \ln\frac{r_u}{r_n}]. \tag{8.109}$$

The plastic moment M_{ps} of the same I beam with straight axis is

$$M_{ps} = 2lth\,\sigma_Y + a\frac{h_a^2}{4}\sigma_Y, \tag{8.110}$$

with $h = R_u - R_l + t$ and $h_a = r_u - r_l$.

Fig. 8.59.

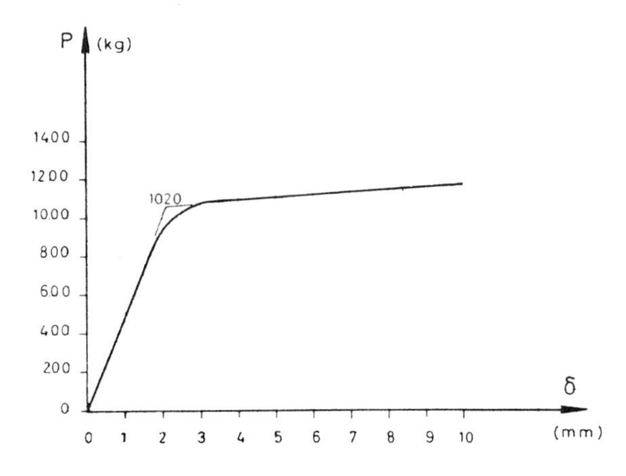

Fig. 8.60.

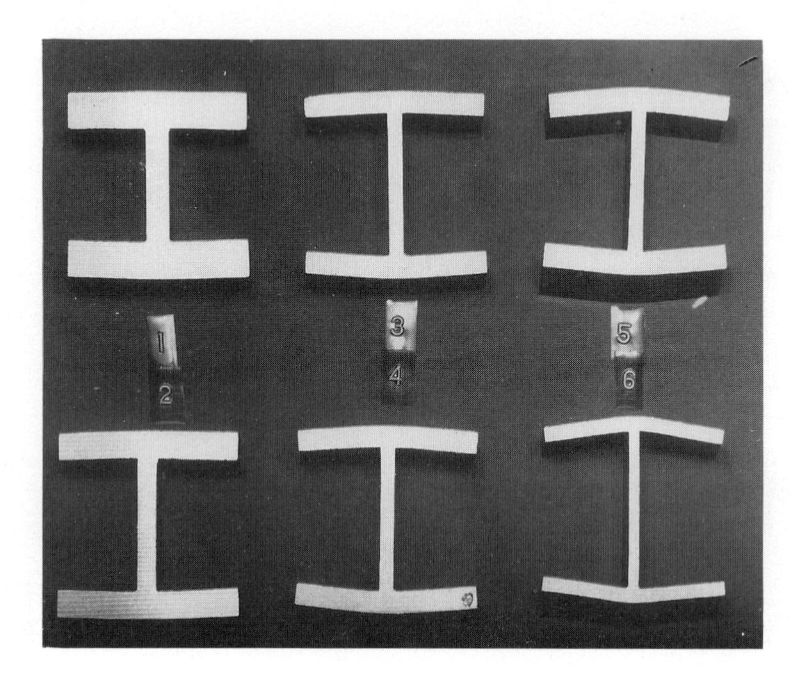

Fig. 8.61.

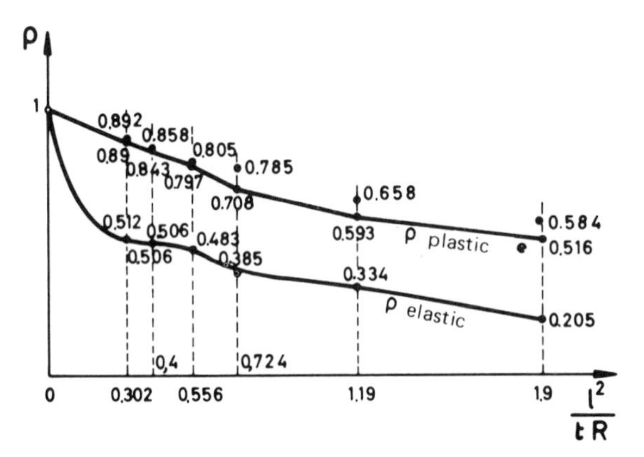

Fig. 8.62.

The overall efficiency ρ of the cross section is thus obtained as $\rho = \frac{M_p}{M_{ps}}$ from eqs. (8.109) and (8.110).

The theoretical predictions given (eqs. (8.91), (8.100), (8.109)) have been tested experimentally on six small-scale models of mild steel [8.62]. The testing device is shown in fig. 8.59. Fig. 8.60 shows a typical moment versus curvature diagram, with the experimental value of M_p. In fig. 8.61, plastically deformed cross sections can be seen, whereas the diagram of fig. 8.62 enables comparison of theoretical efficiencies with experimental values. Experimental values are larger than theoretical ones because the yield condition used is wholly internal to the realistic von Mises condition.

8.7.3. *Circular beam with box shape cross section.*

The circular beam with the cross section shown in fig. 8.63, subjected to uniform bending moment, can be analyzed in a manner similar to that used in Section 8.7.2 for the I beam. It is assumed that the webs are not bent transversally. To achieve this goal, one must have $l \geq \sqrt{0.548tR}$. The stress field is then continued in a statically admissible manner in the central part of the flange. The stress profile is either FA', FAB' or even FAB. If we let

$$\alpha_c = L\sqrt{2/tR}, \tag{8.111}$$

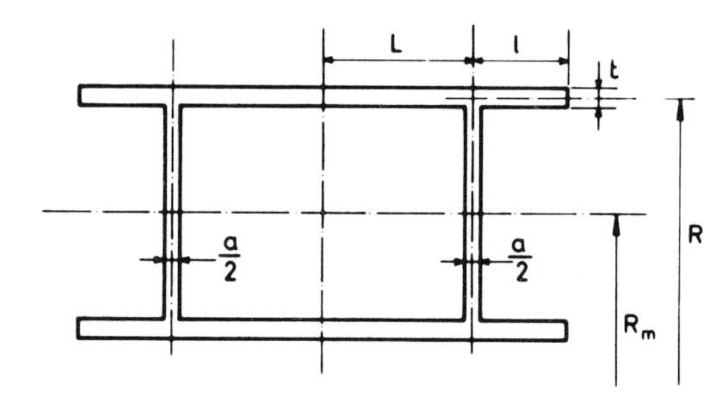

Fig. 8.63.

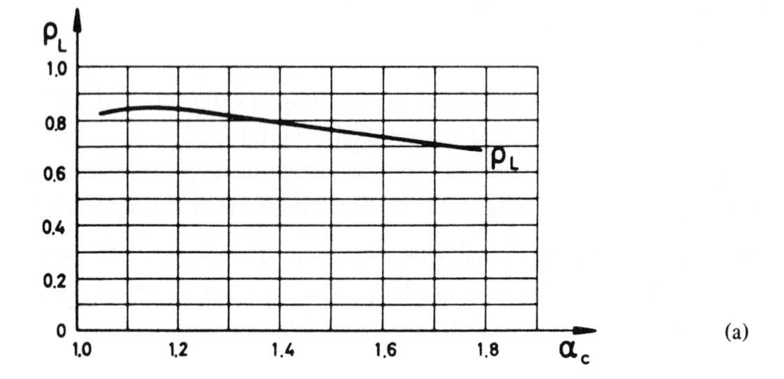

ρ_L
1.0
0.8
0.6
0.4
0.2
0
1.0
1.2
1.4
1.6
1.8
α_c
ρ_L

(a)

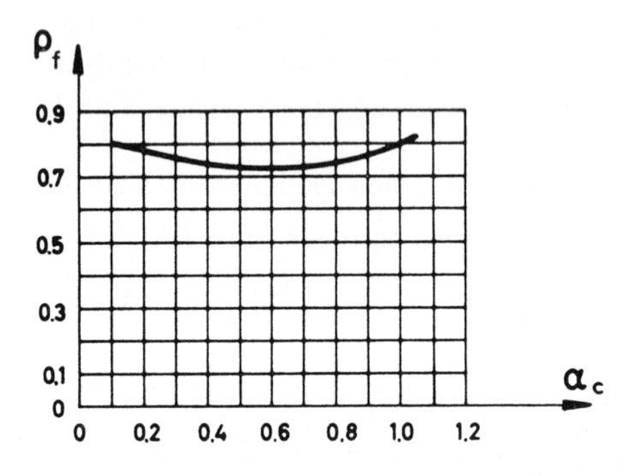

ρ_f
0.9
0.7
0.5
0.3
0.1
0
0
0.2
0.4
0.6
0.8
1.0
1.2
α_c

(b)

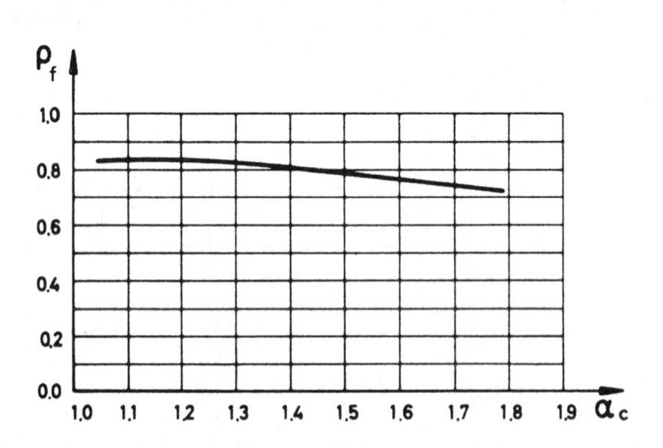

ρ_f
1.0
0.8
0.6
0.4
0.2
0.0
1.0
1.1
1.2
1.3
1.4
1.5
1.6
1.7
1.8
1.9
α_c

Fig. 8.64.

the efficiency ρ_L of the central part is given in fig. 8.64(a) as a function of α_c [8.63]. The efficiency ρ_f of a flange is then given by

$$\rho_f = \frac{l\,\rho_l + L\rho_L}{l + L}. \tag{8.112}$$

Table 8.7.

N°	2L	R_m	ρ_u	ρ_l	ρ_{th}	ρ_{exp}	Differences
1	30	120	0.802	0.758	0.787	0.844	6.1
2	48	120	0.746	0.652	0.713	0.777	8.2
3	44	170	0.811	0.754	0.791	0.858	7.8
4	40	170	0.826	0.772	0.806	0.858	6.1
5	36	170	0.821	0.785	0.810	0.951	14.8
6	30	170	0.756	0.750	0.769	0.887	13.3
7	20	70	0.737	0.672	0.729	0.850	14.2

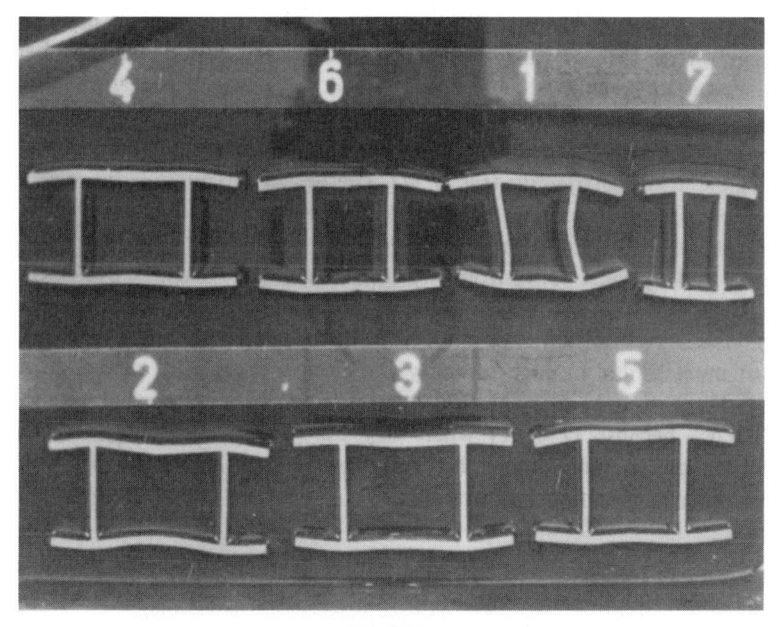

Fig. 8.65.

where ρ_l is the efficiency of the cantilever part of the flange. For the particular case where $l = \sqrt{0.548tR}$, the ρ_f versus α_c curves are given in fig. 8.64(b). An experimental verification of the theory has been conducted on seven small-scale models. Dimensions and results are summarized in table 8.7. The yield stress of the material was 35.150 psi (mild-steel) except for a region 2 in. wide near the welds where its average value was 43.200 psi. In fig. 8.65, the deformed cross sections can be seen. In model 1, buckling of the webs occurred as soon as a plastic collapse mechanism was formed.

8.8. Shakedown analysis.

The remarks of Section 6.8. (plate shakedown problems), concerning the so-called concept of elastic yield criterion and the use of numerical methods, are also valid for shells.

A method of bounding the shakedown domains for shells is proposed by Sawczuk in [4.23]. The method, based on the kinematical theorem (see Section 4.4.5.), gives upper bounds of the shakedown multiplier. Some analytical examples of shakedown interaction curve for a two-parameter loading of a cylindrical shell are analyzed in [4.22] and [4.23].

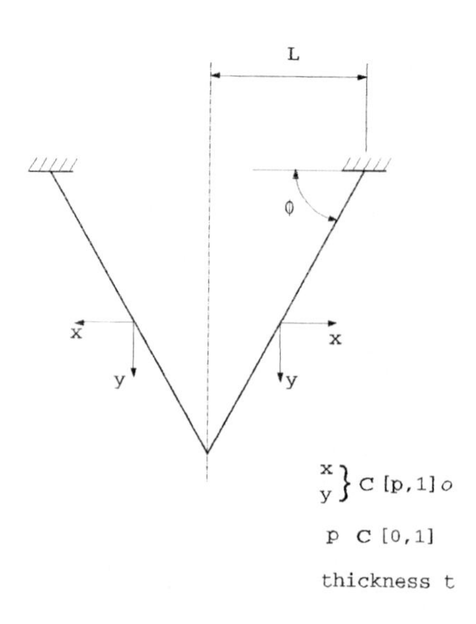

Fig. 8.66. Conical shell subjected to concentrated forces [6.48].

Numerical solutions are given by Morelle and Nguyen Dang Hung [6.48], [6.12]. The method uses the statical theorem and equilibrium finite elements (see Section 6.8). These authors consider the problem of a built-in conical shell subjected to a ring of concentrated forces (fig. 8.66). We consider that each force (X and Y) varies independently, as follows

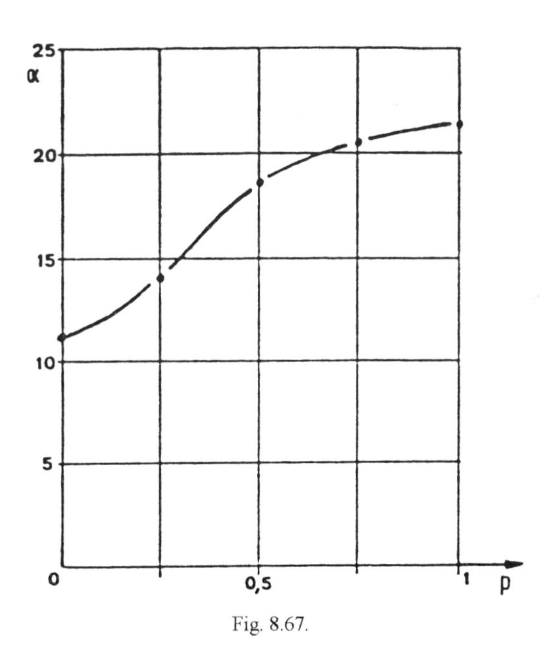

Fig. 8.67.

$$p\alpha \leq X \leq \alpha, \qquad p\alpha \leq Y \leq \alpha$$

$$\text{with} \qquad 0 \leq p \leq 1$$

The computations are performed for the cases p = 0 ; 0.25 ; 0.5 ; 0.75 ; 1. (this last case is that of limit analysis).

Fig. 8.67 shows the variation of the shakedown multiplier α with respect to p, for particular datas.

$$(L = 125 \text{ mm},\ t = 5 \text{ mm},\ N_p = t\sigma_p = 5 \text{ kg/mm},\ \frac{X_{appl}}{N_p} = \frac{Y_{appl}}{N_p} = 0.02)$$

The examination of results display two different behaviours :

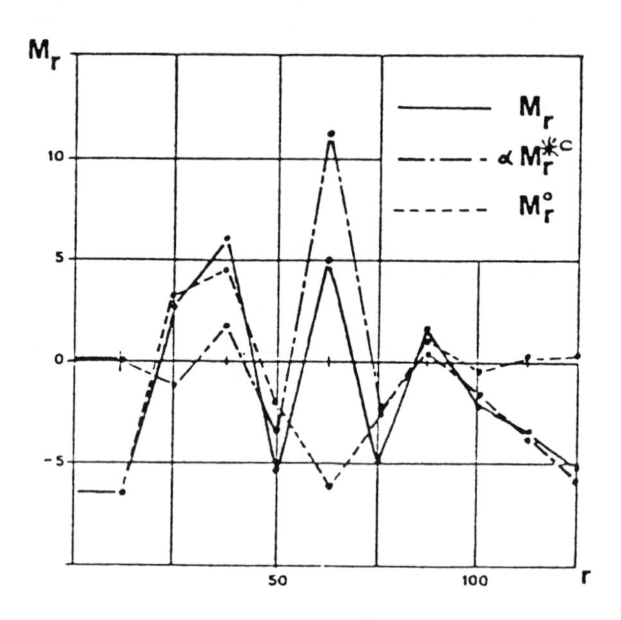

Fig. 6.68. Distribution of M_r, αM_r^{*c}, M_r^o, case p = 0 [6.48]

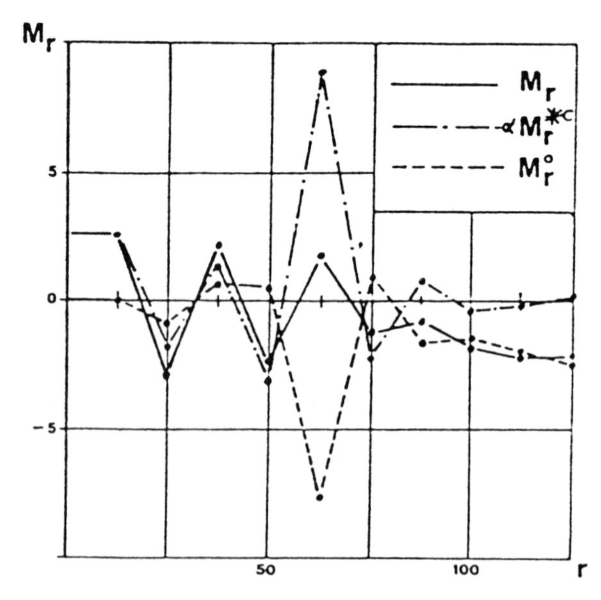

Fig. 6.69. Distribution of M_r, αM_r^{*c}, M_r^o, case p = 1 [6.48]

- for p = 0 ; 0.25 ; 0.5, the shell ruins by plastic fatigue. For the case p = 0, fig. 8.68 shows the collapse radial moment (M_r) and its decomposition in elastic moment $\left(\alpha M_r^{*o}\right)$ and residual moment $\left(M_r^o\right)$

- for p = 0.75 ; 1, the shells ruins by ratcheting (ratchet mechanism). Fig. 8.69 shows, in the case p = 1 the moments M_r, αM_r^{*o} and M_r^o.

From these results, we conclude that (1) the collapse behaviour depends on the variation range of applied loads ; the fatigue plastic mechanism is obtained for loadings lower than those causing the ratcheting mechanism ; (2) a design based on limit analysis may be dangerous if the loading is variable.

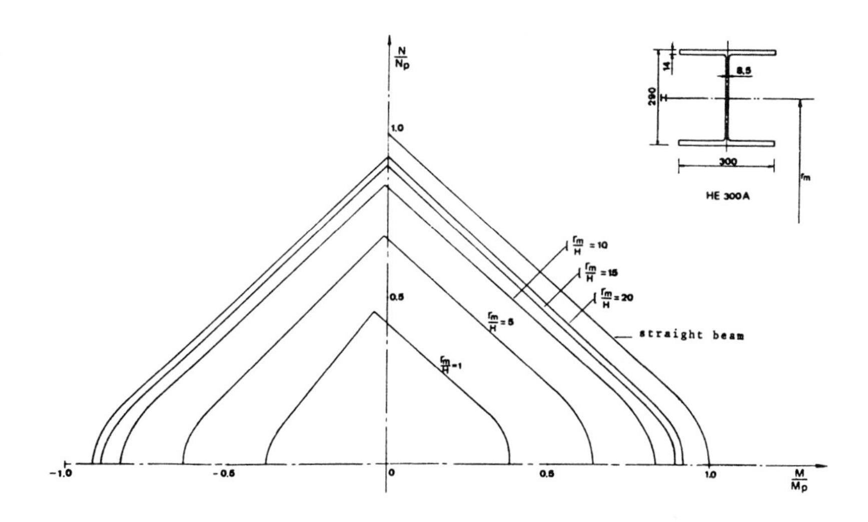

Fig. 8.70.

8.9. Indications on other problems.

Cylindrical shells with circular cut-outs, either with an annular reinforcing ring or without reinforcement have been studied by Hodge [8.64] and by Coon, Gill, and Kitching [8.65], respectively. Ring-reinforced radial branches in cylindrical and spherical vessels have also been considered [8.65].

Shallow shells are treated in the book [8.27] by Hodge.

The reader is referred to the original papers for more information and to the survey book by Olszak and Sawczuk [8.67].

The results given in Section 8.7 have been extended by Lamblin and Guerlement to the combined action of bending moment and axial force ([8.68],[8.69]). Typical interaction curves for a given I profile are shown in fig. 8.70, where r_m is the radius of curvature of the axis of the beam with height H. Experimental verification proved satisfactory [8.69].

Systematic application to industrial rolled I beams showed that a strength reduction of 10% or more is obtained when $\frac{r_m}{H} < 5$ for narrow-flange beams and up to $\frac{r_m}{H} < 20$ for HEA type wide-flange beams.

Hence, it is recommended to take the effect of curvature of the beam into account when this situation is met.

A review of limit analysis and design of containment vessels can be found in [8.70], whereas cylindrical shells with various loadings, boundary conditions and reinforcing rings or ribs have been treated by Guerlement and co-authors ([8.71] to [8.74]). Again, results of these works and of many others, together with references enabling deeper study, can be found in the Atlas of Limit Loads [7.26].

8.10. Problems.

8.10.1. Prove relations (8.28).

8.10.2. Obtain the following bounds for the limiting value of a ring force $2F_o$ applied to an infinitely long cylinder :

$$(2F_o)_+ = 4N_p \sqrt{t/R},$$

$$(2F_o)_- = 2\sqrt{2}\ N_p \sqrt{t/R},$$

using the yield locus consisting of a rectangle inscribed in the polygon ABCDEF of fig. 5.24.

8.10.3. Obtain eq. (8.42) by detailed calculation.

8.10.4. Apply the kinematic theorem to the mechanism found in Section 8.3.2, eq. (8.14), to verify that it gives the limit load (8.27).

References.

[8.1] W.A. NASH, "Recent Advances in the Buckling of Thin shells", Applied Mechanics Reviews, **13** : 3, March 1960.

[8.2] Y.C. FUNG and E.E. SECHLER, "Instability of Thin Elastic Shells", Structural Mechanics, Proceedings of the First Symposium on Naval Structural Mechanics, Pergamon Press, New York, 115-168, 1960.

[8.3] S.B. BARDORF,"A Simplified Method of Elastic Stability Analysis of Thin Cylindrical Shells", N.A.C.A. Tech. Rep., 874, 1947.

[8.4] S.C. BATTERMANN, "Plastic Buckling of Axially Compressed Cylindrical Shells", A.I.A.A. Journal, **3** : 2, 316, January 1965.

[8.5] A.F. KIRSTEIN and E. WENK Jr., "Observation of Snap-through Action in Thin Cylindrical Shells under External Pressure", Proc. S.E.S.A., **XIV** : 1, 1956.

[8.6] W.A. NASH, "An Experimental Analysis of the Buckling of Thin Initially Imperfect Cylindrical Shells Subject to Torsion", Proc. S.E.S.A., **XVIII** : 1, 1961.

[8.7] R.H. HOMEWOOD, A.C. BRINE and A.E. JOHNSON Jr., "Experimental Investigation of the Buckling Instability of Monocoque Shells", Proc. S.E.S.A., **XVIII** : 1, 1961.

[8.8] H. SCHMIDT, "Ergebnisse von Beulversuchen mit doppelt gekrummten Schalenmodellen aus Aluminium", Proc. Symp. Shell Research, Delft, August-September 1961, North-Holland Publ. Comp., Amsterdam, 1961.

[8.9] W.R. OSGOOD, Residual Stresses in Metals and Metal Constructions, Reinhold Publ. Corp. New York, 1954.

[8.10] M.E. SHANK, Control of Steel Construction to Avoid Brittle failure, Plasticity Committee of the Welding Research Council, New York, 1957.

[8.11] J.T.P.YAO and W.H.MUNSE, "Low-cycle Fatigue of Metals-Literature Review", Welding Journal, 182, April 1962.

[8.12] W. FLUGGE, Stresses in Shells, Springer, 1960.

[8.13] C. MASSONNET, "Tension dans les réservoirs sous pression", Technical note B.II,I of the C.E.C.M., Brussels.

[8.14] M. SAVE, "le calcul à la flexion des coques de révolution chargées de façon quelconque", Bulletin of the C.E.R.E.S., **XI**, Liège, 1960.

[8.15] R.L. CLOUD, "Interpretative Report on Pressure Vessel Heads", Welding Research Council Bulletin, 119, 1, January 1967.

[8.16] M.E. LUNCHICK, "Yield Failure of Stiffened Cylinders under Hydrostatic Pressure", Proc. 3rd U.S. Nat. Cong. Appl. Mech., Providence, 1958.

[8.17] M.E. LUNCHICK and J.A. OVERBY, "Yield Strength of Machined Ring-stiffened Cylindrical Shell under Hydrostatic Pressure", Proc. S.E.S.A., **XVIII** : 1, 1961.

[8.18] R.C DEHART and N.L. BASDEKAS, "Yield Collapse of Stiffened Circular Cylindrical Shells", (report) Southwest Research Institute, San Antonio, Texas, September 1960.

[8.19] H.H. DEMIR and D.C. DRUCKER, "An Experimental Study of Cylindrical Shells under Ring Loading", Progress in Applied Mechanics (The Prager anniversary volume), Macmilan, New York, 205-220, 1963.

[8.20] M. SAVE, "Vérification expérimentale de l'analyse limite des plaques et des coques en acier doux", (Experiments on limit loads of mild steel plates and shells), Report MT 21, C.R.I.F., 21 rue des Drapiers Brussels, 1966.

[8.21] B. PAUL, "Limit Load of Clamped Shells with a Reinforcing Ring", PIBAL Report, 424, December 1958.

[8.22] B. PAUL, "Collapse Loads of Rings and Flanges under Uniform Twisting Moment and Radial Force", J. Appl. Mech., **26** : Series E, 2, June 1959

[8.23] B. PAUL and P.G. HODGE Jr., "Carrying Capacity of Elastic-plastic Shells under Hydrostatic Pressure", Proc. 3rd U.S. Nat. Cong. Appl. Mech., 631, 1958.

[8.24] E.T. ONAT, "The Plastic Collapse of Cylindrical Shells under Axially Symmetrical Loading", Quart. Appl. Math., **XIII** : 63, 1955.

[8.25] R. SANKARANARAYANAN, "Plastic Interaction Curves for Circular Cylindrical Shells under Combined Lateral and Axial Pressures", J. Franklin Inst., **270** : 5, November 1960.

[8.26] P.G. HODGE Jr. and J. PANARELLI, "Interaction Curves for Circular Cylindrical Shells According to the Mises or Tresca Yield Criteria", J. Appl. Mech., **29** : 375, 1962

[8.27] P.G. HODGE Jr., Limit Analysis of Rotationally Symmetric Plates and Shells, Prentice-Hall, Englewood Cliffs, N.J., 1963.

[8.28] P.G. HODGE Jr., "Plastic Design of a Closed Cylindrical Structure", J. Mech. Phys. Solids, **12** : 1, 1964.

[8.29] N.A. FORSMAN, "On the Carrying Capacity of Cylindrical Vessels with Flat Closures" (in Russian), Isv. An. U.S.S.R. Mekh. Mashin., 3, 106, 1964.

[8.30] M. SAYIR, "Kollapsbelastung von rotationsymmetrischen Zylinderschalen", Zeit. Ang. Math. Phys. Z.A.M.P., **17** : 353, 1966.

[8.31] M. JANAS and A. KÖNIG, "Limit Analysis of Shells : Shell Roofs and Vessels" (in Polish), Arkady, Warsaw, 1967.

[8.32] Y.V. NEMIROVSKY and Y.N. RABOTNOV, "Limit Analysis of Ribbed Plates and Shells", Nonclassical Shell Problems, pp. 786-807, North-Holland Pub. Co., Amsterdam, 1964.

[8.33] A. BIRON and A. SAWCZUK, "Plastic Analysis of Rib-reinforced Cylindrical Shells", Trans. A.S.M.E., J. Appl. Mech., **89** : Series E, 1, 37, March 1967.

[8.34] M. CARPUSO and A. GANDOLFI, "Sul collasso rigido-plastico dei gusci nervati di revoluzione", (Plastic Collapse of Ribbed Shells of Revolution) (Internal report), Istituto di Tecnica delle Costruzioni, Napoli, 1967.

[8.35] M. CARPUSO and A. GANDOLFI, "Sul collasso plastico dei tubi circolari nervati soggetti a pressione e sforzo assiale" (Plastic Collapse of Ribbed Cylinders under Pressure and Axial Force) (Internal report), Istituto di Technica delle Costruzioni, Napoli, 1967.

[8.36] G. EASON and R.T. SHIELD, "The influence of Free Ends on the Load-carrying Capacities of Cylindrical Shells", J. Mech. Phys. Solids, **4** : 17, 1955.

[8.37] G. EASON, "The Load-carrying Capacities of Cylindrical Shells Subjected to a Ring of Force", J. Mech. Phys. Solids, 7 : 169, 1959.

[8.38] H.H. DEMIR, "Cylindrical Shells under Ring Loads", Proc. A.S.C.E., **91** : ST3 (J. Struct. Div., Part 1), 71, June 1965.

[8.39] W. OLSZAK and A. SAWCZUK, "Die Grenstragfähigkeit von cylindrischen Schalen bei verschiedenen Formen der Plastizitätsbedingung", Acta Technica Academiae Scientiarum Hungariae, **XXVI** : 1, Budapest, 1959.

[8.40] B. PAUL, "Carrying Capacities of Elastic-Plastic Shells with Various End Conditions, under Hydrostatic Compression", J. of Appl. Mech, **26** : Series E, 4, 553, December 1959.

[8.41] P.G. HODGE Jr., "Displacements in an Elastic-Plastic Cylindrical Shell", J. of Appl. Mech, **23** : 73, 1956.

[8.42] P.G. HODGE Jr. and S.V. NARDO, "Carrying capacity of an Elastic-Plastic Cylindrical Shell with Linear Strain-hardening", J. of Appl. Mech, **25** : Trans. A.S.M.E., **80** : 79, 1958.

[8.43] P.G. HODGE Jr., "Yield Conditions for Rotationally Symmetric Shells under Axisymmetric Loading", J. of Appl. Mech, **27** : Trans. A.S.M.E., **82**: Series E, 323, June 1960.

[8.44] W. FLUGGE and T. NAKAMURA, "Plastic Analysis of Shells of Revolution under Axisymmetric Loads", Ing. Archiv., **34** : 4, 238, 1965.

[8.45] D.C. DRUCKER and R.T. SHIELD, "Limit Analysis of Symmetrically Loaded Thin Shells of Revolution", J. of Appl. Mech, **26** : Trans. A.S.M.E., **81** : Series E, 61, 1959.

[8.46] D.C. DRUCKER and R.T. SHIELD, "Limit Strength of Thin-Walled Pressure Vessels with an A.S.M.E. Standard Torsipherical Head", Proc. 3rd U.S. Nat. Cong. Appl. Mech. A.S.M.E., 665, 1958.

[8.47] P.G. HODGE Jr. and C. LAKSHMIKANTHAM, "Limit Analysis of Shallow Shells of Revolution", Trans. A.S.M.E., J. of Appl. Mech, **30** : Series E, 2, 215, June 1963.

[8.48] P.G. HODGE Jr., "Plastic Analysis of Circular Conical Shells", J. of Appl. Mech, December 1960.

[8.49] R.H. LANCE and E.T. ONAT, "Analysis of Plastic Shallows Conical shells", J. of Appl. Mech, **30** : 199, 1963.

[8.50] J. DERUNTZ and J. HODGE, "The Significance of the concentrated Load in the Limit Analysis of Conical Shells", J. of Appl. Mech, **33** :Trans. A.S.M.E., **88** : 93-101, March 1966.

[8.51] U.S. CHWALA and A. BIRON, "Limit Analysis of Shells of Revolution of Arbitrary Shape under pressure", Report 1775, Lab. de Rech. et d'Essai de Matériaux, Ecole Polytechnique de Montreal, 1969.

[8.52] R.W. KUECH and S.L. LEE, "Limit Analysis of Simply Supported Conical Shells Subjected to Uniform Internal Pressure", J. Franklin Inst., **280** : 1, 71, July 1964.

[8.53] D.C. DRUCKER and R.T. SHIELD, "Design of Thin-walled Torsipherical and Toriconical Pressure Vessels Heads", J. of Appl. Mech, June 1961.

[8.54] R.A. STRUBLE, "Biezeno Pressure Vessel Heads", J. of Appl. Mech, **23** : Trans. A.S.M.E., **78** : 642, 1956.

[8.55] G.A. HOFFMAN, "Minimum-weight Proportions of Pressure-Vessel Heads", Trans. A.S.M.E., **29** : Series E, 4, 662, December 1962.

[8.56] G.D. GALLETLY, "Torispherical Shells : A Caution to Designers", J. Eng. Industry, Trans. A.S.M.E., **81** : Series B, 51, 1959.

[8.58] P.R. GAJEWSKI and R.H. LANCE, J. Engng. Ind., **91** : 636, 1969.

[8.59] S.S. GILL, "The limit Pressure of a Flush Cylindrical Nozzle in a Spherical Pressure Vessel", Int. J. Mech. Sci., **6** : 1, 105, 1964.

[8.60] K.S. DINO and S.S. GILL, "The Limit Analysis of a pressure Vessel Consisting of the Junction of a Cylindrical and a Spherical Shell", Int. J. Mech. Sci., **7** : 1, 21, January 1965.

[8.61] K.S. DINO and S.S. GILL, "An Experimental Investigation into the Plastic Behaviour of Flush Nozzles in Spherical Pressure Vessels", Int. J. Mech. Sci., 7 : 12, 817, 1965.

[8.62] C.E. MASSONNET and M.A. SAVE, "Résistance limite d'une poutre courbe en double té à parois minces soumises à flexion pure" (Ultimate Strength of a I-curved Thin-wall Beam Subjected to Uniform Bending), Proc. Int. Ass. Bridge. Str. Eng., **23** : 245, 1964.

[8.63] C.E. MASSONNET and M.A. SAVE, "Résistance limite d'une poutre courbe en caisson soumise à flexion pure" (Ultimate Strength of a Curved Box-shaped Beam Subjected to Uniform Bending), Amici et Alumni (The Campus anniversary volume) Liège, 1964.

[8.64] P.G. HODGE Jr., "Full-Strength Reinforcement of a Cutout in a Cylindrical Shell", Trans. A.S.M.E., J. Appl. Mech., **31** : Series E, 4, 667, December 1964.

[8.65] M.D. COON, S.S. GILL and R. KITCHING, "A Lower Bound to the Limit Pressure of a Cylindrical Pressure Vessel with an Unreinforced Hole", Int. J. Mech. Sci., **9** : 2, 69, 1965.

[8.66] C. RUIZ and S.E. CHUKWUJEKWU, "Limit Analysis Design of Ring-reinforced Radial Branches in Cylindrical and Spherical Vessels", Int. J. Mech. Sci., **9** : 11, 1967.

[8.67] W. OLSZAK and A. SAWCZUK, "Inelastic Behaviour in Shells", Noordhoff, Groningen, 1967.

[8.68] D. LAMBLIN and G. GUERLEMENT, "Interaction Curves for Bending and Axial Forces of Perfectly Plastic Curved I Beams", J. Struct. Mech., **1** (2) : 187-212, 1972.

[8.69] D. LAMBLIN and G. GUERLEMENT, "Effect of Curvature on Plastic Limit Analysis of Curved I Beams", J. Struct. Mech., **3** (3) : 301-315, 1974-1975.

[8.70] M. SAVE,"Limit Analysis and Design of Containment Vessels", Nuclear Eng. and Design, 79, 343-361, 1984.

[8.71] G. GUERLEMENT, "Contribution à l'analyse limite des coques cylindriques "; Doctoral thesis, Faculté Polytechnique de Mons, Belgium, 1975.

[8.72] G. GUERLEMENTand D.O. LAMBLIN, "Limit Analysis of Ring-reinforced Cylindrical Shells", Bull. Polish Acad. Sci., **24**, 1, 111-118, 1976.

[8.73] G. GUERLEMENT, D.O. LAMBLIN and M. SAVE, "Limit Analysis of Cylindrical Shell with Reinforcing Rings", Eng. Struct., **9**, 3, 146-156, 1987.

[8.74] Y.A. HASSANEAN, D.O. LAMBLIN and G. GUERLEMENT, "Contribution of Limit Analysis to Design of Cylindrical Shells", Proceeding First World Conf. on Constructional Steel, 715-718, Elsevier Sci. Pub., 1992.

[8.75] F. ELLYIN and A.N. SHERBOURNE, Nuclear Struct. Engng., **2**, 169, 1965.

9. Reinforced Concrete Plates.

9.1. Introduction.

Mechanical behaviour of reinforced concrete structures is much more complex than that of steel structures. In Chapter 11 of Com. V., we restricted ourselves to a very brief introduction to limit analysis and design of reinforced concrete beams and frames. This extremely short treatment of the question was justified by two reasons :

1. Many problems have not yet found satisfactory solutions, and the applicability of limit analysis and design to reinforced concrete beams and frames is still open to discussion [9.1] ;

2. A sufficiently exhaustive treatment of the subject, with due discussion of the conditions of applicability of the theory, would have required a volume by itself (see [9.2] for more information).

On the other hand, the situation is somewhat less complicated for plates and shells because the percentage of reinforcement is always relatively small and plastic behaviour is more generally and more clearly established. Moreover, analysis and design of plates on the basis of the so-called "yield lime method" [9.3] has long been used by practicing engineers and is now accepted by the European Committee for Concrete [9.4]. As will be emphasized, this method is nothing but a special use of the kinematical approach of limit analysis [9.5, 9.6]. Hence, limit analysis and design of reinforced concrete plates appears to be of considerable practical importance.

Consider a reinforced concrete plate subjected to a system of loads that is defined to within a single load parameter P. Under increasing P, the following phenomena are observed : firstly, a very limited elastic range (OA in fig. 9.1) ; secondly, where the concrete is subjected to tensile stresses, it cracks on account of its small tensile strength (range AB in fig. 9.1) ; thirdly, plastic flow of the reinforcement : during this stage, the bending moment in the most stressed sections is kept at a constant value and a redistribution of bending moments will in general take place until sufficient reinforcement is plastified to allow further opening of cracks, resulting in a collapse mechanism of the plate. The value of the load parameter P reached at the end of the third stage is called the *limit load* (or sometimes, incorrectly, the fracture load).

The behaviour described above implies that a reinforced concrete plate subjected to simple flexure may be regarded as a perfectly plastic structural element.

This is verified by the moment versus curvature diagram of fig. 9.1 [9.7], where the three ranges of mechanical behaviour are indicated. In the present situation however, because of the homogeneous moment field, no redistribution occurs and point B corresponds to collapse.

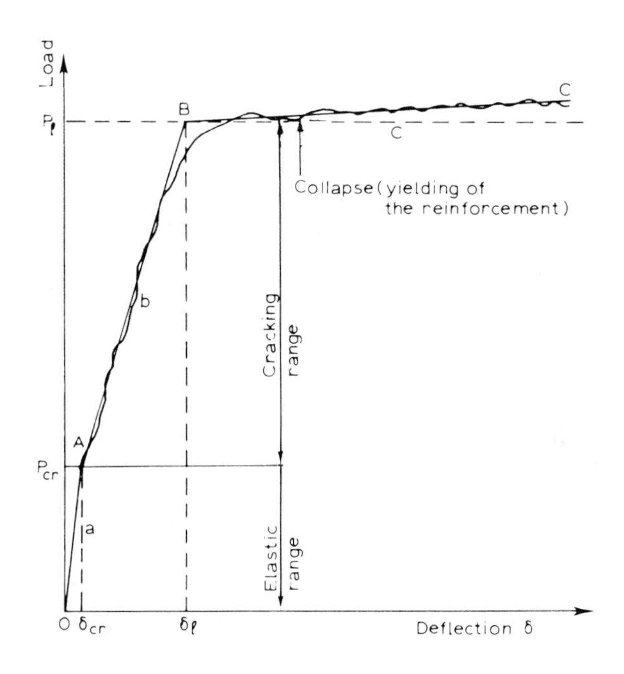

Fig. 9.1. Load versus deflection diagram.

The necessary condition for such a perfectly plastic behaviour are threefold :

1. The percentage of reinforcement is small enough for the plate not to fail by crushing of the compressed concrete before the collapse mechanism is formed ;

2. Shear forces have no appreciable effect on the failure of the plate ;

3. In-plane (membrane) forces have negligible influence on the collapse of the plate.

The theory developed in Sections 9.3 to 9.6 will be based on the three conditions above. In Sections 9.7 and 9.8, we shall discuss of the effects of shear and membrane forces, when not negligible. Predictions of limit loads and collapse mechanisms have been widely verified experimentally [7.17, 9.3].

We adopt the experimental concept of "fracture lines" (according to Johansen [9.3]) that we shall preferably call "hinge lines" (see fig. 9.3). Our criterion will have to be compared with this experimental fact.

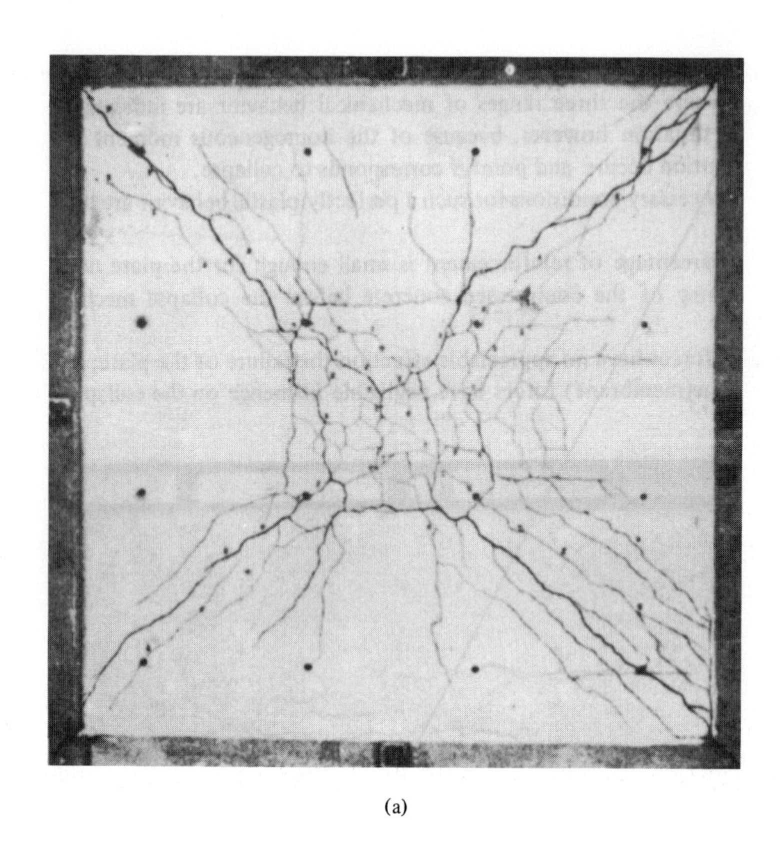

(a)

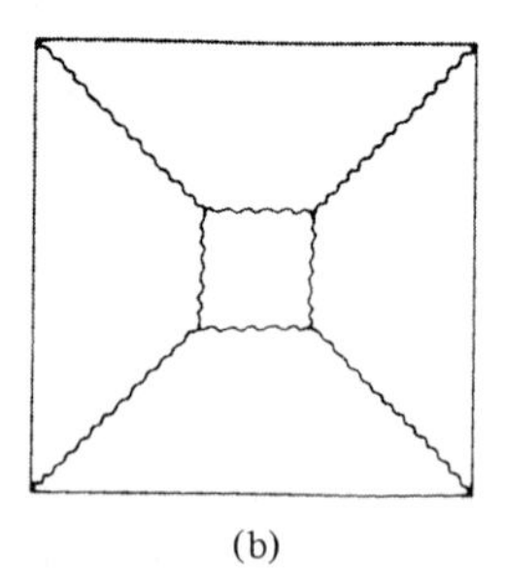

(b)

Fig. 9.2 (British Crown copyright reserved . Reproduced by permission of The Controller of Her Britannic Majesty's Stationary Office)

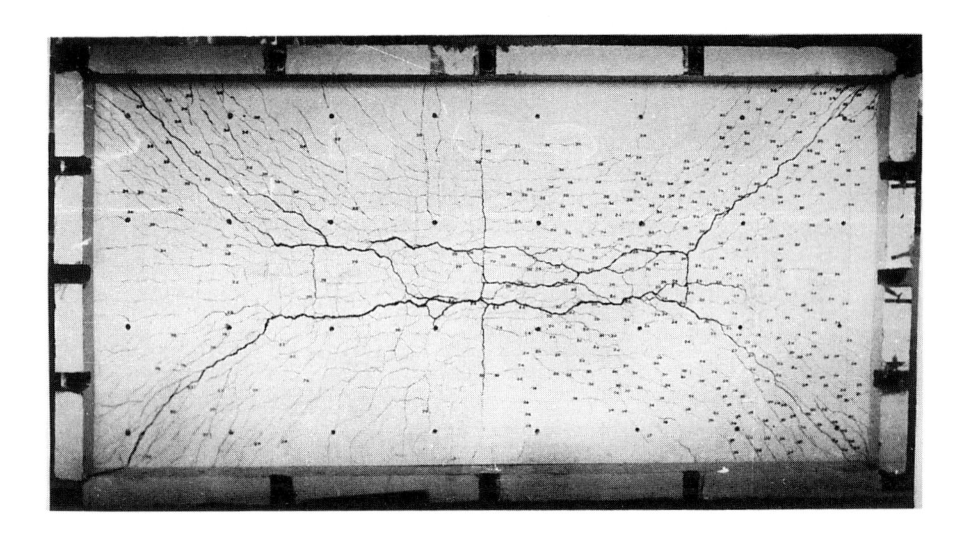

Fig. 9.3 (British Crown. copyright reserved . Reproduced by permission of The Controller of Her Britannic Majesty's Stationary Office)

9.2. Yield condition and flow rule.

9.2.1. *Introduction.*

Accepting the rigid-plastic moment versus curvature diagram of fig. 9.4, we assume that the reinforcement is placed in two orthogonal directions x and y (fig. 9.5). If the reinforcement is not identical in both directions, the corresponding plastic moments will also differ and the plate will be orthotropic with regard to its ultimate strength.

Ultimate moments for simple bending are denoted by M_{px}, M_{py} for positive bending (tension of the lower layer), M'_{px}, M'_{py} for negative bending (all positive values).

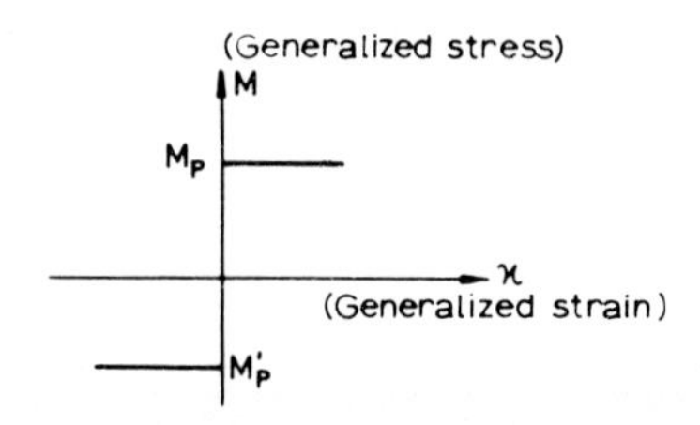

Fig. 9.4.

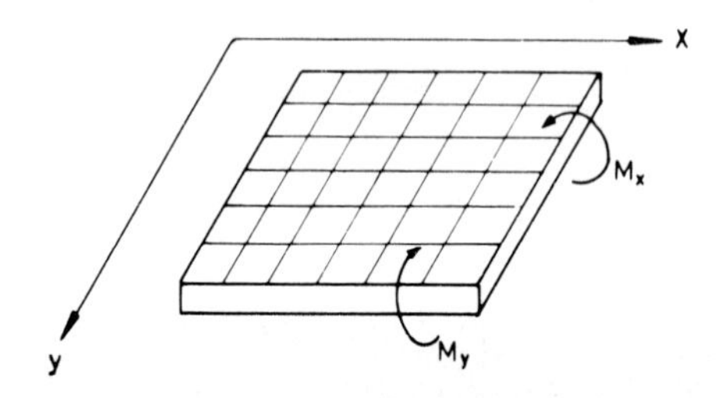

Fig. 9.5.

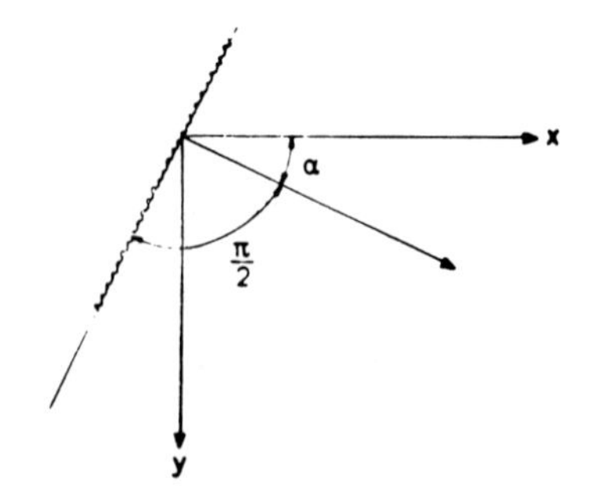

Fig. 9.6.

We accept the following physical criterion : Yielding occurs when the bending moment M_α on the cross section with angle α (fig. 9.6) reaches a certain value $M_{p\alpha}$ ($M'_{p\alpha}$) that depends only on M_{px}, M_{py} (M'_{px}, M'_{py}) and α.

9.2.2. *Yield condition.*

First, we consider only slabs with simple reinforcements. The ultimate moment for positive bending is obtained by putting n = 0 in eq. (5.104).

Thus, we have

$$m = 2\,\gamma\,(1 + \rho - \gamma)\,.$$

Assuming that γ si much lower than unity and owing to (5.99) and (5.101), we obtain

$$M_{p\alpha} = \frac{\sigma_Y t}{2}\,(1+\rho)\,\Omega_s.$$

The quantity Ω_s is additive. Taking account of (5.92), we have for reinforcements placed in two orthogonal directions

$$\Omega_s = \left(\frac{A}{p}\right)_x \cos^2\alpha + \left(\frac{A}{p}\right)_y \sin^2\alpha.$$

Denoting

$$M_{px} = \frac{\sigma_Y t}{2}\,(1+\rho)\left(\frac{A}{p}\right)_x, \qquad M_{py} = \frac{\sigma_Y t}{2}\,(1+\rho)\left(\frac{A}{p}\right)_y,$$

we obtain the curve proposed quite intuitively by Johansen [9.3] and others [9.8, 9.9, 9.10] (*).

$$Y_L, \qquad M_{p\alpha} = M_{px}\cos^2\alpha + M_{py}\sin^2\alpha \tag{9.1}$$

(*) (for a detailed discussion of criterion (9.1), the reader is referred e;g; to the book by R.H. Wood [9.16]).

Hence, the yield strength of the plate at a given point can be described, for positive bending by the *polar diagram* of fig. 9.7 (curve Y) : the magnitude of segment OA represents

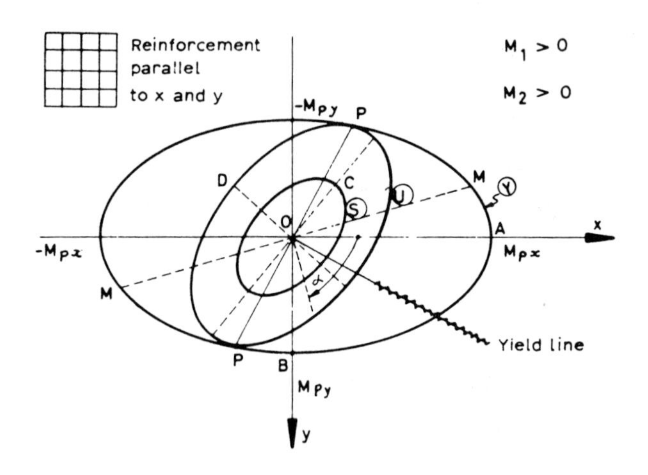

Fig. 9.7.

M_{px}, that of OB represents M_{py}, and generally, OM represents the ultimate bending moment on a cross section normal to the direction of OM (normal stresses on the section having the direction of OM), denoted $M_{p\alpha}$.

The ultimate moment for negative bending can be deduced in a similar way from table II Section 5.6.4. Denoting

$$M'_{px} = \frac{1-\rho}{1+\rho} M_{px}, \qquad M'_{py} = \frac{1-\rho}{1+\rho} M_{py},$$

we have

$$Y_U, \qquad M'_{p\alpha} = M'_{px} \cos^2 \alpha + M'_{py} \sin^2 \alpha. \tag{9.2}$$

The curve Y_L for the lower layer must be used in the region of α where $|M| = M$ and the curve Y_U for the upper layer in the region where $|M| = - M$.(fig. 9.8)

Note that the shapes of curves Y in figures 9.7 and 9.8 have been drawn schematically for more clarity. Real curves may be slightly different.

We now want to obtain the yield condition in terms of the components M_x, M_y, M_{xy} of the moment tensor or, in other words, the yield surface in (M_x, M_y, M_{xy}) space.

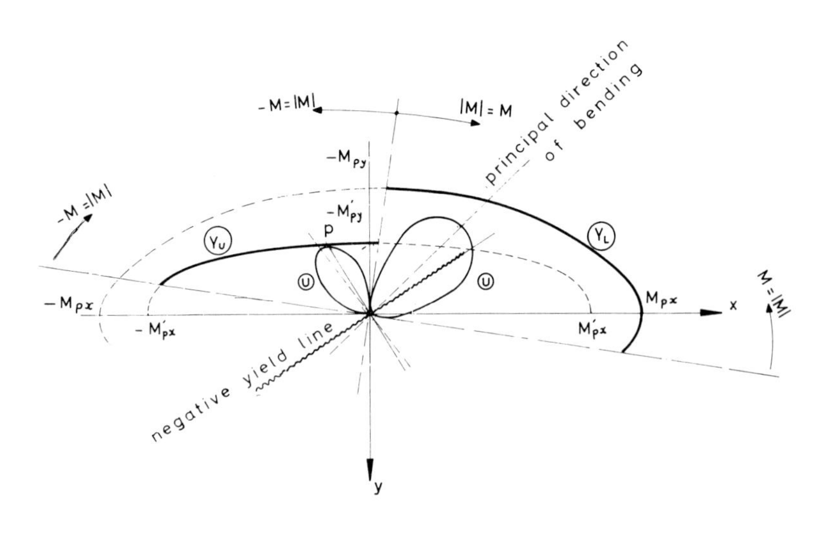

Fig. 9.8.

From the two-dimensional tensorial nature of the moment state it follows that (with convention of fig. 5.6 and with $M_{xy} = -M_{yx}$) (*).

$$.M_\alpha = M_x \cos^2 \alpha + M_y \sin^2 \alpha - M_{xy} \sin^2 \alpha. \tag{9.3}$$

Consider the case

$$M_\alpha = M_{p\alpha}. \tag{9.4}$$

Substitution of expression (9.1) and (9.3) for $M_{p\alpha}$ and M_α into eq. (9.4), yields

$$M_x \cos^2 \alpha + M_y \sin^2 \alpha - M_{xy} \sin^2 \alpha - M_{px} \cos^2 \alpha - M_{py} \sin^2 \alpha = 0. \tag{9.5}$$

(*) See Section 1.1.1 for a similar formula in plane stress [(eq. 1.7)].

With the use of elementary trigonometry (*), eq. (9.5) may be written as

$$\frac{M_{px} - M_x}{2 \tan \alpha} + \frac{M_{py} - M_y}{2} \tan \alpha + M_{xy} = 0. \tag{9.6}$$

This is the equation of a plane in (M_x, M_y, M_{xy}) space.

Varying α, we obtain a set of planes that will envelope a surface, the equation of which is derived by elimination of α from eq. (9.6) and the following equation :

$$-\frac{M_{px} - M_x}{\sin^2 \alpha} + \frac{M_{py} - M_y}{\cos^2 \alpha} = 0, \tag{9.7}$$

which is obtained by differentiating eq. (9.6) with respect to α.

We readily obtain

$$(M_{px} - M_x)\,(M_{py} - M_y) = M_{xy}^2 \tag{9.8}$$

Similarly, from $M_\alpha = -M'_{p\alpha}$ it follows that

$$(M'_{px} + M_x)\,(M'_{py} + M_y) = M_{xy}^2. \tag{9.9}$$

Eqs. (9.8) and (9.9) were obtained, in a slightly different manner, in [9.11] and can also be found in Nielsen [9.8], Wolfensberger [9.12], Kemp [9.9], and Morley [9.10]. These equations represent two cones (fig. 9.7). Their intersection lies on the surface with equation

$$(M_{px} - M_x)\,(M_{py} - M_y) = (M'_{px} + M_x)\,(M'_{py} + M_y), \tag{9.10}$$

which is obtained by eliminating M_{xy} from (9.8) and (9.9). This surface actually reduces to a plane with equation

$$M_{px} M_{py} - M_y (M_{px} + M'_{px}) - M_x (M_{py} + M'_{py}) - M'_{px} M'_{py} = 0. \tag{9.11}$$

(*) Division by sin α cos α would not be acceptable if sin $\alpha = 0$ or cos $\alpha = 0$. However, examination of these two cases easily reveals that the results obtained above remain valid.

This plane is parallel to the M_{xy}-axis and contains the straight lines with equations :

$$M_x = M_{px} \ ,$$
$$M_y = M_{py} \ , \tag{9.12}$$

and

$$M_x = -M'_{px} \ ,$$
$$M_y = M_{py} \ , \tag{9.13}$$

Since the yield surface bounds the common part of the half spaces that are bounded by the tangent planes and contain the origin, it consists only of the parts of the cones [eqs. (9.8) and (9.9)] that are represented on fig. 9.7.

Every tangent plane touches the surface along a straight line that can be assigned a given value of α. Indeed, all points on such a contact line satisfy both eqs. (9.6) and (9.8) (or similar ones for negative yielding). Solving eq. (9.6) for $\tan\alpha$ and using eq. (9.8) we obtain

$$\tan\alpha = -\frac{M_{xy}}{M_{py} - M_y} \ . \tag{9.14}$$

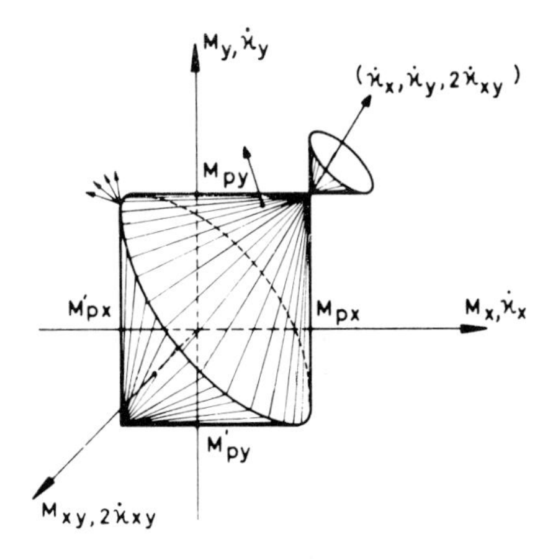

Fig. 9.9.

Because

$$\tan 2\alpha = \frac{2\tan\alpha}{1-\tan^2\alpha},$$

we have

$$\tan 2\alpha = \frac{-2M_{xy}}{(M_{py}-M_y)-(M_{px}-M_x)}. \tag{9.15}$$

For the cone with eq. (9.9), we would obtain (with notation α' to indicate yielding in negative bending)

$$\tan\alpha' = \frac{M_{xy}}{M'_{py}+M_y}, \tag{9.16}$$

and

$$\tan 2\alpha' = \frac{2M_{xy}}{(M'_{py}+M_y)-(M'_{px}+M_x)}. \tag{9.17}$$

Eq. (9.15) can be rewritten as

$$(M_{py}-M_{px})\sin 2\alpha = (M_y-M_x)\sin 2\alpha - 2M_{xy}\cos 2\alpha. \tag{9.18}$$

But we know that, with $\beta = \alpha + \frac{\pi}{2}$

$$M_{\alpha\beta} = \frac{1}{2}(M_x-M_y)\sin 2\alpha + M_{xy}\cos 2\alpha. \tag{9.19}$$

Comparing eqs. (9.18) and (9.19), we deduce that

$$M_{\alpha\beta} = \frac{1}{2}(M_{px}-M_{py})\sin 2\alpha \equiv T_{\alpha\beta}. \tag{9.20}$$

The twisting moment $M_{\alpha\beta}$ associated with the bending moment at yield $M_\alpha = M_{p\alpha}$ is thus determined if the stress point is to be on the yield surface or, equivalently, if curve U is to be *tangent* to curve Y_L (or Y_U). This was shown by Wolfensberger [9.12] and Kemps [9.9].The twisting moment associated with $M_\alpha = -M'_{p\alpha}$ is

$$M_{\alpha\beta} = \frac{1}{2} (M'_{py} - M'_{px}) \sin 2\,\alpha \equiv T'_{\alpha\beta}. \tag{9.21}$$

9.2.3. *Flow rule.*

After discussing the statical conditions for plastic flow, we must now investigate the precise manner in which this flow occurs. Consider a moment state represented by a point on the yield surface, with coordinates M_x, M_y, M_{xy}. According to Prager's terminology, M_x, M_y, M_{xy} are our generalized stresses (Section 5.2). The associated generalized strain rates must be chosen to furnish the specific rate of energy dissipated in plastic flow when the products of the corresponding generalized variables are added. Hence, these generalized strain rates are the rates of curvature $\dot{\kappa}_x$, $\dot{\kappa}_y$ and twice the rate of twist : 2 $\dot{\kappa}_{xy}$.(*).The rate of energy dissipated per unit midsurface of the plate is :

$$D = M_x\,\dot{\kappa}_x + M_y\,\dot{\kappa}_y + 2\,M_{xy}\,\dot{\kappa}_{xy}\,.$$

If we superimpose the systems of orthogonal cartesian coordinates M_x, M_y, M_{xy} and $\dot{\kappa}_x$, $\dot{\kappa}_y$, 2 $\dot{\kappa}_{xy}$ and if we consider a point on the yield surface, the fundamental *flow rule* states that the corresponding generalized strain rates are components of a vector directed along the outward normal to the yield surface at the considered stress point. This is called the *normality rule*. Because of perfect plasticity, the magnitude of the generalized strain rate vector remains arbitrary. Application of the normality law to the yield condition [eq. (9.8)] gives :

$$\begin{aligned} \dot{\kappa}_x &= \lambda\,(M_{py} - M_y)\,, \\ \dot{\kappa}_y &= \lambda\,(M_{px} - M_x)\,, \\ 2\,\dot{\kappa}_{xy} &= 2\,\lambda\,M_{xy}\,, \end{aligned} \tag{9.23}$$

(*) where $\dot{\kappa}_{xy}$ is the twist component of the rate of curvature *tensor*.

where λ is a nonnegative scalar. As the state of stress satisfies eq. (9.8), we have

$$\dot{\kappa}_x \cdot \dot{\kappa}_y = \dot{\kappa}_{xy}^2 \ . \tag{9.24}$$

Application of the normality rule to eq. (9.9) would result in the same relation [eq. (9.24)]. As can be seen on Mohr's circle, fig. 9.8, this relation holds when *one of the principal rates of curvature vanishes.* Hence, *local yielding consists of an elementary yield line* (of length dl). The angle γ of the normal to this yield line with the x-direction is given by

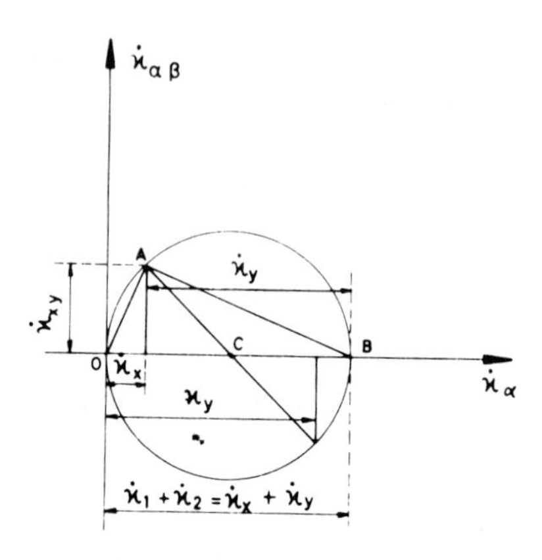

Fig. 9.10.

$$\tan 2\gamma = \frac{2\,\dot{\kappa}_{xy}}{\dot{\kappa}_y \cdot \dot{\kappa}_x} \ . \tag{9.25}$$

Substituting the expression (9.23) for $\dot{\kappa}_x$, $\dot{\kappa}_y$, $\dot{\kappa}_{xy}$ in eq. (9.25), we obtain

$$\tan 2\lambda = \frac{2M_{xy}}{(M_{px} - M_x) - (M_{py} - M_y)} \ . \tag{9.26}$$

Comparison of eq. (9.26) with eq. (9.15) gives $\lambda = \gamma \pm n\,\pi$. Similar results hold for negative yielding. Hence, *the direction of the yield line is that of the section subjected to the yielding moment.*

From inspection of fig. 9.8, it si easily seen that :

1. For an isotropic plate ($\frac{M_{py}}{M_{px}} \equiv k = 1$, $\frac{M_{py}'}{M_{px}'} \equiv k' = 1$, but possibly $M'_{px} \neq M_{px}$) the directions of the yield lines coincide with the principal directions of the moment tensor ;

2. For an orthotropic plate, yield lines no longer follow the principal directions of the moment tensor ; when a positive and a negative yield line occur at a point, they are not orthogonal except when $\frac{(k'-1)}{(1-k)} = \frac{M_{px}}{M'_{px}}$ [9.9], and even in this case they differ from the principal directions of the moment tensor.

The last remark suggests that a discussion of the *singular points* and the *singular line* of the yield surface is in order.

Consider first a singular point characterized by $M_x = M_{px}$, $M_y = M_{py}$, $M_{xy} = 0$. The normality rule requires the strain-rate vector (with components $\dot{\kappa}_x$, $\dot{\kappa}_y 2\,\dot{\kappa}_{xy}$) to lie anywhere on or inside the cone of the outward normals of the yield surface at the singular point (fig. 9.7). Accordingly, we may have any combination of $\dot{\kappa}_1 \geq 0$ and $\dot{\kappa}_2 \geq 0$ with any principal directions. At the singular point $M_x = -M'_{px}$, $M_y = -M'_{py}$, $M_{xy} = 0$ we have any combination of $\dot{\kappa}_1 \leq 0$ and $\dot{\kappa}_2 \leq 0$. As shown in fig. 9.8, at both singular points the curve U coincides with the corresponding curve Y and all directions are potential yield lines (which may or may not become effective). Thus, the singular point corresponds to the intersection of any number of yield lines of the same sign. If this state of stress extends over a region, this region may flow arbitrarily as long as both principal curvatures are everywhere nonnegative.

At the intersection of the two cones we have a singular line. Any point of this line is the intersection of two straight contact lines of the two tangent planes with the two cones. Let $\dot{\kappa}_\alpha > 0$ and $\dot{\kappa}_\alpha' < 0$ denote the nonvanishing principal curvature rates corresponding to these two tangent planes. The normality rule requires that any nonnegative linear combination of these two vectors of curvature rate is admissible. The orientation of the two directions α and α' are well known when the moment tensor is given. Indeed, using eqs. (9.14) and (9.16), and the relation

$$\tan(a-b) = \frac{\tan a - \tan b}{1 + \tan a \,.\, \tan b},$$

we have

$$\tan(\alpha - \alpha') = \frac{M_{xy}(M_{py} - M_y) + M_{xy}(M'_{py} - M_y)}{(M_{py} - M_y)(M'_{py} + M_y) - M_{xy}^2}. \tag{9.27}$$

Linear combination of the two curvature rate tensors results in a new tensor with principal directions that depend on the magnitudes of the coefficients of the combination.

It is worth recalling now that the specific rate of dissipation D (per unit area of midplane) can be viewed as the scalar product of vector $\boldsymbol{\sigma}$ with components M_x, M_y, M_{xy} (point P being on the yield surface) and vector $\dot{\boldsymbol{\varepsilon}}$ with components $\dot{\kappa}_x$, $\dot{\kappa}_y$, $2\,\dot{\kappa}_{xy}$ (fig. 9.11).

$$D = \boldsymbol{\sigma} \cdot \dot{\boldsymbol{\varepsilon}} \ . \tag{9.28}$$

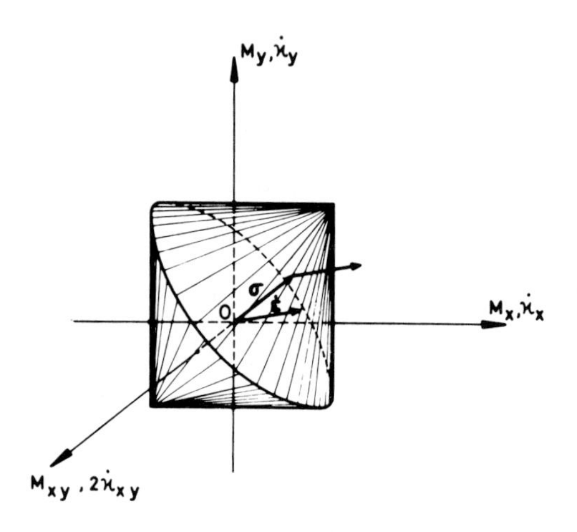

Fig. 9.11.

As is shown in Section 2.4, D is a *single-valued function* of $\dot{\boldsymbol{\varepsilon}}$, even if the yield surface is not regular (that is, if it has flats and vertices, fig. 9.12). In the general case, where there are two nonvanishing principal curvature rates $\dot{\kappa}_1$ and $\dot{\kappa}_2$ with orientation given by the angle α, we have (with $\beta = \alpha + \frac{\pi}{2}$) :

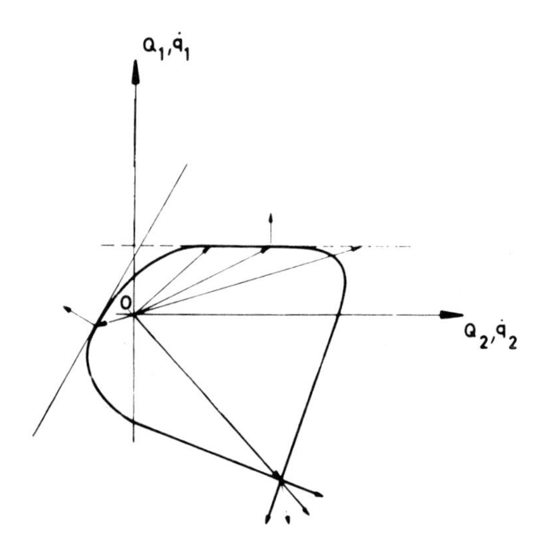

Fig. 9.12.

$$D = M_\alpha \dot{\kappa}_1 + M_\beta \dot{\kappa}_2 . \tag{9.29}$$

Denoting by $\dot{\kappa}_1$ the algebraically largest value of the principal curvature rates, we have

$$D = (M_{px} \cos^2 \alpha + M_{py} \sin^2 \alpha)\dot{\kappa}_1 + (M_{px} \cos^2 \beta + M_{py} \sin^2 \beta)\, \dot{\kappa}_2 \tag{9.30}$$

if $\dot{\kappa}_1 \geq \dot{\kappa}_2 \geq 0$. On the other hand, if $\dot{\kappa}_1 \geq 0 \geq \dot{\kappa}_2$, we have

$$D = (M_{px} \cos^2 \alpha + M_{py} \sin^2 \alpha)\dot{\kappa}_1 + (M'_{px} \cos^2 \beta + M'_{py} \sin^2 \beta)\, |\dot{\kappa}_2| . \tag{9.31}$$

Finally, if $0 \geq \dot{\kappa}_1 \geq \dot{\kappa}_2$, we have

$$D = (M'_{px} \cos^2 \alpha + M'_{py} \sin^2 \alpha)|\dot{\kappa}_1| + (M'_{px} \cos^2 \beta + M'_{py} \sin^2 \beta)\, |\dot{\kappa}_2| . \tag{9.32}$$

The various expressions for D clearly are single-valued functions of the components of the curvature rate tensor, which is given by $\dot{\kappa}_1$, $\dot{\kappa}_2$, and α.

9.3. Discussion of the yield condition.

Theoretical justifications of the chosen yield criterion are presented by Nielsen [9.8], Wolfensgerger [9.12], Kemp [9.9], and Morley [9.10], based on assumed behaviour of steel and concrete.

Indirect experimental verification of the adequacy of the criterion is provided by numerous tests on transversally loaded plates (see, for example, [9.3, 7.16, 9.13]), but little direct experimental support exists for this [9.7], as quoted in [9.10, 9.14 and 9.15].

We shall now summarize the results of experiments made by the coworkers of Massonnet at Liège University, Belgium [9.7].

Square plates 51.2 in. (1.3 m) wide were subjected to biaxial homogeneous bending by sixteen levers (four on each side), acted upon by hydraulic jacks and applying the edge moments by jaws with rubber protection (fig. 9.13). The eight jacks of set A are connected in parallel to the same pump, the eight jacks of set B to a second pump, and the sixteen jacks of set C to a third pump. Principal directions 1 and 2 are obviously parallel to the edges.

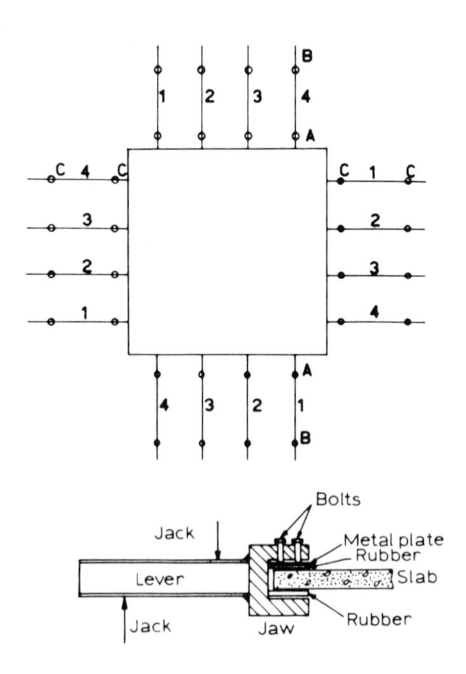

Fig. 9.13. Sketch of the test device.

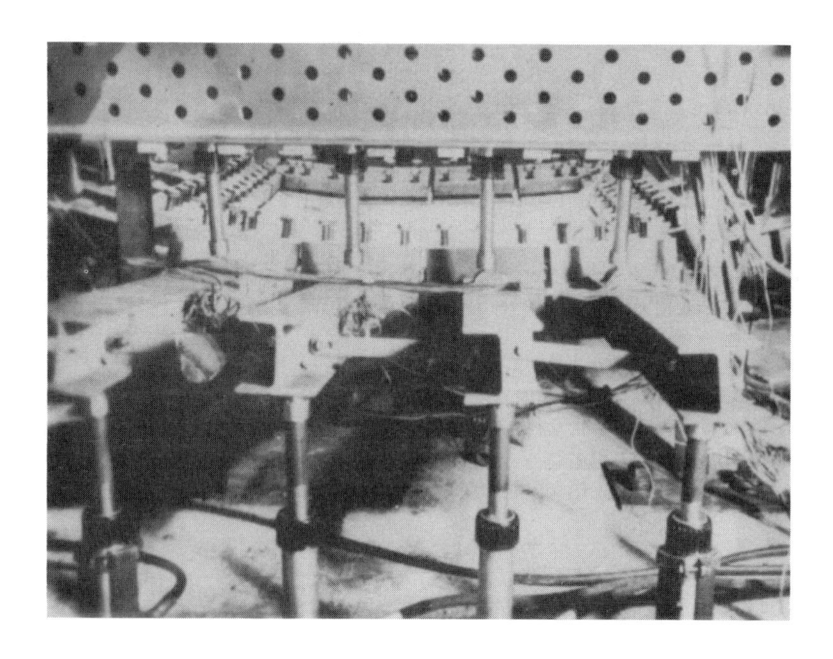

Fig. 9.14.

Jacks C apply the principal moment that will be labelled M_2, whereas jacks A and B apply the moment labelled M_1 and balance the dead weight of the slab and the levers. A photograph of the experimental device is shown in fig. 9.14, where levers and jacks are clearly seen. As is indicated by the photograph, free deformation of the edges is allowed. The moments M_1 and M_2 can be given arbitrary values of either the same signs or opposite signs. There is no shear force, except the negligible one caused by the dead weight. Bending moments are to all intents uniformly distributed along the edges, the largest discrepancies being less than 10%. From recorded pressure versus deflection diagrams analogous to the diagram in fig. 9.1, ultimate values of couples (M_1 , M_2), corresponding to point B in fig. 9.1, have been determined. The characteristics of the tested plates are :

1. Composition of concrete :
   Crushed gravel 3/8 cm, 2760 lb
   Rhine sand 0/2 mm, 1660 lb
   High-strength Portland cement (PHR), 663 lb
   Water, 915 cm^3

2. Metal formworks, vibration by vibrating table.

3. Compressive rupture stress of concrete at 28 days :
   For cubes $6.3 \times 6.3 \times 6.3$ in. : 4840 psi.
   For cylinders ($\varphi = 5.94$ in., h = 11.9 in.) 3970 psi.

4. Tensile rupture stress, on flexion prisms ($3.94 \times 3.94 \times 19.7$ in., span 15.8 in.) 497 psi.

5. Mild steel A 37, 0.394 in. diameter smooth rods, average tensile rupture stress 54400 psi, average yield stress 40200 psi (relative scattering 3.8%).

Fig. 9.15.

6. Slab dimensions : 51.2 in. side, 3.15 in. thickness.

7. Positions of reinforcing rods : layers of 0.802 in^2. of steel per foot, 2.36 in. apart : the effective(*) height is 2.58 in. when there is only one layer near each face of the slab ;

(*) See Section 9.7, fig. 9.77 for definition of effective height.

when two layers exist near the same face, the effective height is 2.58 in. for one and 2.18 in. for the other ; edges of the slab are specially reinforced (fig. 9.15).

The reinforcing rods on top and bottom of any one slab are either unidirectional or follow two orthogonal direction. Their orientation with respect to the edges is defined by the smallest angle α they make with any edge.

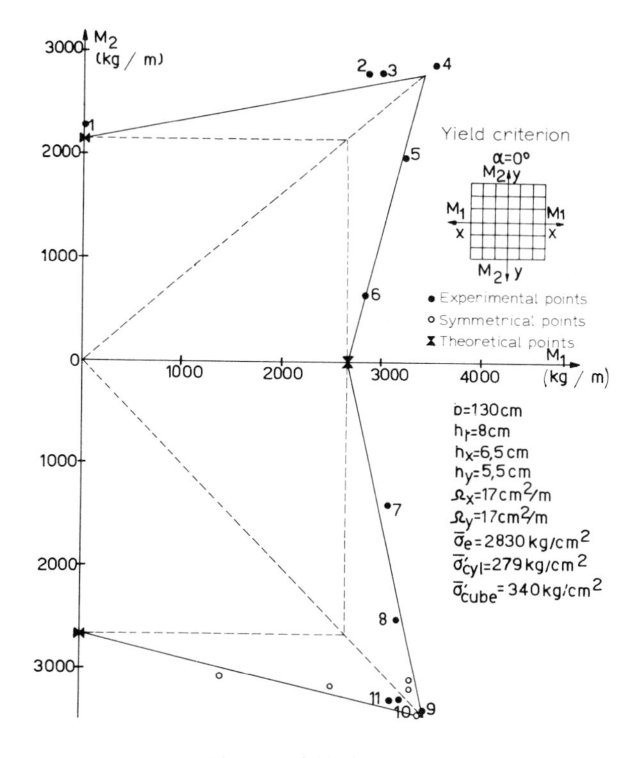

Fig. 9.16. Yield criterion.

The following cases have been considered :

1. $\alpha = 0°$: reinforcement parallel to the edges : 11 slabs. Experimental results are given in fig. 9.16. The theoretical points indicated by   have been calculated in uniaxial bending, as for a beam. For $M_1 \geq 0$ and $M_2 \geq 0$, a difference between M_{p1} and M_{p2} is noted, arising from the difference in effective heights : it is moreover seen that, for $M_1 \approx M_2$ as well as for $M_1 \approx -M_2$, the ultimate strength is larger than in uniaxial bending by about 30% ; finally, we observe that some parts of the yield curve turn their convexity toward the origin.

2. $\alpha = 45°$; results are shown in fig. 9.17. Slabs simply reinforced subjected to bending moments of opposite signs, have only one reinforcement layer near each face. Hence, they are extremely orthotropic, with coefficients $k = 0$ and $k' = \infty$. They are appreciably less strong than the other slabs, but appreciably stronger than predicted by the yield criterion of Johansen (relations 9.1 and 9.2).

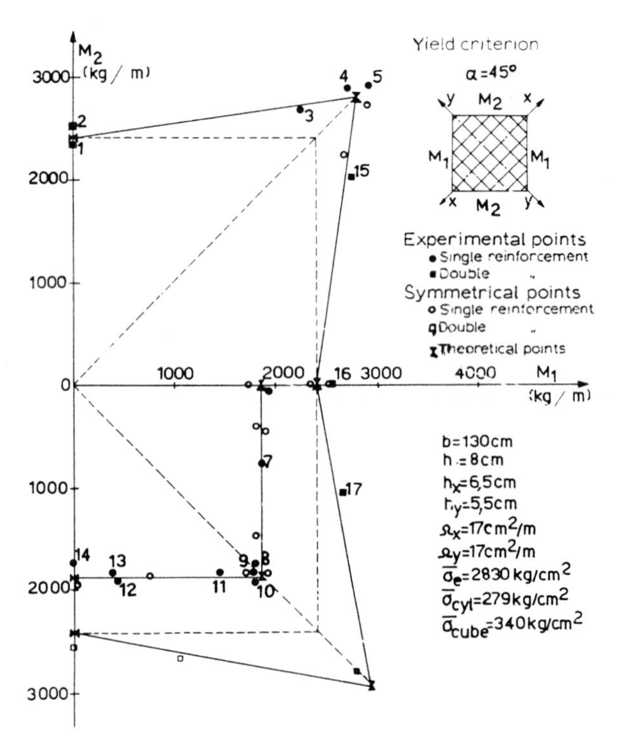

Fig. 9.17. Yield criterion.

All other slabs behave as isotropic plates because they have either two identical orthogonal layers near each face or two such layers near the sole face subjected to tension. Again, pointed corners are found in the diagonal directions of the M_1 , M_2 diagrams (fig. 9.17).

It has been observed that cracks present a general orientation specified by an angle β but are locally formed of segments with two distinct orientations, both of which are different from β. Fig. 9.18 shows, for example, the tension face of the slab subjected to $M_1 = M_2$ with simple reinforcement together with the idealized cracking scheme, with average orientation of 18°30', made of segments parallel and at 45° to the side subjected to M_1.

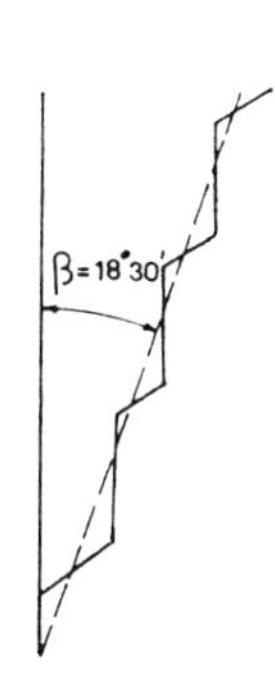

Fig. 9.18.

Theoretical points and connecting segments have been obtained by interpreting the crack patterns and using some simple assumptions. [9.7].

3. $\alpha = 22°30'$ and $\alpha = 30°$; results are shown in fig. 9.19. General features are the same as in the preceding figures. A detailed discussion can be found in the original paper [9.7].

Here, we only point out the following features :

1. *For isotropic plates,* the yield locus of Johansen is entirely inside the locus of the experimental points. Hence, if the ductility of the plate is sufficiently large, any solution based on Johansen's condition gives a lower bound for the limit load (see Section 5.5.2).

When using the condition of Johansen, one can allow for the influence of nonuniform effective height.

The margin of safety in the regions $M_1 \approx M_2$ and $M_1 \approx -M_2$ may be partially cancelled by the fact that, as a rule, a kinematical approach is used. Accordingly, for practical use it seems justified to accept the considered yield condition.

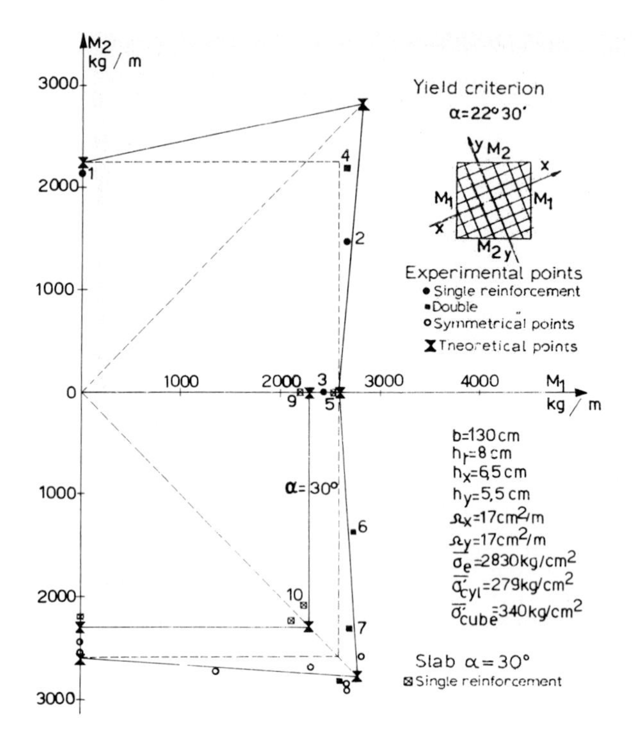

Fig. 9.19. Yield criterion.

2. *For orthotropic plates* (simple reinforcements) :

$$M_{py} = 0, \qquad M_{px} \neq 0 \qquad (k = 0),$$

$$M'_{py} \neq 0, \qquad M'_{px} = 0 \qquad (k' = \infty).$$

In the case where $M_1 \neq 0$, $M_2 = 0$, $\alpha = 45°$, we have $M_x = M_y = M_{xy} = \dfrac{M_1}{2}$.

According to relation (9.8), we then find

$$M_1 = \frac{2\, M_{px}\, M_{py}}{M_{px} + M_{py}}.$$

With $M_{py} = 0$, we obtain $M_1 = 0$, and the normality law predicts $\dot{\kappa}_x = \dot{\kappa}_{xy} = 0$, $\dot{\kappa}_y \neq 0$. Hence, yield lines parallel to the reinforcement rods should be obtained for arbitrarily small values of M_1, whereas experiments give $M_1 \approx 0.7\ M_{px}$. We conclude that Johansen's condition does not apply for very pronounced orthotropy.

Experiments by Lenschow end Sozen [9.14] do not show the increase of yield moments obtained at Liège University [9.7] when $|M_1| = |M_2|$. On the other hand, tests by Lenkei [9.15] not only have shown the existence of twisting moments in the yield line of an orthotropic slab when the principal directions of orthotropy differ from those of the moment tensor, but also indicate a smaller plastic moment than that predicted by Johansen's criterion.

In the absence of more precise information on the subject, it seems reasonable to continue to accept Johansen's yield condition, keeping in mind its approximate character.

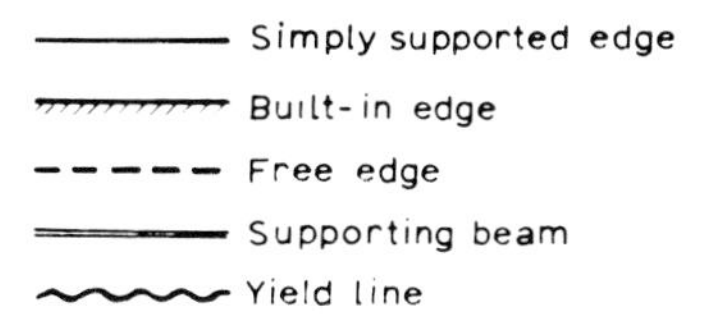

Fig. 9.20.

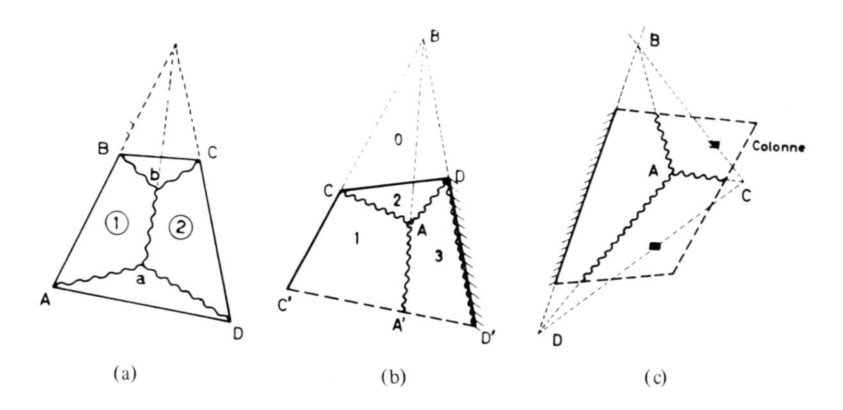

Fig. 9.21.

9.4. The kinematic method (Johansen's fracture line theory).

9.4.1. *Kinematics of the mechanisms.*

In accordance with our yield condition and with experimental evidence, we shall hereafter consider collapse mechanisms made of "hinge lines" (or "yield lines") about which adjacent parts of the plate will experience exclusively a relative rotation. Most often the yield lines will be straight because they will devide the slab into *rigid* parts hinged along the yield lines [9.3].

In the following, boundary conditions will be represented graphically as shown in fig. 9.20. At collapse, the plate is divided into various plane portions by the straight yield lines and thus bends into a polyhedron, or possibly into a ruled surface when an infinite

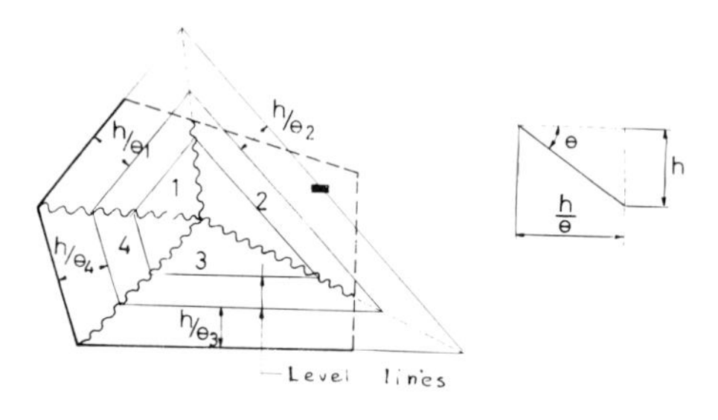

Fig. 9.22.

number of infinitely close yield lines occur. A yield line will be called *positive* when subjected to a *positive* bending moment, and *negative* when subjected to a *negative* bending moment.

Consider first kinematically admissible mechanisms for the plates shown in fig. 9.21. The axes of rotation of the various portions of the plates coincide with either the simply supported edges, or the built-in edges, or are to pass through the points of support [fig. 9.21 (c)].

From inspection of fig. 9.21, we deduce that *a yield line separating two portions of the plate passes through the point of intersection of their axes of rotation.*

As we are only considering incipient collapse under constant load, all velocities of the collapse mechanism are only defined to within a common positive factor, the magnitude of which is irrelevant (except that it must be infinitesimal to justify disregarding changes of geometry).

Thus, in the specification of a mechanism, one angle of rotation, or one deflection, can be assigned an arbitrary value (*).

Conversely, *the mechanism is completely determined by giving the axes of rotation of the various portions of the plate and the ratios of the angles of rotation.* Indeed, as can seen from fig. 9.22, a (polygonal) contour line of arbitrary level h of the deflected plate consists of segments that are parallel to the axes of rotation at distances h/θ_i, these segments intersecting on the yield lines. The latter thus are the loci of the vertices of the polygonal contour lines. It follows directly from this property that the number p of geometric parameters needed completely to determine a mechanism is

$$p = n + a - 1 , \tag{9.33}$$

where n is the number of regions into which the plate is divided by the yield lines and a the degree of indeterminacy that remains in our knowledge of the axes of rotation.

In fig. 9.21 (a), n=4, a=0 and hence p=3. This actually constitutes a three-parameter

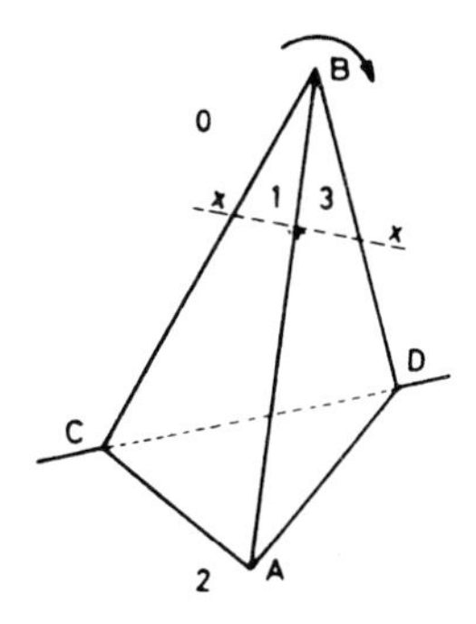

Fig. 9.23.

"family" of mechanisms. In fig. 9.21 (b), n=3, a=0, and p=2. In fig. 9.21 (c), n=3 but a=2 (two axes of rotation have only one fixed point each, namely, the supporting points) and thus p=3+2-1=4.

(*) For brevity, we shall use the terms *rotation, deflection,* and *work* in the following, instead the more accurate terms *rate or rotation, rate of deflection,* and *rate of work..*

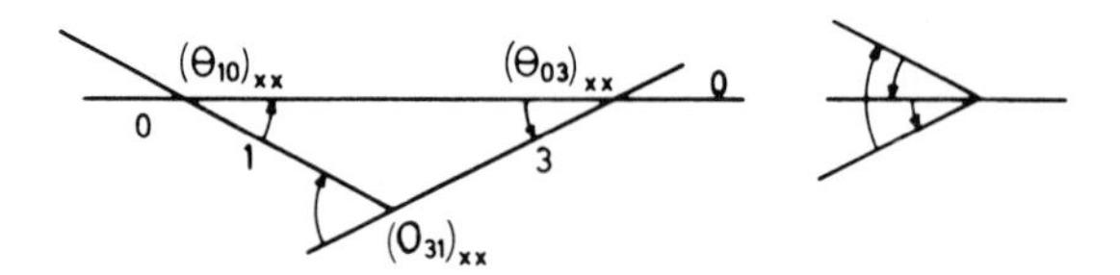

Fig. 9.24.

The values of the angles of relative rotation in a given mechanism can be obtained by a graphical procedure, which is based on the following remark. On any hinge line or its extension, it is always possible to find two points through which pass either the axes of rotation of the adjacent rigid parts [points C, D, B in fig. 9.21 (b), for example] or two other hinge lines [point A, fig. 9.21 (b)]. With adequate sign conventions, the rotation vectors along the rotation axes and yield lines intersecting at the considered points form, at each point, a system equivalent to zero, by their very definition. Hence, intersections may be regarded as joints of a fictitious truss the bars of which are the axes of rotation and the hinge lines, the vectors of rotation playing the role of forces. Consequently, a Maxwell-Cremona diagram can be drawn for the rotation vectors. We shall call it "rotation diagram". To begin with, the magnitude of one rotation is assigned an arbitrary value. This rotation vector is represented by two forces in equilibrium, applied to the truss, equivalent to an internal force in the bar representing the considered axis.

In the example of fig. 9.21 (b), the bars will be AD, AC, BD, BC, and BA. The arbitrary rotation of axis CD, corresponding to a force in bar CD, will be represented by opposite external forces at C and D directed along CD (fig. 9.23). The regions of the plane of the truss are then labeled in correspondence with those determined by the plate boundary and its collapse mechanism [see fig. 9.21 (b) and fig. 9.23].

We then describe closed circular paths about each joint, in a sense of rotation that is fixed once for all.

Suppose for example that we begin at point B, fig. 9.23, moving in the clockwise sense, and successively cross segments BD, AB, and CB, that correspond to hinge lines DD' and AA', and to the axis of rotation CC', respectively. The intersections of the planes 0, 1, and 2 with a plane xx orthogonal to AB are shown in fig. 9.24. If the smallest angles θ_{xx} between the obtained intersection lines (fig. 9.24) are *positive* from one plane to the next *when turning about point B*, we have,

$$(\theta_{10})_{xx} + (\theta_{03})_{xx} + (\theta_{31})_{xx} = 0 \ . \tag{9.34}$$

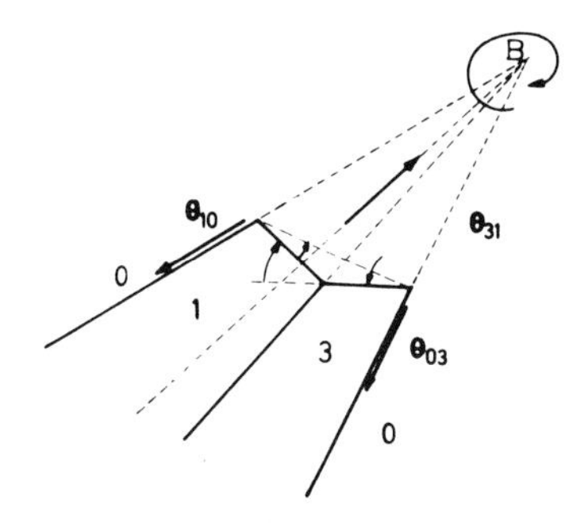

Fig. 9.25.

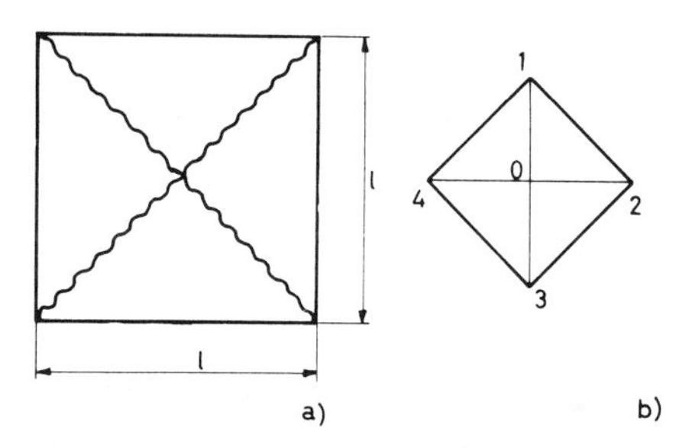

Fig. 9.26.

Two similar relations are obtained from the consideration of intersecting planes normal to CB and DB. We thus arrive at the following result. With a given sign convention for rotation vectors (for example the right-hand screw rule) and if regions i, j, k are successively met when circulating about the joint where bars i, j, and k meet, we have

$$\boldsymbol{\theta}_{ij} + \boldsymbol{\theta}_{jk} + \boldsymbol{\theta}_{ki} = 0. \tag{9.35}$$

This property is illustrated in fig. 9.25. Consider, for example, a square, simply supported, and uniformly loaded isotropic plate. We adopt the collapse mechanism of fig. 9.26, which tends to deform the flat plate into a pyramid. If we give unit displacement to the centre, the rotations about the edges are 2/l. The rotation diagram is shown in fig. 9.26 (b). "Equilibrium" of joint A gives triangle 014, of joint B, 102, etc. In this manner, we obtain

$$D = M_p \,.\, 2l\sqrt{2}\;(2l)\,.\;\sqrt{2} = 8\,M_p. \tag{9.36}$$

The work W of the applied load p is given by the product of p by the volume of the pyramid, that is

$$W = p\,.\,\frac{1}{3}l^2\,.\,1. \tag{9.37}$$

Equating D and W, we find

$$p_+ = \frac{24\,M_p}{l^2}. \tag{9.38}$$

With $P = pl^2$, eq. (9.38) can be written

$$P_+ = 24\,M_p. \tag{9.39}$$

If the plate was built-in and had the plastic moment M'_p in negative bending, relation (9.37) would remain unchanged, whereas the term

$$4\,l\,.\,M'_p\frac{2}{l} = 8\,M'_p$$

would have to be added to the expression (9.36) of D. We therefore would obtain

$$P_+ = 24\,(M_p + M'_p). \tag{9.40}$$

For a rectangular plate (AB=DC=l, AD=BC=L>l, fig. 9.27) it is readily verified that, if yield lines at corners make an angle of 45° with the edges, the rotations diagram is the same as that for the square plate (provided unit displacement is prescribed for segment EF, fig. 9.27). The terms $(L\text{-}l)\frac{4\,M_p}{l}$ and $\frac{pl}{2}(L\text{-}l)$ must be added to the expressions (9.36) and (9.37) of D and W, respectively. We then have

$$p_+\left(\frac{l^2}{3} + \frac{lL}{2} - \frac{l^2}{2}\right) = 8\,M_p + 4\,\frac{L\,.\,l}{l}M_p,$$

or

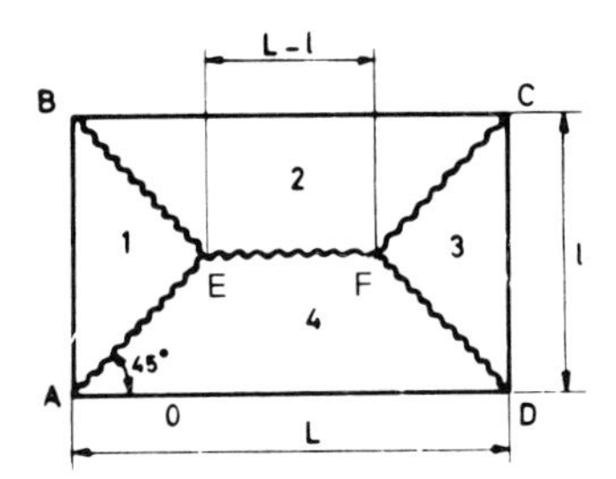

Fig. 9.27.

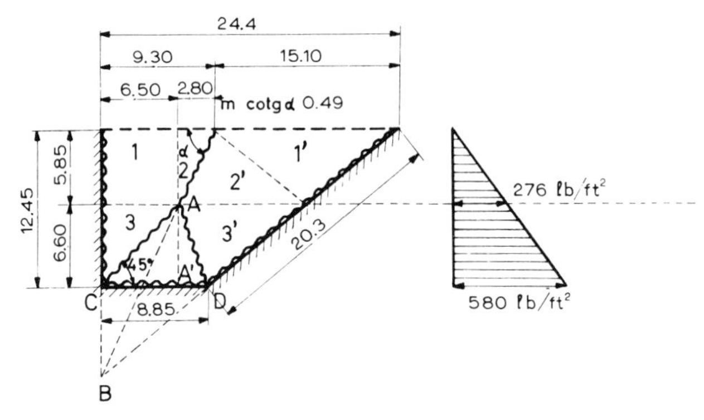

Fig. 9.28.

$$p_+ = \frac{24\,M_p}{l^2}\,\frac{1+\dfrac{L}{l}}{\dfrac{3L}{l}-1}, \tag{9.41}$$

which coincides with formula (7.35) obtained in Section 7.5.2.

As a third example, consider the isotropic plate of fig. 9.28, built-in on three sides, free on the fourth, and subjected to a linearly varying pressure as shown. Assume the mechanism shown in fig. 9.28 and defined by the position of point A and the angle α. Give to point A unit transversal displacement. The fictitious truss consists of the bars AC, CB, BD, DA, and AB, and is subjected to external forces at the joints C and D. The corresponding regions of the plate and the truss are labeled as shown in fig. 9.28 and 9.29. The label 0 refers to the region external to the plate. Forces at C and D are represented by segments 03

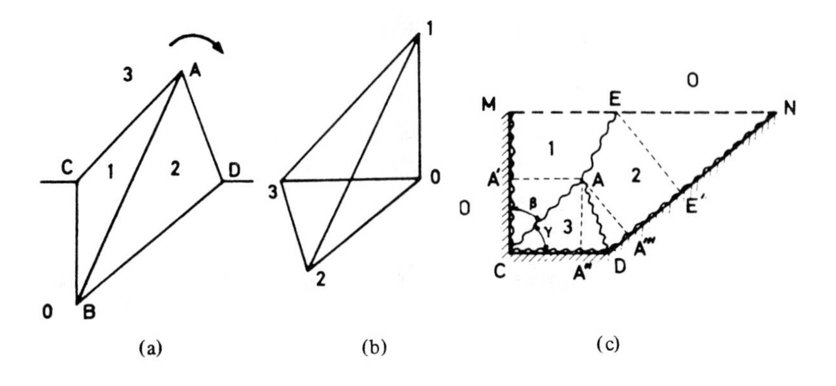

Fig. 9.29.

and 30, with common magnitude $\frac{1}{AA'} = \frac{1}{2}$. The following values of the rotations are obtained from the rotation diagram of fig. 9.29 (b) :

Rotations about

$$\text{BC:} \quad \theta_{BC} = |\overline{01}| = 0.500,$$

$$\text{BD:} \quad \theta_{BD} = |\overline{02}| = 0.500.$$

Rotations of adjacent parts about

$$\text{BA:} \quad \theta_{BA} = |\overline{12}| = 0.900,$$

$$\text{CA:} \quad \theta_{CA} = |\overline{31}| = 0.706,$$

$$\text{AD:} \quad \theta_{AD} = |\overline{32}| = 0.330.$$

With AC=2.83 m, AD=2.10 m, AF=2.00 m, the work dissipated is

$$D = M_p \ (0.76 \times 2.83 + 0.330 \times 2.10 + 0.9 \times 2.00) + M'_p \ (3.8 \times 0.5 + 2.7 \times 0.5 + 6.15 \times 0.5) = 4.494 \ M_p + 6.32 \ M'_p.$$

To obtain the work of the load p, we must evaluate the resultant forces acting on parts 1, 2, and 3. For convenience, we compute partial resultant forces and their distances δ to the corresponding axes of rotation. One finds (fig. 9.29)

$p_{1a} = 2.52$ t, $\delta_{p_{1a}} = 1$ m,

$p_{1b} = 0.38$ t, $\delta_{p_{1b}} = 2.23$ m,

$p_{1c} = 2.80$ t, $\delta_{p_{1c}} = 0.67$ m,

$p'_{1c} = 0.96$ t, $\delta_{p'_{1c}} = 0.50$ m.

(p_{1c} corresponds to the uniform part of the pressure, p'_{1c} to the linearly varying part.)

$p_{2a} = 1.93$ t, $\delta_{p_{2a}} = 0.71$ m,

$p_{2b} = 2.69$ t, $\delta_{p_{2b}} = 1.44$ m,

$p_{2c} = 4.48$ t, $\delta_{p_{2c}} = 0.66$ m,

$p'_{2c} = 1.54$ t, $\delta_{p'_{2c}} = 0.50$ m,

$p_3 = 3.78$ t, $\delta_{p_3} = 0.67$ m,

$p'_3 = 2.57$ t, $\delta_{p'_3} = 0.50$ m.

The work of the pressure thus is

$$W = \sum p_i \, \delta_i \, \theta_i = 0.5(2.52 \times 1 + 0.38 \times 2.23 + 2.8 \times 0.67 + 0.96 \times 0.5)$$
$$+ 0.5(1.93 \times 0.71 + 2.69 \times 1.44 + 4.48 \times 0.66 + 1.54 \times 0.5)$$
$$+ 0.5(3.78 \times 0.67 + 2.57 \times 05) = 8.708 \, .$$

Equation W=D gives

$$4.494 \, M_p + 6.32 \, M'_p = 8.708.$$

With $M'_p = 0$ (simple supports) we would obtain

$$M_p = 19.4\text{kN}.$$

9.4.2. *Work equation.*

Consider (fig. 9.30) a yield line with length l, adjacent to the parts ***i*** and ***j*** of the plate that rotate about the axes i and j, respectively. Denote by $\boldsymbol{\theta}_i$ and $\boldsymbol{\theta}_j$ vectors with magnitude equal to the angles of rotations of the two parts about their respective axes i and j, and directed along these axes, both pointing away from the intersection I of the axes.

Denote by $\boldsymbol{\theta}$ a vector with a magnitude equal to the *relative* rotation of parts ***i*** and ***j*** about the yield line (rotation "in" the yield line), and directed along this yield line. We have [9.30 (b)]

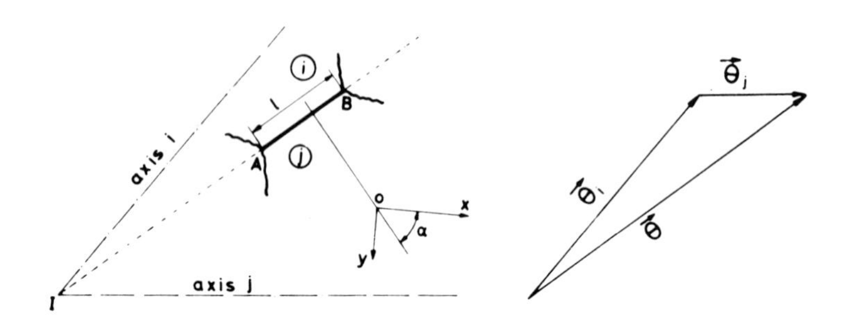

Fig. 9.30.

$$\boldsymbol{\theta} = \boldsymbol{\theta}_i + \boldsymbol{\theta}_j. \tag{9.42}$$

Denote by α the angle of the normal to the yield line with the x-axis, the latter coinciding with the direction of one of the reinforcing bars. According to relations (9.1) or (9.2), the plastic moment is constant along the whole yield line. In a positive yield line we thus have

$$M_\alpha = M_{p\alpha} = M_{px}\cos^2\alpha + M_{py}\sin^2\alpha, \tag{9.43}$$

and the work dissipated in this yield line is

$$D_\alpha = |M_\alpha|\,|\boldsymbol{\theta}|\,|\mathbf{l}| \tag{9.44}$$

where $\mathbf{l}$ is a vector directed along the yield line, oriented as $\boldsymbol{\theta}$, and with a magnitude equal to the length of the line. We have

$$|\boldsymbol{\theta}|\,|\mathbf{l}| = \frac{l_x\,\theta_x}{\sin^2\alpha} = \frac{l_y\,\theta_y}{\cos^2\alpha}, \tag{9.45}$$

where l_x, l_y, θ_x, θ_y are the projections of $\mathbf{l}$ and $\boldsymbol{\theta}$ on the x- and y-axes, respectively.

Substitution of expressions (9.45) and (9.43) for $|\boldsymbol{\theta}|\,|\mathbf{l}|$ and $|M_\alpha|$ into eq. (9.44), and use of $M_{py} = kM_{px}$, give

$$D_\alpha = M_{px}\,(l_y\,\theta_y + kl_x\,\theta_x). \tag{9.46}$$

A negative yield line would have furnished

$$D'_\alpha = M'_{px}\,(l_y\,\theta_y + k'l_x\,\theta_x). \tag{9.47}$$

Summing the works dissipated in all the yield lines we obtain the total work dissipated as

$$D = M_{px}\sum_{+}(l_y\,\theta_y + kl_x\,\theta_x) + M'_{px}\sum_{-}(l_y\,\theta_y + k'l_x\,\theta_x), \tag{9.48}$$

where the first sum is extended to all positive yield lines and the second to all negative yield lines.

When the plate is isotropically reinforced, we have k=k'=1. Hence, eq. (9.48) reduces to

$$D = M_p\sum_{+}(l_y\,\theta_y + l_x\,\theta_x) + M'_p\sum_{-}(l_y\,\theta_y + l_x\,\theta_x). \tag{9.49}$$

As we know that

$$l_x\,\theta_x + l_y\,\theta_y = \mathbf{l}\cdot\boldsymbol{\theta},$$

eq. (9.42) gives

$$l_x\,\theta_x + l_y\,\theta_y = \mathbf{l}\cdot\boldsymbol{\theta}_i + \mathbf{l}\cdot\boldsymbol{\theta}_{j'}, \tag{9.50}$$

and eq. (9.49) can be rewritten :

$$D = M_p \sum_{+} (\mathbf{l} \cdot \boldsymbol{\theta}_i + \mathbf{l} \cdot \boldsymbol{\theta}_j) + M'_p \sum_{-} (\mathbf{l} \cdot \boldsymbol{\theta}_i + \mathbf{l} \cdot \boldsymbol{\theta}_j). \tag{9.51}$$

If we denote by l_i and l_j the projections of $\mathbf{l}$ on θ_i and θ_j, eq. (9.51) becomes

$$D = M_p \sum_{+} (l_i\,|\boldsymbol{\theta}_i| + l_j\,|\boldsymbol{\theta}_j|) + M'_p \sum_{-} (l_i\,|\boldsymbol{\theta}_i| + l_j\,|\boldsymbol{\theta}_j|). \tag{9.52}$$

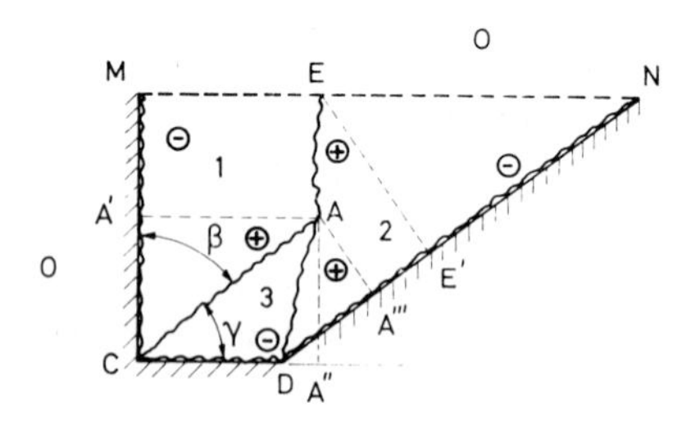

Fig. 9.31.

Eq. (9.52) may be expressed as follows : *in an isotropic plate, the work dissipated in the positive yield lines separating various portions of the plate is the product by* M_p *of the sum of the products of the rotations of these portions about their respective axes by the projections on these axes of the considered yield lines ; the work dissipated in the negative yield lines is obtained similarly, replacing* M_p *by* M'_p *and summing along negative yield lines.* Note that some values l_i or l_j might be negative. In the example of fig. 9.31 of a plate built-in along the edges MCDN and free along MN, we have

$$D = M_p \left[\theta_1 \,(\overline{CA'} + \overline{A'M}) + \theta_2 \,(\overline{DA'''} + \overline{A'''E}) + \theta_3 \,(\overline{CA''} + \overline{A''D}) \right]$$

$$+ M'_p \,(\theta_1 \cdot \overline{CM} + \theta_3 \cdot \overline{CD} + \theta_2 \cdot \overline{DN}) .$$

$\overline{CA''}$ and $\overline{A''D}$ have opposite signs and their sum is CD. The work dissipated is

$$D = M_p(\theta_1 \cdot \overline{CM} + \theta_2 \cdot \overline{DE} + \theta_3 \cdot \overline{CD})$$

$$+ M'_p(\theta_1 \cdot \overline{CM} + \theta_2 \cdot \overline{DN} + \theta_3 \cdot \overline{CD}). \qquad (9.53)$$

9.4.3. *Orthotropic plates : affinity method.*

The use of eq. (9.48) implies the calculation of the projections θ_x and θ_y on the axes (parallel to the reinforcing bars) of the *relative* rotations of the portions adjacent to the yield lines whereas in eq. (9.51) applicable to isotropic plates, *absolute* rotations only are needed. These latter being much easier to evaluate, we want to reduce the study of an orthotropic plate to that of a *fictitious equivalent isotropic plate*, that is an isotropic plate with same limit load P_l (or with same plastic moment M_p under the given loads). We must therefore define the fictitious isotropic plate such that the work dissipated in its mechanism and the work of the loads be the corresponding works in the actual plate multiplied by a common factor. Consider a plate obtained multiplying all dimensions of the original plate parallel to O_y by an "affinity" coefficient γ, without altering the dimensions along O_x. Call the resulting plate the "affine plate", with respect to the given one [9.3]. A point P with coordinates x_1 and y_1 on a segment of yield line becomes P_γ with coordinates x_1 and γy_1 (fig. 9.32).

Suppose that one of the two rigid parts to which point P belongs rotates about an axis RR, fig. 9.32, by an angle θ. The displacement w of P is given by the magnitude of the vectorial product of vectors $\boldsymbol{\theta}$ and $\mathbf{d} \equiv \mathbf{PN}$. We thus have

$$w_p = \theta_y d_x - \theta_x d_y .$$

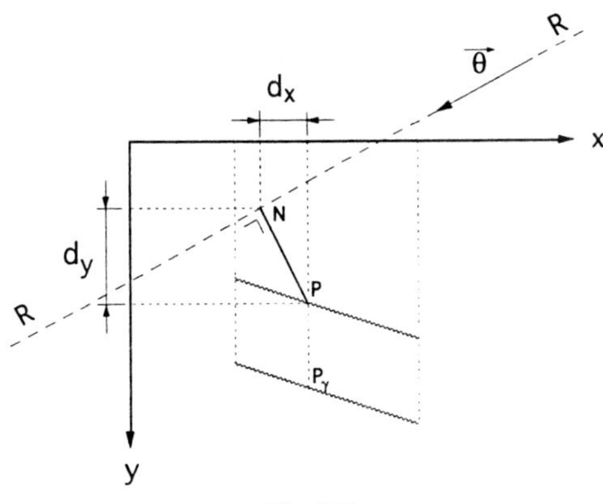

Fig. 9.32.

Similarly, in the affine plate

$$w_{p\gamma} = (\theta_y)_\gamma d_x - (\theta_x)_\gamma d_y \cdot \gamma$$

because $(d_x)_\gamma = d_x$ and $(d_y)_\gamma = \gamma d_y$.

If we let

$$w_{p\gamma} = w_p ,$$

in order, as will be seen later, to maintain the same distributed load q (lb/in^2) on the affine plate as on the given plate, we obtain

$$(\theta_x)_\gamma = \frac{\theta_x}{\gamma}, \tag{9.54}$$

and

$$(\theta_y)_\gamma = \theta_y . \tag{9.55}$$

The affine plate is isotropically reinforced to exhibit moments M_p and M'_p equal to the moment M_{px} and M'_{px} of the given plate, respectively. Application of eq. (9.48) to the affine plate, setting k=k'=1, gives

$$D_\gamma = \sum_{+} \left(M_{px} \gamma l_y \theta_y + M_{px} l_x \frac{\theta_x}{\gamma} \right) + \sum_{-} \left(M'_{px} \gamma l_y \theta_y + M'_{px} l_x \frac{\theta_x}{\gamma} \right). \tag{9.56}$$

We now restrict ourselves to the study of orthotropic plates with the same orthotropy coefficients in both the upper and the lower layers of bars. Hence, k=k'. Let

$$\frac{1}{\gamma^2} = k . \tag{9.57}$$

Substitution of expression (9.57) of γ in eq. (9.56) and comparison with eq. (9.48) yield :

$$D_\gamma = \gamma D . \tag{9.58}$$

The work w of the loads is the sum of the contributions of the distributed load q(x,y) on a certain area A and of the concentrated loads P_i at various points i. Hence,

$$W = \int_A qw \, dA + \sum_i P_i w_i \, . \tag{9.59}$$

For the affine plate, $w_\gamma = w$, $A_\gamma = \gamma A$. If we let $q_\gamma = q$, we have

$$\int_{A\gamma} q_\gamma w_\gamma dA_\gamma = \gamma \int_A qw \, dA \, .$$

Beside, if we let $(P_i)_\gamma = \gamma P_i$, we finally obtain

$$W_\gamma = \gamma W \, . \tag{9.60}$$

Applying the kinematic theorem to the affine plate we write $W_\gamma = D_\gamma$. Substitution of expressions (9.58) and (9.60) for W_γ and D_γ in this relation gives $\gamma W = \gamma D$, and consequently, the affine plate will exhibit the same limit load (or the same plastic moment M_{px}) as the given orthotropic plate. To sum up, the affine plate of an orthotropic plate with given M_{px}, M'_{px}, k and with k'=k, is obtained by multiplying the dimensions in the Oy-direction and the concentrated loads by the affinity coefficient $\gamma = \dfrac{1}{\sqrt{k}}$. Dimensions in the ox-direction and distributed loads are unaltered. If a load is distributed along a line, its resultant must be multiplied by γ.

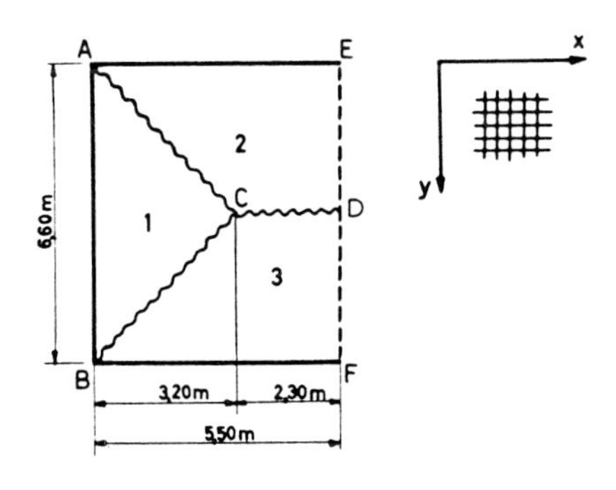

Fig. 9.33.

Consider for example the plate in fig. 9.33, simply supported along the edges EABF and free along EF, uniformly loaded with q=1 (unit of load per unit of area), and assumed to have the collapse mechanism represented on the drawing.

The reinforcement will ensure $k = \frac{M_{py}}{M_{px}} = 2$, and we want to determine M_{px}.

Let the displacement of point C be unity. We have immediately $|\theta_1| = \frac{1}{3.2}$, $|\theta_2| = |\theta_3| = \frac{1}{3.3}$. Obviously $|\theta_{CD}| = \frac{2}{3.3}$ and $(\theta_{CD})_y = |\theta_{CD}|$, whereas $(\theta_{CD})_x = 0$. $\boldsymbol{\theta}_{AC}$ is obtained graphically in fig. 9.34. Obviously, $(\theta_{AC})_x = \frac{1}{3.3} = (\theta_{BC})_x$, $(\theta_{AC})_y = \frac{1}{3.2} = (\theta_{BC})_y$. Application of eq. (9.49) (with no negative yield line) gives

$$D = M_{px}\left[2 \times \frac{3.3}{3.2} + 2\left(2.3 \times \frac{2}{3.3} + 2 \times \frac{3.2}{3.3}\right)\right] = 8.73\ M_{px}\ .$$

The work of the load is

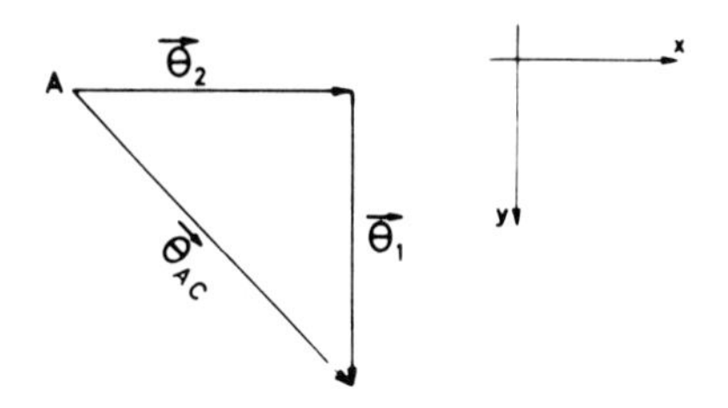

Fig. 9.34.

Hence, from W=D we obtain $M_{px} = 1.68$.

Let us now apply the affinity method to the same problem. The affinity coefficient is $\gamma = \frac{1}{\sqrt{2}} = 0.707$. Consequently, we obtain the plate and the mechanism in fig. 9.35,

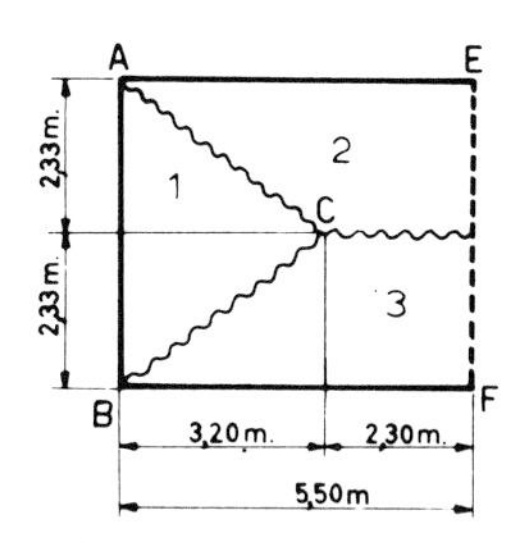

Fig. 9.35.

where $|\theta_1| = \frac{1}{3.2}$, $|\theta_2| = |\theta_3| = \frac{1}{2.33}$. Applying eq. (9.52) we obtain

$$D_\gamma = M_{px}\left(\frac{4.66}{3.2} + 2\times\frac{5.5}{2.33}\right) = 6.162\ M_{px}\ .$$

The work of the loads is

$$W = q\left(4.66\times 3.2\times\frac{1}{2}\times\frac{1}{3} + 2\times 3.2\times\frac{2.33}{2}\times\frac{1}{3} + 2\times 2.33\times 2.3\times\frac{1}{2}\right) = 10.435\ ,$$

whence $M_{px} = 1.692$. The difference between the two results is less than the error due to the numerical approximation. Note also that the value obtained for M_{px} is a lower bound because we have used exclusively the kinematic theorem.

In their book cited in ref. [9.16], Wood and Jones have extended the affinity method to plates with nonorthogonal reinforcing bars.

9.4.4. *Obtaining the actual collapse mechanism.*

Application of the kinematical theorem essentially consists of assuming a type of collapse mechanism, that is a p-parameter family of mechanisms.

For this family, the corresponding load P_+ is readily obtained from the work equation, as a function of the p parameters α , β , γ, δ... which , according to the kinematic theorem, must be chosen to minimize the limit load. This minimum is not necessarily analytic. However, if it is analytic, and if the function $P_+ (\alpha , \beta , \gamma ,...)$ is continuous, the minimum condition is

$$\frac{\partial P}{\partial \alpha} = 0 , \qquad \frac{\partial P}{\partial \beta} = 0 , \qquad \frac{\partial P}{\partial \gamma} = 0 , \dots . \tag{9.61}$$

The system of p eqs. (9.61) furnishes the "best" values of the parameters, and the corresponding mechanism is the actual collapse mechanism provided the assumed family does contain the actual collapse mechanism.

Consider, for example, an infinitely long rectangular plate of width a, that is simply supported along its infinitely long parallel edges and loaded by a concentrated load P at a point of its axis. The plate is isotropic, with plastic moments M_p and M'_p. If we first suppose that the collapse mechanism is of the type represented in fig. 9.36 (a), and if we give a unit displacement to the point of application of the load P, the equation W=D takes the form

$$1.P = M_p \left(8 \frac{\alpha}{a} + \frac{2a}{\alpha} \right) + M'_p \, 2 \frac{a}{\alpha} . \tag{9.62}$$

The system (9.61) reduces here to the single equation $\frac{\partial P}{\partial \alpha} = 0$, which furnishes

$$\alpha = \frac{a}{2} \left(1 + \frac{M'_p}{M_p} \right)^{1/2} . \tag{9.63}$$

Substitution of eq. (9.63) for α into (9.62) yields

$$P = 8 \, M_p \left(1 + \frac{M'_p}{M_p} \right)^{1/2} . \tag{9.64}$$

Expression (9.64) furnishes the smallest value of P that can be obtained from the assumed type of collapse mechanism, but we are by no means certain that this family is the correct one.

This certainty may be achieved only if we are able to apply the combined theorem, indicating for the mechanism which gives the minimum value [eq. (9.64)] of P a corresponding licit moment field. On the other hand, to ascertain that the mechanism obtained above is *not* the correct collapse mechanism, et is *sufficient* to find one mechanism of another family yielding a smaller load than eq. (9.64). Consider, for example, the mechanism shown in fig. 9.36 (b). This mechanism consists of two "fans" of infinitely close positive yield lines, bounded by two negative circular yield lines. This kind of fan may be considered as the limit of the mechanism shown in fig. 9.37, in which the number of positive yield lines is finite, and the negative yield lines form a polygon.

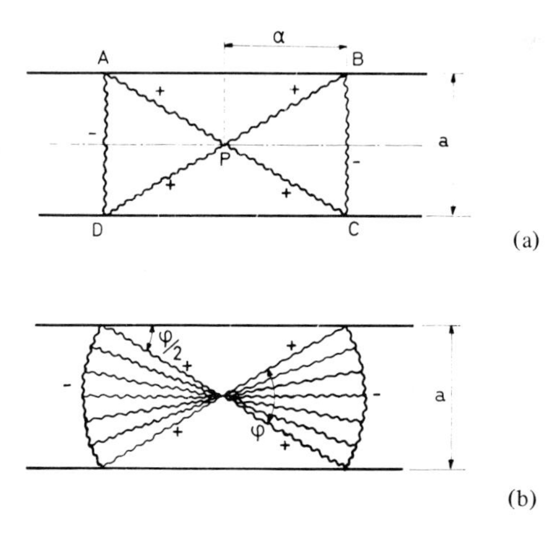

Fig. 9.36.

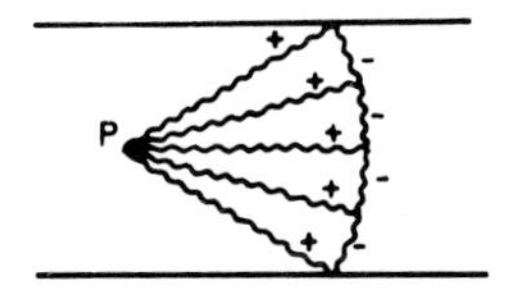

Fig. 9.37.

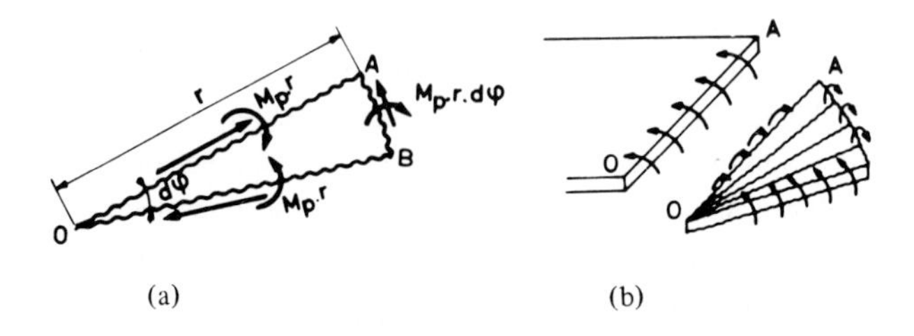

Fig. 9.38.

The work dissipated in a fan, for a unit displacement of its centre point, is obtained as follows. The positive moments M_p acting on the sides OA and OB of an elementary triangle [fig. 9.38 (a)] have the resultant $M_p \cdot r \cdot d\,\varphi$ with the direction AB of the tangent at A to the circular negative yield line. Its angle of rotation is 1/r. Hence, the elementary work dissipated by the moments applied to OA and OB is $M_p\, r \frac{d\varphi}{r} = M_p \cdot d\varphi$. In the negative yield line AB, it is $M'_p\, r \frac{d\varphi}{r} = M'_p \cdot d\varphi$. Summing and integrating, we obtain

$$D = \int_{\varphi} (M_p + M'_p)\, d\varphi = (M_p + M'_p)\, \varphi. \tag{9.65}$$

In considering the mechanism of fig. 9.36 (b), we must also take account of the work done by the moments applied by the fan to its radial edges, as shown in fig. 9.38 (b). Applying the work equation, we obtain

$$1.P = 2\,(M_p + M'_p)\,\varphi + 4\,M_p \cot\frac{\varphi}{2}. \tag{9.66}$$

The only parameter is φ and the minimum condition $\frac{\partial P}{\partial \varphi} = 0$ yields

$$\cot\frac{\varphi}{2} = \left(\frac{M'_p}{M_p}\right)^{1/2}. \tag{9.67}$$

In the particular case where $M_p = M'_p$, eq. (9.67) yields $\varphi = \frac{\pi}{2}$; substituting this value of φ into eq. (9.66), we obtain

$$P = 4\,M_p + 2\,\pi\,M_p = 10.28\,M_p.$$

whereas eq. (9.64) gives

$$P = 8\sqrt{2}\,M_p = 11.3\,M_p.$$

Comparison of the two values of P reveals that the mechanism of fig. 9.35 cannot be the actual collapse mechanism. A number of collapse mechanisms for various loading conditions are found in refs. [7.16, 9.3, 9.16, 9.17 and 9.18]. We shall restrict ourselves to briefly describing some particularly interesting simple cases.

9.4.5. *Some examples of fan mechanisms caused by concentrated loads.*

9.4.5.1. *Circular isotropic plate, simply supported and loaded in its centre.*

The circular fan mechanism transforms the plate into a cone. It gives the limit load

$$P_+ = 2\,\pi\,M_p \tag{9.68}$$

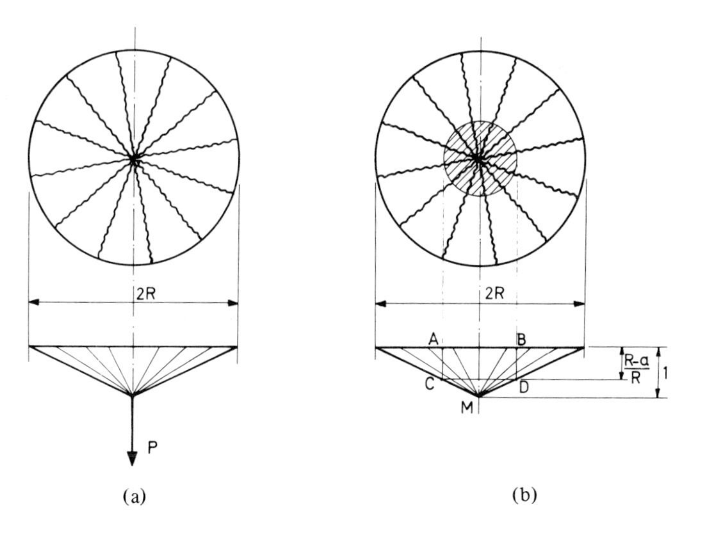

Fig. 9.39.

If the load is distributed over a small central circle of radius a (fig. 9.39), the same mechanism gives the same dissipated work $2\,\pi\,M_p$. But the work of the load here is the product of the intensity p by the combined volumes of the cylinder ABCD and the cone CMD. Thus,

$$W = p\left[\frac{R-a}{R}\,\pi\,a^2 + \pi\,a^2\,\frac{1}{3}\left(1 - \frac{R-a}{R}\right)\right],$$

or

$$W = p\,\pi\,a^2\left(\frac{R - a + \frac{1}{3}a}{R}\right).$$

Introducing the notation $p\,\pi\,a^2 = P$, the equation W=D yields

$$P_+ = \frac{2\,\pi\,M_p}{1 - \frac{\frac{2}{3}a}{R}}. \tag{9.69}$$

9.4.5.2. *Circular isotropic plate, loaded in its centre, and built-in along the edge.*

We now have a circular negative yield line at the edge. We only have to add the term $2\,\pi\,M'_p$ in the expression for D, and thus obtain

$$W = 2\,\pi\,(M_p + M'_p)\,, \tag{9.70}$$

and

$$P_+ = \frac{2\,\pi\,(M_p + M'_p)}{1 - \frac{\frac{2}{3}a}{R}}. \tag{9.71}$$

9.4.5.3. *Isotropic plate of arbitrary shape, built-in along the edge, and subjected to a load distributed over a small circle of radius a.*

A collapse mechanism corresponding to a cone whose axis, normal to the plate, passes through the centre of the circle where the load is applied, and limited by a negative circular yield line of radius R, furnishes the load [eq. (9.71)]. It is easily seen that the smallest value of P_+ is obtained for R as large as possible, i.e., for a yield circle that is tangent to the edge of the plate (fig. 9.40).

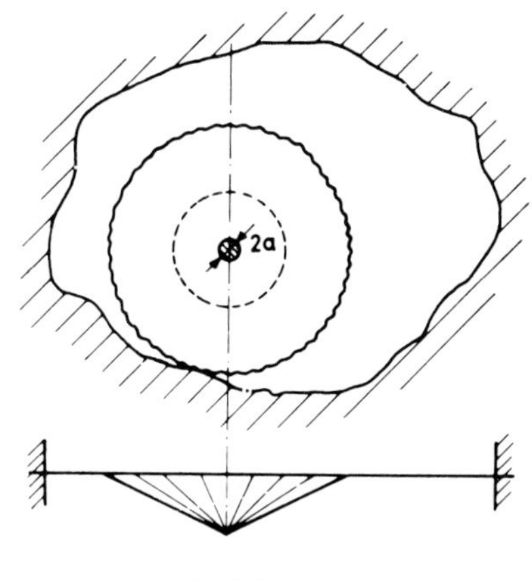

Fig. 9.40.

9.4.5.4. *Circular isotropic plate loaded at the centre and simply supported along a part of its edge.*

For $\beta < 116°$, the collapse mechanism is conical in the supported part of the edges, with two planes in the free part, which are separated by a yield line along the axis of symmetry (fig. 9.41, [9.18]). One obtains

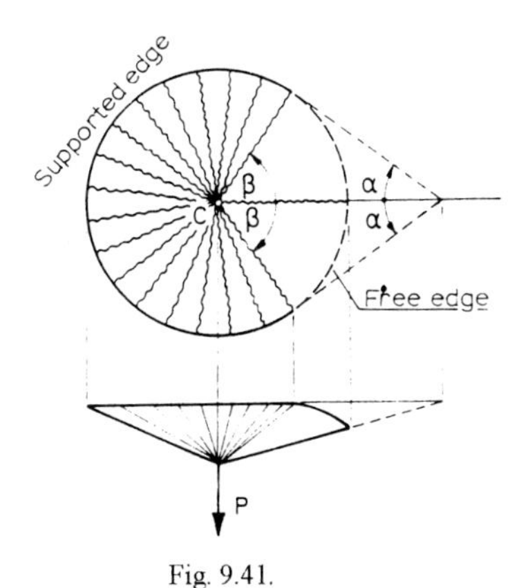

Fig. 9.41.

$$P_+ = 2\,M_p\,(\pi - \beta + \sin\beta). \tag{9.72}$$

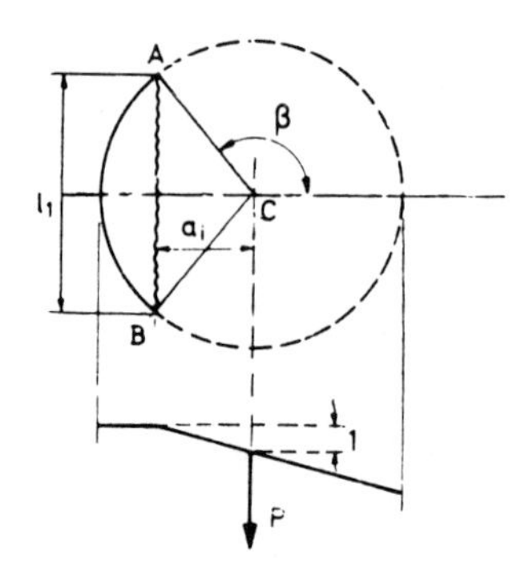

Fig. 9.42.

For $\beta < 116°$, this load is larger than that corresponding to a single straight line joining the extreme points of the supported part of the edge (fig. 9.42). Eq. (9.72) must then be rejected and replaced by

$$P_{+} = 2\, M_p \tan \beta\,. \tag{9.73}$$

9.4.5.5. *Concentrated load applied near the straight edge of an isotropic plate.*

If the edge is simply supported, the smallest load is obtained with

$$\tan \varphi = \left(\frac{M'_p}{M_p}\right)^{1/2}, \tag{9.74}$$

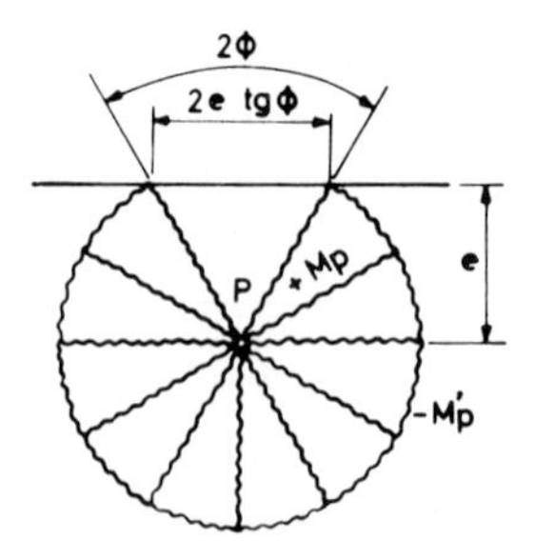

Fig. 9.43.

which, for $M'_p = M_p$, yields

$$P_+ = (3\pi + 2)M_p, \tag{9.75}$$

with $\varphi = 45°$ (fig. 9.43). When M'_p approaches zero, we obtain

$$P_+ = 2\pi M_p, \tag{9.76}$$

with $\varphi = 0$. If the edge is built-in, one returns to the case discussed in Section 9.4.2.3.

9.4.6. *Corner and edge effects.*

9.4.6.1. *Corner effects : introduction.*

Consider a yield line passing through the intersection of two straight simply supported edges (fig. 9.44). It is well known that, when the plate is loaded, the corner point C tends to lift from the support.

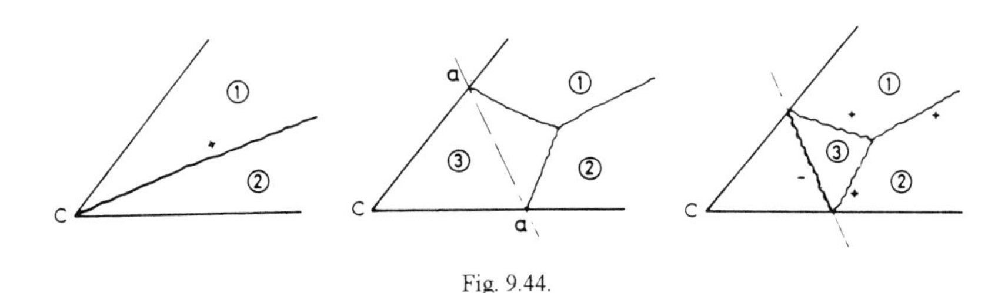

Fig. 9.44.

If it is free to do this (unilateral support), a corner region 3 will actually rotate about a certain axis aa (fig. 9.44) and the point C will move upwards. Thus, the yield line divides into two new branches passing through the intersections of aa and the supporting edges. If the *corner point C is prevented from leaving the support* ("anchored" corners or "bilateral" support), the axis aa is replaced by a negative yield line, and the positive initial yield line must branch into two positive yield lines as shown in fig. 9.44. In both cases, the modification of the mechanism results in a decrease of the load P_+ below that given by the original mechanism. The magnitude of the reduction depends on the shape of the plate and on the ratio $\frac{M_p}{M'_p}$. Detailed discussion of this problem can be found in the book by Wood [9.16]. We shall restrict ourselves to illustrating these considerations by two examples.

9.4.6.2. *Square isotropic plate, simply supported and uniformly loaded.*

Assume bilateral supports. Neglecting corner effects, the collapse mechanism is of the type shown in fig. 9.45. Formula (9.38) furnishes the load $P_+ = \frac{24\,M_p}{l^2}$. The corner effect transforms the mechanism of fig. 9.45 into that of fig. 9.46 (a) which contains two parameters a and b. Provided that the centre O is given a unit displacement, one has, for part 1, $\theta_y = 0$, and $\theta_x = \frac{2}{l}$.

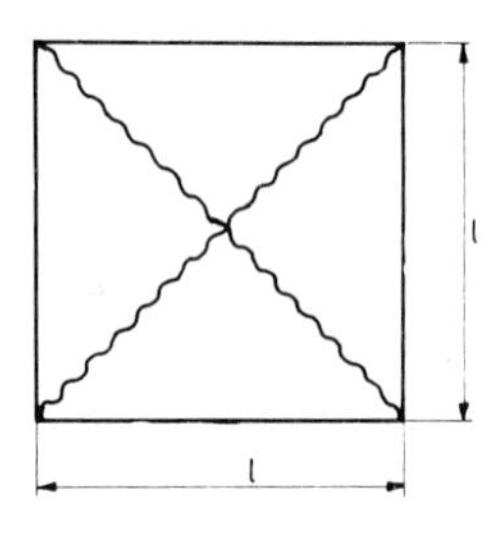

Fig. 9.45.

The triangular part 2 rotates around the negative yield line, whereas the corner itself remains at rest. The vertex S of 2 has the displacement

$$1 - \frac{\frac{a}{\sqrt{2}}}{\frac{l}{2}} = 1 \frac{\sqrt{2a}}{l},$$

and the height SS' of this triangle has the value $\frac{l}{\sqrt{2}} - (a+b)$. The angle of rotation about the negative yield line thus is

$$\frac{1 - \frac{\sqrt{2}\,a}{l}}{l\sqrt{2} - (a+b)}.$$

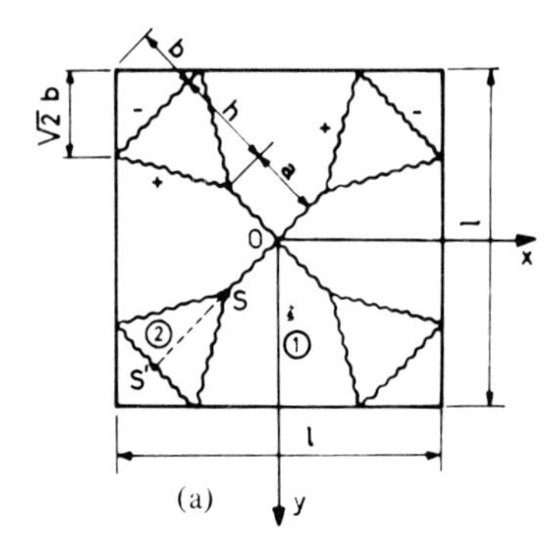

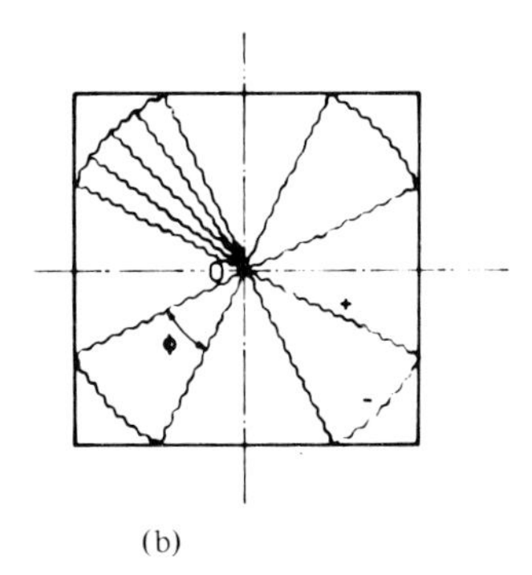

Fig. 9.46.

The work dissipated is

$$D = 4\,M_p \frac{2}{l}\left(l - 2b\sqrt{2}\right) + 4\,(M_p + M'_p)\,2b\,\frac{1 - \dfrac{\sqrt{2}\,a}{l}}{\dfrac{l}{\sqrt{2}} - (a+b)}.$$

Since the corner triangles do not move, the work of the loads is

$$W = \frac{1}{3}\,pl^2 - \frac{4}{3}\,pb^2\left(1 - \sqrt{2}\,\frac{a}{l}\right)$$

Equating W and D, one obtains p as a function of the parameters a end b, which must be calculated to minimize p. Thus,

$$\frac{\partial p}{\partial a} = 0, \qquad \frac{\partial p}{\partial b} = 0.$$

These nonlinear equations must be solved numerically.

Introducing the values of a and b evaluated in this manner in the expression of p, we obtain, if $M'_p = 0$,

$$P \equiv pl^2 = 22\,M_p, \tag{9.77}$$

which compares with $P = 24\ M_p$. If we replace the mechanism of fig. 9.46 (a) by that of fig. 9.46 (b), where the corner effect is obtained by a fan mechanism, we obtain

$$P = 21.7\ M_p \tag{9.78}$$

for $M'_p = 0$.

9.4.6.3. *Triangular plate.*

The case of the equilateral triangular plate (fig. 9.47) is particularly interesting. For a concentrated load at the centre, the simple mechanism of fig. 9.47(a) furnishes

$$P_+ = 6\sqrt{3}\ M_p = 10.4\ M_p\,. \tag{9.79}$$

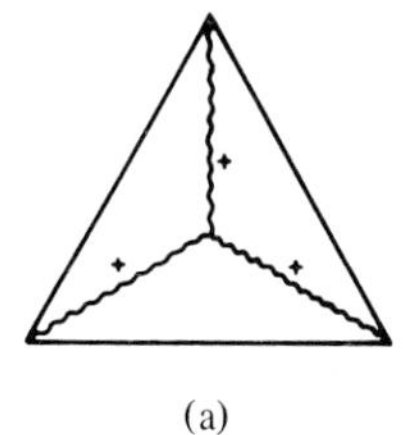

(a)

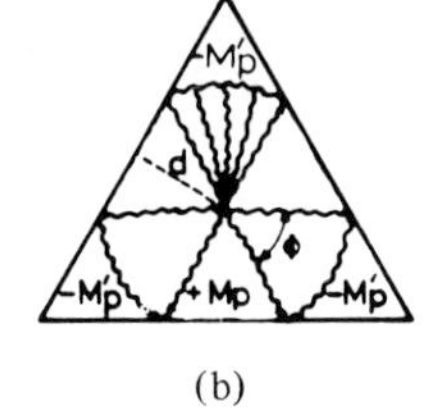

(b)

Fig. 9.47.

For $\varphi = 30°$, the mechanism with three fans, shown in fig. 9.47 (b), yields

$$P_+ = 3\left[\frac{\pi}{6}(M_p + M'_p) + M_p\,\frac{1}{d}\,2d\right].$$

With $M'_p = M_p$ this furnishes

$$P_+ = (\pi + 6)\ M_p = 9.14\ M_p. \tag{9.80}$$

The load reduction thus is $[(10.4\text{-}9.14)/9.14]\times 100 = 13.8\%$, even if the negative reinforcement is identical to the positive one.

9.4.6.4. *Intersection of yield lines and the edges.*

When the kinematic theorem is used, only considerations of kinematical admissibility are relevant [9.6, 9.19].

Because it derives from considerations of statical admissibility, the condition (given, for example, by Wood in ref. [9.16])

$$\psi \geq \cot^{-1}\left(\frac{M'_p}{M_p}\right)^{1/2} \tag{9.81}$$

(see fig. 9.48 for notations) should be disregarded except when one aims at a complete solution (based on the combined theorem). In this case, and for an isotropic plate, when $\psi \neq \pm\pi/2$, twisting moments must exist along the edge. Because n and t are principal

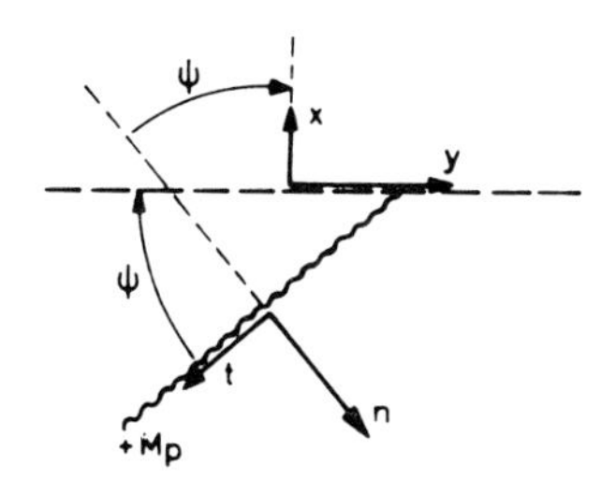

Fig. 9.48.

directions, it follows from $M_x = 0$ that

$$M_{xy} = M_n \cot\psi . \tag{9.82}$$

It is sometimes useful to replace M_{xy} by two shear forces V as shown in fig. 9.49. Since $M_n = M_p$, we have

$$V = M_p \cot\psi . \tag{9.83}$$

Positive values of M_{xy} and V are shown in fig. 9.49.

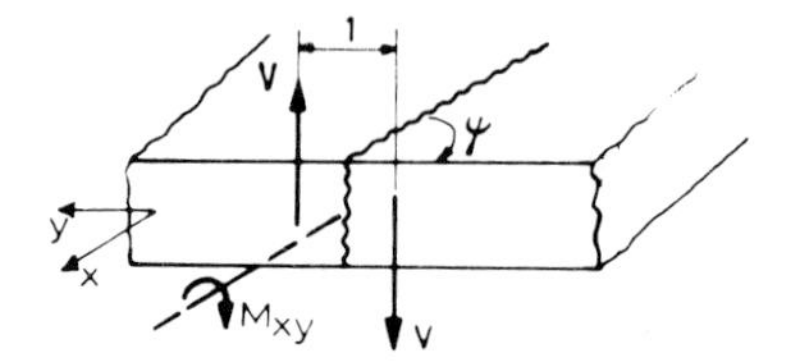

Fig. 9.49.

9.4.7. *Superposition method.*

The principle of superposition, extensively used in elasticity, does not hold in plasticity. However, if we want to construct solutions of more complex problems from those of simpler problems, a (restricted) superposition method may be used.

Consider a plate with given geometry and boundary conditions and assume that the exact plastic moments M_{p1}, M_{p2},..., M_{pn} are known for n distinct loading conditions symbolically represented by P_1, P_2,..., P_n. The plate is then subjected to the simultaneous action of the various loads each of which is multiplied by a positive scalar factor a_i. The resulting loading condition will be symbolically denoted by

$$P_\Sigma = a_1 P_1 + a_2 P_2 + \ldots + a_n P_n . \tag{9.84}$$

We shall prove the following theorem : *the plate with plastic moment*

$$M_{p\,\Sigma} = a_1 M_{p1} + a_2 M_{p2} + \ldots + a_n M_{pn} , \tag{9.85}$$

has a collapse load $P_{\Sigma 1}$ *larger than or equal to* P_Σ . Suppose first that the mechanisms that have yielded M_{p1} , M_{p2} ,..., M_{pn}, all differ from the collapse mechanism for P_Σ. For this mechanism, denote by $\mathcal{P}'$ the powers of the loads and by D' the rates of dissipation. We have, by definition,

$$\mathcal{P}'_{P\,\Sigma} = a_1 \mathcal{P}'_{P\,1} + a_2 \mathcal{P}'_{P\,2} + \ldots + a_n \mathcal{P}'_{P\,n} . \tag{9.86}$$

For the individual loads P_1, P_2,..., P_n, this mechanism is kinematically admissible. Hence, the plastic moments M_i defined by

$$P'_{P_i} = D'(M_i), \qquad i=1, 2, ..., n, \tag{9.87}$$

are such that

$$M_i \leq M_{pi}, \qquad i=1, 2, ..., n. \tag{9.88}$$

From relation (9.88) we immediately obtain

$$a_1 M_1 + a_2 M_2 + ... + a_n M_n \leq a_1 M_{p1} + a_2 M_{p2} + ... + a_n M_{pn}. \tag{9.89}$$

We also know that, for a given mechanism, the total rate of dissipation is proportional to the value of the plastic moment. We thus have

$$\begin{aligned} &a_1 D'(M_1) + a_2 D'(M_2) + ... + a_n D'(M_n) \\ &= D'(a_1 M_1 + a_2 M_2 + ... + a_n M_n). \end{aligned} \tag{9.90}$$

The exact plastic moment under P_Σ is M_Σ given by

$$P'_{P_\Sigma} = D'(M_\Sigma). \tag{9.91}$$

From eqs. (9.86), (9.87), and (9.90), it follows that

$$M_\Sigma = a_1 M_1 + a_2 M_2 + ... + a_n M_n. \tag{9.92}$$

The relations (9.89), (9.85), and (9.92) thus furnish

$$M_\Sigma \leq M_{P_\Sigma}. \tag{9.93}$$

It follows from relation (9.93) that the exact plastic moment for P_Σ is not larger than $M_{P\Sigma}$. Alternatively, the limit load $P_{\Sigma 1}$ of the plate with the plastic moment $M_{P\Sigma}$ is not smaller that P_Σ, and the theorem is proved.

Suppose now that for the loads P_1, P_2,..., P_n the plates with the plastic moment M_{P1}, M_{P2},..., M_{Pn} have the same collapse mechanism, and denote the powers of the loads for this mechanism by P and the rate of the dissipation by D. This mechanism is only kinematically admissible for P_Σ. Hence, the work equation

$$\mathcal{P}_{P\Sigma} = D\,(X) \tag{9.94}$$

furnishes a plastic moment satisfying

$$X \leq M_\Sigma . \tag{9.95}$$

On the other hand,

$$\mathcal{P}_{P\Sigma} = a_1 \mathcal{P}_{P1} + a_2 \mathcal{P}_{P2} + \ldots + a_n \mathcal{P}_{Pn} ,$$

where

$$\mathcal{P}_{Pi} = D\,(M_{Pi}) , \qquad i=1, 2, \ldots, n ,$$

because the mechanism is exact for all P_i. From the two preceding equations we have

$$\mathcal{P}_{P\Sigma} = a_1\, D\,(M_{P1}) + a_2\, D\,(M_{P2}) + \ldots + a_n\, D\,(M_{Pn}) ,$$

or,

$$\mathcal{P}_{P\Sigma} = = D\,(a_1 M_{P1} + a_2 M_{P2} + \ldots + a_n M_{Pn}) \cdot \tag{9.96}$$

Comparison of eqs. (9.94) and (9.96) yields

$$X = M_{P\Sigma} . \tag{9.97}$$

From eqs. (9.97) and (9.95), we have

$$M_{P\Sigma} \leq M_{\Sigma} . \tag{9.98}$$

Comparison of relations (9.98) and (9.93) gives

$$M_{\Sigma} = M_{P\Sigma} . \tag{9.99}$$

We thus have proved the following theorem : *If the collapse mechanisms for the individual loads are identical, the collapse mechanism for the load* P_{Σ} *is the same and the plastic moment* M_{Σ} is equal to $M_{P\Sigma}$.

If we finally consider a given plate with plastic moment M_p, we must have $M_{P\Sigma} = M_p$ or

$$a_1 + a_2 + ... + a_n = 1 . \tag{9.100}$$

Hence, the load P_{Σ} defined by eq. (9.84) which satisfies condition (9.100) is smaller than or at most equal to the limit value $P_{\Sigma l}$. It is the exact limit load when all collapse mechanisms for P_1 , P_2 ,..., P_n are identical. The collapse mechanism for P_{Σ} then coincides with this common mechanism.

An example is provided by the circular isotropic simply supported plate of Section 9.4.5.1. When subjected to a concentrated load P at the centre, its plastic moment is

$$M_{p1} = \frac{P}{2\pi} . \tag{9.101}$$

When the plate is subjected to a load p uniformly distributed on a central circle with radius a, the plastic moment is

$$M_{p2} = \frac{p \pi a^2 \left(1 - \frac{\frac{2}{3} a}{R} \right)}{2\pi} . \tag{9.102}$$

In both cases, the mechanisms correspond to the transformation of the plate into a cone. If the plate is subjected to $k_1\, P + k_2\, p$, its exact plastic moment therefore is

$$M_{p(1+2)} = k_1 \frac{P}{2\pi} + \frac{k_2 p \pi a^2}{2\pi}\left(1 - \frac{2}{3}\frac{a}{R}\right). \qquad (9.103)$$

9.4.8. *Interaction between plate and edge beams.*

In almost all practical cases the plates are monolithic with edge beams. The corresponding problem of collapse is exceedingly difficult. To simplify it, we first consider the simpler problem of a plate simply supported by edge beams which are themselves supported on columns. The edge conditions influence the collapse mode of the plate in a manner that we shall study for the frequently encountered cases of square and rectangular plates [9.16].

Consider first a square plate resting on edge beams with axes in the middle surface of the plate, the torsional rigidity of the beams being neglected. These beams are in turn supported by columns at the four corners of the plate. If the beams are very strong, their deformations will remain elastic (and therefore negligible in the rigid-plastic scheme up to the collapse of the plate). This collapse will take place according to the mechanism of fig. 9.45, called "diagonal mechanism", under the total load

$$P \equiv pl^2 = 24\, M_p. \qquad (9.104)$$

However, if the beams are less strong, they may participate in the collapse of the plate, which presents, then, two median yield lines terminating in plastic hinges in the beams.

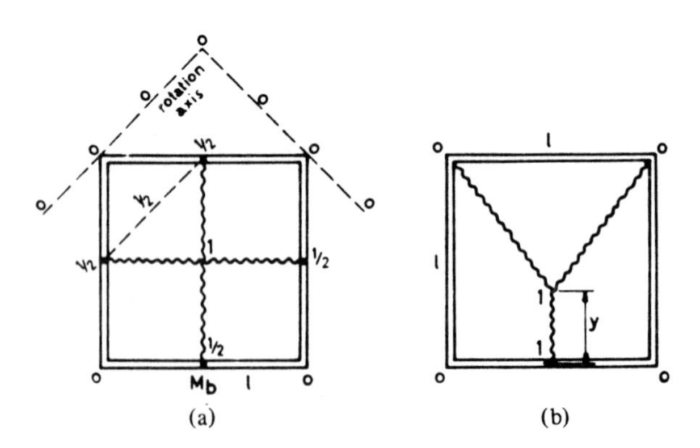

Fig. 9.50.

This mechanism is represented in fig. 9.50 (a). If the centre is given a unit displacement, the hinges displacements are $\frac{1}{2}$. The relative rotation at the yield hinges is $\frac{2\frac{1}{2}}{\frac{1}{2}} = \frac{2}{l}$, which is also the relative rotation of two portions adjacent to a yield line. We thus have

$$D = (2\,M_p\,l + 4\,M_{pb})\,\frac{2}{l} = 4\,M_p + 8\,\frac{M_{pb}}{l},$$

where M_{pb} represents the plastic moment of the edge beams. The work of the loads is $W = \frac{pl^2}{2}$, whence

$$(pl^2)_+ = 8\,M_p\left(1 + \frac{2\,M_{pb}}{M_p\,l}\right). \tag{9.105}$$

If we introduce

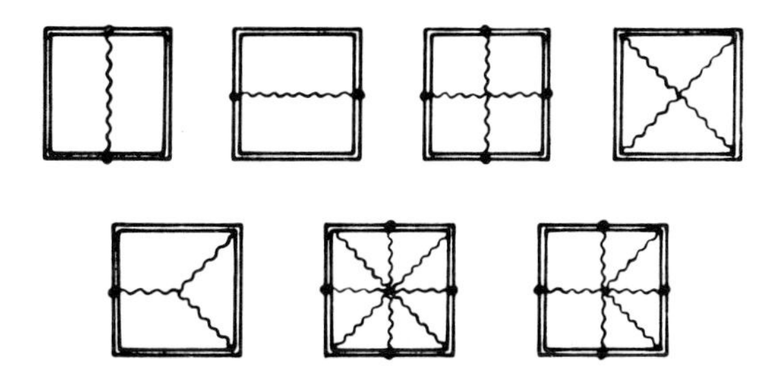

Fig. 9.51.

$$\gamma_p = \frac{M_{pb}}{\frac{M_p\,l}{2}},$$

which will be called the coefficient of plate-beam interaction, eq. (9.105) may be written in the form

$$(pl^2)_+ = 8\, M_p\,(1+\gamma_p)\,. \tag{9.106}$$

Comparing eq. (9.104) with eq. (9.106), we see that, for $\gamma_p = 2$, the "diagonal" collapse mode (plate only) and the "median" mode (plate and beams) result in the same load, under which both of them may occur. All combinations of these two modes are possible, in particular those shown in fig. 9.51, which all yield the same load $(pl^2)_+ = 24\, M_p$. If $\gamma_p > 2$, then the collapse mode is the diagonal one, and the applicable equation is eq. (9.104). If, on the other hand, $\gamma_p < 2$, the "median" mode is relevant, and the governing equation is eq. (9.106). For $\gamma_p = 0$, one returns to the case of the plate supported at the four corners. Eq. (9.106) then gives $(pl^2)_+ = 8\, M_p$.

If one of the beams is weaker than the others, the mechanism of fig. 9.50 (b) furnishes

$$\frac{pl^2}{6\, M_p} = \frac{4 + \dfrac{1}{1 - y/l} + 2\,\gamma_p}{2 + y/l}\,.$$

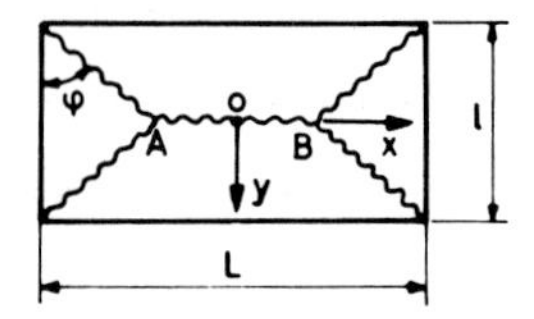

Fig. 9.52.

The value of y / l corresponding to the smallest load is

$$\frac{y}{l} = \frac{10 + 4\,\gamma_p}{8 + 4\,\gamma_p}\left[1 - \left\{1 - \left(\frac{3 + 2\,\gamma_p}{4 + 2\,\gamma_p}\right)\left(\frac{8 + 4\,\gamma_p}{10 + 4\,\gamma_p}\right)^2\right\}^{1/2}\right],$$

γ_p referring to the weak beam ($\gamma_p = 0$ corresponds therefore to a free edge ; it gives $(pl^2)_+ = 14.15\, M_p$).

In the case of a rectangular plate, we shall denote by M_{pb} and M_{pB} the plastic moments of the short and long beams, respectively. If all beams are sufficiently strong, the plate will collapse according to the mechanism represented in fig. 9.52. Under a uniformly distributed load, the work of the load is

$$W = \frac{1}{3} pl^2 \tan\varphi + \frac{1}{2} p (L - l \tan\varphi) l ,$$

if AB takes a unit displacement, and the work dissipated is

$$D = M_p \left[\frac{2l}{\tan \varphi/2} + 2L \frac{2}{l} \right] = 4 M_p \left(\frac{1}{\tan\varphi} + \frac{L}{l} \right).$$

The equation W=D furnishes the expression of p as function of φ. The minimum

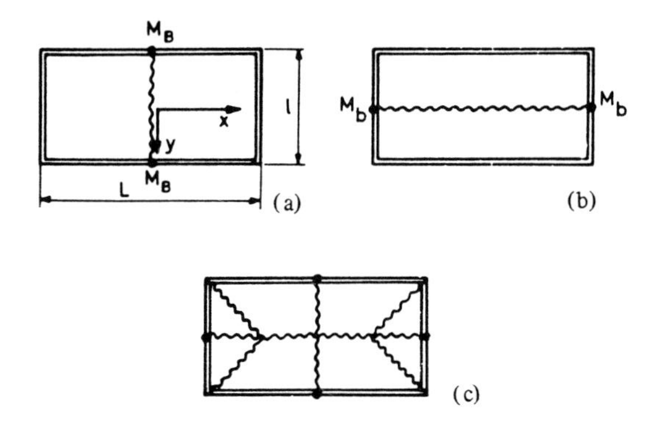

Fig. 9.53.

condition $\frac{\partial p}{\partial \varphi} = 0$ finally yields $\tan\varphi = \left(3 + (l/L)^2\right)^{1/2} - l/L$, and hence

$$(plL)_+ = 24\, M_p \frac{L}{l} \frac{1}{\left[\left(3 + (l/L)^2\right)^{1/2} - l/L\right]^2} . \tag{9.107}$$

It is worth nothing that, for a built-in plate with negative yield lines along the edges, the influence of the additional term in the dissipation is simply to substitute $(M_p + M'_p)$ for M_p in eq. (9.107).

If the long beams are weak in comparison with the short ones, we have the mechanism of fig. 9.53 (a), which gives

$$(plL)_+ = \frac{8}{L}(M_p l + 2 M_{pB}) . \tag{9.107'}$$

If the short beams are weak, we have mechanism (b) of fig. 9.53, which yields

$$(plL)_+ = \frac{8}{l}(M_p L + 2 M_{pb}) . \tag{9.107''}$$

To produce one mechanism rather than the two others, the corresponding load $(plL)_+$ must be the smallest of the three, so imposing two conditions to be satisfied by M_{pb} and M_{pB}. Introducing the notations

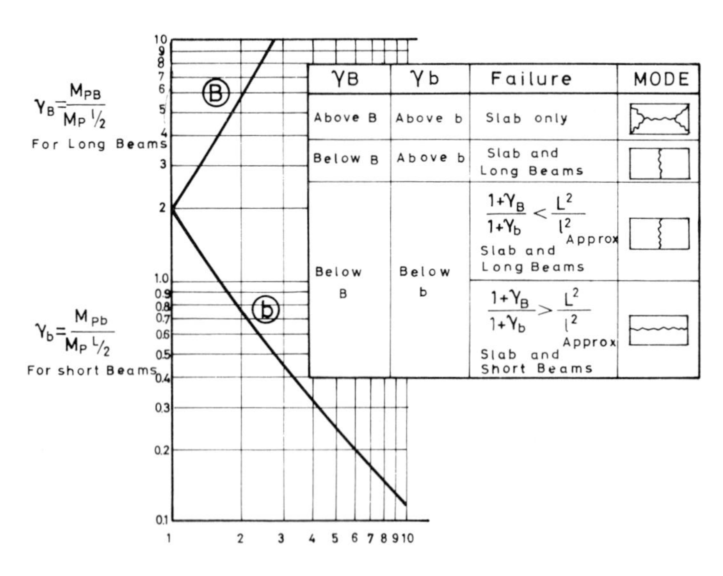

Fig. 9.54.

$$\gamma_B = \frac{M_{pB}}{M_p l/2}, \qquad \gamma_b = \frac{M_{pb}}{M_p L/2}, \tag{9.108}$$

these conditions are easily obtained, by using eqs. (9.107) to (9.108), in the following form :

1. Combined collapse : plate and long beams [fig. 9.53 (a)] if

$$1 + \gamma_B \leq \left(\frac{L}{l}\right)^2 (1 + \gamma_b),$$

and
$$\gamma_B \leq \left(\frac{L}{l}\right)^2 - 1 + \frac{\frac{2L}{l}}{\left[(l/L)^2 + 3\right]^{1/2} - l/L}.$$

Governing eq. (9.107)' .

2. Combined collapse : plate and short beams [fig. 9.53 (b)] if

$$1 + \gamma_B \geq \left(\frac{L}{l}\right)^2 (1 + \gamma_b),$$

and
$$\gamma_b \leq \frac{\frac{2l}{L}}{\left[(l/L)^2 + 3\right]^{1/2} - l/L}.$$

Governing eq. (9.107)'' .

3. Collapse of the plate only (fig. 9.52) if

$$\gamma_B \geq \left(\frac{L}{l}\right)^2 - 1 + \frac{\frac{2L}{l}}{\left[(l/L)^2 + 3\right]^{1/2} - l/L},$$

and
$$\gamma_b \geq \frac{\frac{2l}{L}}{\left[(l/L)^2 + 3\right]^{1/2} - l/L}.$$

Governing eq. (9.107).

If certain of the preceding inequalities are replaced by equalities, mixed collapse modes are obtained such as the one in fig. 9.53 (c). The conditions determining the type of collapse have been represented graphically in fig. 9.54.

9.4.9. *General fan mechanisms.*

In Sections 9.4.4 and 9.4.5, the only fan mechanisms used were circular fans. In order to obtain smaller upper bounds, more general fan patterns can be used. Janas has discussed

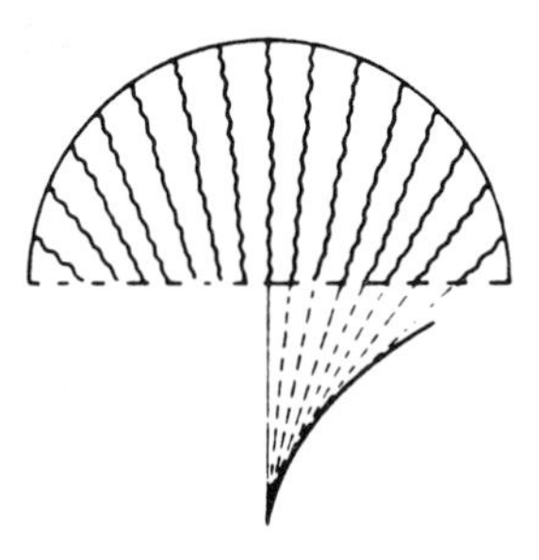

Fig. 9.55.

this question [9.19]. It must first be pointed out that there is no kinematic objection to using a bounding negative yield line that is not orthogonal to the rays of a polar fan, as done by Mansfield [9.20], A. Sawczuk [9.21], and Johansen [9.3] ; The lowest upper bounds are obtained from a bounding logarithmic spiral.

Numerous examples of applications can be found in the book by Sawczuk and Jaeger [7.16]. Obviously, for isotropic plates there exist no licit moments fields corresponding to such fans, because principal directions of curvature rates and moments coincide and positive and negative lines must therefore intersect orthogonally. The better upper bounds obtained from fans with noncircular boundaries simply prove that the complete solutions do not belong to the considered family of mechanisms with polar fans. Indeed, the actual collapse mechanisms might even not belong to the whole yield line class ; they might need completely plastified regions corresponding to the vertices and edges of the yield locus. For this reason, orthogonality of bounding yield line to rays, and in general any condition of statical origin, should be excluded from a purely kinematic approach. Only when a complete solution is desired, should these conditions be allowed to affect the choice of mechanisms.

Improved upper bounds may be obtained from the consideration of more general, nonpolar fans, the straight yield lines of which are tangent to some (curved) envelope (see fig. 9.55, for example). Janas [9.19] has given dissipation formulas for fans of this type, and for their combinations.

The best mechanism of the family is obtained by the differentiation process applied to the choice of the envelope line, or by using the nodal force method (see Section 9.4.10),

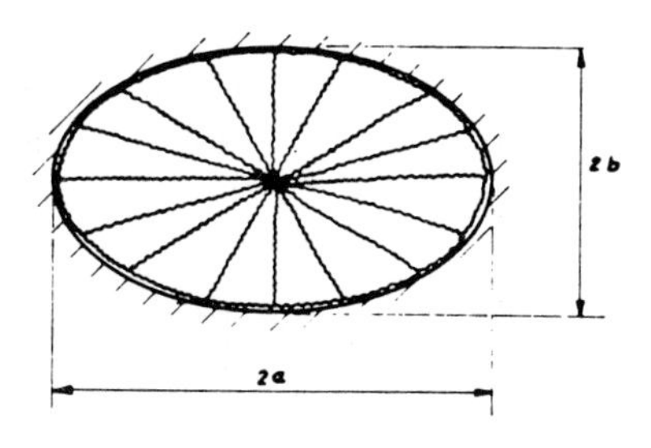

Fig. 9.56 (a).

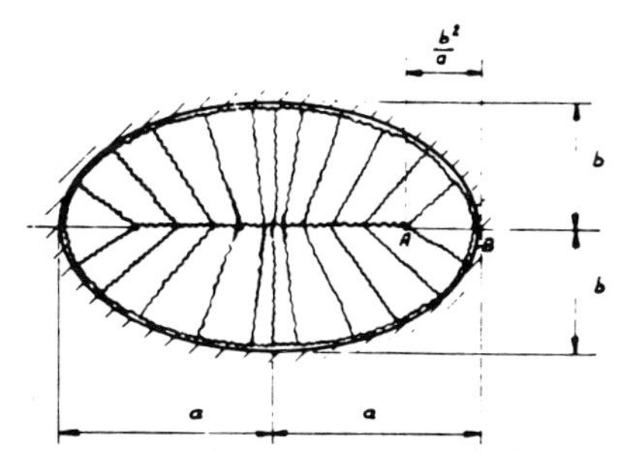

Fig. 9.56 (b).

as done by Nielsen [9.22], who obtained $p_+ = \dfrac{4.40\,M_p}{R^2}$ for the simply supported uniformly loaded, semicircular slab in fig. 9.55. In the case of the elliptic built-in isotropic plate in fig. 9.56, the best collapse pattern is the simple nonorthogonal polar fan of fig. 9.56 (a) when the plate is subjected to a uniform load. It gives

$$p_+ = 3\,(M_p + M'_p)\,\frac{a^2 + b^2}{a^2\,b^2}\,. \tag{9.109}$$

For a uniform line load $\bar{p}$ (lb/in.) along the longer axis, the mechanism of fig. 9.56 (b) gives a lower load. This load $\bar{p}_+$ is given in fig. 9.57.

9.4.10. *Nodal forces.*

According to the kinematic theorem, the best mechanism of a given family is the

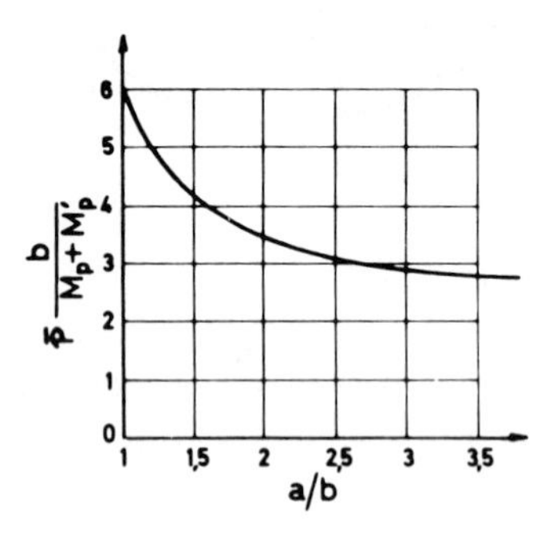

Fig. 9.57.

mechanism that corresponds to the smallest load P_+. This smallest value might well not be an analytic minimum as shown by the example of the built-in beam, fig. 9.58. In this case, $\frac{\partial P_+}{\partial \alpha} = 0$ yields $\alpha = \infty$. Smallest P_+ is obtained for $\alpha = \frac{1}{2}$, largest physically admissible value of the parameter α.

In the following, we *assume* that the differentiation with respect to the parameters will effectively furnish the best mechanism. The resulting set of equations, eqs. (9.61), will

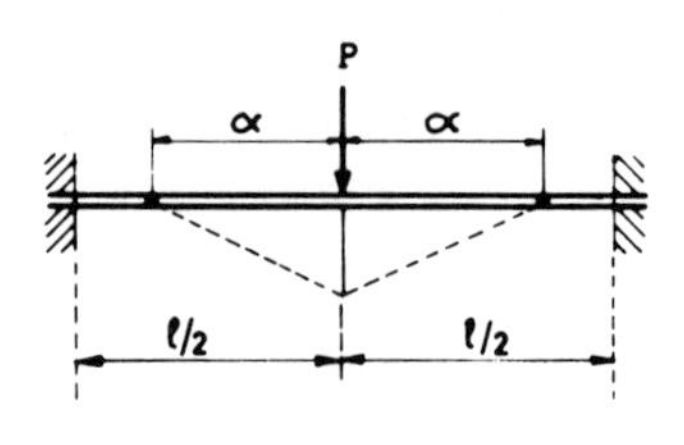

Fig. 9.58. a

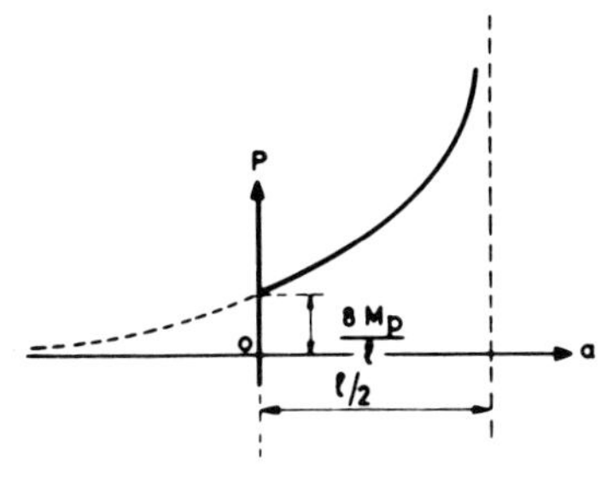

Fig. 9.58. b

sometimes be very hard to solve because these equations are nonlinear. Hence, some substitute method might be searched. One method is the so-called "equilibrium method". If we remark that computing the dissipation corresponding to a given mechanism imposes *only* the value of the bending moment in the yield lines, we can imagine several moment fields that are in equilibrium with the loads and exhibit the imposed moments in the yield lines. These fields will in general violate the yield condition, even in the yield lines where only M_n is determinate.

The work equation can be viewed as an equation of virtual work for a field of this kind. The shear forces, twisting and bending moments acting in the yield lines must be such that every rigid part of the collapsing plate is in equilibrium. Torque and shear forces can be reduced to equivalent transversal forces by the well-known formula of Thomson-Tait, eq. (9.119). Now, if these transversal forces are defined by formulas expressing mathematically that M_{p+} (or P_+) is an analytic extremum, the equilibrium of the rigid parts *acted upon by these transverse forces* will force the parameters to assume values that correspond to vanishing derivatives. Note that this method is nothing but a convenient mathematical substitute for equating the derivatives to zero. Equilibrium of rigid parts is assumed to be possible under *imposed* boundary bending moments and *arbitrary* transversal forces, the latter being then determined from stationarity of M_p (or P_+). The approach is essentially kinematical. No licit stress field is considered here, despite the name of the method. This has not been clearly understood immediately (see [9.23-9.28], and [9.6]). All necessary details of the method can be found in [9.25] and [9.29]. We shall restrict ourselves to an illustrative example. Consider a rectangular built-in plate subjected to a uniformly distributed load p (fig. 9.59). As indicated in fig. 9.59, lower reinforcement is orthotropic, and negative restraining moments along the edge may have different values $-C_1 M_p$, $-C_2 M_p$, $-C_3 M_p$, $-C_4 M_p$.

We consider the three-parameter family of mechanisms (x_1, x_2, y_1) of fig. 9.59. The general formulas [9.25, 9.29] show that all nodal forces K_1 to K_6 vanish in this particular case.

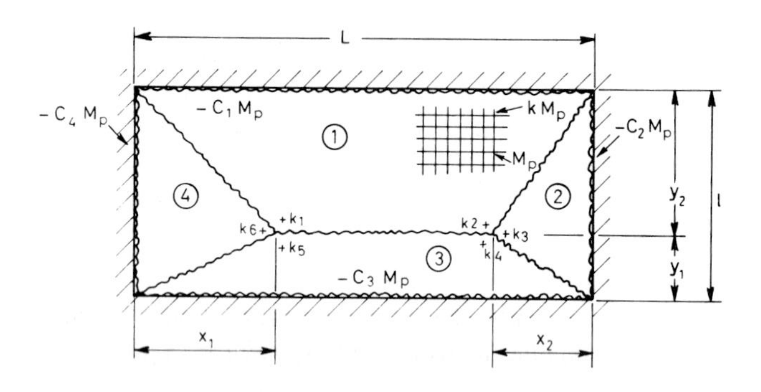

Fig. 9.59. (Reprinted by permission of Thames and Hudson Ltd. from R.H. Wood, Plastic and Elastic Design of Slabs and Plates, 1961.)

Rotational equilibrium of the part 1 about the edge gives

$$M_p L (1 + C_1) = \frac{p x_1 y_2^2}{6} + \frac{p x_2 y_2^2}{6} + \frac{p}{2} (L - x_1 - x_2) y_2^2 \tag{a}$$

and rotational equilibrium of 2 about its edge gives

$$M_p l (1 + C_2) = \frac{p l x_2^2}{6} . \tag{b}$$

Similarly,

$$M_p L (1 + C_3) = \frac{p x_1 y_1^2}{6} + p \frac{x_2 y_1^2}{6} + \frac{p}{2} (L - x_1 - x_2) y_1^2 , \tag{c}$$

$$M_p l (1 + C_4) = p \frac{l x_1^2}{6} . \tag{d}$$

We also have (fig. 9.59)

$$y_1 + y_2 = 1 . \tag{e}$$

We thus have five equations with the five unknowns : x_1, x_2, y_1, y_2, and M_p.

Let $A = \left(1 + C_2\right)^{1/2} + \left(1 + C_4\right)^{1}$

$B = \left(k + C_1\right)^{1/2} + \left(k + C_3\right)^{1/2}$, and solve the system (a) to (e). We obtain [9.16]

$$M_p = \frac{p}{24}\left(\frac{2l}{B}\right)^2 \left[\left\{ 3 + \left(\frac{2l/B}{2L/A}\right)^2 \right\}^{1/2} - \frac{2l/B}{2L/A} \right]^{-2} . \tag{f}$$

It is easily seen that, in the case of the isotropic rectangular plate where k=1 and $C_1 = C_2 = C_3 = C_4 = 1$, relation (f) reduces to eq. (9.107) where 2 M_p is substituted for M_p.

Space limitations prevent us from further developing the nodal force theory. The reader is referred to [9.25, 9.28, and 9.29].

9.5. The static method.

9.5.1. *Introduction.*

The static method consists of finding a licit moment field covering the entire plate. If this field corresponds to a mechanism in accordance with the flow rule, the combined theorem shows that the load is the exact limit load ; otherwise, it is a lower bound P_- for the limit load.

It should be noted that the equilibrium of the rigid parts of a plate mechanism (with no violation of the yield condition in the yield lines) is a necessary but not a sufficient condition for the existence of a licit field. Equivalent transverse forces in the yield lines should be determined from statical admissibility. Verification of the equilibrium of rigid parts, however, is, in general, not a useful step towards obtaining a licit moment field. On the other hand, use of the so-determined tranverse forces (modal forces) in the kinematic method could result in an unnecessarily high upper bound, without at all insuring the existence of a corresponding licit moment field. Consequently nodal forces should not be used in a statical approach.

9.5.2. *Circular plate, simple support.*

We first note that, when both principal moments have same sign, Tresca's yield condition (fig. 9.60) is identical to Johansen's yield condition [eqs. (9.8), (9.9)] for $M_{xy} = 0$.

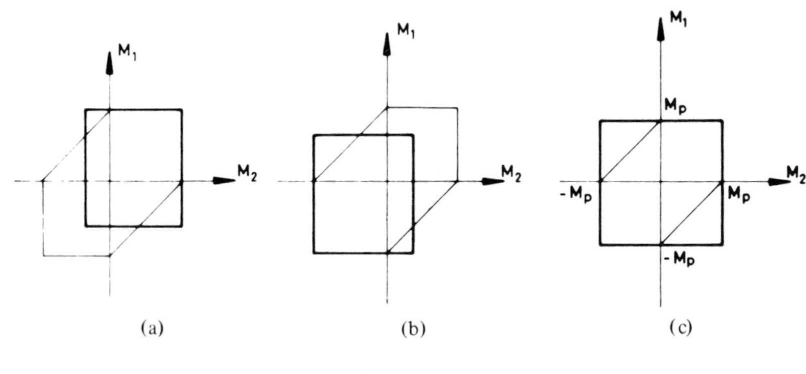

Fig. 9.60.

This remark applies to simply supported circular plates symmetrically loaded, for which the following exact solutions are known :

1. "Circular loading" of total magnitude P on a central circle with radius a concentric to the circular edge. See Section 7.3.2, relations (7.11) to (7.14). The limit concentrated central load is given by eq. (7.16) ;

2. Annual loading : see Section 7.3.3, fig. 7.15 ;

3. Line load : see relation (7.23) ;

4. Annular plate subjected to uniform load : see curve a in fig. 7.17 ;

5. Orthotropic plates : see Section 7.4.2 and ref. [7.26].

Note that all preceding solutions are complete, consisting of both a licit moment field and a corresponding mechanism.

9.5.3. *Built-in isotropic circular plate, circular loading.*

Due to plane polar symmetry, the yield condition in fig. 9.61 can be used. On the basis of the existing solution for the simply supported plate, we consider the moment field.

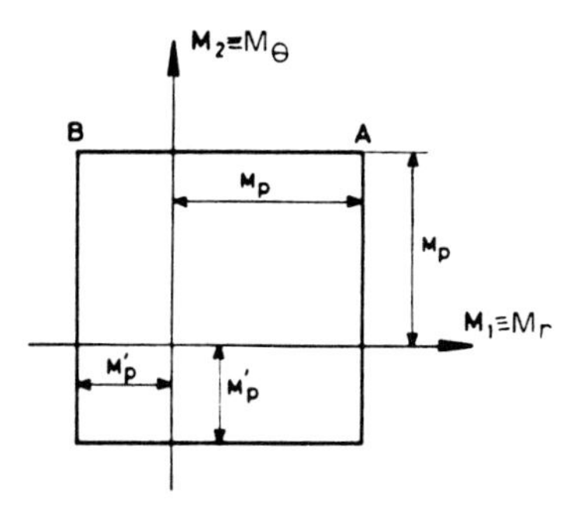

Fig. 9.61.

$$0 \le r \le a \quad \begin{cases} M_\theta = M_p \\ M_r = M_p - (M_p + M'_p) \dfrac{r^2}{a^2} \cdot \dfrac{1}{3 - \dfrac{2a}{R}} , \end{cases}$$

$$a \le r \le R \quad \begin{cases} M_\theta = M_p \\ M_r = M_p - (M_p + M'_p) \dfrac{3 - \dfrac{2a}{r}}{3 - \dfrac{2a}{R}} , \end{cases} \qquad (9.110)$$

where a is the radius of the central loaded circular area. The plastic regime is AB, M_θ being kept constant and equal to M_p, whereas M_r varies in a parabolic manner from M_p at the centre to $-M'_p$ at the edge. Substitution of relations (9.110) into the fundamental equilibrium eq. (7.103) gives

$$p_- = \frac{2\,(M_p + M'_p)}{a^2 \left(1 - \dfrac{2}{3}\dfrac{a}{R}\right)} . \qquad (9.111)$$

With the notation $P = p\,\pi\,a^2$, formulas (9.111) and (9.71) give identical values. Hence, we have obtained the exact limit load

$$P_l = p_l \pi a^2 = \frac{2 \pi (M_p + M'_p)}{1 - \frac{2}{3} \frac{a}{R}} . \tag{9.112}$$

Note that the fields (9.110) and the field (7.12), (7.13) can both be given by

$$0 \leq r \leq a \qquad \begin{cases} M_\theta = M_p \\ M_r = M_p - \frac{pr^2}{6} \end{cases} ,$$

$$a \leq r \leq R \qquad \begin{cases} M_\theta = M_p \\ M_r = M_p - \frac{pa^2}{2} + \frac{pa^3}{3r} \end{cases} , \tag{9.113}$$

where the expression (9.111) or (7.11) for p must be used for the built-in plate and the simply supported plate, respectively. From eq. (9.112) we obtain the exact limit value of the central concentrated load

$$P_l = 2 \pi (M_p + M'_p) . \tag{9.114}$$

9.5.4. *Simply supported isotropic square plate uniformly loaded.*

Assume equal plastic moment for both positive and negative bending (square yield condition, fig. 9.62). In the polar coordinate system of fig. 9.63, with origin at the centre of the plate, consider the field [9.5]

$$\begin{gathered} M_\theta = M_p , \\ M_r = M_p \left(1 - \frac{r^2}{R^2} \right) , \\ M_{r\theta} = 0 , \end{gathered} \tag{9.115}$$

obtained by replacing a by R in eq. (7.12). The radial and circumferential directions are thus assumed to be principal directions. We then change our coordinate system from r, θ to x, y with the formulas

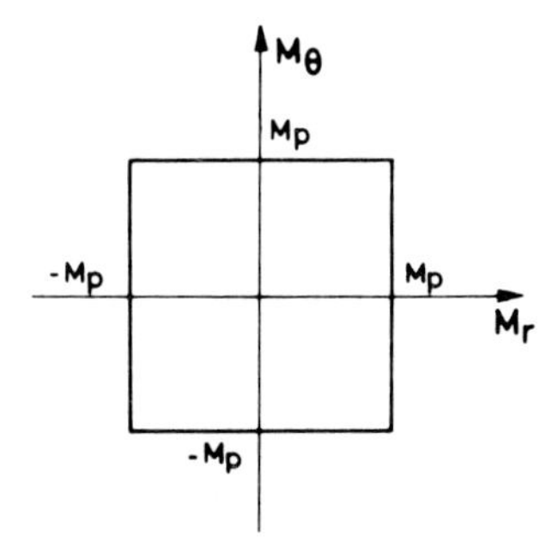

Fig. 9.62.

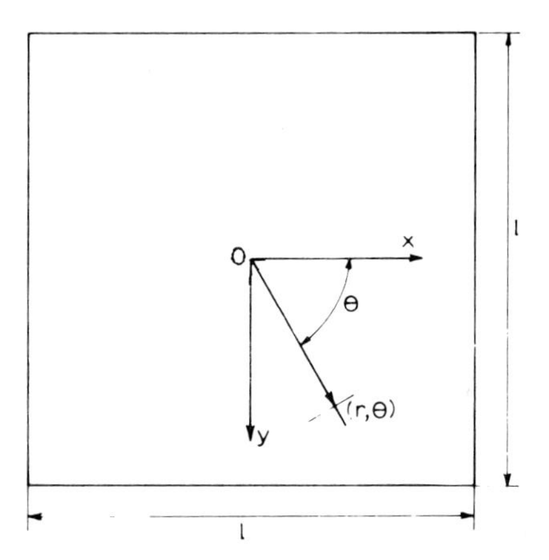

Fig. 9.63.

$$r^2 = x^2 + y^2, \qquad \cos\theta = \frac{x}{r}, \qquad \sin\theta = \frac{-y}{r}, \qquad R = \frac{l}{2},$$

and use the relations (1.10) and (1.11) where M_r, M_θ, M_x, M_{xy}, θ are substituted for σ_1, σ_2, σ, $-\tau$, α, respectively. We transform the field (9.115) into

$$M_x = M_p\left(1 - \frac{4x^2}{l^2}\right),$$

$$M_y = M_p \left(1 - \frac{4y^2}{l^2}\right),$$

$$M_{xy} = 4 M_p \cdot \frac{xy}{l^2} . \tag{9.116}$$

As is readily verified, the yield condition (9.8) is satisfied at every point of the square plate. The boundary conditions also are obviously satisfied.

Substitution of expression (9.116) into the equilibrium eq. (7.37) (where $p=c^t$) yields the lower bound

$$p_- = \frac{24 M_p}{l^2} . \tag{9.117}$$

Comparing with the upper bound, eq. (9.38), we find the exact limit load as

$$p_l = \frac{24 M_p}{l^2} . \tag{9.118}$$

The edge reaction (per unit of length), statically equivalent to the shear force and the torque, is given by [9.16]

$$Q_x = \frac{\partial M_x}{\partial x} - 2 \frac{\partial M_{xy}}{\partial y} . \tag{9.119}$$

Eq. (9.119) applied at $x = \pm \frac{1}{2}$ furnishes $Q_x = \frac{pl}{3}$, which turns out to be constant. Each corner of the plate is subjected to a concentrated vertical downward force :

$$R = \frac{1}{4} \left| pl^2 - \frac{4}{3} pl^2 \right| = \frac{pl^2}{12} .$$

Assume the plate to be simply supported on edge beams with the constant plastic moment M_{pb}, which are in turn simply supported on corner columns. The maximum bending moment in one of these beams is then $M_{max} = \frac{pl^3}{24}$. The beam will remain rigid as long as

$$\gamma_p \equiv \frac{M_{pb}}{M_p(l/2)} > \frac{M_{max}}{(pl^2/24)\,(l/2)} = 2\,, \tag{9.120}$$

a result already found in Section 9.48. For $\gamma_p > 2$, though the plate is completely plastic it collapses according to diagonal mechanism of fig. 9.45.

For $\gamma_p < 2$, we generalize eqs. (9.116) by simply modifying the last equation into

$$M_{xy} = 4\,M_p\,\frac{xy}{l^2}\,(\gamma_p - 1). \tag{9.121}$$

It is readily verified that the obtained moment field is statically admissible for all γ such that $0 < \gamma < 2$, and that the plate is at yield along its median lines (x=y=0). Eq. (9.119) gives the edge reaction

$$Q_x = \frac{4\,M_p}{l}\cdot\gamma_p\,,$$

and the supporting beams are subjected to a maximum bending moment

$$M_{max} = Q_x\,\frac{l^2}{8} = M_{pb}\,.$$

The beams are thus at collapse together with the plate, as shown in fig. 9.50.

We once again have a complete solution, and both equilibrium eq. (7.37) and work equation furnish the exact limit load, namely

$$p_l = \frac{8\,M_p}{l^2}\,(1 + \gamma_p)\,. \tag{9.122}$$

Note that, for $\gamma_p = 0$ (vanishing beams, plate supported at the corners), the plate is plastic at every point as well as for $\gamma_p = 2$ (plate on rigid supports).

9.5.5. *Simply supported isotropic rectangular plate uniformly loaded.*

As in Section 9.4.8, denote by l and L the lengths of the short and long edges, respectively. We generalize the field, eqs. (9.116), into

$$M_x = M_p\left(1 - \frac{4x^2}{L^2}\right),$$

$$M_y = M_p\left(1 - \frac{4y^2}{l^2}\right), \tag{9.123}$$

$$M_{xy} = \frac{4\,M_p xy}{lL}\,.$$

Substitution of expressions (9.123) for M_x, M_y, M_{xy} into equilibrium eq. (7.37) furnishes

$$p_- = \frac{8\,M_p}{l^2}\left(1 + \frac{l}{L} + \frac{l^2}{L^2}\right) \tag{9.124}$$

Though the field (9.123) satisfies the yield condition (9.8) at every point of the plate, there is no corresponding mechanism for rigid support conditions because the trajectories of principal moments are curved lines (except the median lines). Nevertheless, the upper bound, eq. (9.107), does not differ by more than about 1.5% from the lower bound, eq. (9.124), just obtained.

Assume now that the plate is simply supported on edge beams with plastic moment M_{pb} and M_{pB} for the short and long beam, respectively. Substitution of expressions (9.123) for M_x and M_y in eq. (7.37), integration, and use of symmetry, give

$$M_{xy} = p\,\frac{xy}{2} - 4\,M_p\,xy\left(\frac{1}{L^2} + \frac{1}{l^2}\right) \tag{9.125}$$

Reaction on the long beam is [eq. (9.119)]

$$|Q| = p\,\frac{l}{2} - 4\,M_p\,\frac{l}{L^2},$$

and the load corresponding to the collapse of the long beams is determined from condition

$$M_{max} = |Q_{1/2}| \frac{L^2}{8} = M_{pB} \ .$$

We obtain

$$p_{-}(Ll) = \frac{8}{L}(M_p l + 2\,M_{pB})\ . \tag{9.126}$$

Similarly, the load corresponding to collapse of the short beams is

$$p_{-}(lL) = \frac{8}{l}(M_p L + 2\,M_{pb})\ . \tag{9.127}$$

With definitions (9.108) of symbols γ_b and γ_B, eqs. (9.126) and (9.127), respectively become

$$p_{-} = \frac{8\,M_p}{L^2}(1 + \gamma_B)\ , \tag{9.128}$$

$$p_{-} = \frac{8\,M_p}{l^2}(1 + \gamma_b)\ . \tag{9.129}$$

If

$$\frac{1 + \gamma_B}{1 + \gamma_b} < \frac{L^2}{l^2}, \tag{9.130}$$

the load (9.128) is reached before the load (9.129). Long beams are at collapse whereas short beams are still rigid.

Introducing expressions (9.123) of M_x and M_y and expression (9.125) of M_{xy} in the yield condition (9.8), and using eq. (9.128), it is easily seen that the moment field in the plate is licit as long as

$$\gamma_B < \frac{L}{l} + \frac{L^2}{l^2} \tag{9.131}$$

(to compare with a similar condition in the kinematical approach of Section 9.4.8).

When both conditions (9.130) and (9.131) are satisfied, expression (9.128) is the exact limit load because it is identical with the upper bound, eq. (9.107). Similarly, eq. (9.129) is the exact limit load (with collapsing plate and short beams) when

$$\frac{1+\gamma_B}{1+\gamma_b} > \left(\frac{L}{l}\right)^2 \tag{9.132}$$

and

$$\gamma_b < \frac{l}{L} + \frac{l^2}{L^2}. \tag{9.133}$$

When equality signs must be used in relations (9.130) to (9.133), combined mechanisms occur.

9.5.6. *Continuous isotropic rectangular plates subjected to uniform load.*

We consider an isotropic rectangular plate continuous over many transversal simple supports (fig. 9.64) [9.30]. A typical part of this continuous plate, spanning between two adjacent supports apart from the length L, is to be regarded as a plate built-in across the supports, free along the two other edges and subjected to uniform load p.

In order to check the quality of our static solution, we first establish upper bounds as developed in Section 9.4. The assumed mechanisms are shown in fig. 9.64. Upon optimization they give the upper bounds in table 9.1. The plastic moment is M_p for both positive and negative bending. The static solution is constructed by taking polynomials of the second, sixth, and sixth degree for M_y, M_x, and M_{xy}, respectively. After boundary and equilibrium conditions are satisfied, the remaining three parameters are chosen so as not to violate the yield condition (9.8) and (9.9) while maximizing p. This is achieved, by trial and error, for the square plate, where $\rho \equiv l/L = 1$. The resulting values of the parameters turn out to give good lower bounds for other ratios ρ.

Table 9.1. minimum values of $\frac{pl^2}{M_p}$ for various values of ρ.

$\rho = \frac{1}{L}$	Minimum values of $\frac{pl^2}{M_p}$ (upper-bound solution)	Relevant mode
$\to \infty$	$16\,p^2$	
8	1125.75	2'
7	873.63	
6	653.52	
5	465.42	
4	309.36	
3	185.37	
2	93.5 6	
1.5	59.90	
1	34.09	
2/3	21.95	3
1/2	17.02	
1/3	13.27	2
1/4	11.67	
1/5	10.82	
1/6	10.28	
1/7	9.92	
1/8	9.65	
1/12	9.06	
1/16	8.78	
1/20	8.62	
1/2 4	8.51	
1/36	8.34	
1/48	8.25	

The final solution is

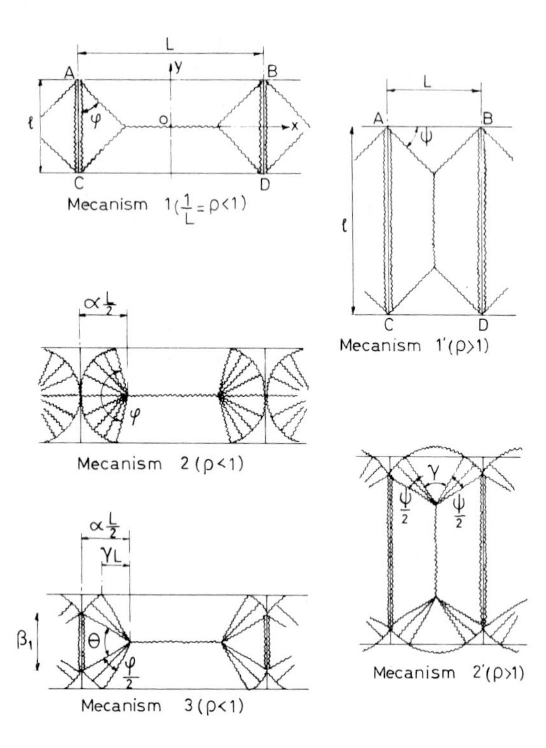

Fig. 9.64.

$$M_x = M_p \left[1 + \frac{1}{l^2} \left(4 - \frac{pl^2}{2\,M_p} + \gamma\rho \right) \frac{x^2}{L^2} + \frac{7\,\gamma x^4}{\rho\, L^4} - 24\,\frac{\gamma x^6}{\rho\, L^6} \right],$$

$$M_y = M_p \left(1 - \frac{4\,y^2}{l^2} \right),$$

$$M_{xy} = \gamma\, M_p\, \frac{xy}{Ll} \left(1 - \frac{4\,x^2}{L^2} \right) \left(1 + \frac{18\,x^2}{L^2} \right), \tag{9.134}$$

where

$$\gamma = \frac{\frac{pl^2}{M_p} - 16\,\rho^2 - 8}{2.5\,\rho}. \tag{9.135}$$

Table 9.2. Upper and lower-bound solutions for various values of ρ

$\rho = \frac{l}{L}$	upper-bound	Values of $\frac{pl^2}{M_p}$ Lower bound	$100\left(\frac{UB - LB}{UB}\right)$
$\to \infty$	$16\,\rho^2$	$16\,\rho^2$	0
8	1125.0	1093.00	2.8
7	873.60	845.00	3.3
6	653.50	630.00	3.6
5	465.40	446.00	4.2
4	309.40	289.00	6.6
3	185.40	175.00	5.6
2	93.60R 59.90	87.80	6.2
1.5	34.09	56.00	6.5
1	21.95	32.10	5.8
2/3	17.02	20.40	7.1
1/2	13.27	15.88	6.6
1/3	11.67	12.23	7.8
1/4	10.82	10.74	8.0
1/5	10 .28	9.96	7.9
1/6	9.92	9. 49	7.7
1/7	9.65	9.18	7.5
1/8	9.06	8.96	7.2
1/12	8.78	8.50	6.2
1/16	8.62	8.31	5.4
1/20	8.51	8.21	4.8
1/24	8.34	8.15	4.2
1/36	8.25R 8.00	8.07	3.2
1/48		8.04	2.5
$1/\infty$		8.00	0

The corresponding values of $\frac{pl^2}{M_p}$ are given in table 9.2 for various values of ρ, together with the upper bound for the same ρ and the difference between the bounds.

The regions where negative reinforcement is needed can be determined from fig. 9.65 in which the locus where the smallest principal moment changes sign is drawn for various values of ρ.

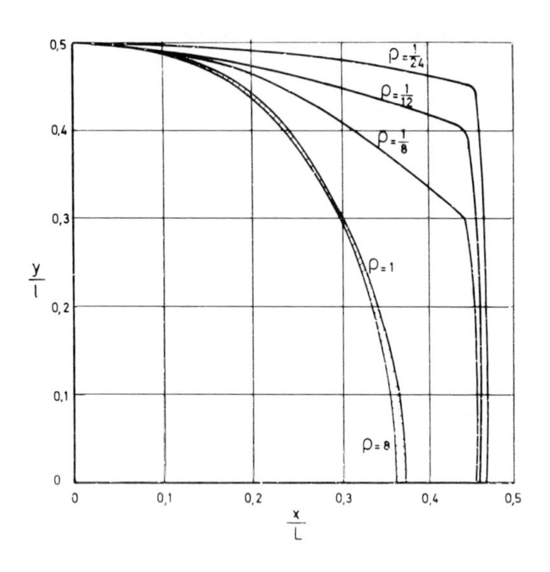

Fig. 9.65.

9.5.7. *Isotropic rectangular built-in plate.*

Even for a square plate with uniformly distributed load, this apparently very simple problem turns out to be extremely difficult. No complete solution is known so far, and presently available good lower bounds were obtained by numerical procedures [9.12, 9.31]. Recently, Sobotka [9.32] has attempted to give a licit moment field with corresponding loads quite close to the upper-bound values given by eq. (9.107) modified by substituting $(M_p + M'_p)$ for M_p. Unfortunately, his moment fields cannot be accepted because they contain inadmissible discontinuities in shear force.

9.5.8. *Orthotropic plates : affinity method.*

The affinity method of Section 9.4.3 will be extended to the statical approach to limit analysis [9.8]. We thus assume that the same orthotropy coefficient k applies to ultimate strength in positive and negative bending. If this situation did not occur, the complete statical procedure should be applied, as was discussed for metal plates in Chapter 7.

If a field M_x, M_y, M_{xy} does not violate the yield condition of an isotropic plate with plastic moments M_p and M'_p, it is readily verified that the field M_x, kM_y and $\sqrt{k}\,M_{xy}$ does not violate the yield condition (9.8) and (9.9) of the plate with plastic moments M_p, M'_p, and orthotropy coefficient k. This is a direct consequence of the very nature of the yield condition (9.8) and (9.9). Now, if so changing the moment field with the orthotropy avoids violating the yield condition, this change does not in general maintain equilibrium if loads are unaltered. To maintain equilibrium, the functions M_x, kM_y and $\sqrt{k}\,M_{xy}$ should be used in a plate obtained from the original one by changing its dimensions by the affine transformation $x^* = x$, $y^* = \sqrt{k}\,y$, where stars refer to the new plate. We then successively have

$$\frac{\partial^2 M_x}{\partial x^{*2}} = \frac{\partial^2 M_x}{\partial x^2},$$

$$\frac{\partial\, kM_y}{\partial y^*} = k\frac{\partial M_y}{\partial y}\cdot\frac{\partial y}{\partial y^*} = \sqrt{k}\;\frac{\partial M_y}{\partial y},$$

$$\frac{\partial\, kM_y}{\partial y^{*2}} = \sqrt{k}\;\frac{\partial^2 M_y}{\partial y^2}\cdot\frac{\partial y}{\partial y^*},$$

$$\frac{\partial^2 \sqrt{k}\;M_{xy}}{\partial x^*\,\partial y^*} = \sqrt{k}\;\frac{\partial}{\partial x}\left(\frac{\partial M_{xy}}{\partial y}\cdot\frac{\partial y}{\partial y^*}\right) = \frac{\partial^2 M_{xy}}{\partial x\,\partial y}.$$

Because the equilibrium equation

$$\frac{\partial^2 M_x}{\partial y^2} + \frac{\partial^2 M_y}{\partial y^2} + 2\frac{\partial^2 M_{xy}}{\partial x\,\partial y} = -p$$

is satisfied in the initial plate, the equilibrium equation

$$\frac{\partial^2 M_x}{\partial x^{*2}} + \frac{\partial^2 k\,M_y}{\partial y^{*2}} + 2\frac{\partial^2 \sqrt{k}\;M_{xy}}{\partial x^*\,\partial y^*} = -p$$

will be satisfied in the affine plate with the same load per unit area. Consequently, to apply a lower-bound solution for an isotropic plate to an orthotropic but otherwise identical plate the following steps are sufficient :

1. Multiply the functions M_y and M_{xy} by k and $\sqrt{k}$, respectively, M_x being unaltered ;

2. Multiply by $\frac{1}{\sqrt{k}}$ all dimensions in the direction of the y-axis, and use the lower-bound value of the load parameter for the resulting isotropic plate.

Consider for example a rectangular, simply supported plate subjected to uniformly distributed load. We immediately transform the field (9.123) of the isotropic plate into

$$M_x = M_p\left(1 - \frac{4x^2}{L^2}\right),$$

$$M_y = k\,M_p\left(1 - \frac{4y^2}{l^2}\right),$$

$$M_{xy} = \sqrt{k}\;\frac{4\,M_p xy}{Ll}\,, \tag{9.136}$$

and the limit load (9.124) into

$$p_- = \frac{8\,M_p\,k}{l^2}\left(1 + \frac{l}{\sqrt{k}\;L} + \frac{l^2}{k\,L^2}\right). \tag{9.137}$$

The upper bound, eq. (9.107), can be transformed in a similar way to give

$$p_+ = \frac{24\,M_p}{l^2}\cdot k\cdot\frac{1}{\left[\left(3 + \frac{l^2}{k\,L^2}\right)^{1/2} - \frac{l}{\sqrt{k}\;L}\right]^2}. \tag{9.138}$$

The two bounds just obtained do not differ by more than 1.5% [9.33, 9.34].

Obviously, the affinity method is not restricted to rectangular plates.

9.5.9. *Other lower bounds and complete solutions.*

Upper-bound solutions, based solely on a yield line pattern, have two main disadvantages :

1. They give too large a value of the limit load and, hence, are unsafe ; this fact is counterbalanced to some extent by the neglected work hardening and the effects of changes of geometry ;

2. Except along the yield lines, they do not give any information on the required reinforcement ; when nonuniform reinforcement is desirable, this drawback is very serious and empirical formulas must be used.

On the other hand, lower-bound solutions (and, obviously, complete solutions) are safe and give all necessary indications concerning economic reinforcement. They are therefore extremely useful but, unfortunately, statically admissible moment fields are much more difficult to obtain than collapse mechanisms. Still more difficult to find are complete solutions, giving the exact limit load.

Presently known complete solutions are few in number. Most of them are found in the books by A. Sawczuk and Jaeger [7.16] and Nielsen [9.8].

Nielsen [9.8] also points out an analogy between some moment fields and slip line fields in plane plastic strain. This analogy furnishes some additional complete solutions, but only for very special problems.

As Prager has shown [7.31], the limit analysis of a framed structure can be formulated as a linear programming problem. Wolfensberger [9.12] has recently applied this remark to find statically admissible solutions in reinforced concrete plates. This method implies suitable linearization of the yield condition and piecewise linearization of the moment field, and requires the use of an electronic computer. For more details see ref. [9.12]. Ceradini [9.35], Gavarini [9.36], and Sacchi [9.37] have worked along similar lines. Massonet has given recently [9.38] some new complete solutions based on combination of radial fields corresponding to plastic regime AB (fig. 9.61).

Following Nielsen [9.8], we give a summary of most (if not all) complete solutions known so far in fig. 9.66 (a) to (t). Moment fields not given in the present book should be found in the references given with the figures. Further lower-bound solutions for rectangular plates with various edge conditions can be found in [9.8].

9.5.10. *Remarks on uniqueness of solution.*

As was seen in Sections 9.4.8, one can obviously find different collapse mechanisms corresponding to the same load. When this load is the exact limit load, the moment fields corresponding to the various mechanisms may possibly differ only in the common rigid regions (see Section 4.4).

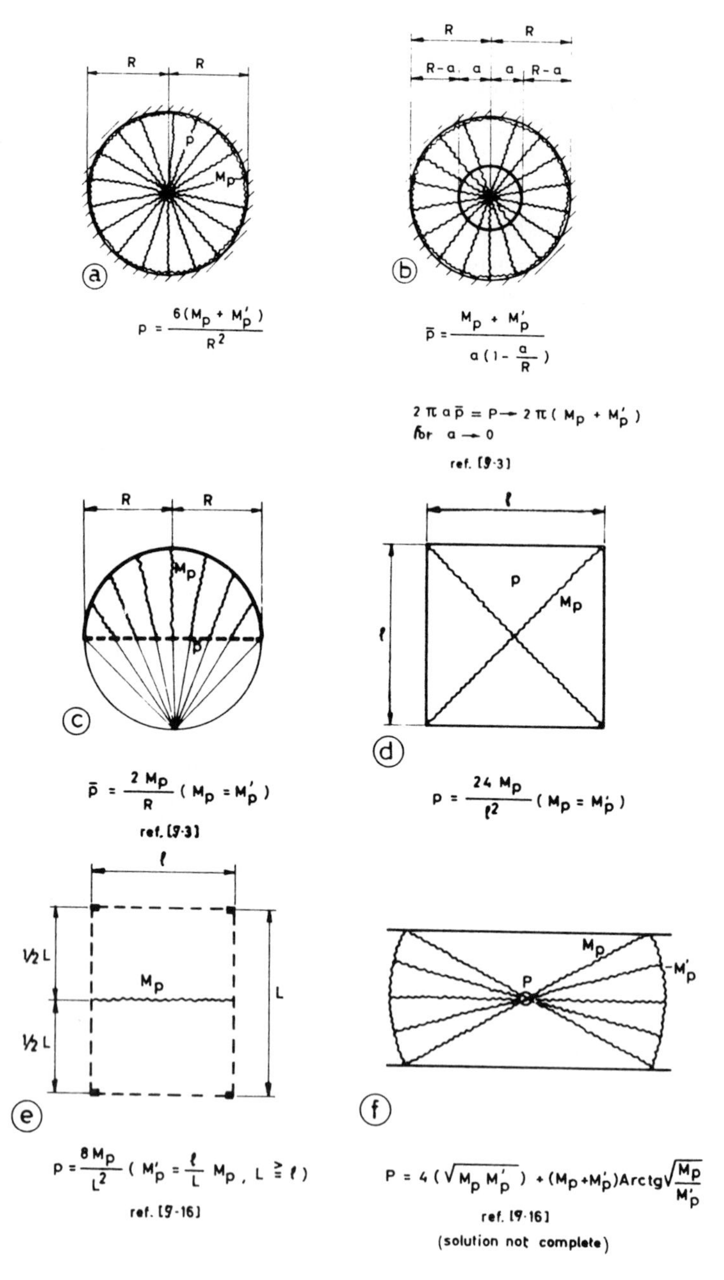

Fig. 9.66.

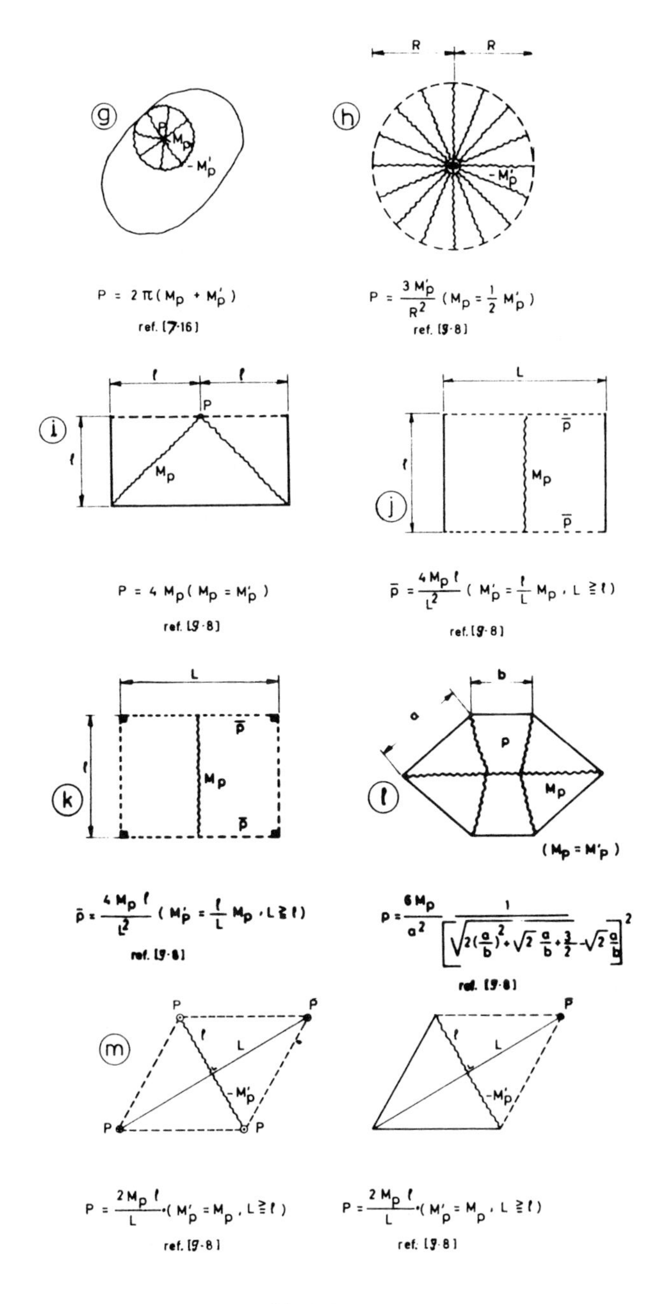

g
P
M_p
$-M_p'$
$P = 2\pi(M_p + M_p')$
ref. [7·16]
h
R
R
$-M_p'$
$P = \frac{3M_p'}{R^2}$ ($M_p = \frac{1}{2} M_p'$)
ref. [9·8]
i
l
l
P
l
M_p
$P = 4M_p$ ($M_p = M_p'$)
ref. [9·8]
j
L
l
$\bar{p}$
M_p
$\bar{p}$
$\bar{p} = \frac{4M_p l}{L^2}$ ($M_p' = \frac{l}{L} M_p$, $L \geq l$)
ref. [9·8]
k
L
l
$\bar{p}$
M_p
$\bar{p}$
$\bar{p} = \frac{4M_p l}{L^2}$ ($M_p' = \frac{l}{L} M_p$, $L \geq l$)
ref. [9·8]
l
b
a
p
M_p
($M_p = M'_p$)
$p = \frac{6M_p}{a^2} \frac{1}{\left[\sqrt{2(\frac{a}{b})^2 + \sqrt{2}\frac{a}{b} + \frac{3}{2}} - \sqrt{2}\frac{a}{b}\right]^2}$
ref. [9·8]
m
P
P
l
L
$-M_p'$
P
P
$P = \frac{2M_p l}{L}$ ($M_p' = M_p$, $L \geq l$)
ref. [9·8]
P
l
L
$-M_p'$
$P = \frac{2M_p l}{L}$ ($M_p' = M_p$, $L \geq l$)
ref. [9·8]

Fig. 9.66.

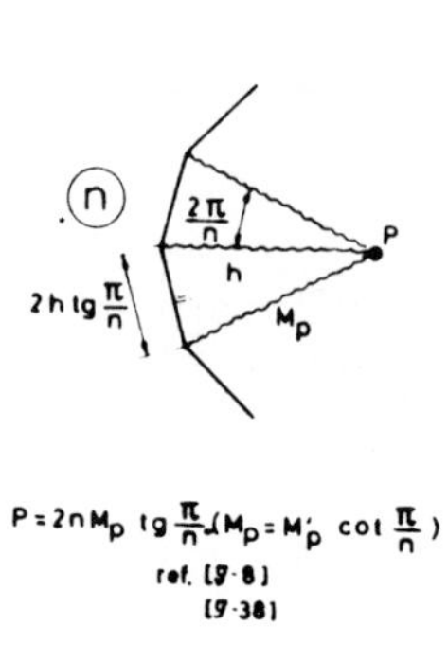

$$P = 2nM_p\ tg\frac{\pi}{n}\ (M_p = M_p'\ \cot\frac{\pi}{n})$$

ref. [9·8]
[9·38]

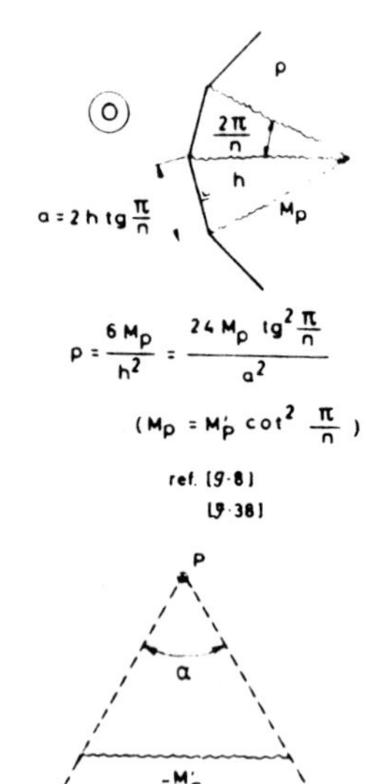

$$p = \frac{6M_p}{h^2} = \frac{24M_p\ tg^2\frac{\pi}{n}}{a^2}$$

$$(M_p = M_p'\ \cot^2\frac{\pi}{n})$$

ref. [9·8]
[9·38]

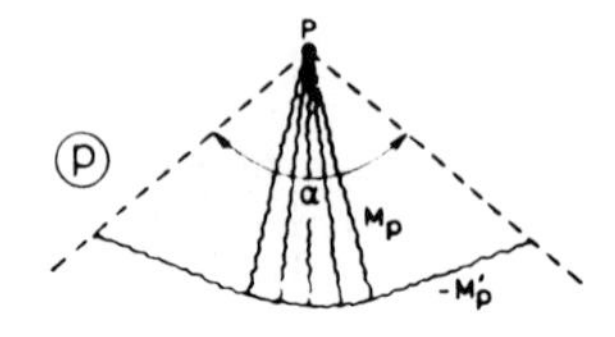

$$P = 2\sqrt{M_p M_p'} + (M_p + M_p')(\alpha - 2\,\mathrm{Arctg}\sqrt{\frac{M_p}{M_p'}})$$

$$(tg^2\frac{\alpha}{2} \leq \frac{M_p}{M_p'})$$

ref. [9·8)

$$P = 2M_p'\ tg\ \frac{\alpha}{2}$$

$$(tg^2\frac{\alpha}{2} \leq \frac{M_p}{M_p'})$$

ref. [9·8]

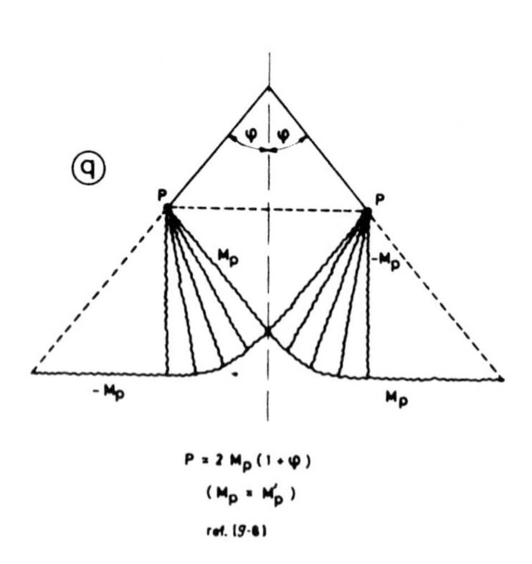

$$P = 2M_p(1 + \varphi)$$

$$(M_p = M_p')$$

ref. [9·8]

Fig. 9.66.

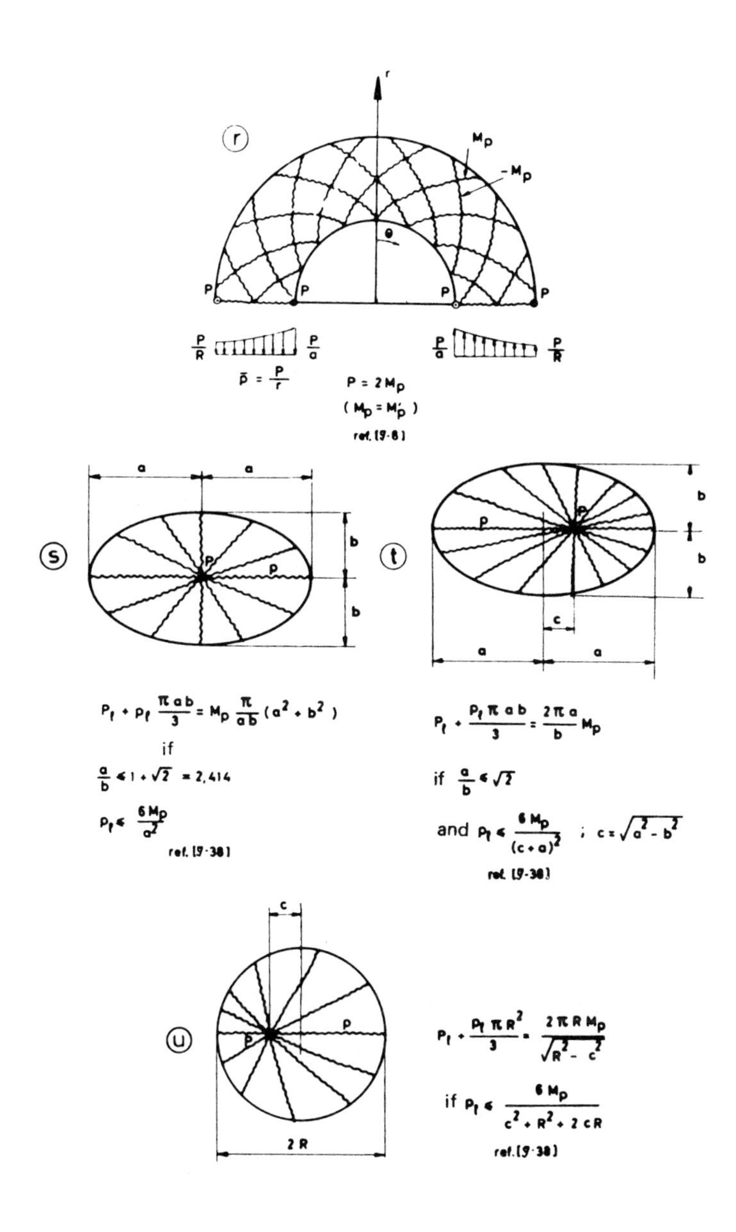

r
Mp
-Mp
θ
P
P
P
P
$\bar{p} = \frac{P}{r}$
P = 2 Mp
(Mp = M'p)
ref. [9-8]
a
a
b
b
p
P
s
t
c
$P_\ell + p_\ell \frac{\pi a b}{3} = M_p \frac{\pi}{ab}(a^2 + b^2)$
if
$\frac{a}{b} \leq 1 + \sqrt{2} = 2{,}414$
$p_\ell \leq \frac{6 M_p}{a^2}$
ref. [9-38]
$P_\ell + \frac{p_\ell \pi a b}{3} = \frac{2\pi a}{b} M_p$
if $\frac{a}{b} \leq \sqrt{2}$
and $p_\ell \leq \frac{6 M_p}{(c+a)^2}$; $c = \sqrt{a^2 - b^2}$
ref. [9-38]
u
2 R
$P_\ell + \frac{p_\ell \pi R^2}{3} = \frac{2 \pi R M_p}{\sqrt{R^2 - c^2}}$
if $p_\ell \leq \frac{6 M_p}{c^2 + R^2 + 2 c R}$
ref. [9-38]

Fig. 9.66.

For example, if, for the simply supported, uniformly loaded square plate, we want to find a licit stress field different from eqs. (9.116), we must cause the principal positive moment acting on the diagonals to reach the value M_p. Such a field was specified by Vallance [9.16]. It gives nonuniform edge reactions. If the plate is supported by edge beams, these reactions cause a maximum moment of $\frac{pl^3}{21}$, instead of $\frac{pl^3}{24}$ with the field (9.116).

If the plate is to be *uniformly* reinforced for $M_p = \frac{pl^2}{24}$, the edge beams can be designed for the smaller maximum moment (here $\frac{pl^3}{24}$). According to the static theorem, the structure (plate + beams) will support the load.

If the reinforcement is adjusted on the moment field, the beams must be designed with the reactions obtained from *this very field* for the yield condition to be nowhere violated. In the case of statical boundary conditions, the limit load for a given loading is also well defined but the collapse mechanism strongly depends on the kinematic boundary conditions. Consider once again the square plate of the preceding example. If the supports are rigid, only the diagonal mechanism can form. But if the supporting beams are designed to become fully plastic at the same time as the plate, the mechanism can be the cone of fig. 9.67, (corresponding to a purely radial moment field). This mechanism is only possible when the supports are free to settle. Nevertheless, it furnishes the same load as the diagonal mechanism. This is a general property.

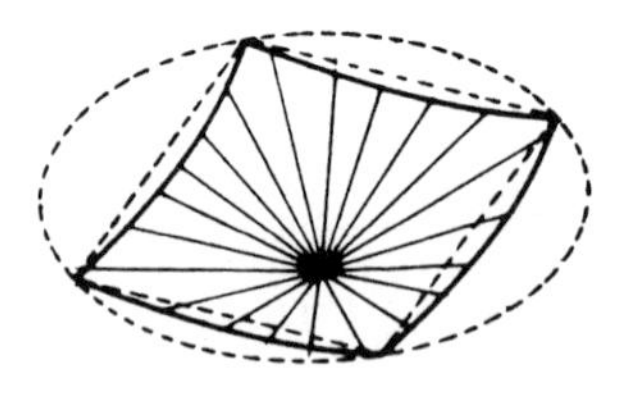

Fig. 9.67.

Indeed, let P_{l_1} and P_{l_2} be the limit loads of two structures differing solely by the kinematic boundary conditions. The stress field of one structure is licit for the other structure. Hence, we simultaneously have $P_{l1} \leq P_{l2}$ and $P_{l_2} \leq P_{l_1}$. Hence, $P_{l_1} = P_{l_2}$.

When different licit stress fields exist for a given structure subjected to given loads, they can be linearity combined to furnish new stress fields. The latter fields are licit for a load equal to the original load multiplied by the sum of the coefficients of the linear combination, provided the yield condition has not been violated [9.38].

9.6. Influence of axial force.

9.6.1. *Introduction.*

When reinforced concrete plates are loaded up to rupture, their load versus deflection relationship does not exhibit a final flat portion as assumed in the simple plastic theory. The shape of the curve may in fact be very different from that predicted by this theory and the maximum obtainable load may be substantially greater than the theoretical limit load. The discrepancy is strongly dependent on the boundary conditions (and, obviously, on the deflection at which rupture is obtained or the test is ended). These facts are illustrated in figs. 9.68 and 9.69.

The increase in carrying capacity stems primarily from axial forces induced by restraints of horizontal displacements, then from geometry changes, and secondarily from work hardening of steel, effects that are disregarded in the simple plastic bending theory. These effects are responsible for the fact that, even when only computed from a good collapse mechanism and thus being upper bounds P_+ to the exact limit load P_l of simple bending theory, most theoretical collapse loads are smaller than the corresponding experimental values (see [7.16, 9.3, 9.16, 9.17, 9.18, 9.53, 9.54, 9.55, 9.56, 9.57]).

The reserve strength of real plate structures with comparison to simple bending estimations is well seen in full-scale tests on complexe slab systems [9.58], which are unfortunately rare. The effects of compressive axial forces in plastic plate response (dome effect) was commented already in 1939 by Gvozdev [9.59] resulting in the allowance by the pre-war Russian code of a reduction of reinforcement for continuous plates. Then, it was accounted for in masonry walls [9.64].

The interaction of bending and axial compression is of an unstable character and, therefore, even small geometry changes are not negligible. The problem was treated by a deformation-type theory by Wood [9.16], Sawczuk [7.16] and Park [9.44, 9.53], and in the framework of the plastic flow theory by Janas [9.43, 9.60], Morley [9.55, 9.61] and Calladine [9.62]. In opposition to the incipient collapse analysis, the two approaches furnish different results at large deflections, even for the rigid-plastic model ; differences are discussed by Baestrup [9.65]. The effects of axial forces including geometry changes will be considered in the framework of the flow theory in Sections 9.6.2 to 9.6.4.

9.6.2. *Finite plastic bending with boundaries restrained against sliding over the supports.*

We consider for illustration a square plate under a uniform load p ; the plate is to consist of a rigid perfectly plastic material, and to be hinged along its supports at the level of the reinforcement.

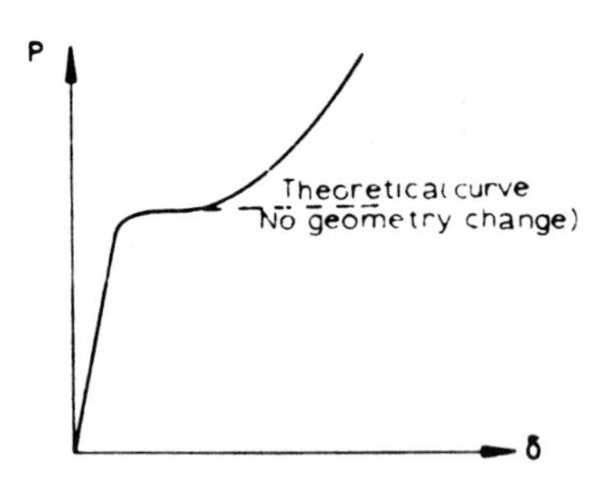

Fig. 9.68. Simply supported plate.

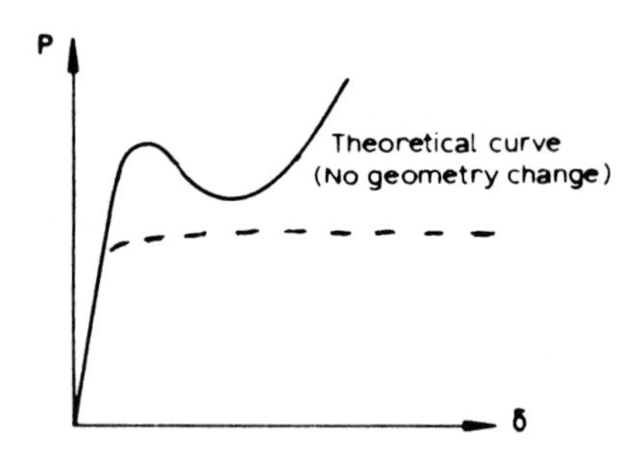

Fig. 9.69. Built-in plate.

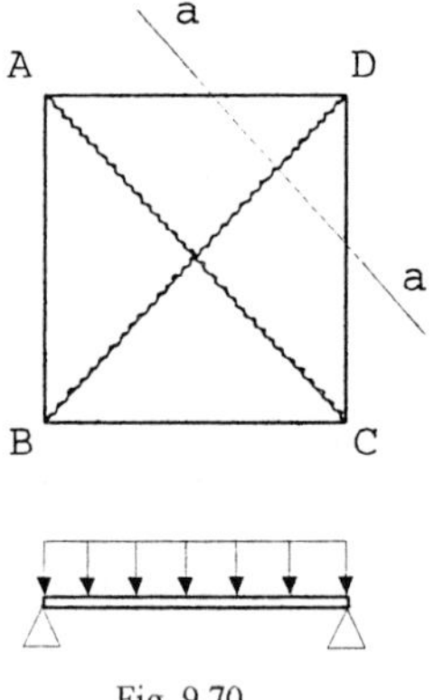

Fig. 9.70.

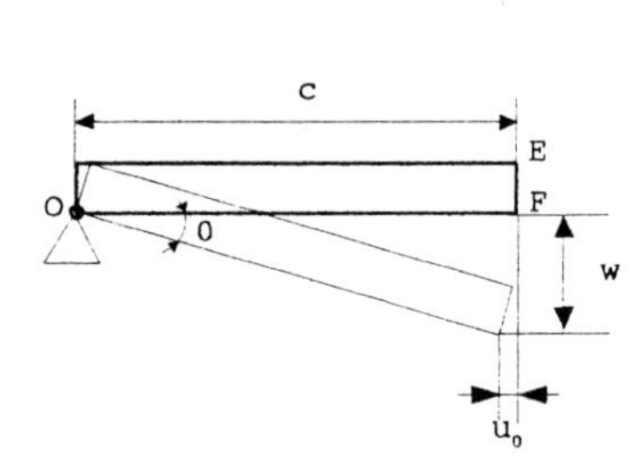

Fig. 9.71.

For the sake of simplicity it is considered that the reinforcement layer coincides with the bottom face of the plate and it does not extend up to the supports.

This plane is taken as the reference plane. The (already known) collapse mechanism of fig. 9.70 is not kinematically admissible for nonzero deflection with the present boundary conditions. Indeed, fig. 9.71 shows the deflected position of a cross section normal to a yield line (as cross section aa in fig. 9.70). Section EF in yield line undergoes a horizontal displacement (see fig. 9.71)

$$u_o = c(1 - \cos\theta) = \frac{w}{\sin\theta}(1 - \cos\theta) \approx w\frac{\theta}{2} \tag{a}$$

(w is small compared to c). The small rotation θ si related to w by

$$\theta = \frac{w}{c}, \tag{b}$$

when

$$u_o = \frac{w^2}{2c}. \tag{c}$$

Differentiating eqs. (b) and (c) with respect to time we obtain

$$\dot{\theta} = \frac{\dot{w}}{c}, \tag{d}$$

$$\dot{u}_o = \frac{w\dot{w}}{c}. \tag{e}$$

Use of eq. (d) in eq. (e) yields

$$\dot{u}_o = w\,\dot{\theta}. \tag{f}$$

Because normals to the reference plane remain normal to the polyhedron into which it transforms, the zero strain rate level is at the distance $\frac{\xi t}{2}$ fo the bottom face such that $\dot{\theta}\,\frac{\xi t}{2} = \dot{u}_o$ or, using eq. (f)

$$\frac{\xi t}{2} = w. \tag{g}$$

Consequently, for vanishing w (impending collapse), boundary conditions enforce the zero strain surface to coincide with the restraining plane. The yield lines must therefore be subjected both to a moment M and an axial force N for such a mechanism to develop. This kinematic condition of internal compatibility at zero deflection is discussed by Janas in ref. [9.19] ; it is responsible for the increase in the load at incipient collapse [9.63].

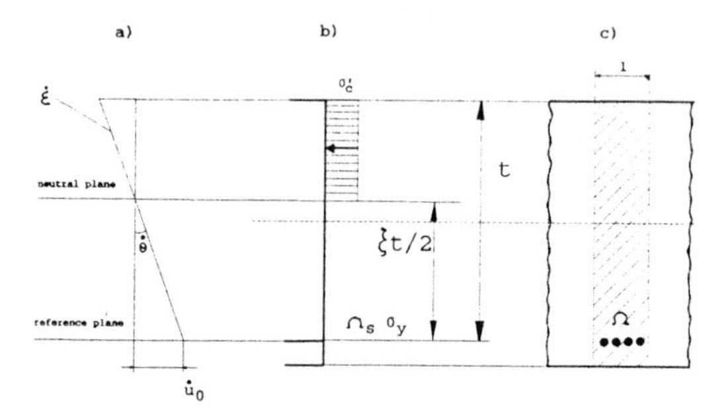

Fig. 9.72.

The assumptions and notations for concrete yield criterion and steel reinforcements are the same as those of Section 5.6. Besides, we supposed that the reinforcements are in the lowest layer of the concrete ($\rho = 1$).

If it is noted that the moment is refered in this Section to the lower plane instead of the middle one in the Section 5.6, the expression $\left(M - N\frac{t}{2}\right)$ must be substituted for the moment M.

let

$$M_* \equiv \frac{\sigma_c' t^2}{2}, \qquad N_* \equiv \sigma_c' t. \tag{9.139}$$

With these new notations, the ratios $\frac{N}{N_*}$ and $\frac{M}{M_*}$ must be substituted respectively to n and (m/2 - n). So, we deduce easily from eq. (5.104) the following MN interaction curve :

$$M = M_* \left[\gamma(2-\gamma) - \left(\frac{N}{N_*}\right)^2 - 2\frac{N}{N_*}(1-\gamma) \right] \tag{9.140}$$

For vanishing w (initiation of collapse mechanism), eqs. (e) and (g) enforce vanishing γ and ξ. With the notations of fig. 5.38, the value $\xi = 0$ corresponds to $\eta = \rho = 1$. Hence, taking into account eq. (5.99), eq. (5.103) gives

$$N = -N_*(1-\gamma), \tag{9.141}$$

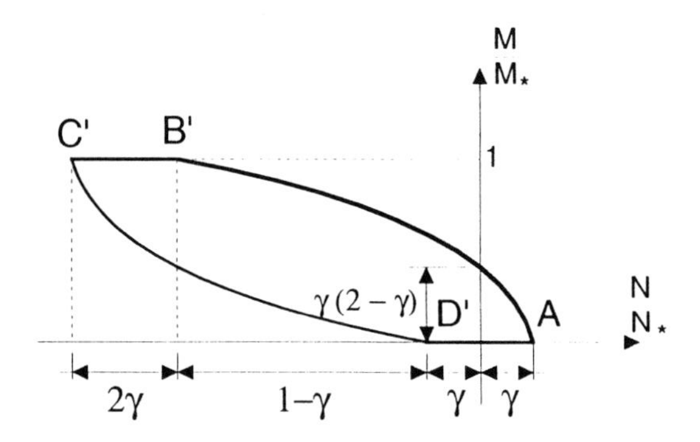

Fig. 9.73.

$$M = M_*\tag{9.142}$$

On the other hand, with no restraint at the boundaries, the plastic moment is

$$M_p = M_* \gamma (2 - \gamma)\tag{9.143}$$

obtained by letting N = 0 in eq. (9.140).

The moment in the yield lines is increased by the factor

$$\frac{M_*}{M_p} = \frac{1}{\gamma (2 - \gamma)},\tag{9.144}$$

and so is the limit load p_l. The maximum tensile axial force is

$$N_c = \sigma_Y \Omega_s = N_* \gamma .$$

For example, with $\sigma_Y =$ 300 N/mm^2, $\sigma_c' =$ 30 N/mm and $\frac{\Omega_s}{t} = 0.01$, we have $\gamma = 0.1$.

With N = 0 and t = 10 cm, we would obtain

$M_p \equiv \frac{\sigma_c' t^2}{2} \gamma (2 - \gamma) = 27.10$ Nm/m. The corresponding $\frac{M}{M_*}$ versus $\frac{N}{N_*}$ parts of the interaction curve being of interest are given in fig. 9.73. Note that eq. (9.140) is represented by the parabolic part AB', whereas segment B'C' corresponds to $\xi = 0$ (neutral layer ar the level of the reinforcement). The values of M and N at point B' are given by eq. (9.141) and (9.142).

Knowing the relation between moments and axial forces in the yield line, we can now study the collapse of the considered slab (fig. 9.70). The rate of work of the external loads is the same as for the simple bending theory. For the collapse mode shown in fig. 9.70, it is

$$W = \frac{pl^2}{12}\dot{w}_o \tag{9.145}$$

whereas the rate of dissipation is now expressed by the formula

$$D = \sum_{i=1}^{n} \int_0^{l_i} \left[M_i(l)\dot{\theta} + N_i(l)\dot{\lambda}_i \right] dl \tag{9.146}$$

where summation extends on all n yield lines with the respective lengths l_i and rotation rates $\dot{\theta}_i$. The internal forced M, N depend on the position of the current neutral plane governed by η through eqs. (5.103). Note that, because of relation (f), dissipation is now deflection dependent.

For the position of edge restraints shown in fig. 9.71, the rotation axes of the collapse mechanism must lie in the bottom plane of the slab, otherwise horizontal displacements of point O occur. Since the neutral axis in a hinge must be the axis of reciprocal rotations of adjacent parts, initial positions are defined, in agreement with eq. (g), by $\xi = 0$. Internal forces can be found from eqs. (5.103) and the rate of dissipation can be computed. The work equation

$$W = D \tag{9.147}$$

yields the following upper bound for the initial collapse load.

$$p_{01} = \frac{24M_*}{l^2}\gamma = p_1 \frac{M_*}{M_p}\gamma \tag{9.148}$$

whereas the solution derived from the simple bending theory with the limit load

$$p_l = \frac{24\, M_p}{l^2} \tag{9.149}$$

is no longer kinematically admissible.

Since restraints enforce ξ=0, large compressive forces $N = -N_*\,(1-\gamma)$ exist in the hinges, and must be supported by the restraints. It can be shown that, in absence of special reinforcements at supports, plastic deformations must develop there. Since there is a zone of concrete crushed in compression (fig. 9.74 a) the collapse mode including rotations around O' can satisfy the zero horizontal displacements condition at the supports, and appears to be admissible. Now, the position of the neutral axis in the hinges is $\xi = \eta$.

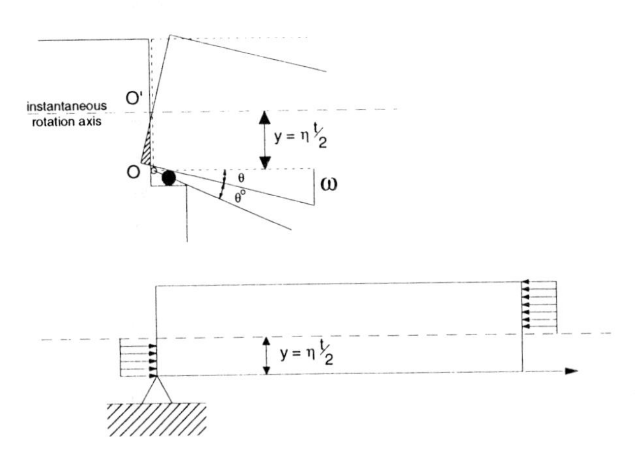

Fig. 9.74.

The axial force acting in the hinges cannot be balanced by horizontal reactions in the punctual contact O and internal forces and dissipation change while *W* remains the same as for the preceding collapse mode. Note, moreover, that the total rate of dissipation must now include a part due to plastic deformation at the supports, where negative rotation rate $\dot{\theta}$ appears. The work equation gives a collapse load dependent on the unknown parameter η which must be found (as in the simple bending theory) from the minimum principle. The minimization procedure is equivalent to satisfying the horizontal equilibrium of rigid portion cut off by the hinges (fig. 9.74 b) and gives

$$\eta = 1 - \gamma \tag{9.150}$$

Finally, a new bound for the collapse load is found to be

$$p_{02} = \frac{p_1}{2}\left[1 + \frac{M_*}{M_p}\right] \tag{9.151}$$

For plastic flow to continue at finite values of the central deflection w_0 the actual collapse load must not be higher than those derived from the work equation [eq. (9.147)] established for the deformed structure. Assuming that the deflection pattern does not change, the current deformations are defined by one parameter, w_0. Normal strains being dependent on the actual deflections, they are seen to vary along the hinges. If collapse follows the mode of fig. 9.71, relation (f) can be introduced into eqs. (5.103) and, after the appropriated integration of eq. (9.146), dissipation can be computed. External work rate is still unchanged [eq. (9.145)].

The work equation gives

$$p_1 = p_1 \frac{M_*}{M_p}\left[1 - \delta_0 (1 - \gamma) + \frac{\delta_0^2}{3}\right], \tag{9.152}$$

where $\delta_o = \frac{w_o}{t}$. For $\delta_o = 0$ we obtain eq. (9.148), as expected. When initial yielding starts with the mode of fig. 9.74, simultaneous plastic deformations develop both in the inner hinges and at supports. According to fig. 9.74 a, we now have $\dot{u}_o = \dot{\theta}\ (y + w)$ and $\dot{u}_o = -\dot{\theta}\ y$ in diagonal and support yield lines respectively. Instantaneous neutral axis at supports lies at $\xi = \eta$ whereas its variable position in diagonal hinges is

$$\xi = \eta + \frac{\frac{w}{t}}{2} \tag{9.153}$$

Relation between the positions of the two axes represents the internal kinematical compatibility conditions [9.19]. The foregoing procedure gives

$$\eta = 1-\gamma-\frac{\delta_o}{2} \tag{9.154}$$

and leads to the solution

$$p_2 = \frac{p_1}{2}\left\{1+\frac{M_*}{M_p}\left[1-\delta_o(1-\gamma-\frac{5}{12}\delta_o)\right]\right\}, \tag{9.155}$$

the unknown parameter η being found in the same way as for initial yielding.

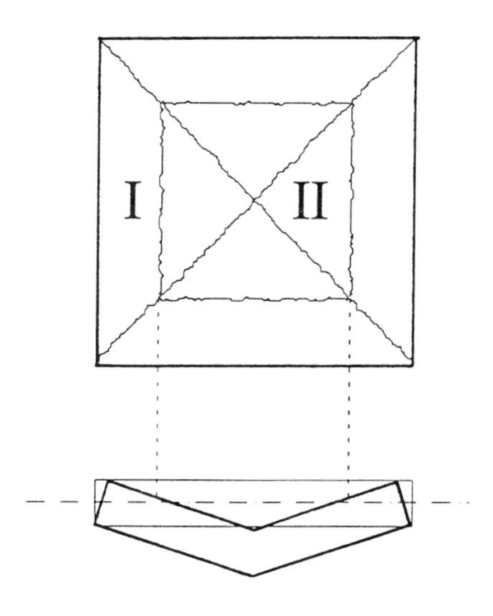

Fig. 9.75.

Both solutions, eqs. (9.152) and (9.155), are valid only if the central deflection remains sufficiently small. If the upper face of the slab descends below the plane of the rotation axes, pure membrane response develops in the hinges of a central zone (II of fig. 9.75). Relations (5.103) are not valid there, the internal forces being M=O, $n=N_p$. In zone

I however, eq. (5.103) still holds. Accounting for this new situation, we may derive the appropriate formula for the collapse loads for both considered modes. They are, respectively,

$$p_{m1} = p_1 \frac{M_*}{M_p}\left(\delta_o \gamma + \frac{1}{3\,\delta_o}\right) \quad , \qquad \delta_o \leq 1 \tag{9.156}$$

$$p_{m2} = \frac{p_1}{3} \frac{M_*}{M_p} [\delta_o \gamma + (1 + 2\gamma)(3 + 2\delta_o)$$

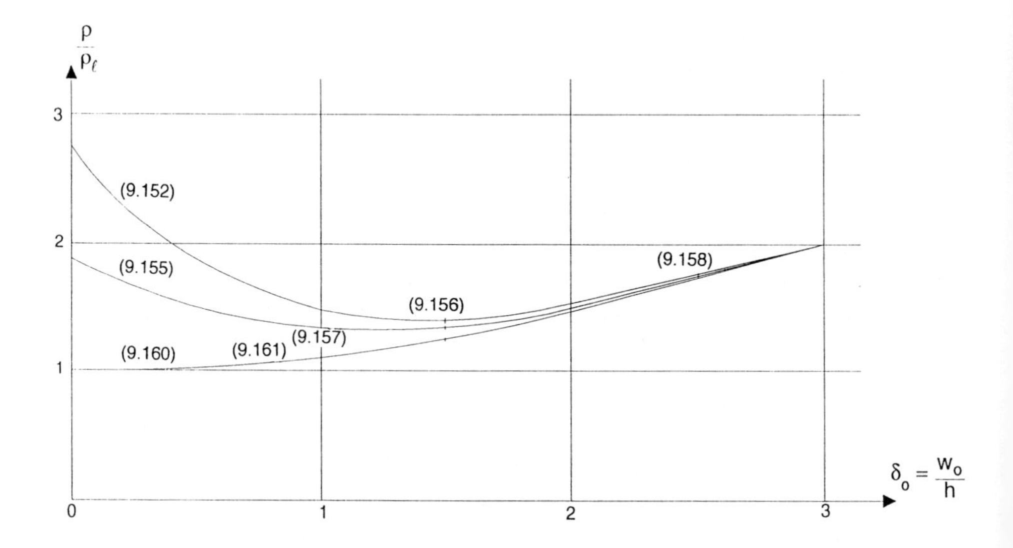

Fig. 9.76.

$$+2(\delta_o+2+2\gamma)\{\alpha-[\alpha(\alpha+2+2\gamma)]^{1/2}\}] \quad ,$$

$$\frac{2}{3}(1+\gamma)\leq\delta_o\leq\frac{1}{2\gamma} \tag{9.157}$$

$$p_{m2}=p_1\frac{M_*}{M_p}\left\{\delta_o\gamma+\frac{1}{\delta_o}\left[\frac{1}{3}-\frac{(1-2\delta_o\gamma)^2}{(1+2\gamma)}\right]\right\},$$

$$\delta_o\leq\frac{1}{2\gamma}. \tag{9.158}$$

The latter formula concerns free inward sliding at supports, because in the mode 2 no tensile reaction is exerced there.

It appears [9.60] that the solution concerning the mode 2 (9.155, 9.157, 9.158) is valid also for a circular restained slab with bottom reinforcement.

We see that a continuous transition from the initial compression (even with crushing of the concrete at supports) to membrane tensile response has been obtained. Collapse load is plotted against deflections in fig. 9.76 for the reinforcement ratio γ=0.2. It is seen that, at the first stage, the collapse load decreases as deflections increase. This unstable situation was also observed experimentally [9.16, 9.53, 9.55].

It may be shown [9.43] that if the position of the neutral axis is a free parameter, the minimum of local dissipation corresponds to pure bending (N=0). Therefore, if the rigid plastic flow theory is followed, the minimum p cannot be inferior to p_1 (see fig. 9.76). In absence of compatibility requirements imposed by edge restraints or large deflections (eq. 9.153), simple bending load p_1 will be attained either at initial collapse (for unrestrained plates - see below) or at a deformed configuration (finite bending of restrained strips [9.43]).

In the case of *unrestrained supports*, free horizontal displacements of the slab edges, arbitrary positions can be adopted for the rotation axes :

$$y=\eta\frac{t}{2}>0$$

without inducing plastic deformations at the supports.

At the initiation of collapse, the neutral axes are defines by $\xi=\eta$ and eqs. (5.103) can be applied to determine the rate of dissipation [eq.(9.146)]. Since the rate of external work is always as for simple bending theory [eq. (9.145)], the appropriate expression for collapse load can be derived from the work equation, eq. (9.147). The parameter η can be derived from the minimum principle or from the equilibrium of horizontal forces applied to the rigid portion ABE (fig. 9.70). The last condition imposes zero total stress resultant in each hinge. With $\eta = 2(1-\gamma)$ there are no normal forces in the hinges [eqs.(5.103)] and the simple bending theory is valid here, giving $p = p_1$ ([eq. (9.149)]. When deflections increase, the work equation [eq. (9.147)] must be established for the deformed structure, as for unrestained supports. The position of the neutral layer varies along the hinge following (9.153).

Repeating the preceding procedure and after appropriate integration of eq. (9.146), the collapse load is found to be dependent upon w_o and η. The minimum principle yields

$$\eta = 2(1-\gamma) - \delta_o \tag{9.159}$$

and the current collapse load is expressed by

$$p = p_1 \left(1 + \frac{\delta_o^2}{12} \frac{M_*}{M_p} \right) \tag{9.160}$$

It can be noticed that since the absolute minimum point occurs at the initial collapse, there is no instability phenomenon, contrary to what was observed for restrained edges.

For the central deflection w_o large enough ($\frac{\eta}{2} + \delta_o > 1$), the membrane response zone appears (see fig. 9.75), with M=0, $N=N_p$, whereas formula (5.103) still holds for the outer region. The limit load is now

$$p_m = p_1 \left\{ \gamma + 1 \left[\delta_o + \gamma - \frac{4}{3} - (2\gamma\delta_o)^{1/2} \right] \frac{M_*}{M_p} \right\}, \qquad \frac{1}{2\gamma} > \delta_o > 2\gamma . \tag{9.161}$$

To minimize the collapse load, the position of the rotation axes η must furnish a vanishing total stress resultant in the hinge. A peripherical compressed zone must therefore exist in the slab as in the mode 2 (fig. 9.74). For $\delta_o < \frac{1}{2\gamma}$ the problem becomes identical to that of unilaterally restrained edges (9.158) since in the latter case inward displacements are also free. The problem of finite plates bending may be studied using a deformation theory. The approach was proposed by Wood [9.16] and developped by Park [9.53] and others.

Since, however, the internal forces are there related to the finite strains rather than the strain rates, the results can not be accepted within the frame-work of the flow theory of plasticity. For comparison of the two approaches, see the survey by Baestrup [9.65].

9.6.3. *Elastic-plastic analysis.*

Elastic deformations and early crackings transform the theoretical curve of fig. 9.76 as shown in fig. 9.77. The peak A was found experimentally by Park [9.44] to correspond approximately to $w_o = w_o^u = 0.5$ t, and with this value the peak load can be estimated using the rigid-plastic load-deflection curves. However, further tests (see [9.53, 9.54, 9.55]) give

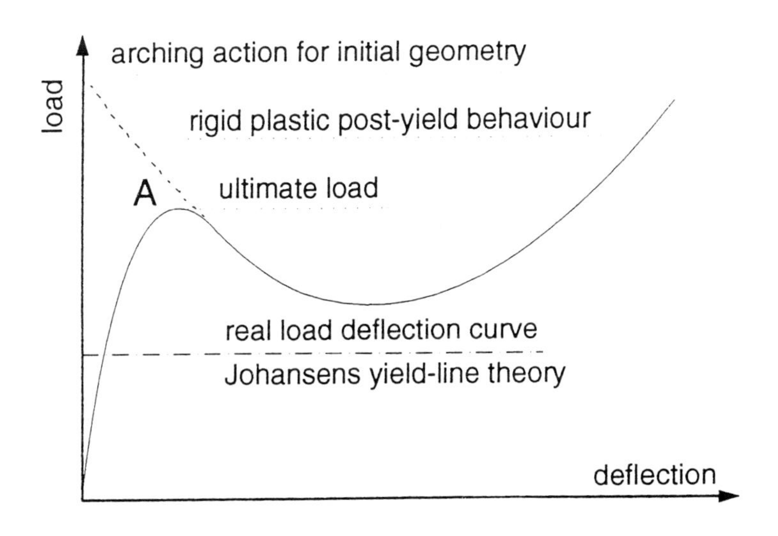

Fig. 9.77.

rather divergent values, and the results depend strongly on the stiffness of support constraints.

Elastic effects change qualitatively the structure response at early stage of deformation process. To take them into consideration, but avoiding the cumbersome elastic-plastic incremental analysis, membrane elastic deformations are accounted for, whereas the elastic flexibility is disregarded. Panels composing a yield line mechanism are thus considered to deform, but remain plane as in the rigid-plastic mode. Such an approach was proposed in the framework at the flow theory by Janas [9.60].

In the deformed configuration of the restrained square plate considered in the preceding section (mode 2, fig. 9.74 a), the kinematical relation (9.153) between the positions of neutral axes at supports (η) and in diagonal hinges (derived from the rigid-body motion) has to be modified by introducing into the deformation rate $\dot{u}_o$ an elastic component $\dot{u}_o^e$ proportional to the rate of evolution of the axial force $\dot{N}$. one obtains thus the kinematical compatibility condition in differential form

$$\xi = \eta + \frac{w}{\frac{t}{2}} + \frac{1}{\varepsilon}\frac{dN}{dw_o}\frac{2}{N_* \frac{t}{2}} \tag{9.162}$$

Where N is taken the mean force along a strip normal to the support, and ε is an average elastic stiffness of the triangular panel (in the direction normal to the hinge) :

$$\varepsilon = \frac{8Et^2}{\sigma'_c L^2} \tag{9.163}$$

depending on the elastic modulus E and the span-to-thickness ratio. In the case of compliant support elements a reduced stiffness ε' should be adopted

$$\varepsilon' = \left(\varepsilon^{-1} + \varepsilon_s^{-1}\right)^{-1} \tag{9.164}$$

where ε_s is a horizontal stiffness of restraining elements. For unrestrained edges we have $\varepsilon_s = 0$ and, therefore, $\varepsilon' = 0$.

Relation (9.162) introduced to the equilibrium equation for axial forces acting on the triangular panel gives the position of the rotation plane η

$$\eta = \frac{1}{2}(\delta_o + 2\gamma - \frac{1}{\varepsilon} - Ce^{-\varepsilon\delta_o}) \tag{9.165}$$

With a constant of integration C which should be found from an initial condition for axial forces. Because the latter are induced by the plastic flow, this condition can be taken N=0 for $\delta_o = 0$. Introducing the above results into the work equation (9.147) or into the moment equilibrium around the support line one obtains the load-central deflection relation :

$$p = p_1 + \frac{12\,M_*}{L^2}\left\{(\gamma+1)^2 - \delta_o\left(1-\gamma--\frac{5}{12}\,\delta_o\right)\right.$$

$$\left. -\left[1-\mu-\left\{1-e^{-\varepsilon\,\delta_o}\left(1-\gamma+\frac{1}{2\,\varepsilon}\right)\right]^2 - \frac{1}{6\,\varepsilon^2}\left(1-e^{-\varepsilon\,\delta_o}\right)^2 \right. \qquad (9.166)$$

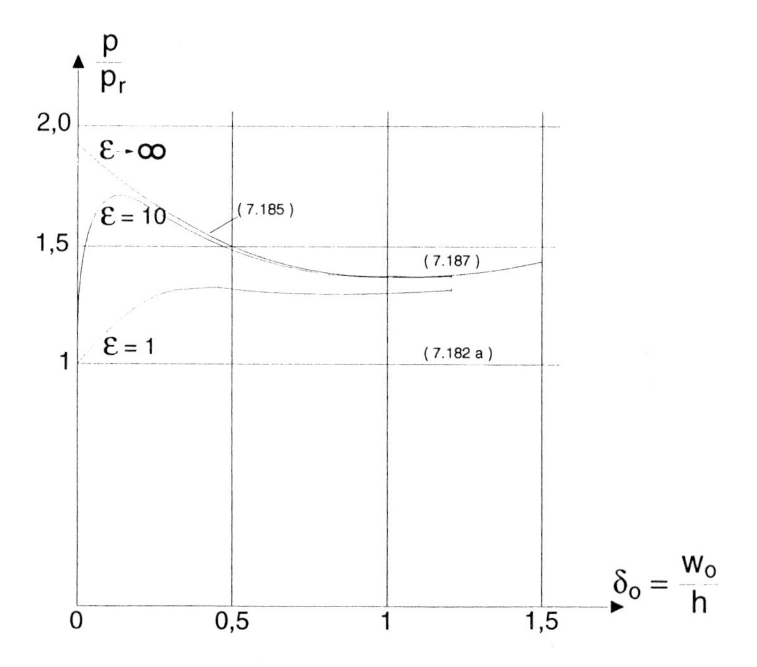

Fig. 9.78.

which is represented, for $\gamma = 0.2$ in the fig. 9.78. The rigid-plastic solution (9.155) and the simple bending for unrestrained supports (9.149) are particular cases with $\varepsilon \to \infty$ and $\varepsilon = 0$ respectively.

The same solution is obtained [9.60] for a restrained circular. Plate using another assumption for the compressiblity of the circular plate (support "springs" modelling the behaviour). Morley [9.61] obtained results slightly different and discussed the range of statical admissiblility of the solution. Changing the initial condition to N=0 for $\delta_o = \delta_i$ permits to account for the initial deflections or the gaps between the plate and supports. An appropriate choice of δ_i permits also (see [9.55]) to account for the pre-collapse flexure deformation.

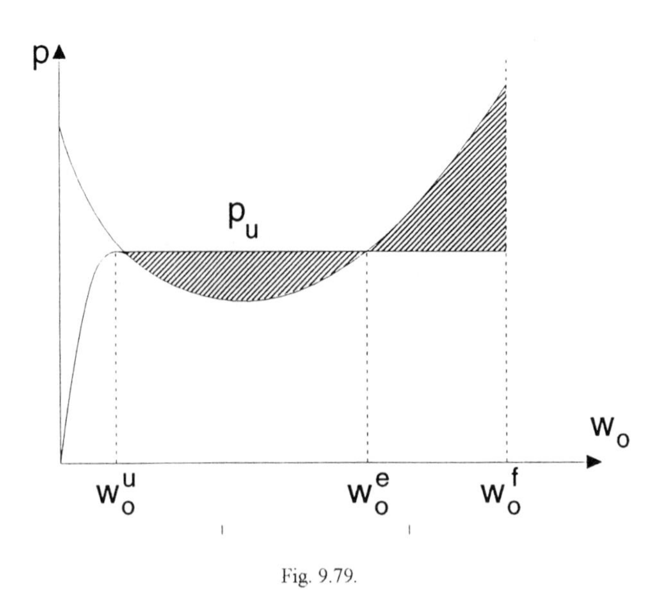

Fig. 9.79.

The above results are directly applicable [9.60] to doubly reinforced plates. If the current load is presented as

$$p = p_l + p_* (\delta_o) \tag{9.167}$$

the excess load p_* appears to depend upon a single reinforcement parameter

$$K = 1 - \gamma_i - \gamma_s + \gamma'_i + \gamma'_s \tag{9.168}$$

where γ_i , γ_s are tension reinforcement ratios in the interior and at supports, respectively ; γ'_i , γ'_s are the corresponding reinforcement at the compressed face. Therefore, if K is the same for different type of reinforcement, the load-deflection rigid-plastic or elastic-plastic curves will differ only by the difference in p_l. This result holds rigorously if $\gamma > \gamma'$, and beyond this limits the error thus introduced remains small.

Comparison with experimental data shows satisfactory correlation with the solution obtained following Section 9.6.3. if the modulus E in (9.163) is taken approximately a half of the initial compression value for concrete E_{co}.

Because of the restabilizing effect of membrane behaviour at large deflections, the peak load attained does not induce necessarily the final collapse. However, if the load is maintained at its peak value the structure snaps violently to a new equilibrium configuration (fig. 9.79).

It is of practical interest to find final deflections of the strucure after snap-through. Because of the finite velocity attained during the accelerated motion in the deflection range corresponding to $p < p_u$, the movement does not finish at w_o^e where the $p = p_u$ on the ascending branch. It can be shown [9.57] that the motion stops at final deflection w_o^f , for which the dashed areas in fig. 9.79 below and above the ordinate p_u are equal. Knowing the rigid-plastic curve and the ultimate peak load (e. g., from the above elastic-plastic analysis) final deflection can be easily estimated in a graphic way.

9.6.4. Practical conclusions.

The preceding theoretical analysis, as well as experimental evidence [9.16, 9.44], show that the carrying capacity can exceed the theoretical limit load P_l.

If the slab is strongly restrained at the boundary against extension of its median plane by surrounding beams or slabs, the factor $\left(1 + \frac{\alpha^2}{4\beta}\right)\rho$ can be applied to the theoretical limit load P_l. The factor ρ is a reduction coefficient accounting for the fact that instability occurs before the maximum load at zero deflection can be reached. According to Wood [9.16], the load factor to use with the carrying capacity $P_c = P_l\left[1 + \left(1 + \frac{\alpha^2}{4\,\beta}\right)\right]\rho$ should be at least 4, and ρ should be taken as :

Reduce reinforcement ratio μ	Reduction coefficient ρ
$\gamma > 0.08$	$\rho = 0.7$
$0.08 > \gamma > 0.04$	
$0.04 > \gamma$	$\rho = 0.5.$

(9.169)

For rectangular plates with L/l >2, and for plates where a combined beam-plate mechanism can occur, the limit load P_l should be used.

If, for unrestrained plates, one is prepared to accept a maximum deflection w_o, Wood [9.16] suggests the following relations.

$$P_c = P_l\left(1 + \frac{0.6\,w_o}{h_t}\right) \quad \text{for} \quad \gamma \approx 0.02\,, \tag{9.170}$$

and

$$P_c = P_l\left(1 + \frac{0.3\,w_o}{h_t}\right) \quad \text{for} \quad \gamma \approx 0.08\,, \tag{9.171}$$

where h_t is the total height.

Relations (9.192) and (9.193) should not be used for rectangular plates with $L \geq 3l$.

9.7. Influence of shear forces. Punching.

Stresses due to shear forces are, as a rule, rather small in plates (see [9.16], p. 11) and their influence is completely negligible, as has been verified by numerous tests up to collapse. The only case to be considered is the punching of the plate by a high concentrated force or by a column reaction.

The punching fracture under a concentrated force is a complicated phenomenon, the interpretation of which remains open to discussion [9.45]. It is influenced by various parameters, among which we have :

1. The quality of the concrete ;

2. The flexural reinforcement (percentage, distribution, adhesion) ;

3. The distribution of bending moments in the vicinity of the concentrated load ;

4. The ratio of the (small) area of application of the load to the effective thickness h of the plate ;

5. The presence of special shear reinforcement (inclined bars, etc.) ;

6. The existence of large compressive forces arising from the constraints at the boundary of the plate in its plane. This last influence is particularly noticeable in the tests described by Guyon [9.46] and Muller [9.47], and related to what was developed in Section 9.6 and explains why the following formula is too conservative.

Indeed, formula (9.172) was empirically deduced by Elstner and Hognestad [9.48] from tests on simply supported square slabs of 1.8 m length, 15 cm thickness, loaded by a column with square section of 15 cm length. They obtained

$$P_V = \frac{7}{8} A_S \sigma'_{r(cyl)} \left(\frac{23.4}{\sigma'_{r(cyl)}} + \frac{0.046}{\varphi_o} \right) \tag{9.172}$$

(units are kg and cm), where P_V is the punching load ; A_S is the "sheared area", product of the perimeter of the punch by the effective thickness of the plate ; $\sigma'_{r(cyl)}$ is the rupture compressive stress of the concrete on cylinders ; $\varphi_o = \frac{P_V}{P_l}$ where P_l is the limit load in bending.

Formula (9.172) implies the absence of shear force reinforcement. When this latter exists, formula (9.172) must be replaced by

$$P_V = A_V \sigma_{Y\,V} \sin\theta, \tag{9.173}$$

where A_V is the cross-sectional area of the inclined shear force reinforcing bars, $\sigma_{Y\,V}$, the yield limit of the shear force reinforcement ; and θ, the angle of the inclined bars with the midplane of the plate.

The safety with respect to P_V must be higher than with respect to P_l because punching is a localized and sudden phenomenon, very sensitive to local imperfections of the concrete and the reinforcement. Despite its limitations, formula (9.172) is one of the most widely used. It gives too large a safety factor when applied to continuous or built-in slabs.

Numerous papers have been devoted to punching of reinforced concrete plates (see, for example [9.49, 7.50]). Moe [9.51] has suggested the following formula :

$$P_V = \tau A_S = A_S \left(\sigma'_{r(cyl)}\right)^{1/2} \frac{15\left(1 - \dfrac{0.075a}{h}\right)}{1 + \dfrac{5.25 \left[A_S \left(\sigma'_{r(cyl)}\right)^{1/2} \right]}{P_l}} , \tag{9.174}$$

where a is the side of the square area of application of the load. In order to avoid punching failure and obtain the bending collapse mechanism, one must have

$$\tau \equiv \frac{P_V}{A_S} \leq \left(9.23 - 1.12\frac{a}{h}\right)\left(\sigma'_{r(cyl)}\right)^{1/2} \qquad \text{for} \qquad \frac{a}{h} \leq 3$$

$$\tau \equiv \frac{P_V}{A_S} \leq \left(2.5 + 1.12\frac{a}{h}\right)\left(\sigma'_{r(cyl)}\right)^{1/2} \qquad \text{for} \qquad \frac{a}{h} \geq 3. \tag{9.175}$$

Formulas (9.174) and (9.175), like formula (9.172), are deduced from tests on simply supported plates.

We finally note with Guerrin [9.52] that punching loads P_V are always very large, especially when the slab is continuous or built-in, even when no special precautions have been taken, and even for very thin plates (4 cm <2 in.). Hence, it is very seldom, except in tests, that multiplication of the actual service loads by the safety factor will result in a load larger than P_V.

9.8. Example of application.

We want to design the slab shown in fig. 9.80 [9.13]. It is supported on the boundary ABCDE and free along EF and FA. In the corner AFE a stair case will be built. Along the edges ABCD the plate is continuous and, hence, can exhibit negative resisting moments. The slab is loaded :

1. On AF by the staircase applying a dead weight of 200 dN/m and a service load of 500 dN/m. With a load factor of s =2, we obtain, at collapse, 1200 dN/m on AF ;

2. On the double line representing a wall, by a line load of 1000 dN/m at collapse ;

3. On its whole area by a uniformly distributed load due to (a) a dead load evaluated at 200 dN/m^2 (approximately 8 cm. thickness) ; (b) a live load of 300 dN/m^2 Etimes s $= 300 \times 2 = 600$ dN/m^2.

The total load thus is 800 dN/m^2.

Solution 1 : isotropic uniform upper and lower reinforcement ($M_p = M'_p$). We begin with the mechanism shown in fig. 9.80. We shall avoid the differentiation process based on the kinematic theorem by considering the equilibrium of rotation of each portion of the slab about its axis of rotation (see Section 9.4.10 dealing with nodal forces). Note that, at the intersection of the yield line from corner B with the side AF, a twisting moment (see Section 9.4.6.4), $M_{xy} = -M_p \cot \alpha$, must be applied by the staircase to the slab. It will be replaced by two forces $V = M_p \cot \alpha$, one on each side of this intersection and a unit length apart. Their directions are shown in fig. 9.80 ; downwards for the force applied to part 3, upwards for that applied to part 2.

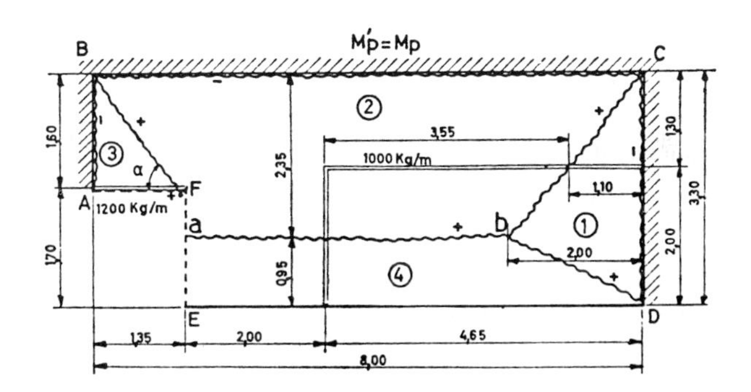

Fig. 9.80.

Equilibrium of rotation gives, for part 1

$$3.3\,(M_p + M_p) = 800 \times 3.3 + \frac{(2.0)^2}{6} + 1000\,\frac{(1.10)^2}{2}\,.$$

Hence $M_p = 360$dN. For part 2

$$8.0\,(M_p + M_p) = 800 \times 1.35 \times \frac{(1.6)^2}{6} + 800\,(6.65 - 2.0)\,\frac{(2.35)^2}{2}$$

$$+ 800 \times 2.0\,\frac{(2.35)^2}{6} + 1000 \times 3055 \times 1.3 + 1000\,(2.35 - 1.30)$$

$$\left(1.3 + \frac{2.35 - 1.30}{2}\right) - \frac{1.35}{1.60} M_p \times 1.60.$$

Hence $M_p = 1080$dN. For part 3,

$$1.6\,(M_p + M_p) = 800\,.\,1.6.\,\frac{(1.35)^2}{6} + 1200\,\frac{(1.35)^2}{2} + \frac{1.35}{1.6} M_p \times 1.35.$$

Hence $M_p = 720$dN. For part 4,

$$(8.00 + 1.35)\,M_p = 800\,(6.65 - 2.0)\,\frac{(3.3 - 2.35)^2}{2}$$

$$+ 800 \times 2.0\,\frac{(3.3 - 2.35)^2}{6} + 1000\,\frac{(3.3 - 2.35)^2}{2}\,.$$

Hence, $M_p = 360$dN.

The very different values of M_p reveal that the mechanism is not the right one. To know how it should be modified, we compute an average value of M_p from the work equation :

$$\sum M_i\,\theta_i = \int pw\,dx\,dy.$$

With a unit displacement for segment ab, we obtain

$$\sum M_i\,\theta_i = \frac{19.8}{6} M_p + \frac{48}{7.05} M_p + \frac{9.6}{5.04} M_p + \frac{19.95}{1.95} M_p = 19.73\,M_p\;,$$

$$\int pw\,dx\,dy = \frac{2365}{2} + \frac{18730}{2.35} + \frac{1480}{1.98} + \frac{2370}{0.95} = 12395.$$

Whence M_p =628dN. Consequently, parts 1 and 4 should be increased to furnish larger M_p, whereas parts 2 and 3 should be decreased. We shall, however, not modify part 3 that exhibits a plastic moment not very different from 680 dN, and shall use the mechanism shown in fig. 9.81.

The equilibrium of rotation now gives for part 1,

$$3.3\ (M_p + M_p) = 800 \times 3.3 \times \frac{(2.65)^2}{6} + 1000 \times \frac{(1.8)^2}{2}.$$

Hence, M_p = 714dN. For part 2,

$$8.0\ (M_p + M_p) = 800 \times 1.35 \times \frac{(1.6)^2}{6} + 800\ (8.0 - 1.35 - 2.65) \times \frac{(1.9)^2}{2}$$

$$+ 800 \times 2.65\ \frac{(1.9)^2}{6} + 1000 \times 2.85 \times 1.3 + 1000\ (1.9 - 1.3)$$

$$\left(1.3 + \frac{1.9 - 1.3}{2}\right) - \frac{1.35}{1.6} M_p \times 1.6.$$

Hence, M_p = 702dN. For part 3, (unmodified), M_p = 720dN. For part 4,

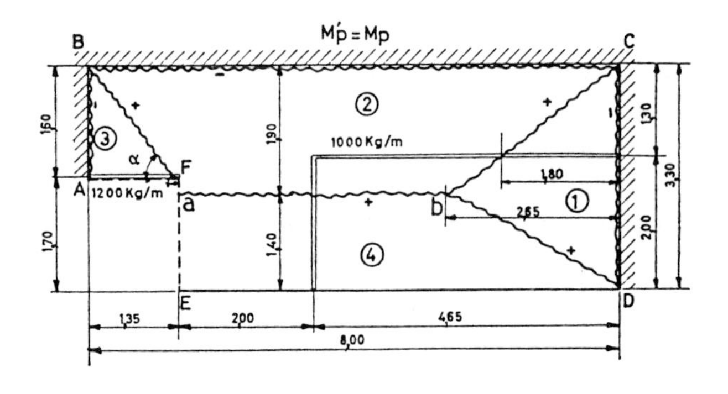

Fig. 9.81.

$$6.65M_p = 800\,(6.65 - 2.65)\,\frac{(1.4)^2}{2} + 800 \times 2.65\frac{(1.4)^2}{6} + 1000\,\frac{(1.4)^2}{2} \; .$$

Hence, $M_p = 724\text{dN}$.

The work equation of the mechanism is

$$\left(\frac{6.6}{2.65} + \frac{7.6}{1.9} + \frac{3.2}{1.6} + \frac{6.65}{1.4}\right) M_p = \frac{4710}{2.65} + \frac{12177}{1.9} + \frac{1482}{1.6} + \frac{4810}{1.4} \; .$$

Hence $M_p = 709\text{dN}$. The variation values of M_p are sufficiently close to accept the mechanism of fig. 9.81 as the best of its family, and take the plastic moment $M_p = 709\text{dN}$.

Assuming that the right family was chosen (a fact that could not be verified with certainty without using the static theorem) it remains to determine the reinforcement. With a total thickness $h_t = 8$ cm., the effective thickness is $h = 6.5$ cm.. Mild steel reinforcing bars have $\sigma_Y = 24\text{kN/cm}^2$ and we shall use a conrete with $\sigma'_r = 240\text{dN/cm}^2$. The lever of internal forces can be approximated to 0.95 h [7.16] and we can write

$$M_p = A_S\, \sigma_Y\, 0.95h. \tag{a}$$

Substituting 709 dN for M_p and the values above for σ_Y and h in relation (a), we obtain

$$A_S = 0.0479\text{cm}^2/\text{cm} = 4.79\ \text{cm}^2/\text{cm} \; .$$

We shall use seven bars of 10 mm diameter per meter. The reinforcement ratio $\frac{A_S}{h}$ is thus 0.84%.

The bars are palced in two orthogonal upper layers and two orthogonal identical lower layers. Their total length is, anchorages excluded,

$$28\ A_{slab} = 675\ \text{m} \; .$$

Solution 2 : exclusively lower isotropic reinforcement ($M'_p = 0$). The work equation of the mechnaism shown in fig. 9.81 now gives, with $M'_p = 0$,

$$\left(\frac{3.3}{2.65}+\frac{8}{1.9}+\frac{1.6}{1.6}+\frac{6.65}{1.4}\right)M_p = 12546 \ ,$$

that is, $M_p = \frac{12546}{11.22} = 1120$dN. Hence, we slightly modify the mechanism to increase the value of M_p in part 4 where M'_p did not enter. With the mechanism of fig. 9.82, we obtain, for part 1,

$$3.3M_p = 800 \times 3.3\,\frac{(2.2)^2}{6} + 1000\,\frac{(1.73)^2}{2}$$

and, hence, $M_p = 1100$dN. For part 2,

$$8.0M_p = 800 \times 1.1\,\frac{(1.6)^2}{6} + 800 \times 0.25\,\frac{(1.6)^2}{2} + 1200 \times 0.25 \times 1.6$$

$$+\, 800\,(8.0 - 1.35 - 2.2)\,\frac{(1.65)^2}{2} + 800 \times 2.2\,\frac{(1.65)^2}{6} + 1000 \times 2.92 \times 1.3$$

$$+\, 1000\,(1.65 - 1.3)\left(1.3 + \frac{1.65 - 1.3}{2}\right) - \frac{1.1}{1.6}M_p \times 1.6 \ .$$

Hence, $9.1\,M_p = 10834$ or $M_p = 1192$dN . For part 3,

$$1.6M_p = 800 \times 1.6\,\frac{(1.1)^2}{6} + 1200\,\frac{(1.1)^2}{2} + \frac{1.1}{1.6}M_p \times 1.1.$$

Hence, $M_p = 1165$dN. For part 4,

$$6.65M_p = 800 \times 4.45\,\frac{(1.65)^2}{2} + 800 \times 2.2\,\frac{(1.65)^2}{6} + 1000\,\frac{(1.65)^2}{2} \ .$$

Hence, $M_p = 1060$dN.

The four values of M_p do not differ appreciably from that given by the work equation. On that basis, we accept the mechanism of fig. 9.82 with $M_p = 1190$dN. We then obtain,

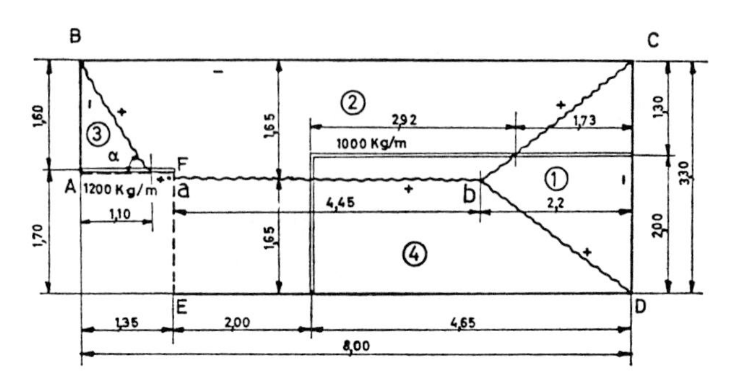

Fig. 9.82.

from relation (a), $A_S = 8.05\ cm^2/m$. We place two orthogonal lower layers of eleven bars of 10 mm. diameter per meter. Their total length is 22 $A_{slab} < 28\ A_{slab}$.

We see that an economy of 25% on the volume of reinforcement is achieved when the slab is treated as simply supported along ABCD. Continuity (that is, $M'_p \neq 0$) proves economic only if upper bars can be placed exclusively where negative moments exist, information that is given by a static approach (or by empirical rules, if reliable). Obviously, with simple supports ($M'_p = 0$) one must be prepared to accept cracks at the upper face in the vicinity of the supports.

9.9. Problems.

9.9.1. On the basis of the example treated in Section 7.9, show that built-in edged "push off" the yield lines, whereas simple supports or free edges "attract" them.

Hint : study the influence of displacing the yield lines on the dissipation and on the work of applied loads.

9.9.2. Determine the approximate limit load p_+ of a square isotropic plate uniformly loaded, simply supported on three edges, and free along the fourth. Show that the mechanism in fig. 9.83 must be rejected

Answer : $p_+ = \dfrac{14.15\ M_p}{l^2}$.

Hint : start with the mechanism of half a rectangular plate with sides l and 2l.

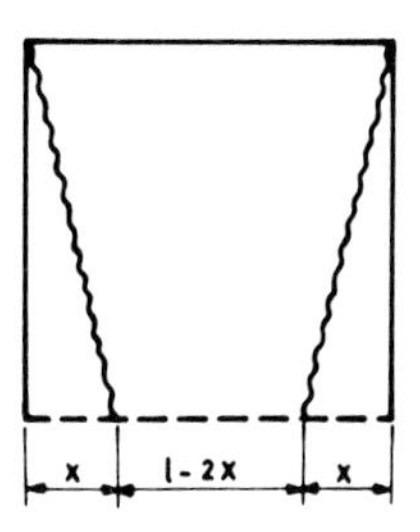

Fig. 9.83.

9.9.3. Determine theapproximative limit load p_+ of a square isotropic plate with side l, uniformly loaded, and supported by four corner columns.

Answer : $p_+ = \dfrac{8\,M_p}{l^2}$.

9.9.4. Determine the approximate limit load p_+ of the triangular, isotropic uniformly loaded plate in fig. 9.84 without accounting for the corner effect.

Obtain the right mechanism by differentiation.

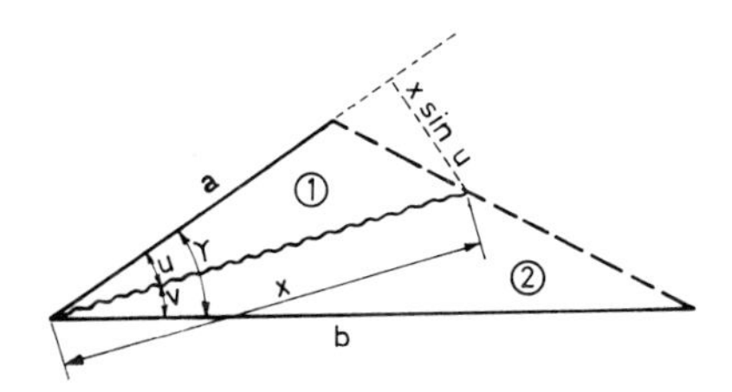

Fig. 9.84.

Answer : $(\text{total load})_+ = \dfrac{6\,M_p}{\tan\dfrac{\gamma}{2}}$.

9.9.5. A simply reinforced isotropic rectangular plate is simply supported on three edges and free along the fourth (fig. 9.85). It is subjected, at collapse, to a uniformly Q = 45 kN at each point

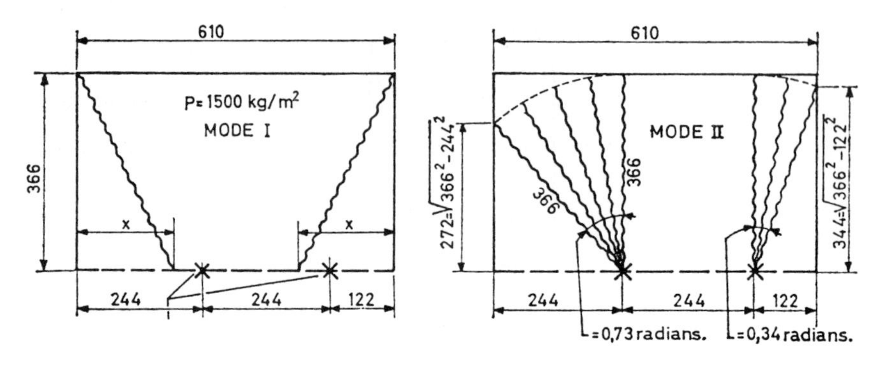

Q = 45 kN at each point lengths in cm

Fig. 9.85. (Reprinted by permission of Thames and Hudson Ltd. from R.H. Wood, Plastic and Elastic Design of Slabs and Plates, London 1961.)

distributed load $p = 156 kN/m^2$ and two concentrated loads of 45 kN acting on the free edge as shown in fig. 9.85. Determine the necessary M_p (by mechanisms).

Answer : $M_p = 43.5$ kN.

Hint : apply the superposition method, using the mechanisms shown in fig. 9.85.

9.9.6. Determine the approximate limit load P_+ of a rectangular isotropic balcony, doubly reinforced, subjected to a concentrated load P at a free corner (fig. 9.86).

$$\textit{Answer} : P_+ = 2\left(M_p \cdot M'_p\right)^{1/2} + (M_p + M'_p)\left(\frac{\pi}{2} - 2 \arctan\left(\frac{M_p}{M'_p}\right)^{1/2}\right)$$

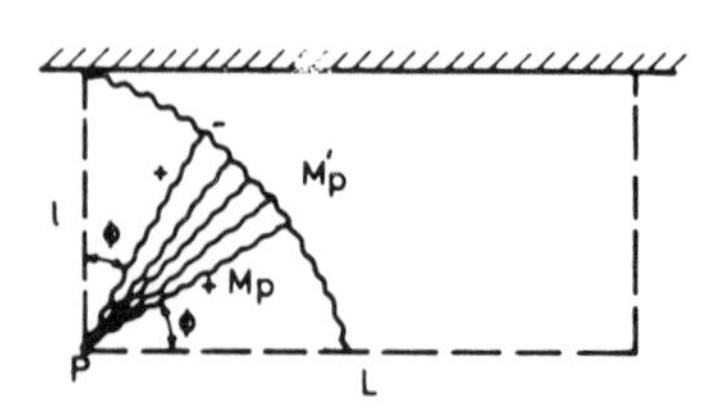

Fig. 9.86.

Hint : use the mechanismof fig. 9.86 differentiate to obtain the best mechanism ; note the influence of the plastic moment M_p for positive bending.

9.9.7. Determine the approximate limit load p_+ of a square isotropic plate, uniformly loaded, simply supported on one side, and supported by two corner columns on the opposite side (fig. 9.87).

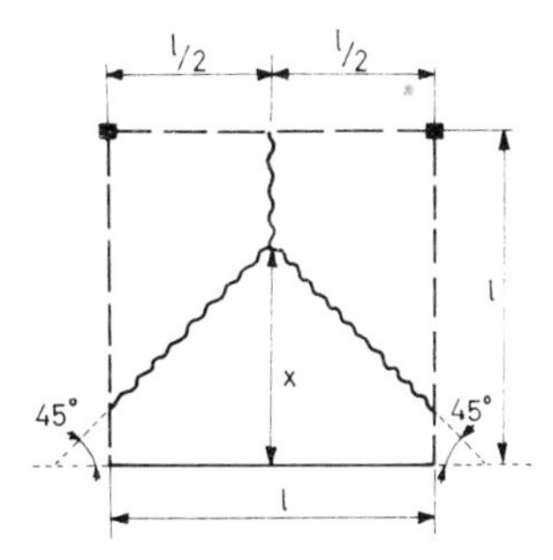

Fig. 9.87.

Answer : $p_+ = \dfrac{8.5\ M_p}{l^2}$.

Hint : use mechanism of fig. 9.87. Differentiate with respect to x.

9.9.8. Same problem as problem 9.9.7, but with a built-in edge with $M'_p = M_p$ replacing the simply supported edge.

Answer : $p_+ = \dfrac{11.2\ M_p}{l^2}$.

9.9.9. Same problem as problem 9.9.7, but two adjacent sides are simply supported and the remaining corner is column-supported. Take l=4 m.

Answer : $p_+ = \dfrac{M_p}{1.5}$.

9.9.10. Determine the approximate total limit load P_+ of a circular isotropic plate (simple reinforcement), uniformly loaded ($P = p \pi R^2$), simply supported on its edge by four columns angularly apart from $\frac{\pi}{2}$.

Answer : $P_+ = 14.1\ M_p$.

9.9.11. Obtain the formula for the dissipation in a circular fan when the reinforcement is orthotropic (fig. 9.98).

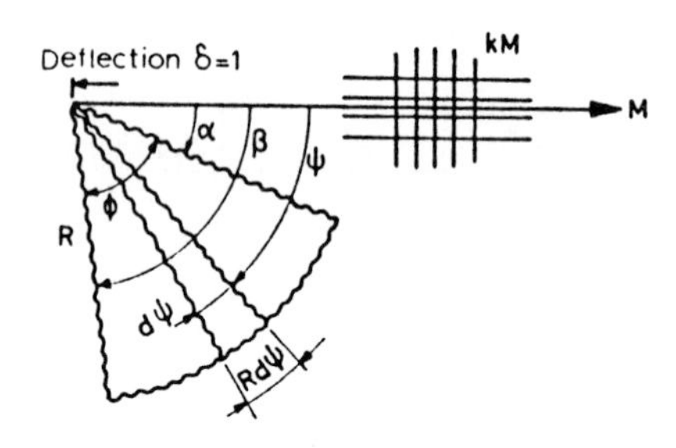

Fig. 9.88.

Answer : $D = (M_p + M'_p) \left\{ \left[\frac{(1-k)}{2}\right] \cos(\beta + \alpha) \sin \varphi + \left[\frac{(1+k)}{2}\right] \right\}$.

With the obtained formula, determine the limit load P_+ of a circular orthotropic plate subjected to a concentrated load at the centre.

Answer : $P_+ = \pi (1 + k) (M_p + M'_p)$.

9.9.12. A very long, isotropic, doubly reinforced rectangular plate is supported on its two sides and subjected to a concentrated load P at mid-span. Its plastic moments M_p and M'_p are different.

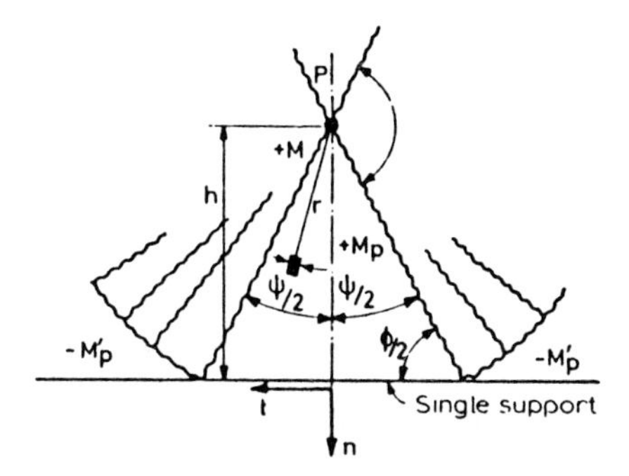

Fig. 9.89. (Reprinted by permission of Thames and Hudson Ltd. from R.H. Wood, Plastic and Elastic Design of Slabs and Plates, London 1961.)

1. Show that the angle $\frac{\varphi}{2}$, fig. 9.89, of the fans with the side must be larger than or equal to $\operatorname{arccot}\left(\frac{M'_p}{M_p}\right)^{1/2}$.

2. With the following equilibrium equations in polar coordinates, applicable to a field without polar symmetry,

$$\frac{1}{r^2}\frac{\partial}{\partial r}\left(r^2\frac{\partial M_r}{\partial r}-\frac{1}{r}\frac{\partial M_\theta}{\partial r}+\frac{1}{r^2}\frac{\partial^2 M_\theta}{\partial \theta^2}-\frac{2}{r^2}\right)\frac{\partial}{\partial r}\left(r\frac{\partial M_{r\theta}}{\partial \theta}\right)=-p,$$

$$V_r=\frac{M_r}{r}+\frac{\partial M_r}{\partial r}-\frac{M_\theta}{r}-\frac{1}{r}\frac{\partial M_{r\theta}}{\partial \theta},$$

$$V_\theta=-2\frac{M_{r\theta}}{r}-\frac{\partial M_{r\theta}}{\partial r}+\frac{1}{r}\frac{\partial M_\theta}{\partial \theta},$$

show that the moment field

$$M_\theta=+M_p,$$

$$M_r=-M_p\tan^2\theta,$$

$$M_{r\theta}=0,$$

is statically admissible (assuming the rigid part outside the fans to be strong enough).

Determine the approximate limit load P_, prescribing that φ be such to give plasticity on the edge.

Answer : $P\_ = 4\left(M_p \cdot M'_p\right)^{1/2} + 2\left(M_p + M'_p\right)\varphi.$

Compare with eq. (9.66), after having used eq. (9.67).

3. Determine the edge reactions between the fans.

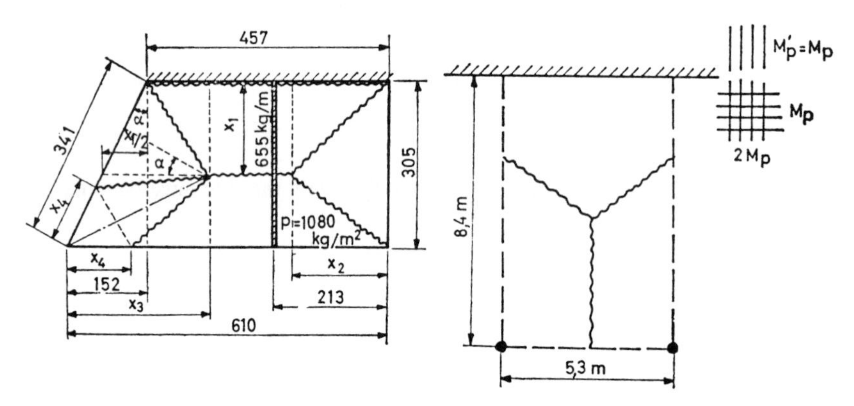

Fig. 9.90. lengths in cm. (Reprinted by permission of Thames and Hudson Ltd. from R.H. Wood, Plastic and Elastic Design of Slabs and Plates, London 1961.)

9.9.13. Determine the the isotropic plate shown in fig. 9.90, with $M_p = M'_p$.

Answer : $M_p = 5.97$ kN. *Hint* : use the mechanism shown in fig. 9.90, and obtain the values of x_1, x_2 , x_3, and x_4 by a "cut and try" process, for largest M_p.

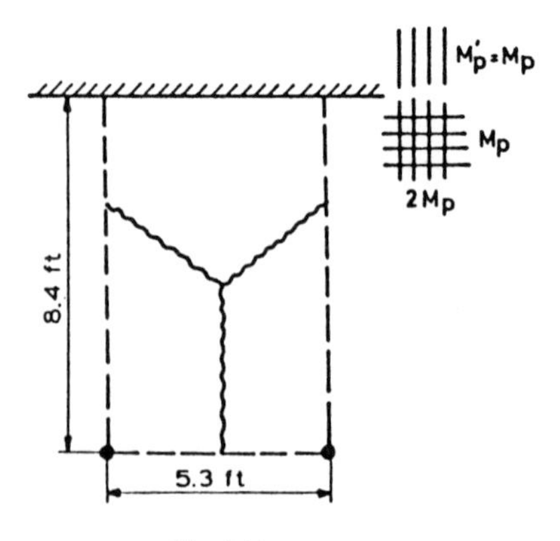

Fig. 9.91.

9.9.14. Determine the plastic moment of the orthotropic rectangular plate in fig. 9.91, built-in on one short side and column-supported on the two opposite corners. The loads are

Dead weight (h_t = 24cm.)	600dN/m^2
Covering	100 dN/m^2
Live load at collapse	1500 dN/m^2
Total p	2200 dN/m^2

Answer : M_p = 68.4kN. *Hint* : use the affinity method.

References.

[9.1] "Flexural Mechanics of Reinforced Concrete", Proc. A.S.C.E.-A.C.I. Int. Symp., Miami, Nov. 1964, pub. by A.S.C.E., 1965.

[9.2] C.E. MASSONNET, "Limit Design Applied to Steel and Concrete Structures" (lecture notes), Univ. of California at Berkeley, Civil Eng. Dept., 1965.

[9.3] K.W. JOHANSEN, "Yiled-line Theory" (translated from the Danish), Cement and concrete Association, London, 1962.

[9.4] Comité Européen du Béton : "Recommandations pratiques unifiées pour le calcul et l'exécution des ouvrages en béton armé" (European committee for Concrete : unified practical recommendations), 1964

[9.5] W. PRAGER, "The General Theory of Limit Design", Proc. 8th Int. Cong. Appl. Mech., Istanbul, 1952, **2** : 65, 1956.

[9.6] M.A. SAVE, "A Consistent Limit-Analysis Theory for Reinforced Concrete Slabs", Magazine of Concrete Research, **19** : 58, 3, March 1967, and discussion in **19** : 61, 252, December 1967.

[9.7] R. BAUS and S. TOLACCIA, "Calcul à la rupture des dalles en béton armé et étude expérimentale du critère de rupture en flexion pure", (Yield-line Theory and Experimental Investigation of the Yield Criterion of Reinforced Concrete Slabs in Pure Bending), Ann. Inst. Tech. Bat. Trav. Pub., Paris, June 1963.

[9.8] M.P. NIELSEN, "Limit Analysis of Reinforced Concrete Slabs", Acta Polytechnica Scandinavia Ci 26, Copenhagen, 1964.

[9.9] K.O. KEMP, "The Yield Criterion for Othotropically Reinforced Concrete Slabs", Int. J. Mech. Sci., **7** : 11, November 1965.

[9.10] C.T. MORLEY, "On the Yield Criterion of an Orthogonally Reinforced Concrete Slab Element", J. Mech. Phys. Solids, **14** : 1, 33, January 1966.

[9.11] C.E. MASSONNET and M.A. SAVE, "Calcul plastique des constructions (Plastic Analysis of Structures), **2**, Centre belgo-luxembourgeois d'information de l'acier, Brussels, 1963.

[9.12] R. WOLFENSBERGER, "Traglast und optimale Bemessung von Platten", Technische Forschungs- und Beratungsstelle der Schweizerischen Zement Industrie, Wildegg, 1964.

[9.13] G.A. STEINMANN, "La théorie des lignes de rupture" (Yield-line Theory), Comité Européen du Béton, Bull. d'Inf., 27, September 1960.

[9.14] R.J. LENSCHOW and M.A. SOZEN, "A Yield Criterion for Reinforced Slabs", A.C.I. Journal, May 1967. Discussion of this paper is in A.C.I. Journal, November 1967.

[9.15] P. LENKEI, "On the Yield Condition for Reinforced Concrete Slabs", Archiwum Inzynierii Ladovej, **XIII** : 1, 5, Warszawa, 1967.

[9.16] R.H. WOOD, "Plastic and Elastic Design of Slabs and Plates", Thames and Hudson, London 1961. See also, by L.L. JONES, R.H. WOOD, "Yield-line Analysis of Slabs", Thames and Hudson, Chatto and Windus, London, 1967.

[9.17] Comité Européen du Béton, (C.E.B.) (European Commitee for Concrete), Bulletins 27, 43, 45.

[9.18] A.R. RJANITSYN, "Calcul à la rupture et plasticité des constructions" (Limit Analysis and Plasticity of Structures), Eyrolles, Paris, 1959.

[9.19] M. JANAS, "Kinematical Compatibility Problems in Yield-line Theory", Mag. Conc. Research, **19** : 58, 33, March 1967.

[9.20] E.H. MANSFIELD, "Studies in Collapse Analysis of Rigid-plastic Plates with a Square Yield Diagram", Proc. Roy. Soc., Series A, **241** : 311, 1975.

[9.21] A. SAWCZUK, "Grenztragfähigkeit der Platten", Bauplanung-Bautechnik, **11** : 7, 8, 1957.

[9.22] M.P. NIELSEN, "On the calculation of Yield-line Patterns with Curved Yield Lines", R.I.L.E.M. Bull. 19, 67, June 1963

[9.23] L.L. JONES, "Recent British Advances in Yield-line Analysis by the Equilibrium Method", in "Flexural Mechanics of Reinforced Concrete", Proc. Int. Symp., Miami, November 1964, A.S.C.E., 1965.

[9.24] R.H. WOOD, "Plastic Design of Slabs using Equilibrium Methods", in "Flexural Mechanics of Reinforced Concrete", Proc. Int. Symp., Miami, November 1964, A.S.C.E., 1965.

[9.25] K.O. KEMP, "The evaluation of Nodal and Edge Forces in Yield-line Theory", Mag. Conc. Research Special Pub., London, May 1965.

[9.26] C.T. MORLEY, "Equilibrium Methods for Least Upper Bounds of Rigid-plastic Plates", Mag. Conc. Research Special Pub., London, May 1965.

[9.27] R.H. WOOD, "New Techniques in Nodal-force Theory for Slabs", Mag. Conc. Research Special Pub., London, May 1965.

[9.28] L.L. JONES, "The Use of Nodal forces in Yield-line Analysis", Mag. Conc. Research Special Pub., London, May 1965.

[9.29] L.L. JONES, "Ultimate load Analysis of Reinforced and Prestressed Concrete Structures", Chatto and Windus, London, 1962.

[9.30] M. HOLMES and K.A. STEEL, "Upper and Lower Bound Solutions to the Collapse of a Continuous Slab under Uniform Load", Mag. Conc. Research, **16** : 47, 83, June 1964.

[9.31] Z. SOBOTKA, "La limite supérieure et inférieure de la capacité portante des dalles rectangulaires encastrées" (Upper and Lower Bounds to the Collapse Load of Rectangular Built-in Slabs), Comité Européen du Béton, Report of the 10th

General Meeting, London, October 1965.

[9.32] Z. SOBOTKA, "La capacité portante plastique des dalles rectangulaires encastrées avec la charge uniforme" (Plastic Carrying Capacity of Rectangular Plates with Uniform Load), Acta Technica Csaw, **6**, 676, 1966.

[9.33] A. SAWCZUK, "Grenztragäfhigkeit der Platten", Bauplanung-Bautechnik, **11** : 7, 315, July 1957 ; 8, 359, August 1957.

[9.34] K.O. KEMP, "A Lower-Bound Solution to the Collapse of an Orthotropically Reinforced Slab on Simple Supports", Mag. Conc. Research, **14** : 41, 79, July 1962.

[9.35] G. CERADINI and C. GAVARINI, "Calcolo a rottura e programmazione lineare", Giornale del Genio Civile, Jan.-Feb., 1965.

[9.36] C. GAVARINI, "I theoremi fondamentali del calcolo a rottura e la dualita in programmazione lineare", Ingegneria Civile, **18**, 1966.

[9.37] G. SACCHI, "Contribution à l'analyse limite des plaques minces en béton armé" (thèse), Fac. Polytechnique de Mons, Mons, Belgium, 1966.

[9.38] C.E. MASSONNET, "Complete Solutions Describing the Limit State of Reinforced Concrete Slabs", Mag. Conc. Research, **19** : 58, 13, March 1967.

[9.39] J. ZAWIDZKI and A. SAWCZUK, "Plastic Analysis of Fiber-Reinforced Plates under Rotationally Symmetrie Conditions", Int. J. Solids and Struc., **3** : 3, 413, May 1967.

[9.40] A. HILLERBORG, "Strimlemethoden för Platter pa Pelare", Vinkelplattor M.M., Svenska Riksbyggen, Stockholm, 1959.

[9.41] R.E. CRAWFORD, "Limit Design of Reinforced Concrete Slabs", Proc. A.S.C.E., J. Eng. Mech. Div., **90** : EM5, 321, October 1964.

[9.42] C.T. MORLEY, "The minimum Reinforcement of Concrete Slabs", Int. J. Mech. Sci., **8** : 4, 305, April 1966.

[9.43] M. JANAS, "Large Plastic Deflections of Reinforced Concrete Slabs", Int. J.

Solids and Struct., **3** : 4, November 1967.

[9.44] R. PARK, "Ultimate Strength of Rectangular Concrete Slabs under Short-term Uniform Loading with Edge Restrained Against Lateral Movement", Proc. Inst. Civ. Eng., **28** : June 1964.

[9.45] "Effort tranchant", Colloquium de Wiesbaden, 1963. Comité Européen du Béton, Bull. d'Inf. 40, 41, 42, Paris, 1964 ("Shear strength", European Commitee for Concrete).

[9.46] Y. GUYON, "Béton précontraint" (Prestressed Concrete), vol. 2, Collection de l'ITBTP, Eyrolles, Paris, 1958.

[9.47] M.J. MULLER "Quelques aspects du comportement des dalles et des poutres précontraintes en phase élastique et à la rupture" (Some aspects of the behaviour of prestressed slabs and beams in elastic and ultimate range), Groupement belge de la précontrainte, publication A.B.E.M. 15, Brussels, March 1959.

[9.48] R.C. ELSTNER, E. HOGNESTAD, "Shearing of reinforced concrete slabs", J. Amer, Concrete Inst., July 1956.

[9.49] C. FORSELL, A. HOLMBERG, "Stampellast pa plattor av betong" (Concentrated loads on concrete slabs), Betong 31 : 2, Stockholm, 1946.

[9.50] G.D. BASE, "Some tests on the punching shear strenth of reinforced concrete slabs, Technical Report", Cement and Concrete Association, July 1959.

[9.51] J. MOE, "Shearing Strength of reinforced concrete slabs and footings under concentrated loads", Portland Cement Association, Research and Development Laboratories, Stokies, Ill., U.S.A., April 1961.

[9.52] A. GUERRIN, Traité de béton armé, t. IV, Dunod, Paris, 1960.

[9.53] R. PARK, W.L. GAMBLE, "Reinforced Concrete Slabs", J. Willey, New York, 1980.

[9.54] E.H. ROBERTS, "Load carrying capacity of slab strips restrained against longitudinal expansion", Concrete 3, 369, 1969.

[9.55] M.W. BAESTRUP, C.T. MORLEY, "Dome effect in reinforced concrete slabs : elastic-plastic analysis", J. Structural div. ASCE, 106, ST6 1255, June 1980.

[9.56] "Cracking delfection and ultimate load of concrete slab systems", ACI special publication, n 30, American Concrete Institute, Detroit, 1971.

[9.57] A.A. GVOZDEV, S.M. KRYLOV, "Recherches expérimentales sur les dalles et plancher-dalles effectuées en Union Soviétique", Comité Européen du Béton, Bull. d'Inf., n 50, 174, Paris, 1965.

[9.58] A.J. OCKLESTON, "Load tests on three-storey reinforced concrete building in Johanesburg", The Structural Engineer, 33, 304, 1955.

[9.59] A.A. GVOZDEV, "0n the basis of the p. 33 of the design code for reinforced concrete", (in Russian), Stroitelnaya Promyshlennost, 17, 1939.

[9.60] M. JANAS, "Arching action in elastic-plates", J. Struct. Mechanics, **1**, 3, 277, 1973.

[9.61] C.T. MORLEY, "Yield line theory of reinforced concrete slabs at moderately large deflections", Magazine of concrete research, **19**, 60, 211, September 1967.

[9.62] C.R. CALLADINE, "Simple ideas in the large-deflection plastic theroy of plates and slabs", Engineering Plasticity, Heyman and Leckie eds., Cambridge University Press, 93, Cambridge, 1968.

[9.63] M. JANAS, A. SAWCZUK, "Influence of position of lateral restraints on carrying capacities of plates", Comité Européen du Béton, Bull. d'Inf., n° 58, 166, Paris, 1966.

[9.64] E.L. Mc DOWELL, K.E. Mc KEE, E. SEVIN, "Arching action theroy of masonry walls", J. Structural Division, ASCE, **82**, ST2, 915-1, 1956.

[9.65] M.W. BAESTRUP, "Dome effect in reinforced concrete slabs : rigid-plastic analysis", J. Structural Division, ASCE, **106**, ST6, 1237, June 1980.

10. Reinforced Concrete Shells.

10.1. Introduction.

Reinforced concrete shell structures, and particularly reinforced concrete shell roofs, have been increasingly used in the last decades, and their development is continuing. There are numerous publications on the subject ; a large number of references can be found in the books [5.1] and [8.12]. Despite this wide use of reinforced concrete shells, their limit analysis is still in a relatively early stage of development [10.1]. In particular, no direct experimental support has been given of the yield surfaces used. Nevertheless, investigations were performed to establish the yield criterion, first by Sawczuk and Olszak for cylindrical shells [10.2], next by Capurso [9.24] and one of the authors [10.25, 10.26] for a more general condition, stated in Section 5.6.

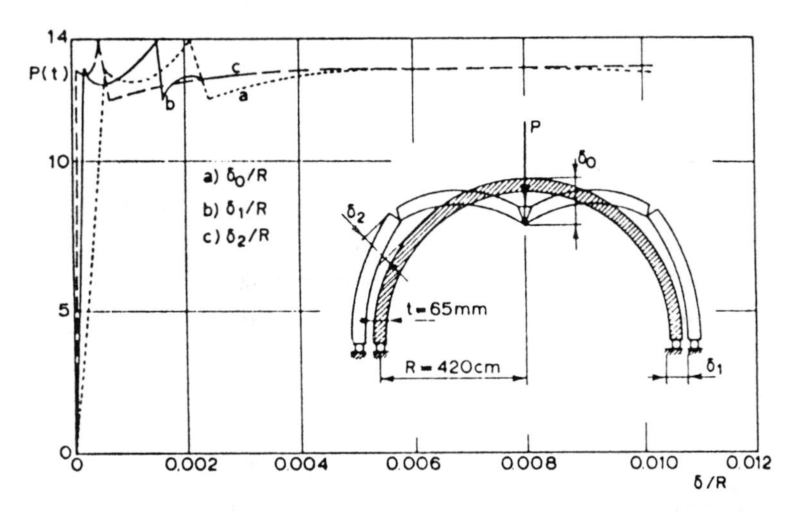

Fig. 10.1. Load-deflection relations for a reinforced concrete dome under concentrated load [10.18].

Numerous tests to collapse, though of high engineering interest [10.5 to 10.9], have been conducted to investigate some peculiar points of the behaviour of some reinforced concrete shells and, hence, cannot give much information on the applicability of limit analysis, though they have influenced the development of approximate methods to evaluate the collapse load [10.10 to 10.12]. These methods are approximate either because they introduce assumptions or concepts that are of dubious value from the point of view of limit analysis [10.10], or because they are purely kinematical and, hence, furnish only an upper bound [10.12 and 10.13].

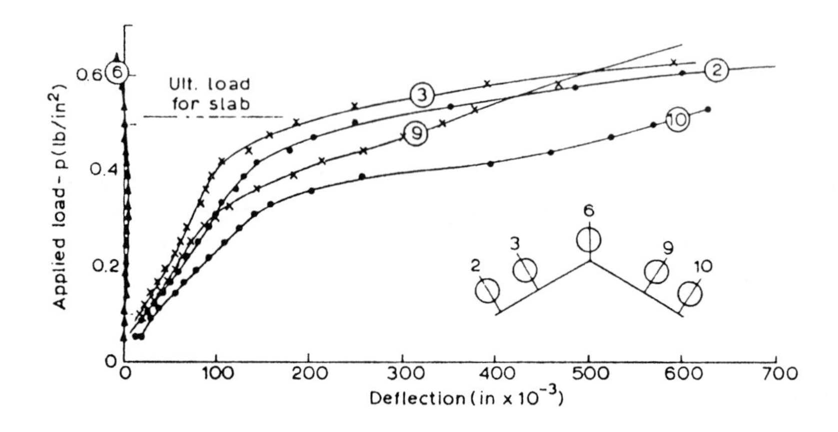

Fig. 10. 2. Model structure 203 (C1) 3 free edges. Deflections of midspan section (collapse test) [10.15].

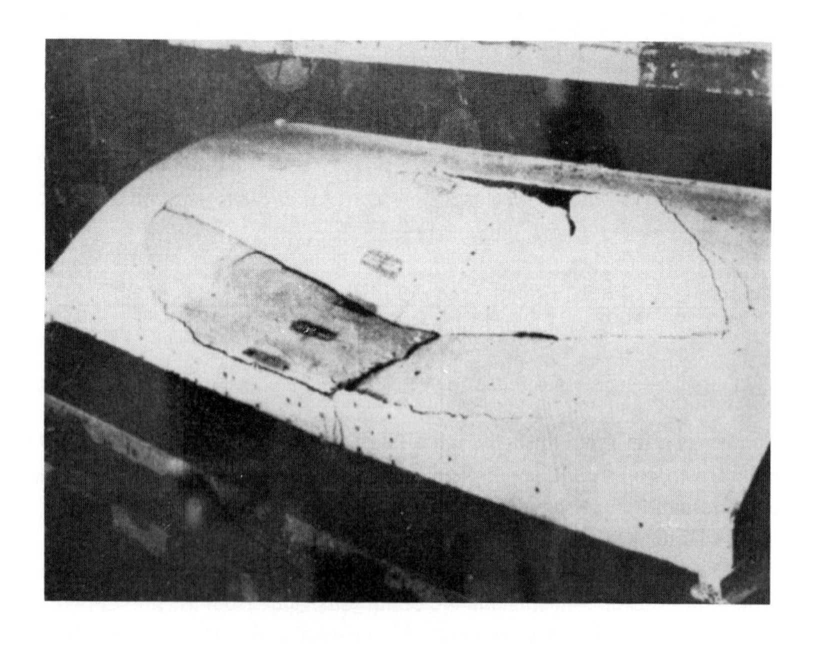

Fig. 10.3. Cylindrical panel after collapse [10.9].

Available experimental data strongly indicates that collapse mechanisms and limit loads (in the sense of limit analysis) do exist, as can be seen in fig. 10.1 [10.14] and fig. 10.2 [10.15]. For the folded structures studied in ref. [10.15], a safe estimate of the collapse load can be obtained by applying the yield line method to the plate. Experiments on cylindrical shell roofs also show [10.5] that plastic deformations are concentrated in "generalized yield lines" (see fig. 10.3) which will be used in Section 10.3. Because obtaining the right collapse pattern is often very difficult (at least as much as in plate problems) a sound trend is to combine experimental and analytical techniques, as strongly advocated by Sawczuk [10.4], Olszak and Sawczuk [8.63], and Haidukov [10.16]. From experimental collapse patterns, and with a sufficiently accurate yield surface, limit loads are obtained by the work equation of the kinematic theorem.

In thin shells, instability, and particularly creep-buckling, may be the relevant cause of collapse. Hence, taking account of the small amount of experimental support, limit analysis should be regarded as *one* theory to evaluate the safety with respect to a particular collapse mode, and should be used with adequate caution.

10.2. Classification of plastic yielding regimes.

The problem of the limit analysis of reinforced concrete shells based on yielding conditions stated in Section 5.6 is, at first sight, very intricate. Nevertheless, for a lot of practical cases, additional approximations are often allowed with regard to the shell geometry. Then, we need to establish a classification of the possible limit states for the yield-lines, founded on two parameters. Parameters $\dot{\theta}$ and $\dot{u}_o$ are those of Section 5.6.

a) *bending ratio* r_c. This parameter is connected directly to the regimes of plastic yielding defined in the previous Sections. Let $\dot{\varepsilon}_e$ be the maximum value of the strain rate $\dot{\varepsilon}_n$,and $\dot{\varepsilon}_m$ the value of $\dot{\varepsilon}_n$ in the mean layer. We define the bending ratio as the ratio of the bending strain rate to the membrane strain rate :

$$r_c = \frac{\dot{\varepsilon}_e - \dot{\varepsilon}_m}{\dot{\varepsilon}_m}$$

or, account taken of eq. (5.108),

$$r_c = \frac{t}{2}\left|\frac{\dot{\theta}}{\dot{u}_o}\right| = 2\left|\frac{\dot{\theta}}{\dfrac{\dot{u}_o}{}}\right| \tag{10.1}$$

Because of the normality law, it is obvious that the bending state predominates in the vicinity of the tops B and D of the parabolas and the membrane state in the vicinity of the vertices A and C (fig. 5.39 and 5.40). More precisely, it is easy to show that the cone of strain rate, at A and C, is such that :

$$r_c \leq 1 \tag{10.2}$$

b) *wavelength parameter* r_λ. It is necessary to introduce the wavelength λ of the membrane strain rate $\dot{u}_o$ such that the ratio $\frac{|\dot{u}_o|}{\lambda}$ is of the same order that the variation of $\dot{u}_o$, along the yield line :

$$\left|\frac{\partial \dot{u}_o}{\partial s}\right| \approx \frac{|\dot{u}_o|}{\lambda} . \tag{10.3}$$

In Chapter 8, we have seen that the choice of the regimes of plastic yielding, for the limit states of steel shells, depends on this wavelength. We assume that this is also valid for the reinforced concrete shells. Now, let R be the minimum curvature radius and t the thickness in the considered part of the shell. The wavelength parameter is defined by :

$$r_\lambda = \frac{\lambda^2}{tR}. \tag{10.4}$$

So, we obtain the following classification for the reinforced concrete shells as given in table 10.1.

It can be observed that the Johansen's yield-line theory for slabs is a particular case of the general theory, associated to :

$$r_c = \frac{1}{1 - 2\gamma}, \qquad r_\lambda < 1$$

Indeed, in this model, we suppose that the axial forces vanish.

Compartment A3 of table 10.1 corresponds to the limit state of slabs, computed by the flexural theory of Wood and Janas (Section 9.6). This theory overestimates the actual limit load, even when Johansen's simplified theory underestimates it.

Column B involves shallow shells. The theory was stated by Rzhanitsin [10.12, 10.13] for the extreme cases of membrane state (B1) and bending state (B3).

Table 10.1.

	(A) $r_\lambda \leq 1$	(B) $r_\lambda \approx 1$	(C) $r_\lambda > 1$
$r_c \leq 1$	(A1) plane stress state - asymptotic membrane state of the slabs (Wood, Janas) (Sect. 9.6)	(B1) membrane theory of shallow shells (Rzhanitsin) [10.12]	(C1) membrane state of shells
$r_c > 1$ ($r_c \approx 1$)	(A2) membrane-slabs - Johansen's theory (Chp. 9) -Post-yield state (Wood-Janas) (Sect. 9.6)	(B2) general theory of shallow shells	(C2) general theory of shells
$r_c \gg 1$	(A3) -bending theory of slabs of Wood - Janas (Sect. 9.6)	(B3) bending theory of shallow shells (Rzhanitsin) [10.3]	(C3) bending state of shells

Let us consider now column C. Either, the membrane state predominates (C1), then we restrict ourselves to vertices A and C (fig. 5.39 and 5.40). Or, the bending state predominates (C3). If the membrane and bending states are of the same order of magnitude (C2), we must work on the whole yield surface. Thus, the previous table permits to indicate the relative place of each particular theory.

Note that different regimes may occur in different yield-lines, and even on different parts of a yield-line.

10.3. Upper bound solutions.

10.3.1. *Circular cylindrical tank.*

Consider a vertical cylindrical tank subjected to the hydrostatic internal pressure

$$p(x) = p_o \frac{x}{l}. \tag{10.5}$$

The origin x=0 is at the top cross section which is free, whereas the bottom end is built-in (fig. 10.4). Because of the symmetry of the problem, the collapse involves one or possibly several circumferential yield-lines and a "fan" of meridional yield-lines. Following Sawczuk and Olszak [10.2], three types of collapse may occur : complete collapse of short tank, complete collapse with compressed ring, partial collapse of long tanks. We examine only the first type.

First, we determine the plastic regime of yield-lines. Let $d\theta$ be the angle between two yield-lines of the fan. By geometrical considerations, we obtain the extension and the rotation rates of a meridional yield-line

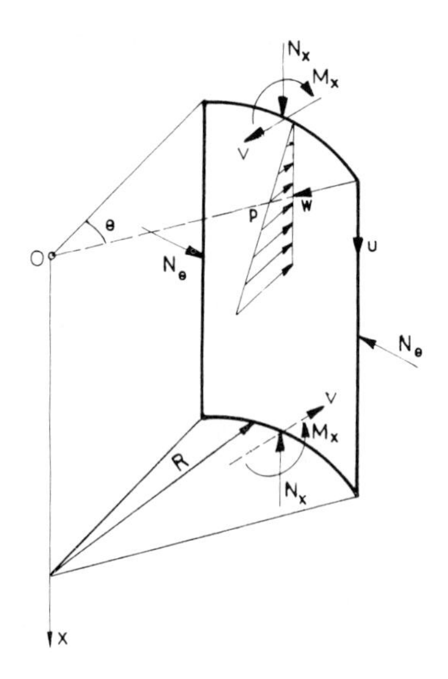

Fig. 10.4.

$$d\dot{u}_o = -\dot{w}\, d\theta \qquad\qquad d\dot{\theta} = -\frac{\dot{w}}{R}\, d\theta. \qquad (10.6)$$

The bending ratio (10.4) is equal to

$$r_c = \frac{t}{2R} \qquad (10.7)$$

which is a negligible quantity. Thus, the yield-lines are in membrane state (vertices A and C of fig. 5.39 and 5.40).

From the normality law, it is easily seen that the generalized strain rate

$$\dot{\kappa}_x = -\frac{d^2\dot{w}}{dx^2}, \qquad (10.8)$$

corresponding to M_x, vanishes. The regimes A and C furnish a conical mechanism :

$$\dot{w}(x) = -\dot{w}_o \left(1 - \frac{x}{l}\right), \qquad \dot{w}_o > 0 \tag{10.9}$$

with circumferential yield-line at the bottom x=l, which is in bending state :

$$\dot{u}_o = 0 \qquad \dot{\theta} = \frac{\dot{w}_o}{l} \;.. \tag{10.10}$$

We assume that the circumferential reinforcement is such that

$$N_p(x) = N_p \cdot \frac{x}{l} \;. \tag{10.11}$$

The rate of dissipation is provided by extension in the meridional yield-lines (*) and rotation of the "hinge" at the bottom :

$$D = -2\pi \int_o^l N_p(x)\,\dot{w}\,dx + 2\pi R M_p \frac{\dot{w}_o}{l} \;.$$

Owing to (10.9) and (10.11), we have :

$$D = 2\pi \frac{N_p\, l\, \dot{w}_o}{3} + 2\pi R M_p \frac{\dot{w}_o}{l} \;.$$

In the other hand, the power of the external load is given by

$$W = -2\pi R \int_o^l p(x)\,\dot{w}\,dx \;.$$

Finally, equating D and W, we find an upper bound of the maximal pressure :

(*) Note that extension in the meridional yield-lines has circumferential direction.

$$p_{0+} = \frac{6\,M_p}{l^2} + \frac{N_p}{R}\,. \tag{10.12}$$

For the other collapse mechanisms, the reader is referred to ref. [10.25].

10.3.2. *Conical shell.*

A truncated conical shell of the type used for foundations of high towers has been studied by S. Kaliszky [10.18]. We consider a conical mechanism bounded by a circumferential hinge circle (fig. 10.5) rotating with the rate $\dot{\psi}$. At the distance s of the hinge, the extension and rotation rates of yield lines, separated by angle d θ, are respectively :

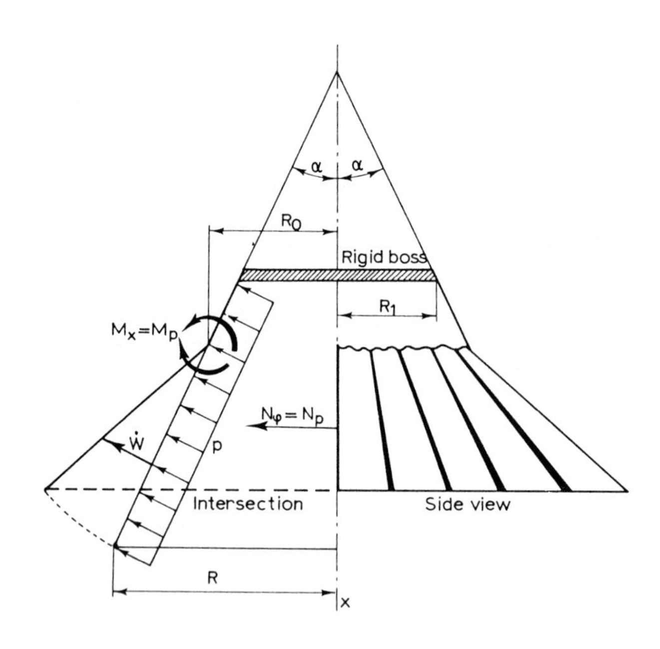

Fig. 10.5. Collapse mode for a trucated cone [10.18].

$$d\,\dot{u}_0 = \dot{\psi}\ s\cos\alpha\, d\,\theta\,, \qquad d\,\dot{\theta} = \dot{\psi}\ \sin\alpha\, d\,\theta\,. \tag{10.13}$$

Thus, the bending ratio (10.1) is negligible :

$$r_c = \frac{e}{2\,s}\,\mathrm{tg}\,\alpha \tag{10.14}$$

except near the hinge.

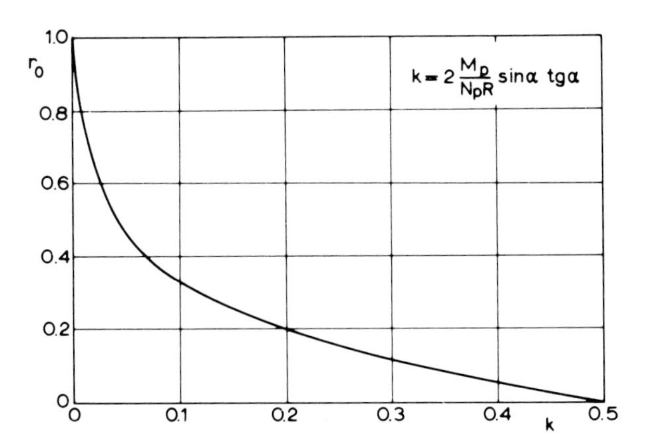

Fig. 10.6. Position of the circumferential hinge in a conical shell [10.18].

The work equation then gives the limit load (see fig. 10.2 for notations) :

$$p = \frac{3}{R^2}\,\frac{N_p\,R\cos\alpha\,(1-r_0)^2 + 2\,M_p\,r_0\sin^2\alpha}{r_0^3 - 3\,r_0 + 2}\,, \tag{10.15}$$

where $r_0 = \frac{R_0}{R}$. The parameter R_0 is obtained from the minimum condition $\frac{dp}{dR_0} = 0$, according to the kinematic theorem. It is given in fig. 10.6 as a function of the shell parameter k.

10.3.3. *Spherical cap.*

In other rotationally symmetric shells, the collapse mechanism consists of one or possibly several circumferential yield-lines, and of a "fan" of meridional yield-lines. In these last lines, it may be easily shown that the bending ratio r_c is not governing.

Indeed, let $\dot{v}$ and $\dot{w}$ be the tangential and normal displacement rates, in the meridian plane (fig. 10.7). The circumferential extension rate in the meridional yield-lines is given by :

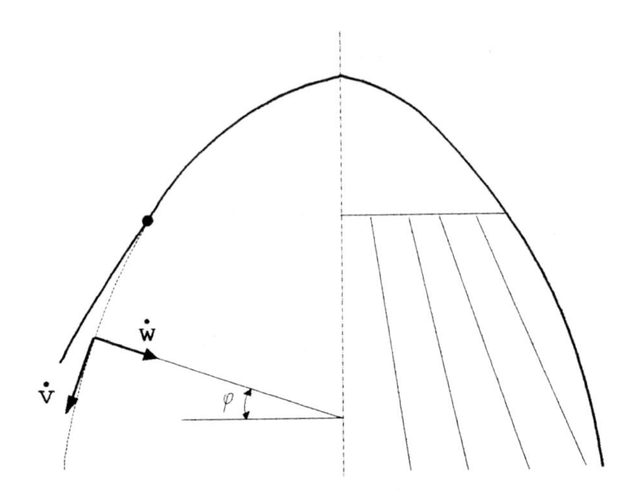

Fig. 10.7.

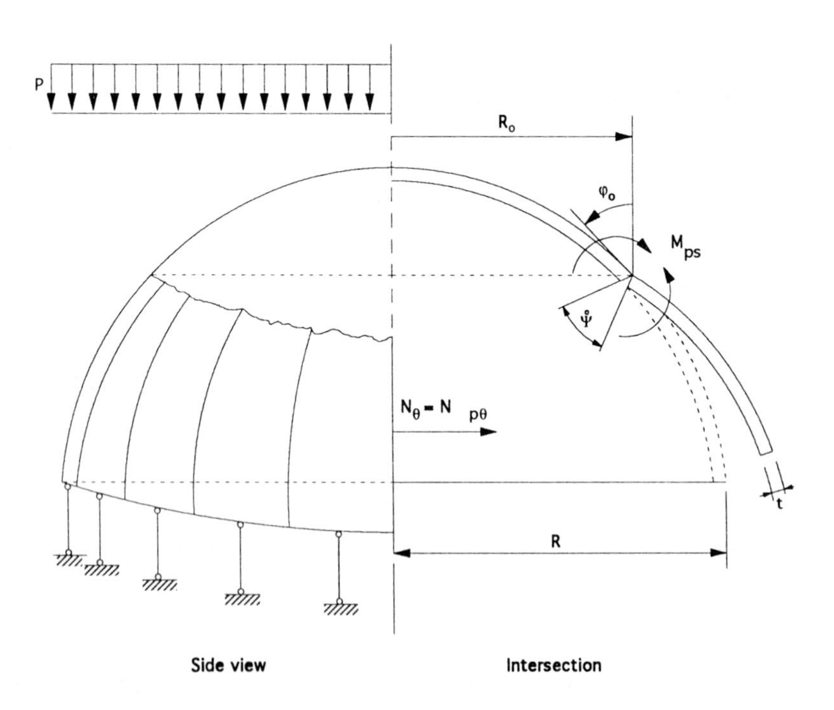

Fig. 10.8. Yielding zones in a spherical cap [10.14].

$$d\dot{u}_o = (\dot{v} \sin \varphi - \dot{w} \cos \varphi)\, d\theta. \tag{10.16}$$

Let be R the curvature radius in the meridian plan. The relative rotation of the sides of the meridional yield-line is :

$$d\dot{\theta} = \left(\frac{\dot{v}}{R} + \frac{1}{R} \frac{\partial \dot{w}}{\partial \varphi} \right) \sin \varphi \, d\theta. \tag{10.17}$$

Owing to (10.1), we see that the bending ratio is negligible :

$$r_c = \frac{t}{2\,R} \frac{\left(\dot{v} + \dfrac{\partial \dot{w}}{\partial \varphi} \right) \sin \varphi}{\dot{v} \sin \varphi - \dot{w} \cos \varphi} \approx \frac{t}{2\,R}. \tag{10.18}$$

Let us examine as an example the problem of a simply supported spherical cap subjected to vertical load, uniformly distributed on the horizontal projection area [10.19, 10.20, 10.25].

The meridional strip rotates by an angle $\dot{\psi}$ with respect to lines tangent to the shell surface (fig. 10.5). According to the previous remark, the energy is dissipated by extension in the meridional yield-lines and rotation of the hinge circle :

$$D = \int_o^{\varphi_o} R\, d\varphi \int_o^{2\pi} N_{p\theta}\, d\dot{u}_o + 2\,\pi\, R \cos \varphi_o\, \dot{\psi}\, M_{ps}. \tag{10.19}$$

An easy computation leads to :

$$D = 2\,\pi\, R \left[R\, N_{p\theta} (1 - \cos \varphi_o - \varphi_o \sin\varphi_o) + M_{ps} \cos \varphi_o \right] \dot{\psi}. \tag{10.20}$$

On the other hand, the power of external loads is given by :

$$P = \frac{\pi\, R^3}{6} p^+ \left[6 \cos^2 \varphi_o (1 - \cos \varphi_o) + \cos 3\, \varphi_o - 3 \cos 2\, \varphi_o + 3 \cos \varphi_o - 1 \right] \dot{\psi}.$$

Finally, an upper bound p^+ of the limit load is obtained from the power equation. The parameter φ_o must be adjusted so that it minimizes the pressure p. Extended computations are given in [10.19, 10.20, 10.25].

10.3.4. *Shallow shells.*

A simplified kinematic method for shallow shells has been developed by A.R. Rzhanitsyn (see [10.12, 10.13 and 8.63]). It corresponds to case B1 of table 10.1. Consider, for example, a shallow shell (the rise f is small with respect to any span AD, BE,...) with plane polygonal base, subjected to a concentrated force P, and hinged at the edges (fig. 10.9). A possible collapse mechanism is shown in fig. 10.9 (b). Lines of strain discontinuities OA, OB, OC, OD, OE, enable the various rigid parts to rotate about the edges AB, BC, etc.

The rate of circumferential extension and rotation in line OA are (see fig. 10.9)

$$\dot{u}_{oA} = \dot{w}\, z\, (\cot\alpha_i + \cot\beta_i)\,, \qquad \dot{\theta} = \dot{w}\, (\cot\alpha_i + \cot\beta_i)$$

where z si taken from the equation of the median surface of the shell.

The bending ratio (10.1) is negligible :

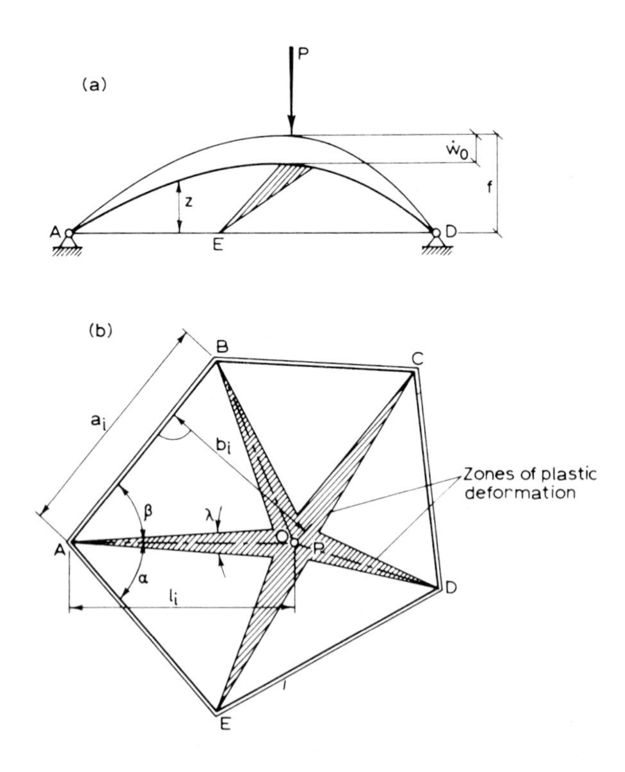

Fig. 10.9 (a), Fig. 10.9 (b). Collapse pattern for a shallow shell [10.12].

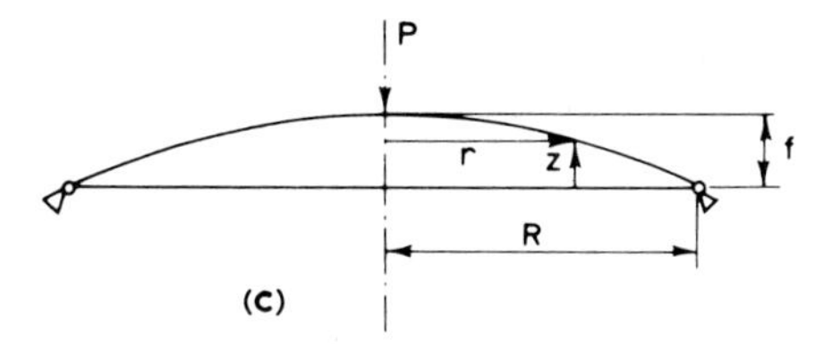

Fig. 10.9 (c).

$$r_c = \frac{t}{2z} \approx \frac{t}{2f},$$

as soon as the height f of the shallow shell (fig. 10.c) is much greater than the thickness t. Then, the shell is in membrane regime (case B1). The total rate of dissipation for the m yield lines is

$$D = N_p \dot{w} \sum_{i=1}^{m} (\cot\alpha_i + \cot\beta_i) \int_0^{l_i} z \, ds,$$

where s is the abscissa along the yield line i. The value of P is then obtained from the work equation, together with the minimum condition that gives the location of point 0.

For example, in the case of a shallow shell of revolution with equation $z = f\left[1 - (r^n/R^n)\right]$, fig. 10.9 (c), the corresponding collapse load is

$$P_+ = 2\pi N_p f \frac{n}{n+1}. \tag{10.22}$$

10.3.5. *Cylindrical shell roofs.*

10.3.5.1. *Introduction.*

Consider a cylindrical shell referred to a cylindrical system of coordinates (fig. 10.10). In the absence of symmetry of revolution, there are six generalized stresses : M_x, M_θ, $M_{x\theta}$, N_θ, N_x, $N_{x\theta}$. The shear forces V_x and V_θ are reactions.

We also have $M_{x\theta} = M_{\theta x}$ and $N_{x\theta} = N_{\theta x}$ because, for all z such that $-t/2 \leq z \leq t/2$, one has $z/R \ll 1$. Hence, the yield condition will be represented by an

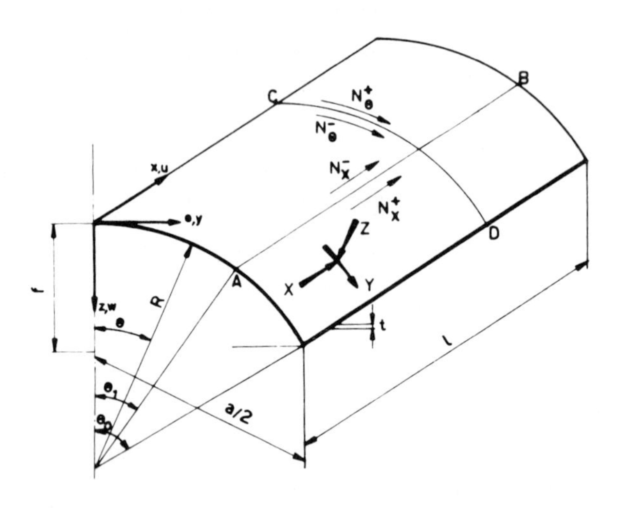

Fig. 10.10.

hypersurface in a six-dimensional stress space. In a kinematic approach with lines of concentrated rate of deformation, the deflection rate $\dot{w}$ normal to the shell is bound to be con»inuous, whereas not only slope discontinuities are admissible but also discontinuities in the rates of tangential displacements $\dot{u}$ and $\dot{v}$. In order to evaluate these discontinuities, we shall need the following relations :

$$\dot{\varepsilon}_x = \frac{\partial \dot{u}}{\partial x}, \qquad \dot{\varepsilon}_\theta = \frac{1}{R}\left(\frac{\partial \dot{v}}{\partial \theta} - \dot{w}\right),$$

$$\dot{\kappa}_x = \frac{\partial^2 \dot{w}}{\partial x}, \qquad \dot{\kappa}_\theta = \frac{1}{R}\left(\dot{w} + \frac{\partial^2 \dot{w}}{\partial \theta^2}\right),$$

$$\dot{\gamma}_{x\theta} = \frac{1}{R}\frac{\partial \dot{u}}{\partial \theta} + \frac{\partial \dot{v}}{\partial x},$$

$$\dot{\kappa}_{x\theta} = \frac{1}{R}\left(\frac{\partial \dot{v}}{\partial x} + \frac{\partial^2 \dot{w}}{\partial x\, \partial \theta}\right). \tag{10.23}$$

10.3.5.2. *Collapse mechanisms.*

Collapse mechanisms formed of generalized yield lines [10.4, 10.21, 10.22, 10.23] have been suggested by the experiments of Sawczuk [10.4], who tested cylindrical shells in reinforced mortar with the following characteristics : $\sigma'_c = 140 dN/cm^2$ (mortar) ; reinforcement : two orthogonal layers with square mesh of 13 mm side (78 wires per meter), at both the upper and lower face, made of wire of 1.2 mm diameter.

The geometry of the various shells is given in table 10.2. The ratio $t/R \approx 0.023$ was chosen to avoid buckling failures. The support conditions are shown in fig. 10.11. There were no edge beams. The shells were subjected to external radial pressure, applied by means of a rubber bag, and measured with an accuracy higher than $0.001\ dN/cm^2$.

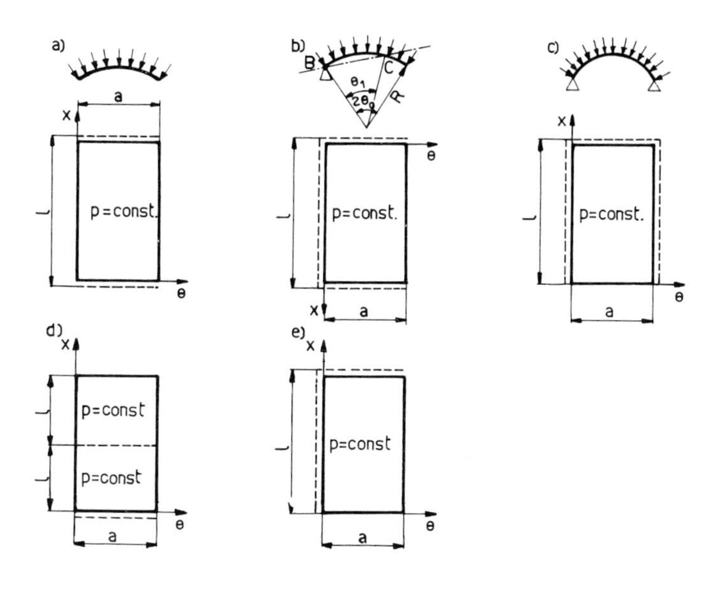

Fig. 10.11.

Table 10.2.

Type	θ_0	R, cm	f, cm	a, cm	L, cm	t, cm
A	30°	93	12.5	93	192	2.2
B	30°	93	12.5	93	95	2.0
C	17°	93	4.1	54.2	192	2.2
D	17°	93	4.1	54.2	95	2.2

Table 10.3.

Type	a, cm	l, cm	p, dN/m^2	Type	a, cm	l, cm	p, dN/m^2
Aa	93	188	1050	Cb	52	188	1.400
Ba	93	90	3500	Db	52	90	1.600
Ca	54	188	700	Ac	89	190	1.250
Da	54	90	1950	Bc	89	91	3.000
Ab	91	188	1550	Ad	93	94	1.200
Bb	91	90	3000	De	54	91	1.500

The collapse pressures given in table 10.3 are the maximum values that could be obtained, with wide opening of the cracks. Each value is the average of two or three measurements that do not differ by more than $\pm 4\%$.

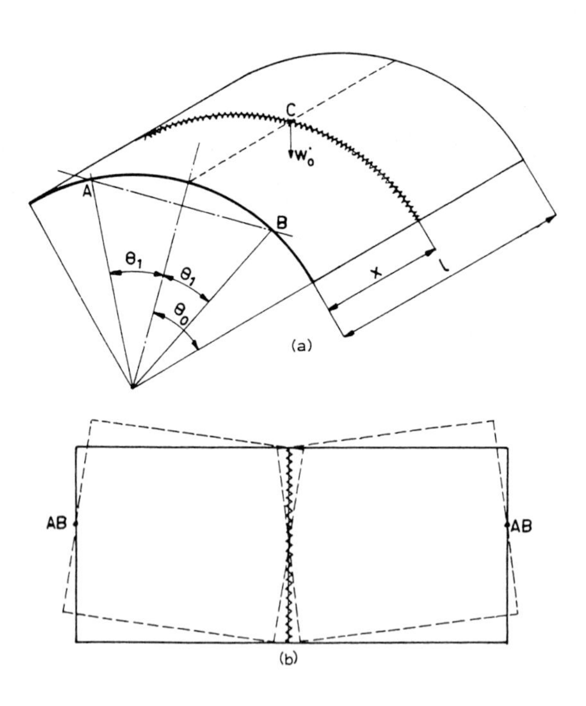

Fig. 10.12.

We now consider some of the observed collapse mechanisms :

1. Long shell, simply supported at two ends, types Aa and Da with $l/a \approx 2$, type Ca with $l/a \approx 3$, fig. 10.12. A "beam mechanism", with a transversal hinge circle at midspan, takes place.

Each rigid portion on both sides of the hinge circle rotates with respect to an axis AB (fig. 10.9) located in the plane of the simple support. Consider the rigid part where $0 \leq x \leq l/2$. In relation (10.23) we use the conditions $\dot{\varepsilon}_\theta = \dot{\kappa}_\theta = \dot{\gamma}_{x\theta} = 0$ and the boundary condition $\dot{u}\,(0\,,\theta\,,R)=0$ that comes from the fact that points A and B do not move. We obtain

$$\dot{w} = x\,\frac{2\,\dot{w}_o}{l}\cos\theta, \tag{10.24}$$

$$\dot{v} = x\,\frac{2\,\dot{w}_o}{l}\sin\theta\,, \tag{10.25}$$

$$\dot{u} = \frac{2\,R\,\dot{w}_o}{l}(\cos\theta - \cos\theta_1)\,, \qquad 0 \leq \theta \leq \theta_0\,. \tag{10.26}$$

As both rigid parts are identical, it is easily seen that $\dot{v}$ is the same on both sides of the yield line, whereas $\dot{u}$ simply changes its sign as does $\frac{\partial\,\dot{w}}{\partial\,x}$. We therefore have the following discontinuities :

$$\dot{u}_o = \dot{u}] = \frac{4\,R\,\dot{w}_o}{l}(\cos\theta - \cos\theta_1)\,, \tag{10.27}$$

$$\dot{\theta} = \frac{\partial\,\dot{w}}{\partial\,x}] = \frac{4\,\dot{w}_o}{l}\cos\theta\,. \tag{10.28}$$

The bending ratio is

$$r_b = \frac{e}{2\,R}\;\frac{\cos\theta}{|\cos\theta - \cos\theta_1|}\,.$$

If f is the height of the roof, we see that the bending ratio is generally much lower than unity :

$$r_b \approx \frac{e}{2f}.$$

The regime of plastic yielding is represented by the vertex C ($N_x = N_p'$) for $|\theta| \le \theta_1$ and A ($N_x = N_p$) for $\theta_1 \le |\theta| \le \theta_o$ (fig. 5.39 and 5.40).

The power equation gives finally :

$$p^+ = \frac{8 N_p R}{L^2 \sin \theta_o}\left[\left(1 + \frac{N'_p}{N_p}\right)\left(\sin \theta_1 - \theta_1 \cos \theta_1\right) + \left(\theta_o \cos \theta_1 - \sin \theta_o\right)\right], \tag{10.29}$$

θ_1 is evaluated from the condition of minimum $\frac{dp_+}{d\theta_1} = 0$:

$$\theta_1 = \frac{\theta_o}{1 + \frac{N'_p}{N_p}}. \tag{10.30}$$

In the case of the tested shells N'_p=308 kN/m, N_p=178 kN/m, θ_o=30°, R=93 cm, l=188 cm. Formula (10.29) gives p=156 kN/m^2, to compare with the experimental value p_{ex}=105 kN/m^2. The same formula, applied to the shell Ca (l/a=3.5) gives p=5 kN/m^2 whereas p_{ex}= 7 kN/m^2. Hence, formula (10.29) can be applied with confidence in the range $2.0 \le l/a \le 3.5$.

2. Shells simply supported on two short edges and on a third edge (Type b, fig. 10.11). When the third supporting edge is long, fig. 10.13 (a), there is complete collapse of the shell with the mechanism shown in the figure. Let $\dot{w}_o$ be the radial displacement rate of the segment DE. For part ABDEA we have the following rates of displacement :

$$\dot{w} = \frac{\dot{w}_o}{\sin 2\theta_o} \sin \theta,$$

$$\dot{v} = \frac{-\dot{w}_o}{\sin 2\theta_o} (\cos \theta - 1),$$

$$\dot{u}_o = 0, \tag{10.31}$$

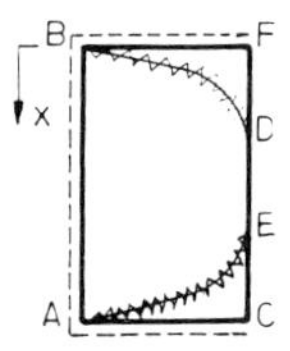

Fig. 10.13.

satisfying the boundary conditions $(\dot{w})_{\theta=0} = 0$, $(\dot{v})_{\theta=0} = 0$, $(\dot{w})_{\theta=2\theta_o} = \dot{w}_o$. Part BFD rotates with respect to axis BC [fig. 10.11 (b)]in the plane x=0. Point C is determined by the angle θ_1. The rates of displacement of this part are

$$\dot{u} = KR(\cos\theta - \cos\theta_1),$$
$$\dot{v} = Kx\sin\theta,$$
$$\dot{w} = Kx\cos\theta, \tag{10.32}$$

where $K = \dfrac{\dot{w}_o}{x_o \cos 2\theta_o}$, x_o being the abscissa of point D, fig. 10.13. Because $\dot{w}$ must be continuous across the yield line BD, the following equation is obtained for this yield line

$$x = x_o \cot 2\theta_o \cdot \tan\theta. \tag{10.33}$$

The following discontinuities thus occur in this yield line

$$\dot{u}] = \frac{\dot{w}_o R}{x_o \cos 2\theta_o}(\cos\theta - \cos\theta_1),$$
$$\dot{v}] = \frac{\dot{w}_o}{\sin 2\theta_o}\left(\frac{\cos 2\theta}{\cos\theta} - 1\right), \tag{10.34}$$

$$\frac{\partial \dot{w}}{\partial x}] = \frac{\dot{w}_o}{x_o \cos 2\theta_o}\cos\theta,$$
$$\frac{\partial \dot{w}}{\partial \theta}] = \frac{1}{R}\frac{\dot{w}_o}{\sin 2\theta_o}\frac{\cos 2\theta}{\cos\theta}. \tag{10.35}$$

The parameter θ_1 and x_o must then be adjusted to render the pressure p obtained from the work equation a minimum.

The bending ratio is also not governing :

$$r_b \approx \frac{t}{2R}.$$

We may then assume that the yield-line is in membrane regime. Determining the upper bound of the collapse pressure by the previous theory, and evaluating numerically the minimum of p, we obtain (see [10.25])

$$p = 21 \text{ kN/m}^2 > p_{ex} = 16 \text{ kN/m}^2$$

for the shell such that : $\theta_o = 17°$, R=93 cm, L=90 cm.

10.3.5.3. *Other work on the subject.*

The method sketched above has been developed by M. Janas [10.3, 10.22] and used by him for the evaluation of the collapse load of a cylindrical shell hinged along heavy edge beams and supported on flexible diaphragms [10.22]. Unfortunately, the results are not directly applicable to reinforced concrete shells because of the choice of the yield surface. The same remark applies to the work of Fialkow [10.23], who obtained upper and lower bounds to the limit radial pressure of cylindrical roofs simply supported on two end diaphragms (the obtained bounds differ from the average value by 2% to 25%).

10.4. Lower bound solutions for circular cylindrical tanks, under axisymmetric loading.

10.4.1. *Yield condition.*

We consider again the problem of a vertical cylindrical tank subjected to the hydrostatic internal pressure (10.5) (see fig. 10.4).

Following Sawczuk and Olszak [10.2], we shall give a whole set of statical solutions.

In this Section, compressive forces will be regarded as positive, as is usual in reinforced concrete practice. We suppose that the reinforcement is simple and placed in the extreme layer of concrete ($\rho = 1$ with the notations of Section 5.6).

When there is no axial force N_x, the ultimate moment for positive bending in the circumferential yield-lines, is obtained by taking $n_x = 1$ in the yield curve (5.114) :

$$m_x = -2\gamma_x^2 + 2\gamma_x \equiv m'_p. \tag{10.36}$$

For negative bending, the ultimate moment m'_p vanishes (see regime A'D' in Table 5.6.2). On the other hand, because of symmetry, we know (see Sections 5.1.4 and 8.3.1) that M_θ is a reaction. The ultimate forces in the meridional yield-lines are obtained (see fig. 5.39)

- in tension, at the vertex A : $n_\theta = -\gamma_\theta \equiv -n_p$;

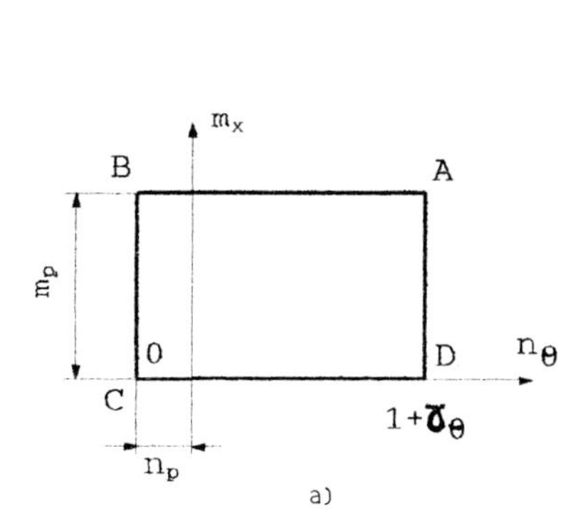

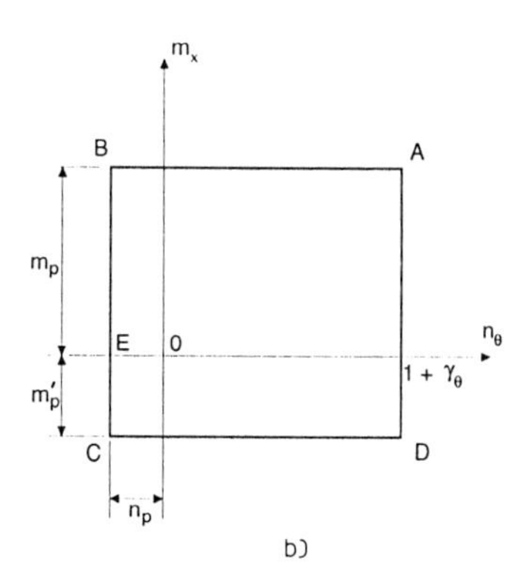

Fig. 10.14. (a) (b)

- in compression, at the vertex C : $n_\theta = \gamma_\theta + 1$.

Finally, the yield criterion in variables m_x and n_θ is represented by fig. 10.14.

10.4.2. *Collapse mechanisms.*

The generalized strain rates $\dot{\varphi}_x$ and $\dot{\lambda}_\theta$ corresponding to m_x and n_θ, respectively are given in table 10.4. By using the normality law, we have seen at Section 10.3 that regimes AB and CD must be rejected and that regimes BC and DA furnish conical mechanisms.

Table 10.4.

stress profile fig.10.14	Yield condition	Strain rates (to within a positive common factor)	
AB	$m_x = m_p \equiv -2\gamma_x^2 + 2\gamma_x$ $-n_p \leq n_\theta \leq 1 + \gamma\theta$	1	0
BC	$n_\theta = n_p \equiv$ $-m'_p \leq m_x \leq m_p$	0	- 1
CD	$m_x = -m'_p \equiv 0$ $-n_p \leq n_\theta \leq 1 + \gamma_\theta$	- 1	0
DA	$n_\theta = 1 + \gamma_\theta$ $-m'_p \leq m_x \leq m_p$	0	1

10.4.3. *Equilibrium equation.*

When compressive stresses and strains are regarded as positive, the fundamental equilibrium equation [eq. (8.10)] can be rewritten as

$$\frac{l^2}{c^2}\frac{d^2 m_x}{dx^2} - n_\theta - p^* = 0, \tag{10.37}$$

where l is the length of the shell, $C^2 = \frac{4 l^2}{Rt}$, R is the radius of the median surface, t the effective thickness (see Section 9.7), and $p_* = p(x)\frac{R}{t\sigma'_r}$. The origin of the abscissa x is located at one end of the shell.

10.4.4. *Cylindrical shell subjected to hydrostatic pressure.*

10.4.4.1. *Nonuniform reinforcement : complete collapse.*

The origin x=0 is at the top cross section which is free, whereas the bottom end is built-in. The boundary conditions thus are (see fig. 10.4 for sign convention, all positive elements)

$$\dot{w}(l) = 0, \qquad \dot{w}(0) = -\dot{w}_o \qquad (\text{with } \dot{w}_o > 0),$$

$$m_x(0) = 0, \qquad v_x(0) \equiv \frac{4 l}{t^2 \sigma'_r} V_x(0) = 0. \tag{10.38}$$

Various collapse mechanisms can occur, depending on the geometry and the reinforcement of the shell. A complete discussion can be found in the paper by Sawczuk and Olszak [10.2]. Having in mind the "complete collapse" of the tank,eq. (10.9), we assume that the axial reinforcement is uniform and exclusively at the inner face : $m_p(x) = \text{const}$, $m'_p(x) = 0$. Pressure is given by (10.5).

With the stress profile CB, fig. 10.14 that is, $n_x = -n_p(x)$, $0 \le m_x \le m_p$, integration of eq. (10.37) with due account of the boundary conditions (10.38), gives

$$m_x = m_p\left(\frac{x}{l}\right)^3 \tag{10.39}$$

and

$$p_o^* = \frac{6\, m_p}{C^2} + n_p . \tag{10.40}$$

It is easily seen that the mechanism (10.9) corresponds to the considered stress field. Indeed, the value (10.40) is the same as the upper bound (10.12), written in reduced variables. Applying the combined theorem, we conclude that the expression (10.40) furnishes the exact limit load. The collapse mechanism and the stress field are shown in fig. 10.15.

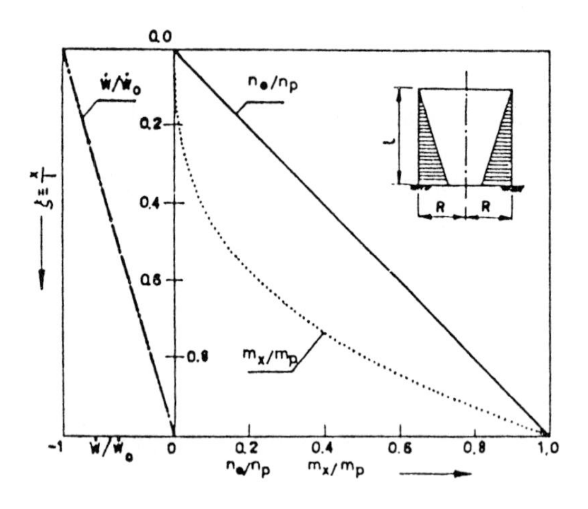

Fig. 10.15.

If a uniform axial reinforcement at the external face of the cylinder ($m'_p \neq 0$) is added to the preceding reinforcement, the yield condition in fig. 10.11 must be used. For sufficiently short shells, complete collapse can still occur. Limiting values of the length parameter C, together with an auxiliary unknown $\frac{x_1}{l}$, are obtained from the conditions that, at $x = x_1$, $m_x = -m'_p$, and $\frac{dm_x}{dx} = 0$ (stress profile ECEB in fig. 10.11, with point C corresponding to abscissa x_1). In the present situation, it can also be shown [10.2] that the stress field remains admissible with nonvanishing circumferential reinforcement at the top of the cylinder. We thus have, in general,

$$n_p(x) = n_{po} + n_p \frac{x}{l} . \qquad (10.41)$$

The limiting values C_1 of C are given in fig. 10.14, for $0 \leq \frac{m'_p}{m_p} \leq 2$. When $m'_p = m_p = 1$ and $n'_p = n_p = 1$, we had found in Section 8.3.4.3 that $C_1^2 = 17.1$.

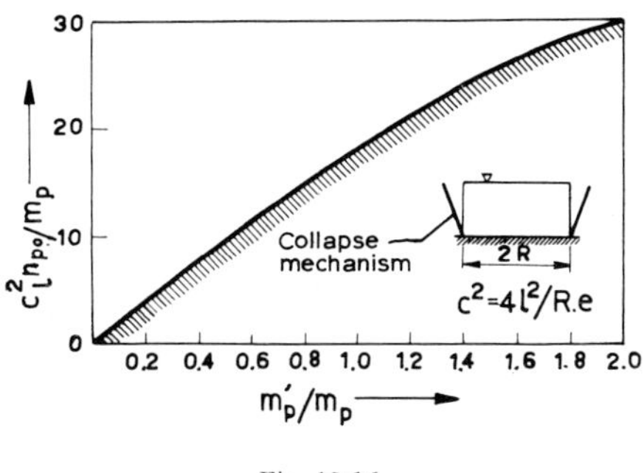

Fig. 10.16.

10.4.4.2. *Uniform reinforcement : complete collapse.*

Consider the same loading as in Section 10.4.4.1, but a reinforcement with $m_p \neq 0$, $m'_p \neq 0$ in the axial direction, $n_p \neq 0$ in the circumferential direction, where all three values m_p, m'_p and n_p do not depend on x. From what was seen in the preceding Section, it can be expected that, with increasing length of the shell, a positive hinge circle at the bottom end will be accompanied by a negative hinge circle at a certain abscissa x_1, with possibly circumferential contraction of an upper region where $0 \leq x \leq x_0$ (fig. 10.14). At the boundary x_0 of this region, n_θ must jump from 1 to $-n_p$ because regimes AB and CD are not possible. Such a discontinuity is admissible (see Sections 5.6 and 8.6). We therefore consider the following stress field :

$$\begin{aligned}
n_\theta &= 1, & -m'_p \leq m_x \leq m_p \quad &\text{for} & 0 \leq x \leq x_0, \\
n_\theta &= -n_p, & -m'_p \leq m_x \leq m_p \quad &\text{for} & x_0 \leq x \leq x_1, \\
n_\theta &= -n_p, & -m'_p \leq m_x \leq m_p \quad &\text{for} & x_1 \leq x \leq 1,
\end{aligned} \tag{10.42}$$

with the boundary conditions :

$$m_x(1) = m_p, \qquad m_x(x_1) = -m'_p, \qquad m_x(0) = 0, \qquad v_x(0) = 0 \tag{10.43}$$

$$\begin{array}{lllll}
m_x & \text{continuous in} & x = x_1 & \text{and} & x = x_0\ , \\
n_\theta & \text{discontinuous at} & x = x_0 & \text{by the amount} & 1 + n_p\ , \\
v_x & \text{continuous at} & x = x_1 & \text{and} & x = x_0\ .
\end{array} \tag{10.44}$$

Integration of eq. (10.37) then gives

$$m_x = C^2 \left(\frac{x^2}{2l^2} + p_0 \frac{x^3}{6l^3} \right) \quad \text{for} \quad 0 \le x \le x_0 \tag{10.45}$$

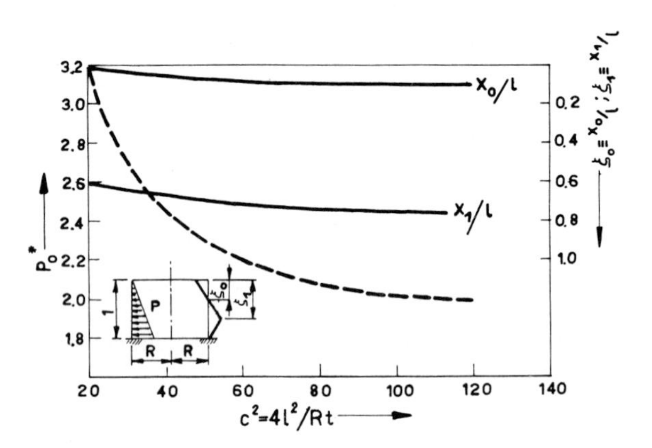

Fig. 10.17.

$$m_x = C^2 \left(-n_p \frac{x^2}{2l^2} + p_o^* \frac{x^3}{6l^3} \right) + C^2 A \frac{x}{l} + C^2 B \ . \tag{10.46}$$

From condition (10.44) at $x = x_o$, one obtains

$$A = \frac{x_0}{l^2} (1 + n_p) \ ,$$

$$2B = -\frac{x_0^2}{l^2} (1 + n_p) \ . \tag{10.47}$$

The first two conditions of (10.43), together with the condition

$$\left(\frac{dm_x}{dx} \right)_{x = x_1} = 0$$

of an analytical extremum at $x = x_1$, give the following equations for the unknown p_o^*, x_1 and x_0 :

$$\frac{6\, m_p}{C^2} + 3\, n_p - p_o^* + 6\,(1 + n_p)\frac{x_0}{l} - 3\,(1 + n_p)\frac{x_0^2}{l^2} = 0 \ ,$$

$$\frac{6}{C^2}(m_p + m'_p) + 3\, n_p\left(1 + \frac{x_1^2}{l^2}\right) - p_o^*\left(1 - \frac{x_1^3}{l^3}\right) = 0 \ ,$$

$$\frac{2\, x_o}{l}(1 + n_p) = 2\, n_p\frac{x_1}{l} - p_o^*\frac{x_1^2}{l^2} \ . \tag{10.48}$$

The corresponding mechanism is

$$\dot{w} = -\dot{w}_o\frac{x - x_0}{x_1 - x_0} \quad \text{for} \quad 0 \le x \le x_1 \ ,$$

$$\dot{w} = \dot{w}_o\frac{x - l}{l - x_1} \quad \text{for} \quad x_1 \le x \le l \ , \tag{10.49}$$

where $\dot{w}_o$ is the modulus of the radial displacement rate at $x = x_1$. The solution given above is statically and kinematically licit and hence complete. It is valid provided the positive moment m_x in the vicinity of the upper edge does not attain m_p at a certain abscissa x_2. To study the limiting case where $m_x = m_p$ at $x = x_2$, we first remark that, to remain within the yield restriction, m_x must exhibit an analytical maximum at $x = x_2$, whereas expression (10.45) shows that m_x is monotonically increasing for $0 \le x \le x_0$. Hence, $x_0 \le x_2 \le x_1$. The condition $m_x(x_2) = m_p$ and $\left(\frac{dm_x}{dx}\right)_{x_2} = 0$ give $x_2 = x_0 = \frac{2\, n_p}{p_o^* - x}$. Substituting this expression for x_0 in eq. (10.46), using eqs.(10.47) and letting $m_x = m_p$, we finally obtain the critical value C_l at which eqs. (10.48) and (10.49) cease to apply. When $m_p = m'_p = n_p = 1$, the considered solution is valid for $17.1 < C^2 < 115$, and the corresponding values of p_o^*, $\frac{x_0}{l}$, $\frac{x_1}{l}$ are given in fig. 9.12. If $C^2 = 17.1$ the mechanism described in Section 10.4.4.1 takes place. If $C^2 = 115$, partial collapse occurs, as will be seen in Section 10.4.4.3. Fig. 10.15 shows the stress profile, the stress field, and the yield mechanism (with a circumferentially compressed upper zone) when $m_p = m'_p = n_p = n'_p = 1$ and $C^2 = 75$.

10.4.4.3. *Uniform reinforcement : partial collapse.*

for large values of C^2, only the bottom region of the tank will collapse, whereas the upper region will remain rigid.

In the collapsing region, the stress profile BCB, fig. 10.7, will be valid, and the mechanism will be as shown in the insert at the right of fig. 10.16. Because the stress field can be continued in a licit manner in the rigid region, the solution is complete. For detailed derivation, the reader is referred to the original paper [10.2]. The limit value of p_o^* is given in fig. 10.16 versus C^2, for the three types of mechanisms, when $m_p = m'_p = n_p = 1$.

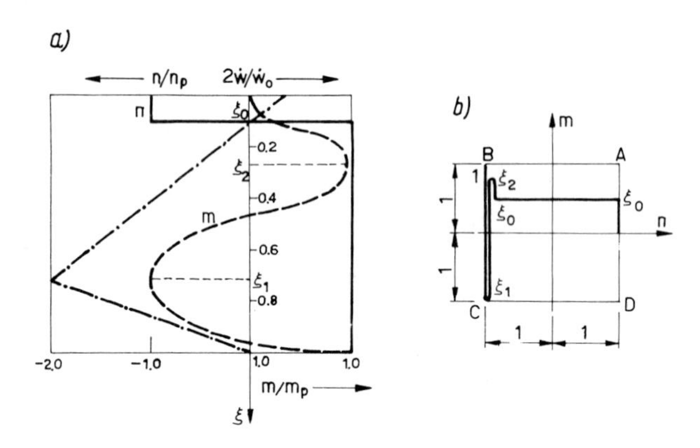

Fig. 10.18.

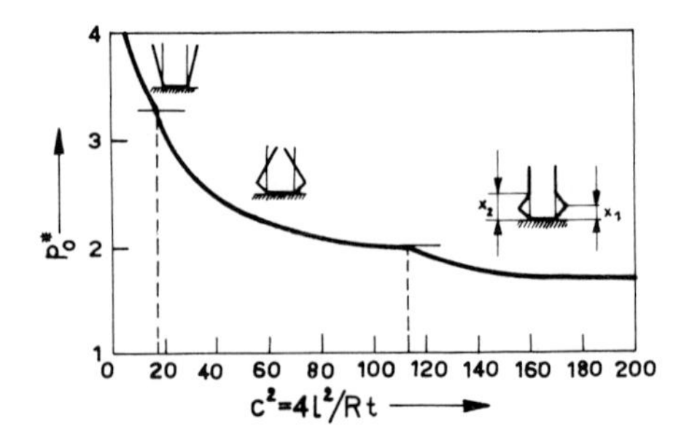

Fig. 10.19.

10.4.5. *Cylindrical silo.*

A vertical cylindrical silo, free at the upper edge, built-in at the bottom edge and containing a medium of density γ, with coefficient of friction μ on the wall and coefficient k of internal friction, has been studied by Sawczuk and Köning [10.17], in a manner similar to that used in Section 10.4.4. The limit load p_o and the collapse mechanisms are given in fig. 10.20 valid when $m_p = m'_p = n_p = 1$.

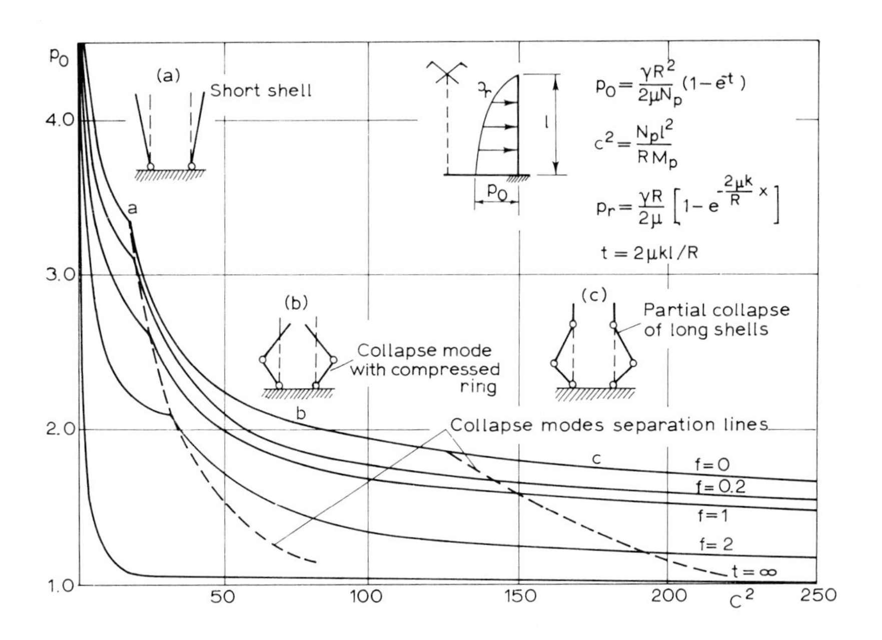

Fig. 10.20. Mechanisms of collapse and collapse pressure for nonuniformly loaded cylindrical shells [10.17].

References.

[10.1] F. LEVI, "Methodes simplifiées de calcul des voiles minces a courbure gaussienne nulle" (Simplified calculation methods of shells with vanishing gaussian curvature), Simplified Calculation Methods for Shell Structures, pp. 445-446, North-Holland Publ. Co., Amstrerdam, 1964.

[10.2] A. SAWCZUK and W. OLSZAK, "A method of Limit Analysis of Reinforced Concrete Tanks", Simplified Calculation Methods for Shell Structures, pp. 416-437, North-Holland Publ. Co., Amsterdam, 1964.

[10.3] M. JANAS, "Limit Analysis of Nonsymmetric Plastic Shells by a Generalized Yield Line Method", Nonclassical Shell Problems, pp. 997-1010, North-Holland Publ. Co., Amsterdam, 1964.

[10.4] A. SAWCZUK, "On experimental foundations of the Limit Analysis Theory of Reinforced Concrete Shells", Shell Research, pp. 217-231, North-Holland Publ. Co., Amsterdam, 1961.

[10.5] A.L.L. BAKER, "Further Research in Reinforced Concrete, and its Application to Ultimate load Design", Proc. Inst. Civil. Eng., **2** : 2, August, 1953.

[10.6] P.B. MORICE, "Research on Concrete Shell Structures", Proc. Ist Symp. Shell Roof Constr., London, 1952, Cement and Concrete Ass. London, 1954.

[10.7] A.C. VAN RIEL, W.J. BERANEK and A.L. BOUMA, "Test on Shell Roof Models of Reinforced Concrete Mortar", Proc. 2nd Symp. Shell Roof Constr., Oslo, 1957.

[10.8] G.R. MITCHELL, "Shell Research at the Building Research Station", Proc. 2nd Symp. Shell Roof Constr., Oslo, 1957.

[10.9] A.L. BOUMA, A.C. RIEL, H. VAN KOTEN and W.J. BERANEK, "Investigations on Models of Eleven Cylindrical Shells made of Reinforced and Prestressed Concrete", Shell Reseach, pp. 79-101, North-Holland Publ. Co., Amsterdam, 1961.

[10.10] H. LUNDGREN, Cylindrical Shells, The Danish Technical Press, Copenhague, 1949.

[10.11] A.L.L. BAKER, "Ultimate Strength Theory for Short Reinforced Concrete Cylindrical Shell Roofs", Mag. Conc. Research, 10, 3, 1952.

[10.12] A.R. RJANITSYN, "The Design of Plates and Shells by the Kinematical Method of Limit Equilibrium", IX Congres Int. de Mec. Appl., Actes, **VI** : 331, Brussels, 1956.

[10.13] A.R. RJANITSYN, "Calculation of Shallow Shells by the Limit Design Methods", Simplified Calculation Methods of Shell Structures, pp. 438-444, North-Holland Publ. Co., Amsterdam, 1961.

[10.14] A.M. OVETCHKIN, "Analysis of Reinforced Concrete Rotationally Symmetric Shells" (in Russian), Gosstroyizdat, Moscow, 1961.

[10.15] A.R. DYKES, "Experimental and Theoretical Studies of Folded Plate Structures", Nonclassical Shell Problems, pp. 941-976, North-Holland Publ. Co., Amsterdam, 1964.

[10.16] G.K. HAIDUKOV, "Limit Equilibrium Design of Shallow shell Panels", Nonclassical Shell Problems, pp. 977-996, North-Holland Publ. Co., Amsterdam, 1964.

[10.17] A. SAWCZUK and J.A. KÖNIG, "Limit Analysis of Reinforced Concrete Silo" (in Polish), Arch. Inz. Lad., **8** : 161, 1962.

[10.18] S. KALISZKY, "Untersuchung einer Kegelstrumpfschale auf Grund der Traglastverfahrens", Acta Tech. Hung., **34** : 159, 1961.

[10.19] N V. AKHVLEDIANI, "On the Calculation of Reinforced Concrete Shells of Revolution According to the Limit Equilibrium Method" (in Russian), Soob. AN Gruz. S.S.R., **18** : 205, 1957.

[10.20] N V. AKHVLEDIANI, "Analysis of Reinforced Concrete Domes According to the Limit Equilibrium Method" (in Russian),Issled. Teor. Sooruzh., **10** : 123, Gosstroyizdat, Moscow, 1961.

[10.21] M. JANAS and A. SAWCZUK, "On Carrying Capacities of Arch Dams", Symp. Arch. Dams (Southampton, 1964), Pergamon Press, Oxford, 1964.

[10.22] M. JANAS, "Limit Analysis of a Cylindrical Shell Roof" (in Polish), Arch. Inz. Lad., **8** : 365, 1962.

[10.23] M.N. FIALKOW, "Limit Analysis of Simply Supported Circular Shell Roofs", J Eng. Mech. Div., Proc. A.S.C.E., June 1958 (with errata in the October 1958 issue).

[10.24] M. CAPURSO, "On the Limit Analysis of Reinforced Concrete Shells" (in Italian), Giornale del Genio Civile, **104** : 83, February 1966, and 167, March 1966.

[10.25] G. DE SAXCE, "Extension de la méthode de Johansen aux coques en béton armé", Laboratoire de Mécanique des Matériaux et Stabilité des Constructions, Int. Report, Univ. of Liège, Belgium, 1984.

[10.26] G. DE SAXCE, "Extension of the Yield-Line Method to the Reinforced Concrete Shells". Proceedint Int. Conf. I.A.S.S.-85, Moscow, September 1985.

11. Plane Stress and Plane Strain.

11.1. Introduction.

In Section 7.1 we have defined a disk as a body with the geometry of a plate but subjected to forces acting in its median plan. We recall that the thickness must be small with respect to the dimensions in the median plane. In this situation, if we refer the disk to the orthogonal coordinate system x, y, z of fig. 11.1 where axes x and y lie in the median plane, stress components parallel to the z-axis may be neglected. We thus set

$$\sigma_z = \tau_{xz} = \tau_{yz} = 0. \tag{11.1}$$

Eqs. (11.1) define a *state of plane stress.* The z-axis is obviously a principal axis of the stress tensor, which is plane and will be completely determined by the components σ_x, σ_y, τ_{xy}. These three stress components are our generalized stresses Q_i, whereas the corresponding generalized strain rates $\dot{q}_i$ are $\dot{\varepsilon}_x$, $\dot{\varepsilon}_y$, $\dot{\gamma}_{xy}$. As noted in Section 2.4, the principal axis z of the stress tensor is also principal for the strain-rate tensor when the body is isotropic. Hence, $\dot{\gamma}_{xz} = \dot{\gamma}_{yz} = 0$. As a rule, $\dot{\varepsilon}_z \neq 0$, but on account of $\sigma_z = 0$, the strain rate $\dot{\varepsilon}_z$ does not appear in the expression for the dissipation and its value is irrelevant. From rotational equilibrium about the axis z, we have (see Section 1.1.1),

$$\tau_{xy} = -\tau_{yx}. \tag{11.2}$$

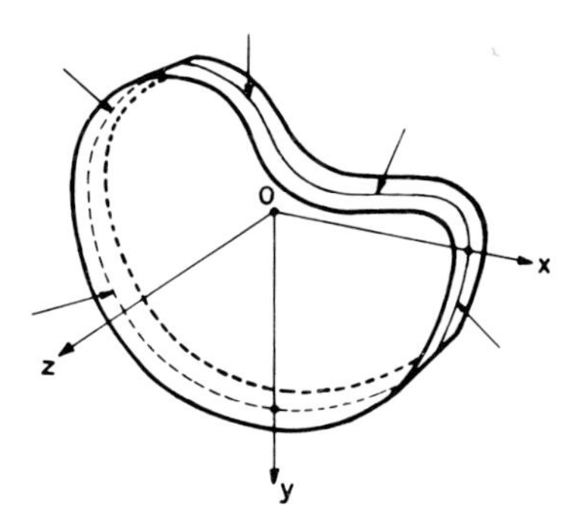

Fig. 11.1.

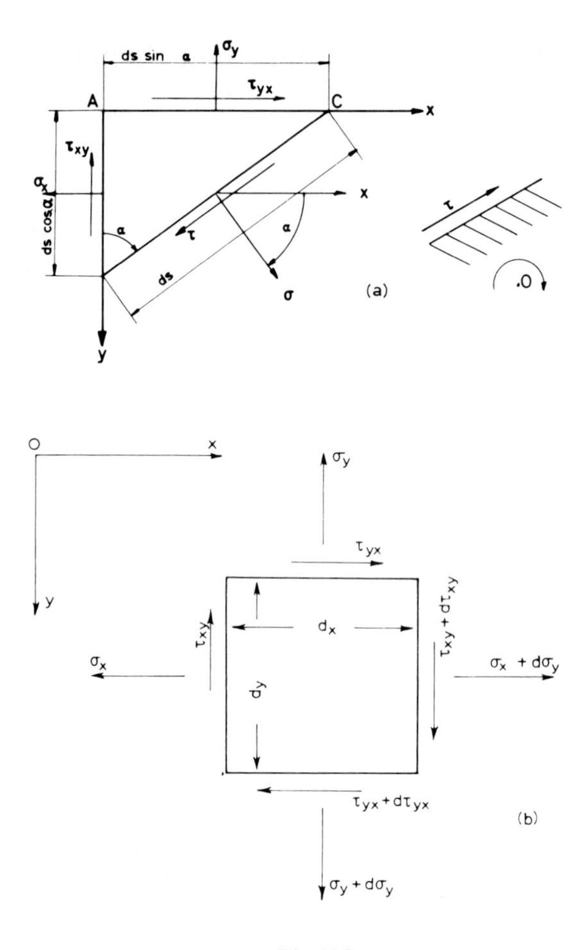

Fig. 11.2.

Positive stresses are shown in fig. 11.2 (a). Equilibrium of translation of the elementary parallelepiped of fig. 11.2 in the x- and y-directions give

$$\text{(a)} \qquad \frac{\partial \sigma_x}{\partial x} + \frac{\partial \tau_{xy}}{\partial y} + X = 0,$$

$$\text{(b)} \qquad \frac{\partial \tau_{xy}}{\partial x} + \frac{\partial \sigma_y}{\partial y} + Y = 0, \qquad (11.3)$$

where X and Y are components of the body force per unit volume. The conditions of equilibrium at the boundary are

$$\text{(a)} \qquad l\,\sigma_x + m\,\tau_{xy} = \overline{X},$$

$$\text{(b)} \qquad l\,\tau_{xy} + m\,\sigma_y = \overline{Y}, \qquad (11.4)$$

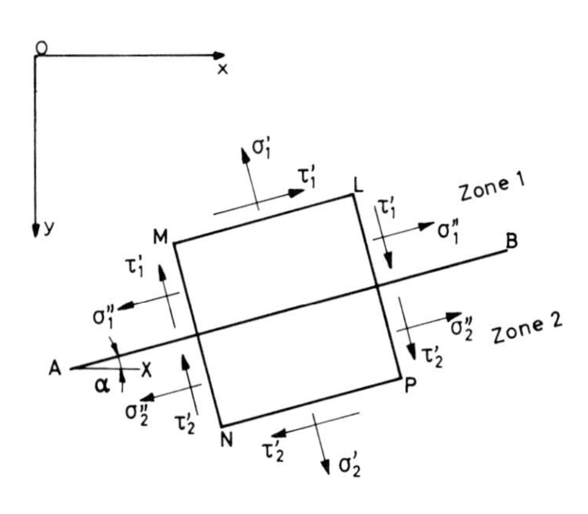

Fig. 11.3.

where $\overline{X}$ and $\overline{Y}$ are the components of the surface traction per unit area, and l and m the direction cosines of the outward pointing normal at the considered point of the boundary.

With expressions (1.12) and (1.13) for the principal stresses σ_1 and σ_2 in the xy-plane, the yield condition (1.32) of Tresca can be written as :

$$\max\left[\left|\frac{\sigma_x+\sigma_y}{2}\right| + \left(\left\{\frac{\sigma_x-\sigma_y}{2}\right\}^2 + \tau_{xy}^2\right)^{1/2} ;\right.$$

$$\left. 2\left(\left\{\frac{\sigma_x-\sigma_y}{2}\right\}^2 + \tau_{xy}^2\right)^{1/2} \right] = \sigma_Y . \qquad (11.5)$$

We have seen in Section 5.4.2, eq. (5.36) that the dissipation per unit volume is given by

$$D_V = \sigma_y \cdot \max|\dot{\varepsilon}_i|, \qquad i=1,2,3, \qquad (11.6)$$

where $\dot{\varepsilon}_i$ denotes a principal strain rate. The principal strain rates $\dot{\varepsilon}_1$ and $\dot{\varepsilon}_2$ in the xy-plane are related to the components $\dot{\varepsilon}_x$, $\dot{\varepsilon}_y$, $\dot{\gamma}_{xy}$ by eqs. (1.24) and (1.25), whereas $\dot{\varepsilon}_3 \equiv \dot{\varepsilon}_z$ is related to $\dot{\varepsilon}_1$ and $\dot{\varepsilon}_2$ by the incompressibility relation (1.26). Stress discontinuities are often used in plane stress problems. We recall from Section 5.6.2 that vectors $\boldsymbol{\sigma}'$ and $\boldsymbol{\tau}'$ normal to the discontinuity line AB (fig. 11.3) must be continuous across AB, whereas vector $\boldsymbol{\sigma}''$ parallel to AB may be discontinuous across AB.

Discontinuity surfaces in the velocity field are also admissible, but it can be shown [3.9] that the component of the velocity vector **V** along the normal to the surface of discontinuity must remain continuous. Hence, the dissipation per unit area in the surface of discontinuity is

$$D = \frac{\sigma_Y}{2} \, |\mathbf{V}]|, \qquad (11.7)$$

where **V**] is the jump in the tangential velocity and $\frac{\sigma_Y}{2}$ the yield stress in pure shear, σ_Y being the yield stress in simple tension.

Consider now the *state of plane strain*. It can be obtained practically, for example, with a very long solid that would be generated by the translation of the disk of fig. 11.1, with its loading, in the direction of the z-axis. Far from the end sections, all cross sections behave identically and, hence, the displacements of any point occurs in its own cross section. We thus have :

$$\varepsilon_z = \gamma_{zx} = \gamma_{zy} = 0, \qquad (11.8)$$

and also

$$\dot{\varepsilon}_z = \dot{\gamma}_{zx} = \dot{\gamma}_{zy} = 0. \qquad (11.9)$$

Eqs. (11.8) and (11.9) define a state of plane strain. From eq. (11.9) it is seen that the generalized stresses of the problem are σ_x, σ_y, τ_{xy} (as in plane stress) because the other stress components do not work. The z-axis is a principal axis of both the strain-rate tensor and the stress tensor. We thus have

$$\tau_{xz} = \tau_{yz} = 0, \qquad (11.10)$$

whereas the magnitude of $\sigma_z \equiv \sigma_3$ depends on the state of stress in the xy-plane and on the yield criterion. With the yield condition of Tresca, it is easily seen in fig. 1.5 that for $\dot{\varepsilon}_z \equiv \dot{\varepsilon}_3 = 0$, the stress point must lie in one of the planes

$$\begin{aligned} &(a) \qquad \sigma_1 - \sigma_2 = \sigma_Y \\ &(b) \qquad \sigma_1 - \sigma_2 = -\sigma_Y \end{aligned} \tag{11.11}$$

and σ_z may vary form 0 to σ_Y or from $-\sigma_Y$ to 0.

Eqs. (11.11) can also be written as

$$(\sigma_1 - \sigma_2)^2 = \sigma_Y^2, \tag{11.12}$$

or

$$(\sigma_x - \sigma_y)^2 + 4\,\tau_{xy}^2 = \sigma_Y^2. \tag{11.13}$$

With the yield criterion of von Mises, the condition $\dot{\varepsilon}_3 = 0$ and the normality law give

$$\sigma_3 = \frac{1}{2}(\sigma_1 + \sigma_2). \tag{11.14}$$

Substitution of expression (11.14) for σ_3 into eq. (1.35) results in

$$(\sigma_1 - \sigma_2)^2 = \frac{4}{3}\sigma_Y^2, \tag{11.15}$$

that is,

$$(\sigma_x - \sigma_y)^2 + 4\,\tau_{xy}^2 = \frac{4}{3}\sigma_Y^2. \tag{11.16}$$

The similarity of relation (11.12) and (11.15) will be noted : they differ only by the magnitude of the constant in the right-hand side. In a (σ_1 , σ_2) space they are both represented by two straight lines making 45° angles with the axes (see Problem 2.6.4, fig. 2.7). The dissipation per unit volume is, for the Tresca condition,

$$D_V = \sigma_Y \, |\dot{\varepsilon}_1| \qquad (\text{with } |\dot{\varepsilon}_1| = |\dot{\varepsilon}_2|), \tag{11.17}$$

and for the Mises condition,

$$D_V = \frac{2\,\sigma_Y}{\sqrt{3}} \, |\dot{\varepsilon}_1|. \tag{11.18}$$

11.2. Plane stress : perforated disks.

11.2.1. *Square disk with a slit.*

It is known from experiments that a perforated disk subjected to uniaxial tension and made of a ductile and negligibly work-hardening material flows plastically when the axial force attains the value

$$N^* = \sigma_Y \, A_n \, , \tag{11.19}$$

where A_n is the "net area" of the smallest cross section through the hole. This empirical concept of "net area" is at the basis of the practical design methods for perforated or notched bars made of ductible metals. As will be seen in the following, plastic limit analysis gives a rigourous theoretical support to the concept of net area, while indicating its limits of applicability.

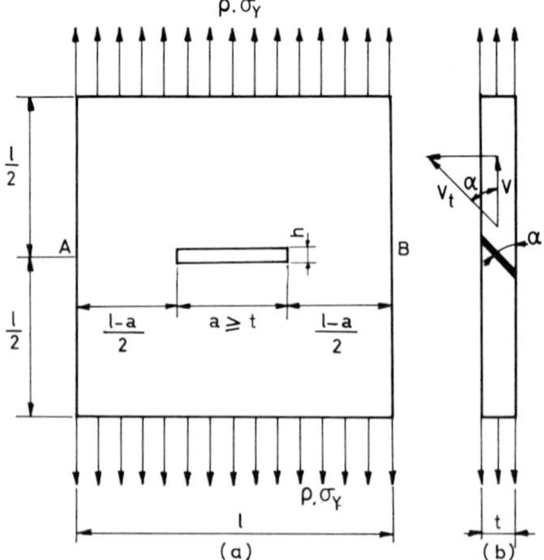

Fig. 11.4. (From Plastic Analysis of Structures by P.G. Hodge Jr. Copyright 1959, Mc Graw-Hill Book Company. Used by permission of Mc Graw-Hill Book Company, Inc.)

Consider the square disk of fig. 11.4 subjected to uniaxial tensile stress σ. The central slit has a length a larger than the thickness t but with negligible width h. From the empirical formula (11.19), the stress at the boundary is

$$\sigma^* = \frac{N}{A} = \frac{\sigma_Y (l - a)\, t}{lt} = \frac{l - a}{l}\sigma_Y .$$

In the absence of a slit, the limit stress would equal σ_Y. Hence, defining the "cutout factor" ρ as

$$\rho = \frac{\sigma^*}{\sigma_Y}, \tag{11.20}$$

we immediately obtain

$$\rho = \frac{l - a}{l}. \tag{11.21}$$

The empirical formula (11.21) has the following theoretical verification. Consider a collapse mechanism with a plane of discontinuity through line AB (fig. 11.4), inclined by an angle α with respect to the median plane of the disk. The area of this surface of slip is $A = \left[\frac{t(l - a)}{\sin\alpha}\right]$. If the upper and lower parts of the disk go apart with a relative velocity V in the direction of the applied load, the relative velocity tangential to the plane of discontinuity is $V_t = \frac{V}{\cos\alpha}$. The power dissipated thus is, according to eq. (11.7) for D,

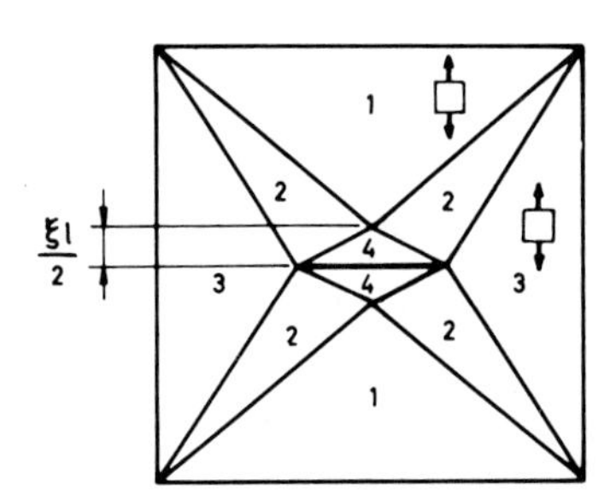

Fig. 11.5. (From Plastic Analysis of Structures by P.G. Hodge Jr. Copyright 1959, Mc Graw-Hill Book Company. Used by permission of Mc Graw-Hill Book Company, Inc.)

$$D = \frac{V\,t\,(l - a)}{\sin 2\,\alpha}\,\sigma_Y\,. \tag{11.22}$$

The power of the loads is

$$P = \sigma\, l\, t\, V\,. \tag{1.23}$$

From the work equation $P = D$, we obtain

$$\sigma = \frac{(l - a)}{l\,\sin 2\alpha}\,\sigma_Y\,.$$

The minimum value of σ is obtained with $\alpha = \frac{\pi}{4}$, and the corresponding upper bound for the cutout factor is

$$\rho_+ = \frac{l - a}{l}. \tag{11.24}$$

A discontinuous statically admissible stress field is then constructed with regions of homogeneous state of stress (fig. 11.5). The parameter ξ defining the extent of the various regions is adjusted to maximize the applied stress σ without violating the yield condition. In this manner, one obtains [3.9]

$$\xi = 1 - \frac{a}{l} + \left(\frac{a}{l}\right)^2,$$

and the corresponding lower bound for the cutout factor is

$$\rho_- = \frac{l - a}{l}\,. \tag{11.25}$$

From comparison of eqs. (11.24) and (11.25) it is concluded that

$$\rho = \frac{1-a}{l} \tag{11.26}$$

in accordance with the empirical expression (11.21).

11.2.2. *Square disk with a central circular hole.*

Consider the square disk of fig. 11.6, subjected to uniaxial tension. The mechanism of fig. 11.4 (b) can also be used in the present case, giving the upper bound (11.24). However, for increasing a, the disk is expected to collapse in the manner of a frame [fig. 11.6 (b)], with four plastic hinges in the weakest sections. With a mechanism of this kind, containing four plastified regions [in grey in fig. 11.6 (a)] similar to plastic hinges, the following upper bound is obtained [3.9] :

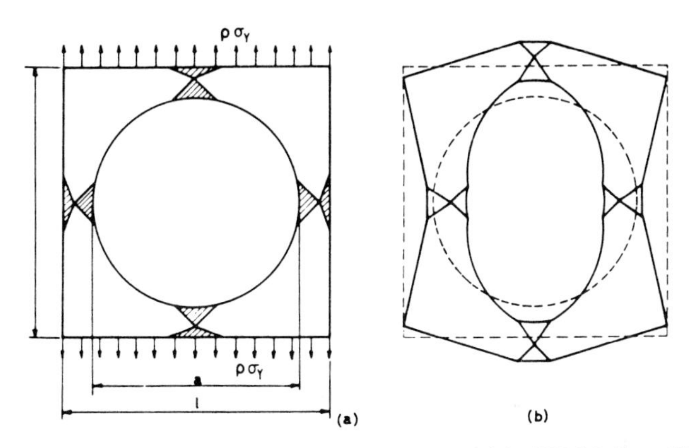

Fig. 11.6. (From Plastic Analysis of Structures by P.G. Hodge Jr. Copyright 1959, Mc Graw-Hill Book Company. Used by permission of Mc Graw-Hill Book Company, Inc.)

$$\rho_+ = \left(2 - 4(a/l) + 3\,(a/l)^2\right)^{1/2} - \frac{a}{l}\;. \tag{11.27}$$

The upper bound (11.27) is smaller than the upper bound (11.24) when $\frac{1}{3} \le \frac{a}{l} \le 1$. With a discontinuous statically admissible stress field made of five kinds of regions, and maximization of the applied stress σ with respect to a parameter ξ, the following bounds are obtained [3.9] :

$$\rho_+ = \left(2 - 4(a/l) + 3\,(a/l)^2\right)^{1/2} - \frac{a}{l}\;. \tag{11.27}$$

The upper bound (11.27) is smaller than the upper bound (11.24) when $\frac{1}{3} \leq \frac{a}{l} \leq 1$. With a discontinuous statically admissible stress field made of five kinds of regions, and maximization of the applied stress σ with respect to a parameter ξ, the following bounds are obtained [3.9].:

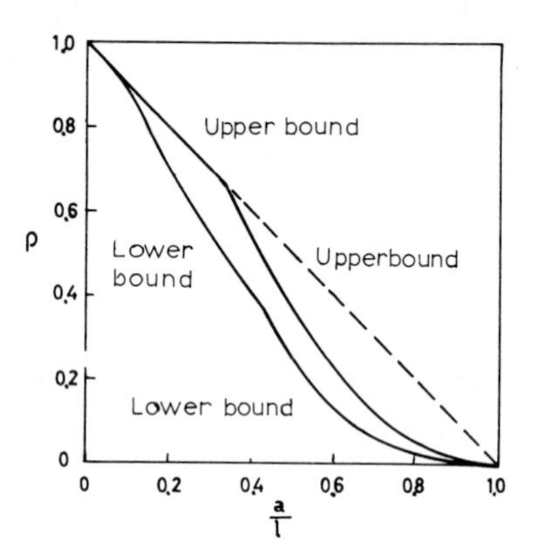

Fig. 11.7. (From Plastic Analysis of Structures by P.G. Hodge Jr. Copyright 1959, Mc Graw-Hill Book Company. Used by permission of Mc Graw-Hill Book Company, Inc.)

$$\rho_{-} = \frac{(1-a)^2}{2\,al} , \tag{11.28}$$

valid when $0.443 \leq \frac{a}{l} \leq 1$. When $0 \leq \frac{a}{l} \leq 0.443$,

$$\rho_{-} = \frac{(1-a\backslash l)(1-\xi)}{\frac{a}{l}}, \tag{11.29}$$

with

$$\frac{\xi(1-\xi)}{\{(3\xi-2)(2-\xi)\}^{1/2}} = \frac{a}{l} .$$

The various bounds are represented in fig. 11.7. It is worth remarking that, in the present case, the concept of "net area" is valid for $a/l < 1/3$ but ceases to be applicable for larger a/l.

A different situation arises when the tensile load is applied by a rigid pulling device that enforces uniform velocity at the upper and lower edge instead of uniform stress. The upper bound (11.27) must then be rejected whereas the upper bound (11.24) applies. On the other hand the stress field of fig. 11.8 is statically admissible for any a/l, and the corresponding lower bound is $\rho_- = (l - a)/l$ which thus turns out to be the exact value.

11.2.3. *Square disk with central circular hole, subjected to biaxial tension.*

Consider the disk of fig. 11.9. Let $\rho_x \sigma_Y$ and $\rho_y \sigma_Y$ be the applied stresses that the disk can support separately at the limit state.

For each ratio a/l, we want to determine, in a space with coordinates ρ_x, ρ_y the locus of the points corresponding to collapse. In the absence of a hole, this locus is identical to the yield locus for $\sigma_Y = 1$.

With the Tresca yield condition, the hexagon of fig. 11.10 is obtained. The desired curves have been bounded by Gaydon [11.1] for various values of a/l. They lie in the shaded regions in fig. 11.10. The bounds for ρ_x are given in fig. 11.11, when $\rho_x / \rho_y = -1$. In the case of equal tensile stresses ($\rho_x / \rho_y = 1$), eq. (11.24) again furnishes an upper bound.

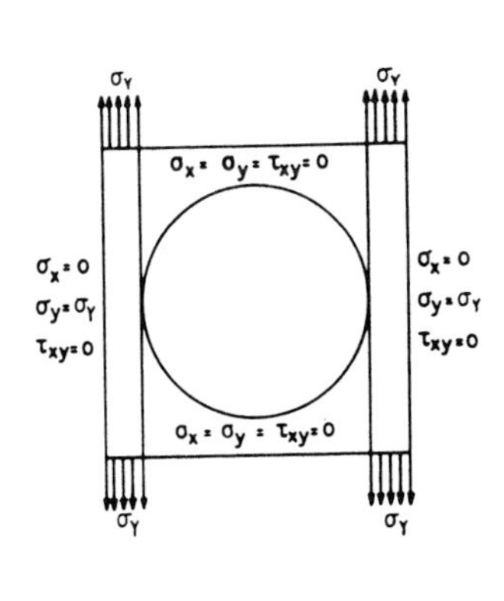

Fig. 11.8.

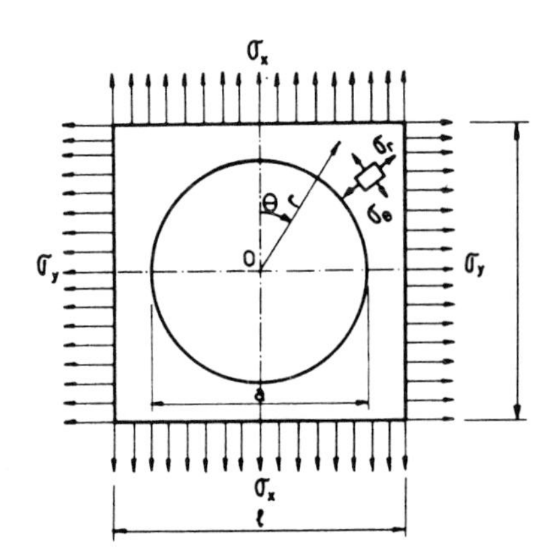

Fig. 11.9.

Consider now the following stress field, referred to the polar coordinate system of fig. 11.9 :

$$\sigma_r = \frac{\sigma_Y \, \rho}{1 - \frac{a}{l}} \left(1 - \frac{a}{2\,r} \right),$$

$$\sigma_\theta = \frac{\sigma_Y \, \rho}{1 - \frac{a}{l}}, \qquad \text{for} \qquad \frac{a}{2} \le r \le \frac{l}{2}, \tag{11.30}$$

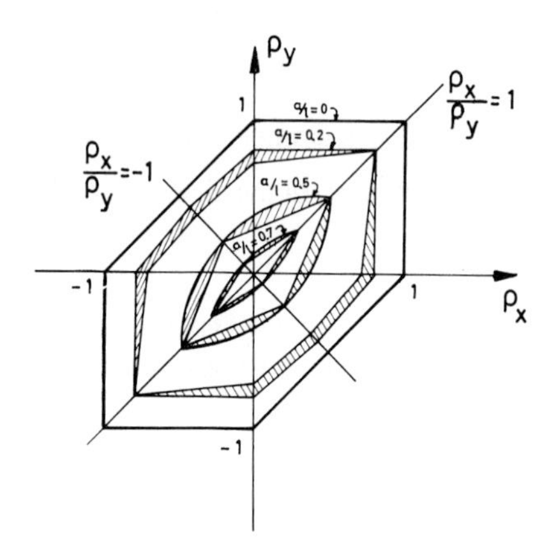

Fig. 11.10.

$$\tau_{r\,\theta} = 0,$$

and

$$\sigma_r = \sigma_\theta = \sigma_Y \, \rho, \qquad \tau_{r\,\theta} = 0, \qquad \text{for} \qquad \frac{l}{2} \le r. \tag{11.31}$$

The reader will easily verify that this field satisfies the equilibrium equation

$$\frac{d(r\,\sigma_r)}{dr} - \sigma_\theta = 0,$$

and the boundary conditions. The yield condition of Tresca is not violated as long as

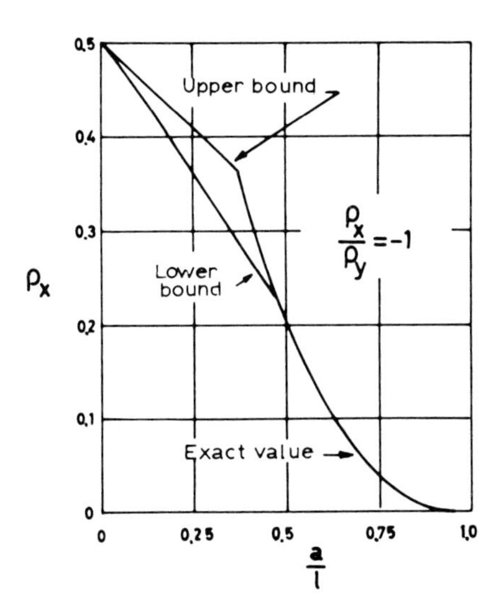

Fig. 11.11.

$$\rho \leq 1 - \frac{a}{l}. \tag{11.32}$$

Hence, when $\rho_x = \rho_y = -1$ the corresponding point of the interaction curve ρ_x versus ρ_y is given by

$$\rho_x = \rho_y = 1 - \frac{a}{l}. \tag{11.33}$$

When the ratio σ_x / σ_y is not known beforehand, it is desirable to evaluate a "general cutout factor" ρ such that, if the disk is able to support the simultaneous stresses σ_x and σ_y it will be able, when perforated, to support simultaneously $\rho\, \sigma_x$ and $\rho\, \sigma_y$. We first assume that buckling does not occur. We next remark that the interaction curves ρ_x versus ρ_y are symmetrical with respect to the rays with eqs. $\rho_x = \rho_y$ and $\rho_x = -\rho_y$. Suppose that, for the considered ratio a / l, the points with coordinates $(\rho_x , 0)$, $(0 , \rho_y)$and the point corresponding to $\rho_x / \rho_y = 1$ are known (point 1, 2, and 3 in fig. 11.12, respectively). It is then possible to draw an hexagon inscribed in the unknown exact curve (dashed in fig. 11.12). The general cutout factor will be given by the smallest value of the ratio of the lengths of segments OA' and OB for all rays like OB, fig. 11.12. The smallest ratio $|\overline{OA}| / |\overline{OB}|$ will give a lower bound to ρ. Because segments $\overline{O1}$ and $\overline{O2}$ are equal, it remains to compare the lengths of segment O1 with $|\overline{O3}| / \sqrt{2}$. From eq. (11.33) we have $|\overline{O3}| / \sqrt{2} = (1 - a/l)\sqrt{2}$, larger that $|\overline{O1}|$as given in fig. 11.7. Hence, the general cutout factor can be evaluated in this figure.

11.2.4. *Rectangular disk subjected to uniaxial tension and perforated by several holes.*

Consider a rectangular flat bar pulled between rigid grips (as in a testing machine). When there is only one row of n holes with diameter a, the mechanism shown in fig. 11.4, together with the generalization of the stress field represented in fig. 11.8, give the exact cutout factor

$$\rho = \frac{l - na}{l} . \tag{11.34}$$

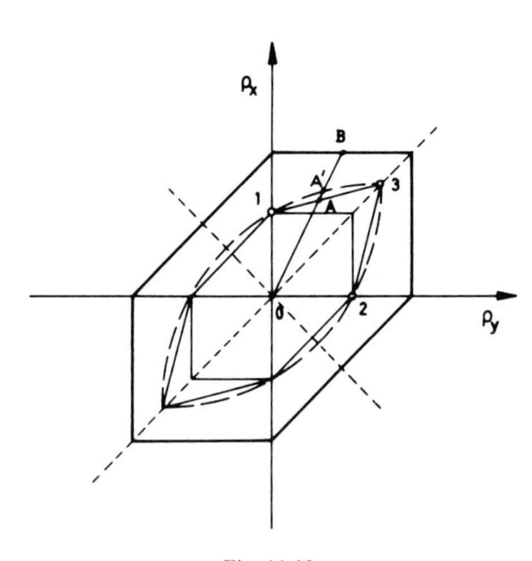

Fig. 11.12.

In analogy with the case of one single hole, relation (11.34) implies that a ≥ t.

When there are two rows with n and n'=n-1 holes, respectively, the same mechanism remains applicable and gives

$$\rho_+ = \frac{l - na}{l} . \tag{11.35}$$

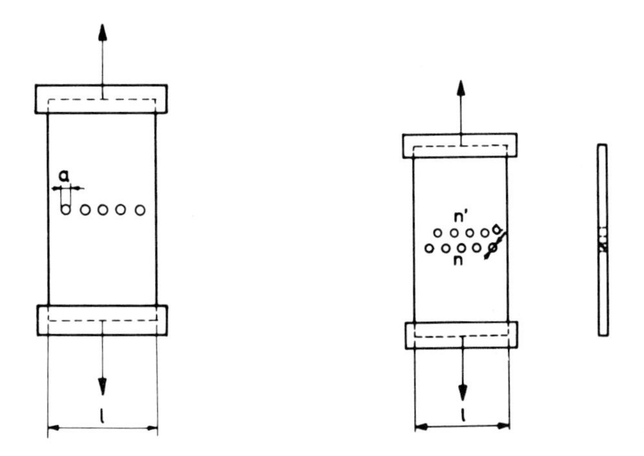

Fig. 11.13 (a).

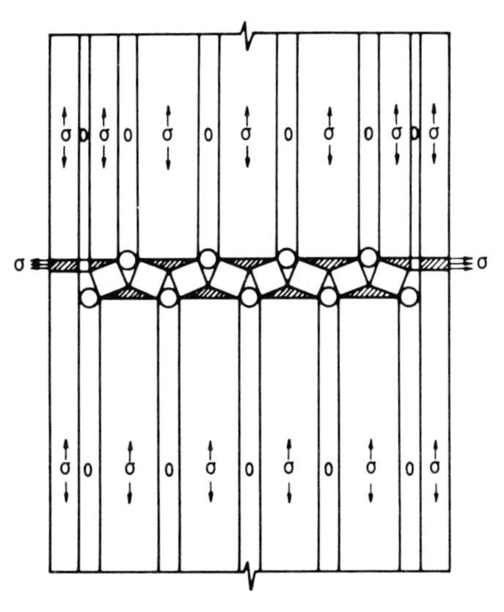

Fig. 11.13 (b).

A more complicated discontinuous stress field is used [11.2], as shown in figs. 11.13 (b), (c), and (d). The dashed areas are in plane hydrostatic state of stress. The tractions on the longitudinal edges in fig. 11.13 (b) are eliminated by the field of fig. 11.13 (d). The parameters C, θ and γ are chosen to maximize the applied load without violating the yield condition. For the particular case n=5, n'=4, $a/l = 1/20$, Brady and Drucker [11.2] have

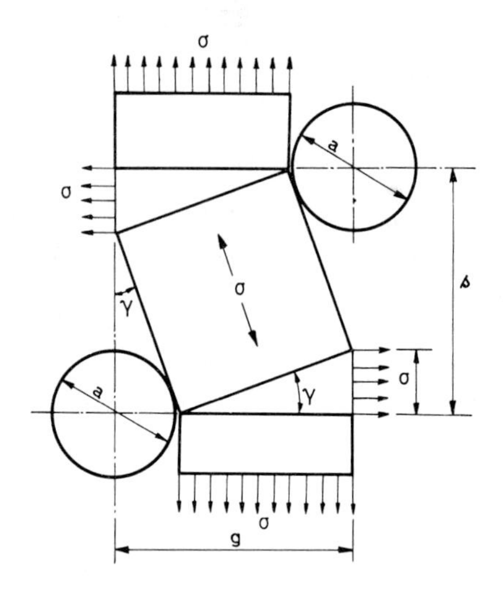

Fig. 11.13 (c).

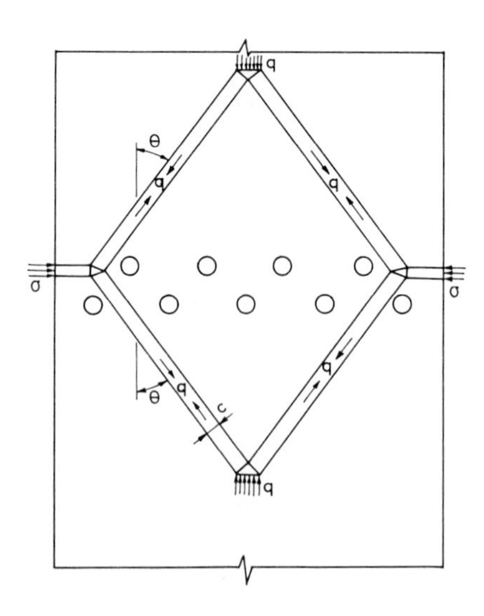

Fig. 11.13 (d).

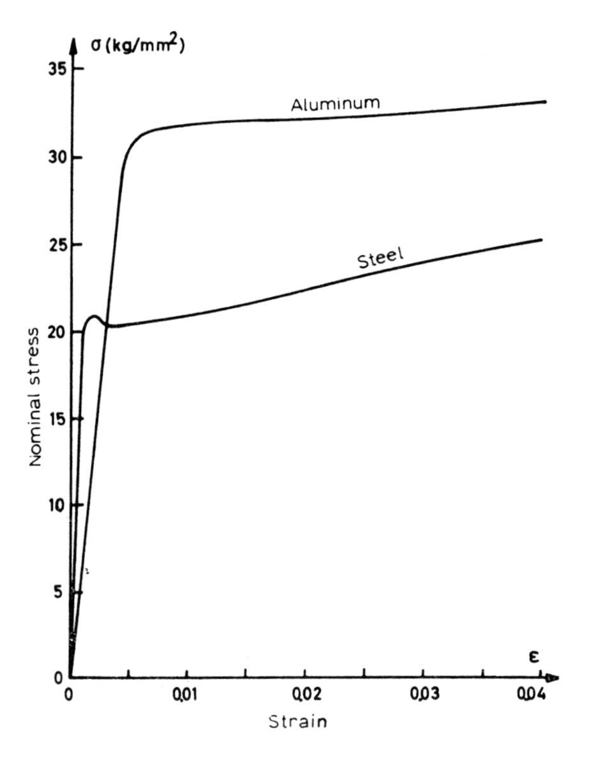

Fig. 11.14. Stress curves for steel and aluminium sheet.

found $\theta = 35°$, $\gamma = 19.9°$ and C as large as possible. The corresponding lower bound for the cutout factor is

$$\rho_{-} = 0.70 , \tag{11.36 a}$$

whereas relation (11.35) gives

$$\rho_{+} = 0.75 . \tag{11.36 b}$$

In a similar manner, the following results have been obtained [11.2] :

1. With three rows of holes (n=5, n'=4, n''=5, a=0.5 in., l=10 in., distance between axes of rows : 1 in.),

$$0.67 \leq \rho \leq 0.72. \tag{11.37}$$

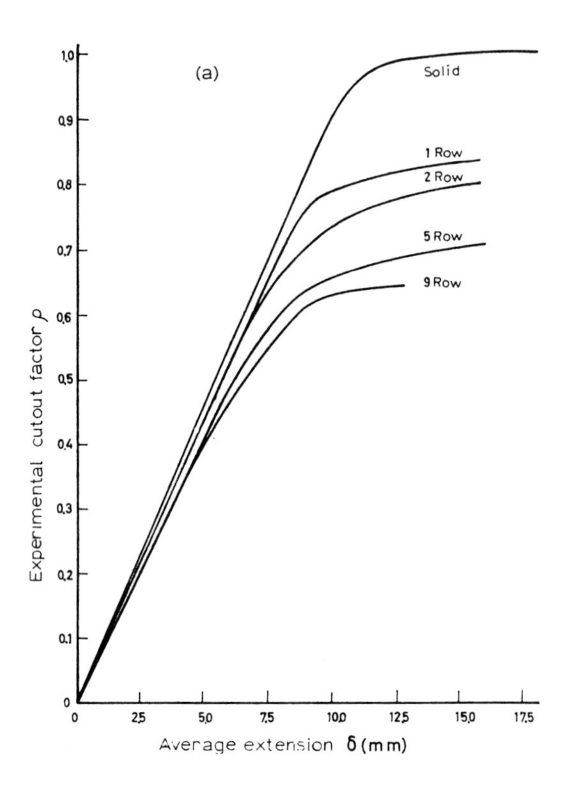

Fig. 11.15 (a) 61S - T6 aluminium sheet.

2. With five rows of holes (same dimensions and distances),

$$0.58 \leq \rho \leq 0.69. \tag{11.38}$$

3. With nine rows of holes (same dismensions and distances),

$$0.55 \leq \rho \leq 0.63. \tag{11.39}$$

In order to apply the "net area" method to the cases studied by Brady and Drucker, the smallest net area must be found, either cross-sectional or following a transversal broken line [11.3]. The geometrical parameters are l, a, n, n', and s and g (fig. 11.16). In the present situation, any segment like AB in fig. 11.16 is larger than $|CD| = \frac{|CE|}{2}$. Hence, the smallest net area is always a cross section though the centres of a row of holes. We thus have An = const = l - 5a, and the corresponding limit load is (with thickness t=1),

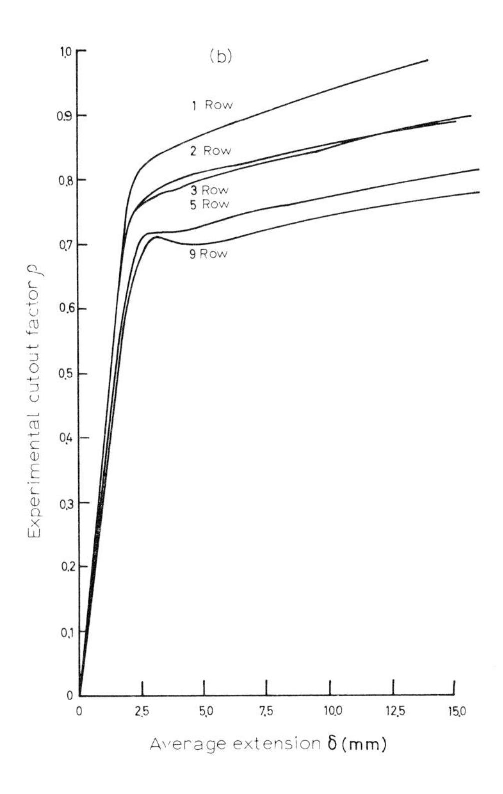

Fig. 11.15 (b). Typical hot rolled steel sheet results.

$$P^* = \frac{\sigma_Y}{S} (1 - 5a),$$

where S is the safety factor. Application of the "$s^2/4g$" rule as prescrived by the A.I.S.C. specification [11.4] would also result in a constant P^*. Because the limit load P_l (theoretical as well as experimental) is smaller than σ_Y (1 – 5a) for more than one row of holes, the practical method gives a decreasing safety with an increasing number of rows. It must be remarked that the net area method is most often used for riveted or bolted connections, where forces are applied in the holes. In this latter case, the discrepancy seems smaller [11.2].

For more information on the subject, we refer the reader to the original paper [11.2] in which an experimental verification of the theroy is also given. Figs. 11.14 and 11.15 summarize the experimental results. It is seen that, in accordance with the theory, the values of ρ are strongly influenced by the configurations of the holes. Experimental cutout factors are not, as a rule, smaller than the theoretical upper bounds probably because the Mises criterion applies better to the steel and aluminium used than does the Tresca criterion.

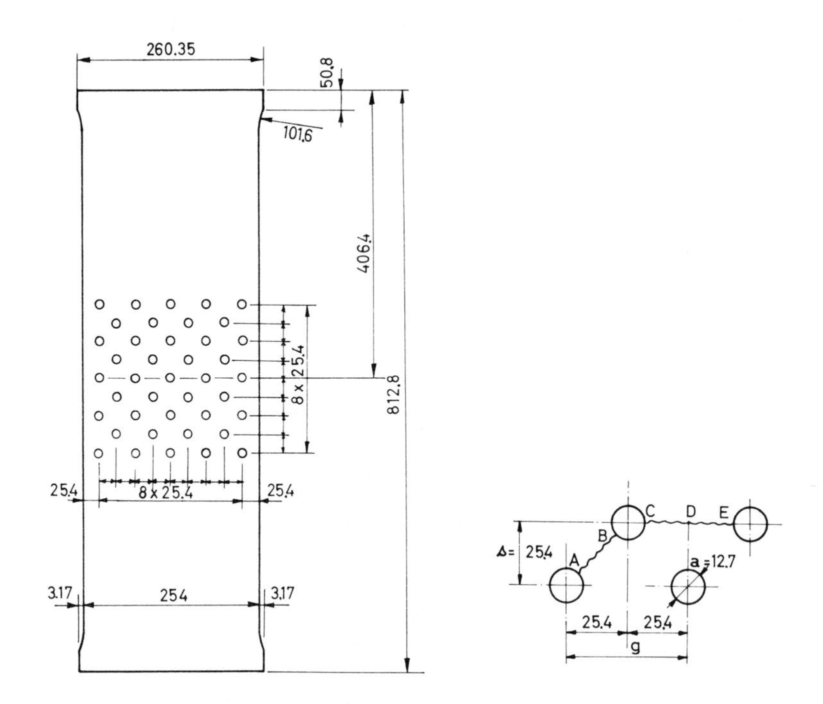

Fig. 11.16. Test piece configuration.

11.2.5. *Square disk with a central circular reinforced cutout.*

In order to restitute, at least partially, the strength lost from the presence of a hole, the disk may be reinforced, for example, as shown in fig. 11.17. Whatever the type of reinforcement, the reinforced disk will be analyzed as formed of a basic disk and a hub of larger thickness. The mechanism shown in fig. 11.4 gives, in the present case,

$$\rho_+ = 1 - \frac{a}{l} + \frac{B}{\frac{lt}{2E}} \tag{11.40}$$

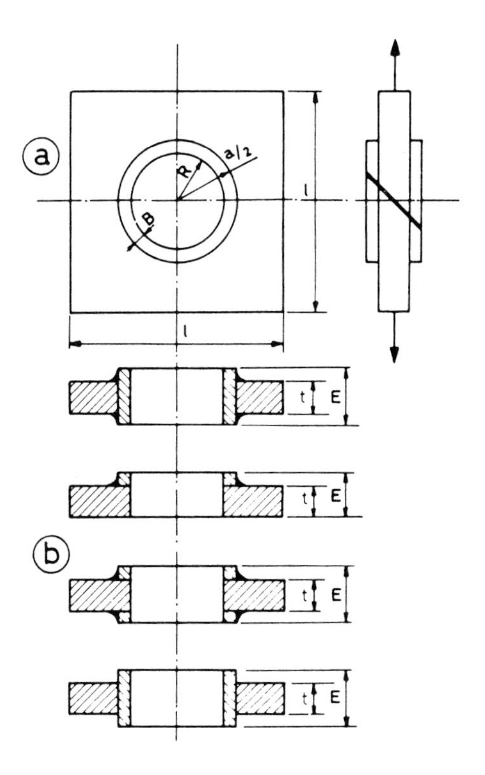

Fig. 11.17. (From Plastic Analysis of Structures by P.G. Hodge Jr. Copyright 1959, Mc Graw-Hill Book Company. Used by permission of Mc Graw-Hill Book Company, Inc.)

(see fig. 11.17 for notations). A lower bound to ρ will be obtained with a discontinuous stress field formed of regions of five types in the disk, and a special region in the hub (fig. 11.18).

The hub is loaded as shown in fig. 11.18 (b) and satisfies the interaction formula (see Com. V., eq. (5.4))

$$\frac{M}{M_p} + \left(\frac{N}{N_p}\right)^2 = 1.$$

With the parameters h=t/E, f=R/B, b=a/l, j=2B/a, one obtains [3.9] :

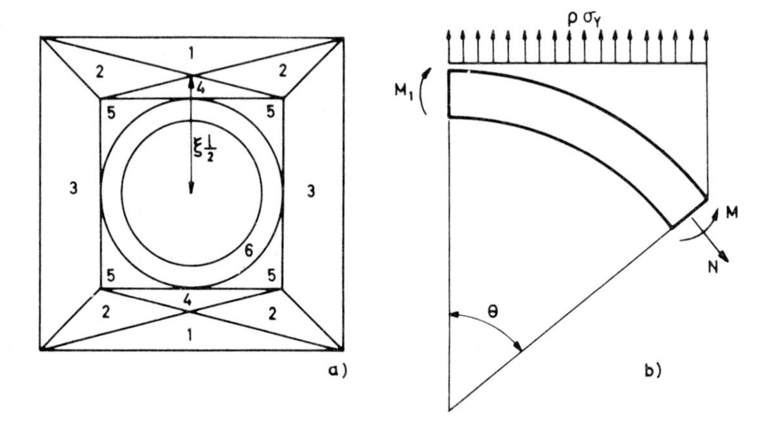

Fig. 11.18. (From Plastic Analysis of Structures by P.G. Hodge Jr. Copyright 1959, Mc Graw-Hill Book Company. Used by permission of Mc Graw-Hill Book Company, Inc.)

$$\rho_- = \min [\rho_1 , \rho_2] , \tag{11.41}$$

where

$$\rho_1 = \rho_h + \frac{(1 - \rho_h)(1 - b)^2}{b(2 - \rho_h)}$$

$$\left[1 - \frac{b(\rho_2 - \rho_h)}{1 - b}\right]^2 = 4 b^2 \frac{(\rho_2 - \rho_h)^2}{(1 - b)^2 - (\rho_2 - b\rho_h)^2} , \tag{11.42}$$

with

$$\rho_h = \frac{-f + (f^2 + 2)^{1/2}}{h(1 + f)} \qquad \text{if} \qquad \frac{1}{2} \le f,$$

$$\rho_h = \frac{1}{h(1 + f)} \qquad \text{if} \qquad 0 \le f \le \frac{1}{2}. \tag{11.43}$$

When the reinforcement must be *designed* to ensure an assigned cutout factor ρ^* (for example, to give the full strength of the disk : $\rho^* = 1$), B and E will be chosen, from relation (11.40), to yield $\rho_+ > \rho^*$. It must then be verified, with eqs. (11.41) to (11.43), that $\rho^* \leq \rho_-$. In the opposite situation, larger E and B must be tried.

When the load is applied by rigid grips, the upper bound (11.40) remains valid, whereas the following lower bound can be obtained [3.9] :

$$\rho_- = 1 - \frac{a}{l} \sin \theta_1, \tag{11.44}$$

where θ_l must satisfy

$$\left[\frac{(1-\gamma f)^2}{2\,(2jf - h)} + 1 + \frac{h}{2}\right] \sin^2 \theta_1 - (1 + jf + h) \sin \theta_1 + \left[jf + \frac{h}{2} - \frac{(1-jf)^2}{h}\right] \leq 0.$$

Hodge and Perrone [11.5] have studied various types of reinforced holes and have compared their results with the experiments of Vasarhelyi and Hechtman [11.6]. The verification is very good for disks with rigid grips. Theoretical lower bounds assuming uniform stress at the boundary are always smaller than experimental values, most often by about 25%.

11.2.6. *Numerical solutions.*

Finite element solutions are proposed by different authors.

In the paper [6.16] of Belytschko and Hodge, bounds were found on the limit load for the four square slabs with sides of length L shown in fig. 11.19. Each slab was weakened by an opening and subjected to a uniaxial loading.

As an example, the finite element representations used to determine the lower and upper bounds for the slab with the square opening are shown in figs. 11.20 and 11.21 respectively. The lower bound model has 18 equilibrium elements and 41 independent parameters (for the principle of the method used, see Section 6.3.1). The upper bound model has 101 displacement elements and 117 independent parameters (see Section 6.4.1).

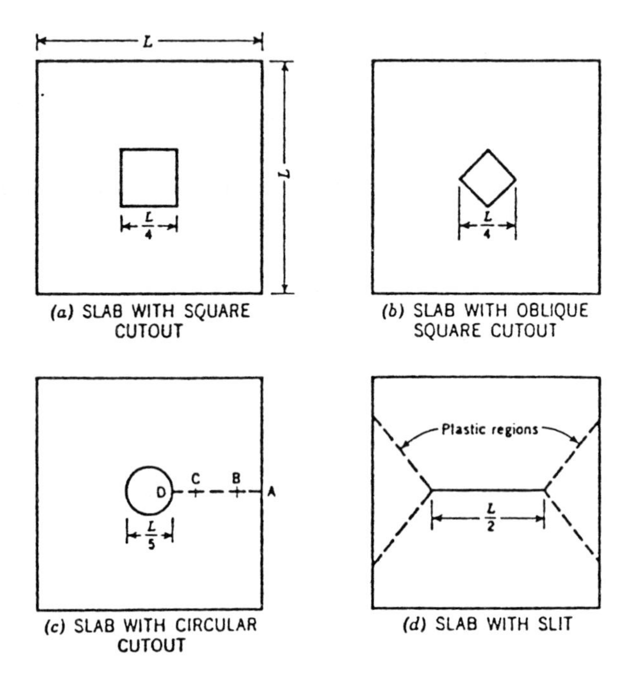

Fig. 11.19. Problem Geometries (All slabs were loaded by uniform tensile load applied in vertical direction). [6.16]

The results obtained, given in Table 11.1 are expressed in terms of cutout factors ρ. In addition, Table 11.1 gives the average of the two bounds, the maximum possible error for this average and, where applicable, other numerical or analytical solutions.

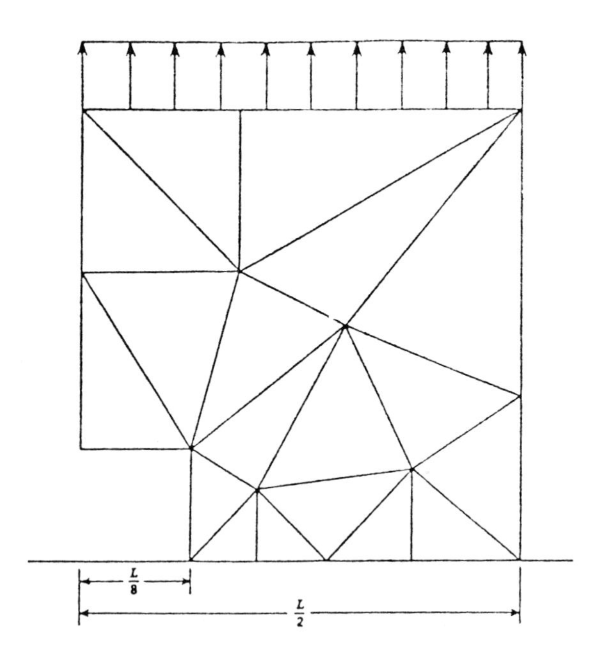

Fig. 11.20. Slab with square opening equilibrium element discretisation [6.16].

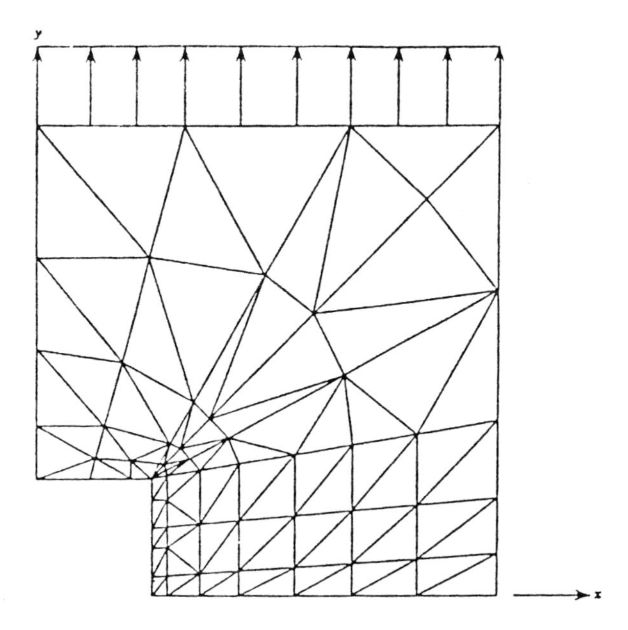

Fig. 11.21. Slab with square opening displacement element discretisation [6.16].

Table 11.1. Results of Belytschko and Hodge [6.16].

	Lower bound	upper bound	average	maximum error	other numerical solutions	analytical solutions (exact)
Slab (1)	ρ^- (2)	ρ^+ (3)	$\rho^a = (\rho^+ + \rho^-)/2$ (4)	$100\,[(\rho^+ - \rho^a)/\rho^a]$ (5)	(6)	
Slab with square cutout	0.693	0.764	0.729	4.8	0.722 [6.36]	
Slab with oblique square cutout	0.740	0.799	0.759	2.5		
Slab with circular opening	0.793	0.824	0.809	1.9	0.806 [6.36] 0.885 [6.67]	0.8
Slab with slit	0.498	0.522	0.510	2.1	0.538 [6.67]	0.5

As the bounds obtained are true bounds in that all conditions of the upper and lower bound theorems are met exactly, the results may be assessed by comparing the upper and lower bound results. As can be seen from Table 11.1, the bounds are within 5% of each other for all of the results except for the slab with a square hole. Therefore the results are suitably accurate for most engineering purposes.

For purposes of checking which of the bounds is more accurate, it would be desirable to check these bounds against an exact solution. However, *for the Mises material no exact solutions are available for weakened slabs*. An exact limit load of 0.5 has been obtained for the slab with a slit and the Tresca yield criterion (6) [see eq. (11.24)]. As the Tresca yield surface inscribes the Mises, this result is immediately applicable as a lower bound for the Mises material ; it may also be used to derive an upper bound of 0.576. As can be seen from Table 11.1 the bounds obtained here are 0.498 and 0.522.

The lower bound result for the slit slab is of interest in that the problem was solved with progressively finer element subdivisions ranging from 5 to 16 elements without any improvement in the bound. This was in marked contrast to the other problems run, in which subdividing the elements invariably improved the bound. On this basis it was conjectured that 0.50 is an exact solution to this problem. However, an attempt to substantiate this conjecture by using a very fine finite element representation for the upper bound problem was not successful. Difficulties were encountered because of the nature of the velocity field for this problem at collapse ; evidently, the slab is plastic in the two triangular regions at the sides of the slit, as shown in fig. 11.19 (d) *and rigid everywhere else*. Therefore, *either very steep velocity gradients or velodity discontinuities must occur near the ends of the slit* along the line separating the plastic and rigid regions.

The two first problems (fig. 11.18')(a) and (b) have been also investigated by Casciaro and Di Carlo in [6.36].

As they use the mixed finite element method (see Section 6.5.2), their model gives only an approximation of the collapse cutout factor, but no true bounds. Upper bounds have also been obtained for the two last problems (fig. 11.19 (c) and (d)) by Hayes and Marcal [6.36], using displacement finite elements.

11.3. Notched bars in tension.

Consider the V-notched bar shown in fig. 11.22. For very small thickness ($t/b \to 0$) we are in plane-stress conditions, and the mechanism of fig. 11.22 (a), with slip planes at 45° of the median plane of the bar, gives the upper bound $P_+ = \sigma_Y\, bt$ for the limit load.

A central band of width b in uniform traction with magnitude σ_Y [fig. 11.22 (d)] is a statically admissible stress field with the same load. Hence, the exact limit load is

$$P_l = \sigma_Y\, bt\,. \tag{11.45}$$

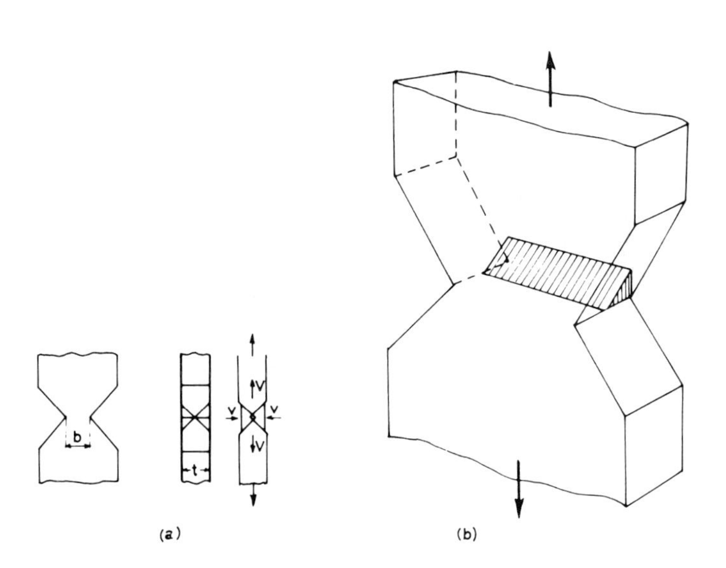

Fig. 11.22.

For an arbitrary value of the ratio t/b, the mechanism shown in fig. 11.22 (a) gives the upper bound [11.7]

$$P_+ = \sigma_Y\, bt\left(1 + \frac{\sqrt{2t}}{4b}\right), \tag{11.46}$$

the slip surfaces having the total area $2bt\sqrt{2}\ +t^2$, and the relative velocity of the sliding blocks in the slip planes being $\sqrt{2}\ v$. A fourth of the total sliding surface is shaded in fig. 11.22 (b).

In plane strain conditions ($t/b \rightarrow \infty$), the exact limit load is [11.8]

$$P_l = \sigma_Y\, bt\left(1 + \frac{\pi}{4}\right) = 1.785\ \sigma_Y\ bt. \tag{11.47}$$

For large but finite t/b ratio, the slip-line field of the plane strain state provides an upper-bound solution, with the load given by eq. (11.47). As shown in fig. 11.23, the upper bound (11.46) gives a higher load than eq. (11.47) (and, hence, should be rejected, according to the kinematic theorem) for $t/b \geq 2.22$. The experimental results of Findley and Drucker [11.9] show that : (a) the aluminium specimen tested seems to satisfy the yield criterion of von Mises better than Tresca's ; the corresponding experimental limit loads, when divided by $2/\sqrt{3}$, come close to the points representing the limit loads of the steel specimen ; and (b) a very large t/b ratio (larger than 6) is required to obtain plane strain conditions. The need of great experimental care (in particular for what regards axiality of the load) is emphasized in [11.9]. The conclusion under (b) above is not confirmed by Sczcepinski and Miastkowski [11.10] who tested forty-eight mild steel flat-notched bars and concluded from their experimental results shown in fig. 11.24 that, for $t/b \geq 2$ the limit load is practically constant and equal to the plane-strain limit load.

Experiments on nonsymmetric notches are reported by Dietrich [11.11], whereas a survey on the problem of notched bars has been presented by Szczepinski [11.12], who cites important literature on the subject.

The collapse of a strip with a V-notch was also computed numerically by Nguyen Dang Hung [6.33] [6.12]. He uses equilibrium finite elements and von Mises' criterion of yield "in the mean" (relaxed problem, see Section 6.3.2). Nguyen Dang Hung finds the approxipation of the limit load P=1.192 bt instead of the exact solution $P_l = 1.155$ bt (see [3.12]).

Fig. 11.25 represents the collapse mechanism obtained by finite elements. The elements for which the yield point is reached, in the sense of the mean, are in black.

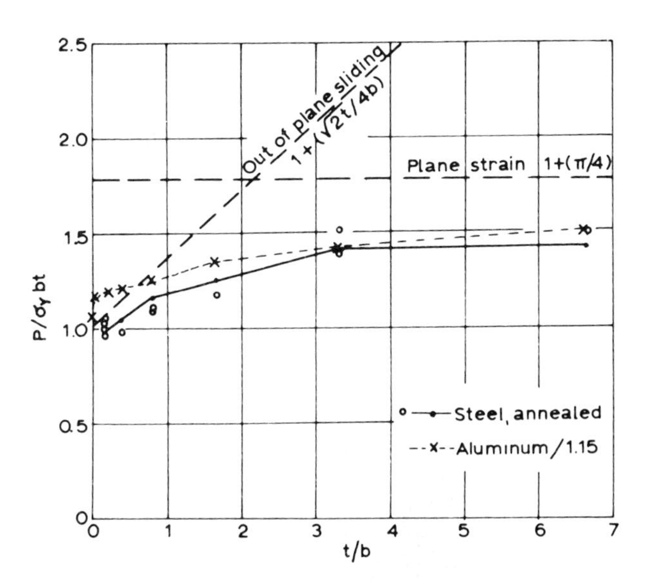

Fig. 11.23.

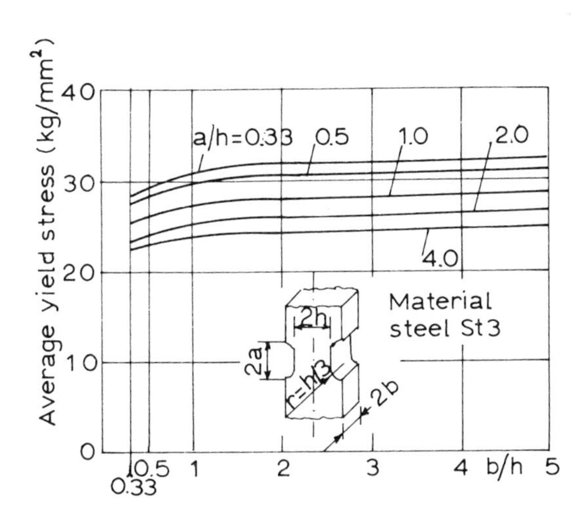

Fig. 11.24.

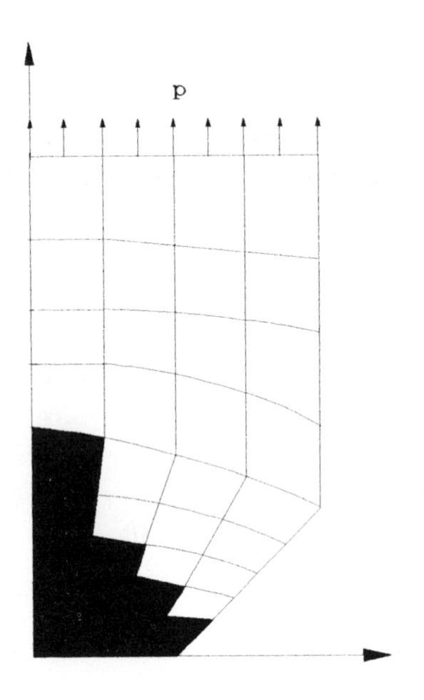

Fig. 11.25.

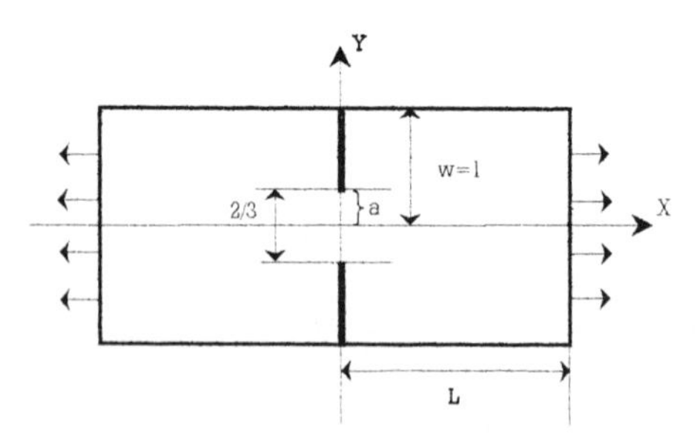

Fig. 11.26. Geometry fo the plane strain problem [6.35].

Another interesting problem is that of a rectangular bar with symmetric infinitely thin external cuts. Numerical solutions are given by Christiansen [6.35].

The bar is assumed large in the z-direction, and the *plane strain* approximation is used. The bar is subject to a uniform tensile force at the ends of density λ per unit length in the y-direction. The length of the bar is 2L, the width 2w, and *each of the cuts* are w - a deep (see fig. 11.26) ; the Mises condition (11.13) is used as yield condition.

Also the Tresca condition (11.5) may be used, but usually the linearization is more complex.

Several values of the width of the intact region have been used. We start with a few comparisons for the case a=1/3, which has been studied most often in the literature. A trivial lower bound for the collapse multiplier γ^* is found by considering only the intact region with the total tensile force 2w γ :

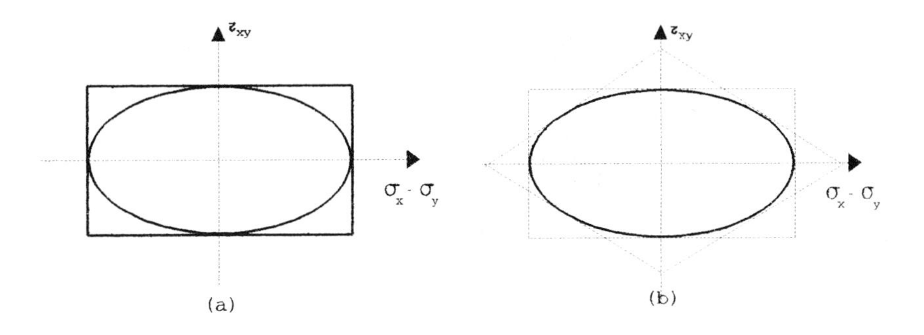

Fig. 11.27. Linearizations of the yield conditic.n (11.13) : (a) condition (11.48) ; (b) condition (11.49) [6.35].

$$\lambda(\text{lower}) \cdot 2w = 2\sigma_0 \cdot 2a$$

or

$$\lambda^* \geq 2/3.$$

To get an upper bound we disregard the cuts :

$$\lambda(\text{upper}) \cdot 2w = 2\sigma_0 \cdot 2w$$

or

$$\lambda^* \leq 2 .$$

For a bar of infinite length (L = ∞) the following bounds are found in Reference [11.25] (p. 482) :

$$1 \leq \lambda^* \leq 4/3 .$$

For a bar of infinite length and width the solution can be found exactly [11.24] [11.25] (L = ∞ , w = ∞):

$$\lambda^* = \frac{(2+\pi)}{3} .$$

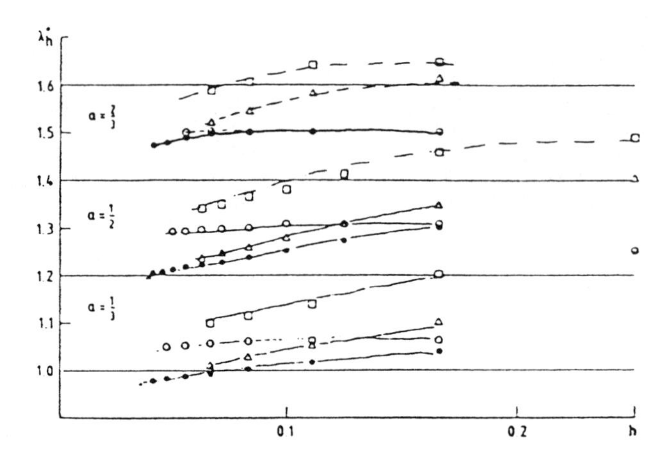

Fig. 11.28. Approximate collapse multipliers for three values of the width of the intact region : a = 1/3, a = 1/2 and a = 2/3 [6.35].

Finite bars are weaker. This classical problem has been used to measure material properties and should be of considerable interest. In addition the problem is mathematically equivalent to the classical punch problem [11.25]. Mixed finite elements (see Section 6.5) are used with the following stress and displacement fields :

$\underline{\sigma}$, piecewise bilinear, $\underline{u}$ piecewise bilinear (elements (a))

$\underline{\sigma}$, piecewise constant, $\underline{u}$ piecewise bilinear (elements (b))

Both of these couples have been combined with two linearizations of the yield condition (11.13) (see fig. 11.27) :

$$\frac{1}{2}\,|\sigma_x - \sigma_y| \leq 1, \quad |\tau_{xy}| \leq 1 \qquad (11.48)$$

$$\frac{1}{2}\,|\sigma_x - \sigma_y| \leq 1, |\tau_{xy}| \leq 1, \quad \frac{1}{2}\,|\sigma_x - \sigma_y| + |\tau_{xy}| \leq \sqrt{2} \qquad (11.49)$$

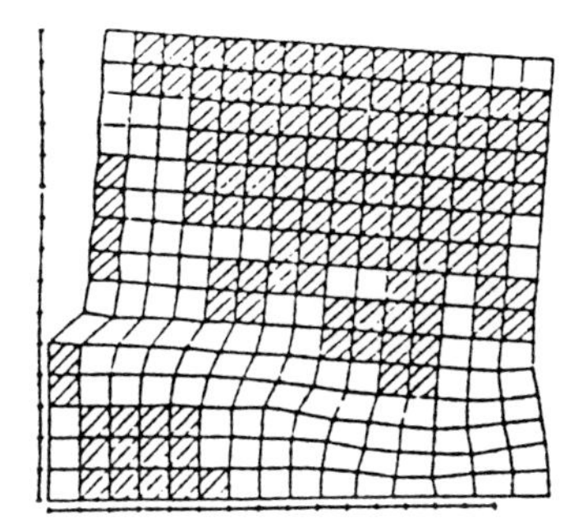

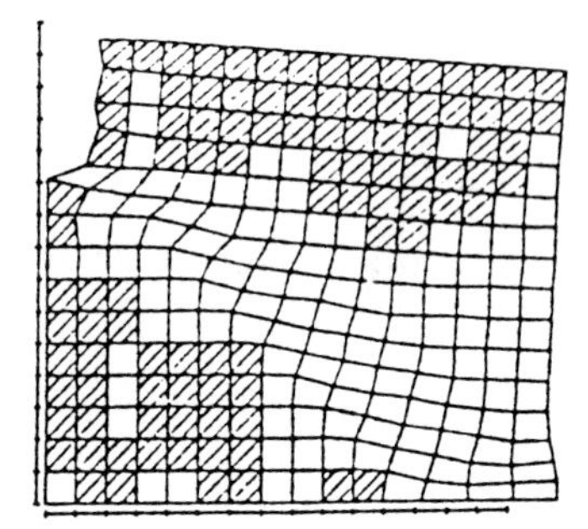

Fig. 11.29. Collapse state for elements (a). Condition (11.49), L = 1, h = 1/15 : (a) a = 1/3 ; (b) a = 2/3. The non-plastic region is hatched.

For L=1, the resulting approximate collapse multipliers λ^* computed for a length h of the element side are shown in fig. 11.28. More complete results are given in ref. [6.35].

Clearly for this problem the elements (b) give the better value for the collapse multiplier. Since the elements (b) also results in a smaller programming problem with fewer primal variables, allowing computations for smaller element size, the piecewise constant elements for $\boldsymbol{\sigma}$ are definitely preferable in this respect.

The error does seem to be a decreasing function of h, although not as a constant times h. The collapse state for particular values of "a" is represented in fig. 11.29.

11.4. Thin rotating disks.

11.4.1. *Introduction.*

Consider a thin circular disk of uniform thickness t, rotating with an angular speed ω (rad/sec). Let γ be the weight per unit volume of the material. The force of inertia per unit volume has the value $F = \gamma\,\omega^2\, r/g$, where r is the radial coordinate of the considered unit volume (fig. 11.30) and g the gravity acceleration. In elastic range, the principal () stresses σ_r and σ_θ (fig. 11.30) are given by [1.1]

$$\text{(a)} \qquad \sigma_r = \frac{3+\nu}{8}\,\frac{\gamma}{g}\,\omega^2 \left(b^2 + a^2 - \frac{a^2 b^2}{r^2} - r^2 \right)$$

$$\text{(b)} \qquad \sigma_\theta = \frac{3+\nu}{8}\,\frac{\gamma}{g}\,\omega^2 \left(b^2 + a^2 + \frac{a^2 b^2}{r} - \frac{1+3\nu}{3+\nu}\, r^2 \right) \qquad (11.50)$$

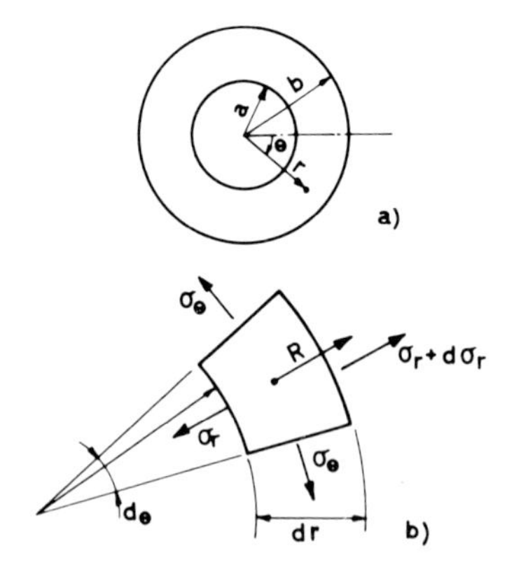

Fig. 11.30.

For thin disks, the third principal stress σ_z can be neglected (plane-stress conditions). The stress distribution (11.50) is shown in fig. 11.31 (a). Both principal stresses are positive throughout the disk. With the Tresca yield criterion, plasticity will first occur at r=a, when

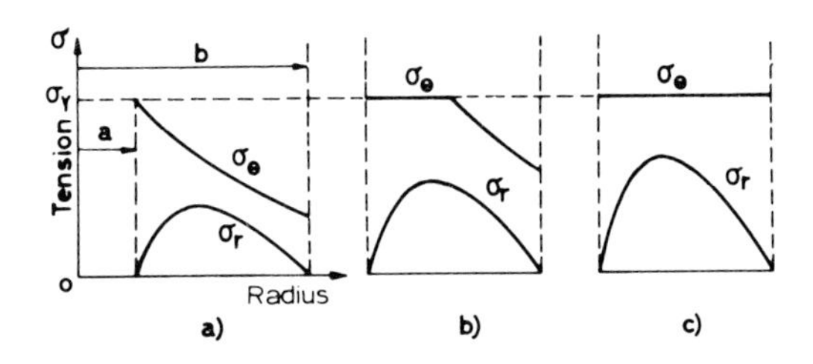

Fig. 11.31.

$\sigma_\theta = \sigma_Y$. With increasing speed ω, plasticity will spread in the disk and the stress distribution becomes the type shown in fig. 11.31 (b).

As long as there remains an outer elastic annulus, displacements remain small (of the order of elastic displacements) and, hence, acceptable. At a certain value ω_l of the angular speed, the plastic zone spreads to the outer radius b and unrestricted plastic flow sets in, with the stress distribution of fig. 11.31 (c). If ω_s is the angular speed under service conditions, the safety factor against plastic collapse by overspeed will be $s = \omega_l / \omega_s$. Note that this safety factor is meaningful only if failure does not occur previously from other causes such as fatigue, excess elastic strains or creep. The influence of the temperature field on the elastic-plastic strains and on the local values of the yield stress σ_Y should also be considered.

At the plastic limit state, the stress field is statically determinate. Indeed, to the equilibrium equation in the radial direction

$$\frac{d}{dr}(r\,\sigma_r\,t) - t\,\sigma_\theta + \frac{\gamma}{g}\,\omega^2\,r^2 = 0, \tag{11.51}$$

it is sufficient to add the yield condition

$$\sigma_\theta = \sigma_Y \tag{11.52}$$

to obtain the function σ_r (r), taking into account the boundary conditions

$$\sigma_r = 0 \qquad \text{at} \qquad r=a, \tag{11.53}$$

$$\sigma_r = \sigma_{rb} \text{ at } \qquad r=b. \tag{11.54}$$

The stress σ_{rb} vanishes when the outer edge is free, and is given by

$$\sigma_{rb} = \frac{T}{t_b} \tag{11.55}$$

when the edge is loaded by radial tensile forces with magnitude T per unit of length (due to turbine blades, for example). It must also be verified that the obtained solution satisfies the inequality

$$\sigma_\theta > \sigma_r > 0. \tag{11.56}$$

11.4.2. *Disk with constant thickness.*

The equilibrium eq. (11.51) will be rendered nondimensional by dividing both sides by σ_Y. If we let

$$s_r = \frac{\sigma_r}{\sigma_Y}, \qquad s_\theta = \frac{\sigma_\theta}{\sigma_Y}, \qquad K = \frac{\gamma}{g}\frac{b^2\omega^2}{\sigma_Y}, \quad \rho = \frac{r}{b}, \tag{11.57}$$

eq. (11.51) is rewritten, with the yield condition $s_\theta = 1$,

$$\frac{d}{d\rho}(s_r\, t\, \rho) = t\,(1 - K\rho^2). \tag{11.58}$$

Integration of (11.58) gives

$$s_r = 1 - \frac{1}{3}K\rho^2 + \frac{C}{\rho}. \tag{11.59}$$

The integration constant C is determined from the boundary condition at the inner edge. In the absence of a central hole, a=0 and, by symmetry, $s_r = s_\theta = 1$ at $\rho = 0$. Hence, C=0 and

$$s_r = 1 - \frac{K}{3}\rho^2. \tag{11.60}$$

When there exists a central hole, $a \neq 0$, then $s_r = 0$ at $\rho = a/b \equiv \rho_a$, and the corresponding expression for s_r is

$$s_r = 1 - \frac{K}{3}\rho^2 - \frac{1}{\rho}\left(\rho_a - \frac{K}{3}\rho_a^3\right). \tag{11.61}$$

To obtain the thickness t of the disk when ω_l is assigned, expression (11.57) for K is substituted into eq. (11.61) or eq. (11.60). If we let $\rho = 1$ in the resulting relation, and use the condition $s_{rb} = \sigma_{rb} / \sigma_Y$, we obtain t.

It is remarked that, in the absence of a central hole, the upper limit for K is 3, according to relation (11.60). On the other hand, when there is a central hole, the situation differs depending on the existence of edge forces T. If there are no such forces (T=0), use of $s_r = 0$ for $\rho = 1$ in eq. (11.61) yields

$$K = \frac{3}{1 + \rho_a + \rho_a^3}. \tag{11.62}$$

From eq. (11.62) and (11.57) we see that neither K nor ω_l depends on the value t of the thickness. When $T \neq 0$ however, for given a, b, σ_Y and ω_l, the thickness t is directly dependent on T. Because a uniform thickness may then become too large (see [11.13], p. 535), it is reasonable to consider disks with an inner part with constant thickness and an outer part with variable thickness.

10.4.3. *Practical shapes of rotating disks.*

We restrict ourselves to disks such that $1 \leq K \leq 3$ (the usual range in turbine disks) and assume that the condition $s_{rb} = 1$ is to be satisfied. For more detailed analysis, in particular, when $s_{rb} < 1$, we refer the reader to Heyman [11.13].

Consider first a disk with a constant thickness for $0 \leq r \leq r_x$ and with a doubly conical outer part, with cone angle α, fig. 11.32. In this outer region

$$t = t_b + A\,(1 - \rho), \tag{11.63}$$

where A is a constant with the dimension of length. Substitution of expression (11.63) for t in eq. (11.58) and integration give

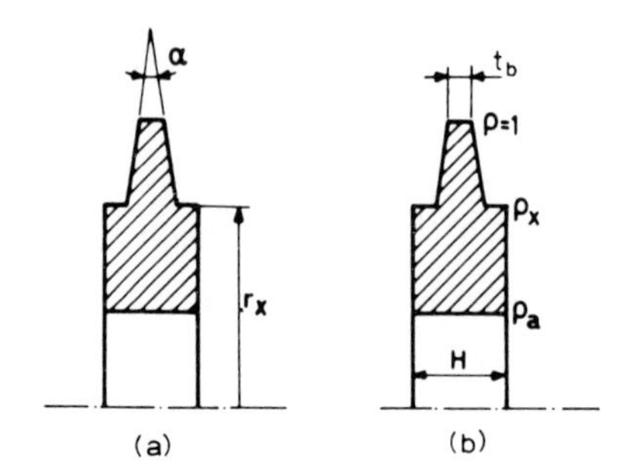

Fig. 11.32.

$$s_r\left[t_b + A\,(1-\rho)\right] = (t_b + A)\left(1 - \frac{K\rho^2}{3}\right)$$

$$-\frac{A}{2}\rho\left(1 - \frac{K\rho^2}{2}\right) - \frac{C}{\rho}\,, \tag{11.64}$$

where C is an integration constant.

The magnitude of A must be such that $(s_r)_{max} \le 1$, and the yield condition $s_\theta = 1$, $s_\theta \ge s_r \ge 0$ is not violated. Differentiating eq. (11.64), it is found that, for s_r to decrease with ρ decreasing from 1, it is necessary that

$$A \ge t_b\,K. \tag{11.65}$$

The constant C is then obtained from the condition $s_r = s_{rb}$ at $\rho = 1$, and we finally have

$$s_r\,t = (t_b + A)\left(1 - \frac{K\rho^2}{3}\right) - \frac{A\rho}{2}\left(1 - \frac{K\rho^2}{2}\right)$$

$$-\frac{1}{\rho}\left[(t_b + A)\left(1 - \frac{K}{3}\right) - \frac{A}{2}\left(1 - \frac{K}{2}\right) - t_b\right]. \tag{11.66}$$

If the radius r_x at which the profile of the disk changes is so chosen as to have the coefficient of A in eq. (11.66) vanish when $\rho = \rho_x$, the forces transmitted from the doubly-conical part to the central part will not depend on the angle α of the outer part. This condition is

$$3\rho^2 + 2\rho - \left(\frac{6}{K} - 1\right) = 0, \qquad (11.67)$$

and its solution is

$$\rho^* = \frac{1}{3}\left[\left(\frac{18}{K} - 2\right)^{1/2} - 1\right]. \qquad (11.68)$$

If $\rho_x = \rho^*$, the radial force per circumferential unit of length transmitted across the section with equation $\rho = \rho_x$ is

$$(s_r\, t)_x = t_b \left(1 - \frac{K\rho_x^2}{3} + \frac{K\rho_x}{3}\right). \qquad (11.69)$$

The thickness H of the constant thickness part is then obtained as explained in Section 11.4.2, regarding r_x as the outer radius with $T = (s_r\, t)_x\, \sigma_Y$.

From the preceding analysis it is seen that the thickness H will be independent of A if $\rho_x = \rho^*$. The smallest admissible value t_b K, relation (11.65), can then be given to A. Even more noticeable is the fact that, if the profile of the outer part is changed, the force $(s_r\, t\, \sigma_Y)_x$ transmitted in the section $\rho = \rho_x$ is practically independent of the chosen profile (and so is H), for $\rho_x \geq \rho^* = 0.7208$. This can be seen in fig. 11.33, and is discussed by Heyman in ref. [11.13]. In fig. 11.34 corresponding to K=1.5, it can also be noted that, if $\rho_x \geq \rho^*$, the force $(s_r\, t\, \sigma_Y)_x$ depends very little on s_{rb} and, hence, the condition $s_{rb} = 1$ will have little influence on the magnitude of H.

For these reasons, the more easily feasible doubly-conical outer part will usually be preferred to other shapes. It will be connected by fillets to the central part with constant thickness. The curves in fig. 11.35 are good design aids. The radius a may be chosen for

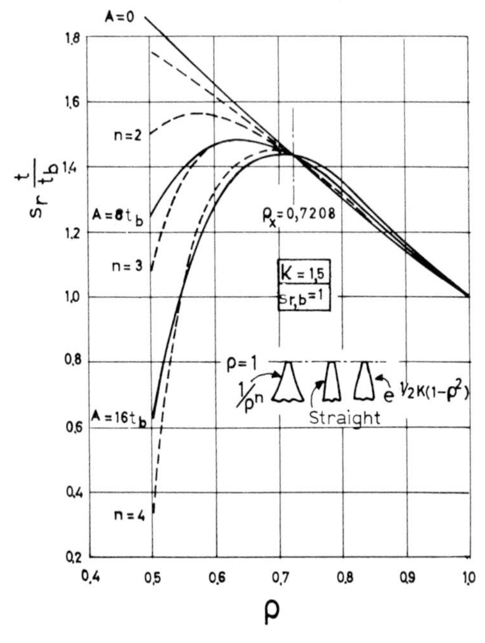

Fig. 11.33.

to coincide with some critical non-dimensional radius ρ_c where s_r is a maximum in a disk with constant thickness (the dashed curve in fig. 11.35). Fig. 11.36 gives ρ_x versus K

Consider, for example, the disk shown in fig. 11.37 [11.13]. The yield stress is $\sigma_Y = 422\ ^N\!/\!_{mm^2}$. The design angular speed is 3.000 rpm. With a safety factor of 1.5, we have $\omega_1 = 4.500$ rpm. At the bursting speed, the edge loading is $T = 10.25\ ^{kN}\!/\!_{mm}$. We compute

$$K = \frac{\gamma}{g}\frac{b^2\,\omega^2}{\sigma_Y}\ ,$$

and

$$t_b = \frac{T}{\sigma_Y} = 24.3\ \text{mm.}$$

The dashed curve in fig. 11.35 gives $H/t_b = 5.5$ and $\rho_a = 0.315$. Hence, $a = 0.315 \times 635 = 200$ mm. Then, $A = t_b\, K = 41.5$ mm. In fig. 11.36 we find $\rho_x = 0.64$. The thickness of the doubly-conical part at $\rho = \rho_x$ thus is

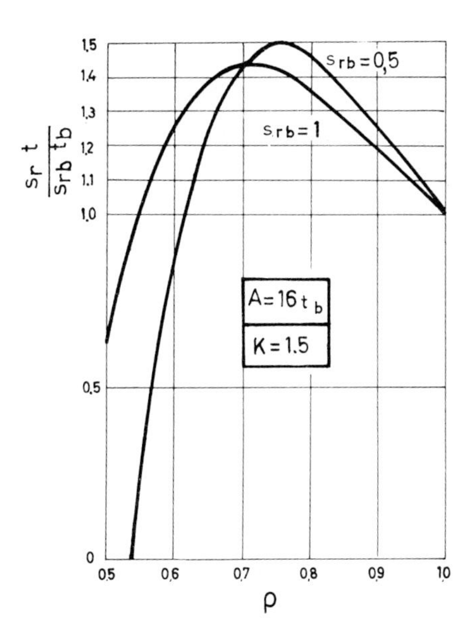

Fig. 11.34.

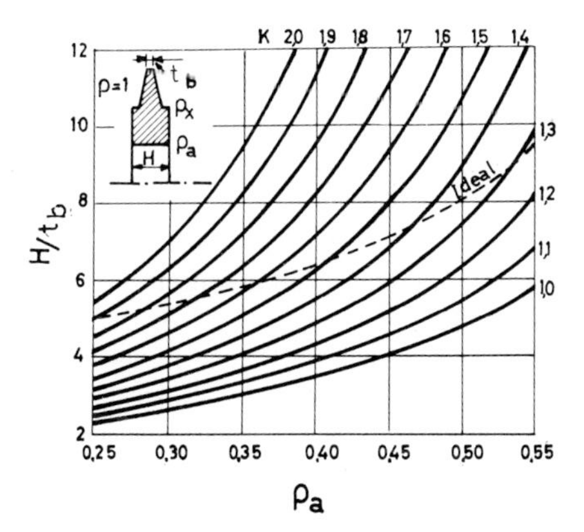

Fig. 11.35.

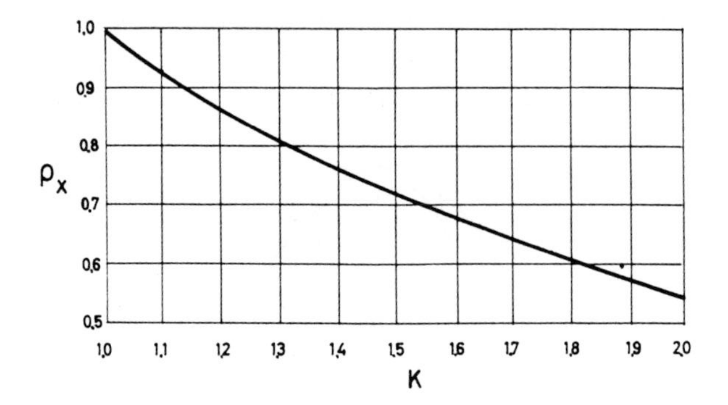

Fig. 11.36.

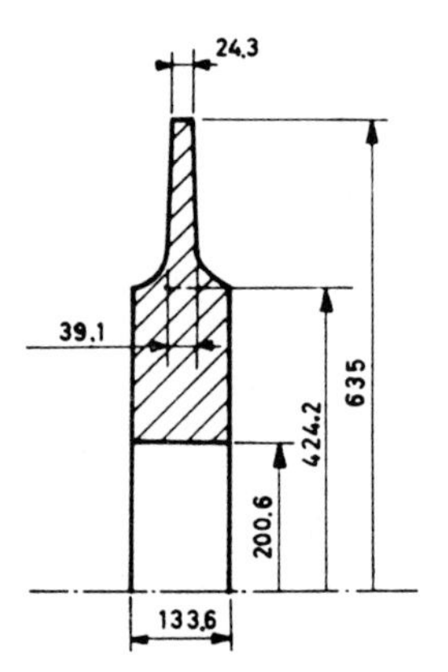

Fig. 11.37. (all mm)

$$t_x = t_b + A\ (1 - \rho_x) = 39.25 \text{ mm},$$

and the uniform thickness is H = 5.5 × 24.3 = 134 in. Fillets will be machined at the junction of the two parts, as shown in fig. 11.37.

11.5. Other planes stress problems.

Nguyen Dang Hung [6.33], [6.12] has studied numerically by displacement finite elements the problem of disk with eccentric circular hole subjected to tension (see Section 6.3.2). The value 0.793 of the limit factor (mean stress at the reduced section divided by the yield stress) is good, by comparison with the experimental volume equal to 0.77 [11.26].

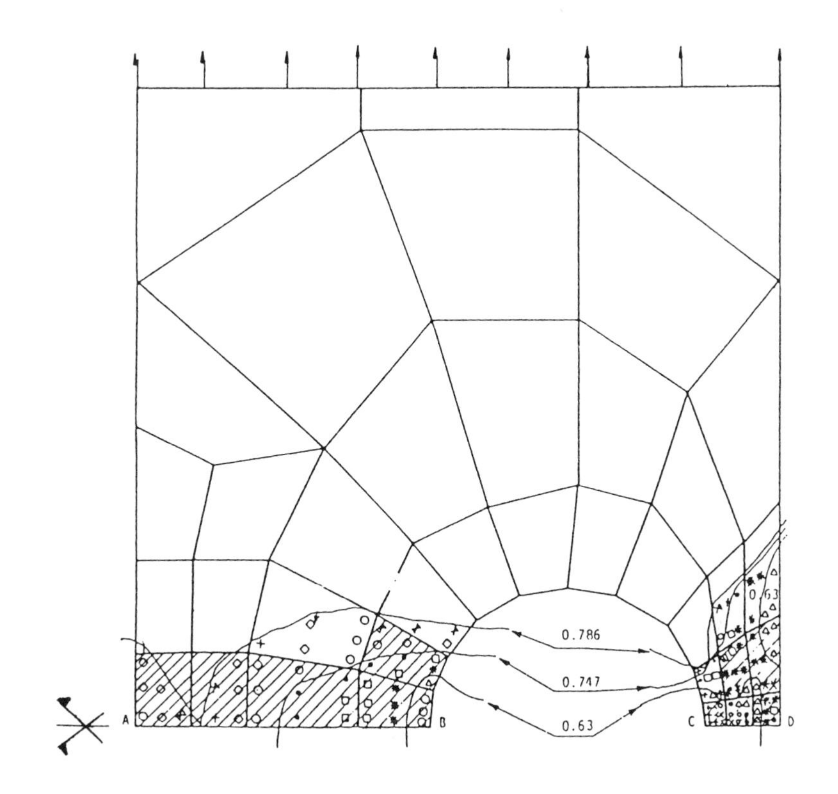

Fig. 11.38. Tension specimen with eccentric hole (267 degrees of freedom) [6.33].

Fig. 11.38 shows the limit state representation.

Another example analyzed by Nguyen Dang Hung in [6.33] concerns the limit pressure of a metal layer between rigid and rough parallel platens. All the curves given correspond to the particular case where w/h = 2 (fig. 11.39).

Table 11.2 compares the results obtained by the three types of approaches : statical (Section 6.3.2), kinematical (Section 6.4.2) and hybrid (Section 6.6). To each given number of degrees of freedom (D.O.F), the value of the nondimensional limit pressure q/2K is associated.

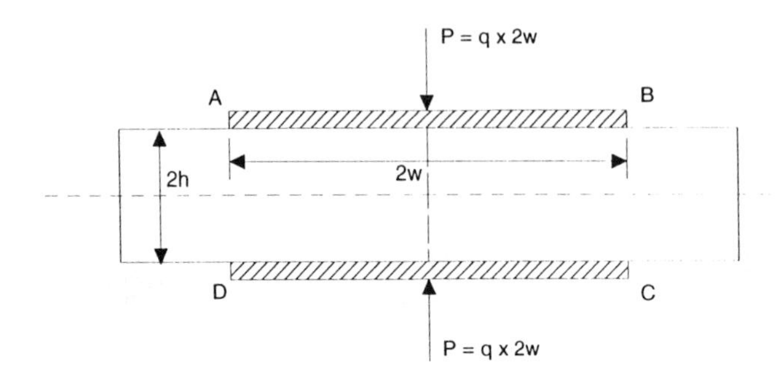

Fig. 11.39. Compression of a metallic plate between rough rigid parallel plates. [6.33]

The theoretical result obtained by Prandtl, namely

$$q/2K = 1.16,$$

corresponds to plane *strain* conditions. It can be roughly converted to *plane stress* conditions by multiplying it by $\sqrt{3/2}$, which gives

$$q/2K = 1.16\sqrt{3/2} = 1.0045.$$

Fig. 11.40 represents the collapse mechanisms corresponding to the various discretizations and approaches.

Table 11.2.

Statical		Kinematical		Hybrid	
D.O.F.	q/2k	D.O.F.	q/2k	D.O.F.	q/2k
24	0.902	30	0.928	30	0.933
80	0.914	90	0.924	90	0.924
168	0.916	182	0.922	182	0.921

Rectangular disks with various boundary conditions, made of a material with the von Mises yield condition, have been treated by Szmodits [11.15]. His lower bounds are obtained with discontinuous stress fields, and Mohr's circle is systematically used to deal with the discontinuity conditions. Discontinuous velocity fields are used in the kinematic

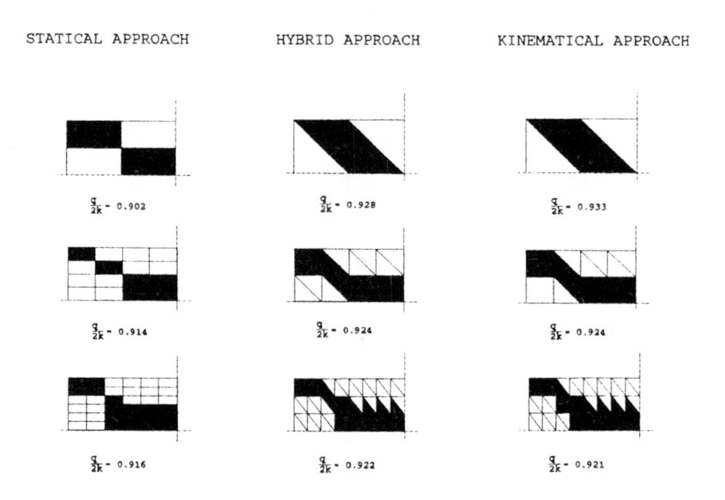

Fig. 11.40. Convergence of the ultimate mechanism [6.33]

analysis. Both types of fields are rather crude and give bounds that are far apart (lower bounds are roughly half of the upper bounds). Problems of the same type can be solved, using a general method of series expansions of the unknown functions that reduces the analysis to a linear programming problem where the coefficients of the expansions are the variables and the limit load is the cost function [11.16].

11.6. Plane strain : thick tube.

The state of plane strain has been widely studied in the mathematical theory of perfectly plastic solids. Prager and Hodge devote approximately half their book to the subject [3.13]. Plane strain occupies also an important part of the books by Hill [3.12], Nadai [1.4], and Sokolovsky [3.14].

Many of the problems treated do not, however, belong to the theory of structures but deal with various deformation processes : indentation, extrusion, etc., and therefore are not included in this volume. Only one structural plane-strain problem will be considered hereafter : the thick tube subjected to internal pressure. This subject is of important practical interest and has been widely studied experimentally [11.19]. For other plane-strain problems the reader is referred to the books cited above, where other interesting questions may also be found, e.g., torsion of prismatic and cylindrical bars with various cross sections, etc.

It is not possible to discuss here the numerous papers dealing with the plastic thick tube subjected to internal pressure. Such a discussion can be found in refs. [11.20] and [11.21]; autofrettage is treated in ref. [3.12]. We restrict ourselves to the following questions : (a) what is the magnitude of the pressure at which plasticity first occurs ; and (b) what is the magnitude of the limit pressure, and what is the physical significance of it.

Let a and b be the internal and external radii of the tube, respectively, and p the internal pressure. The tube is referred to a cylindrical coordinate system, fig. 11.41.

The longitudinal, radial, and circumferential directions are principal directions because of the symmetry of revolution. In elastic range, we have the well-known Lame equations :

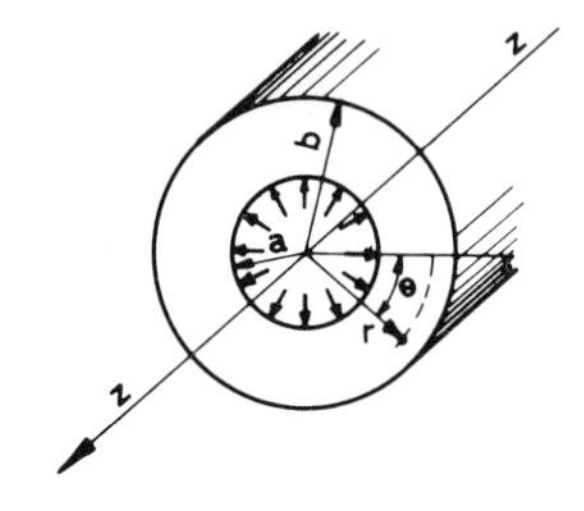

Fig. 11.41.

$$\text{(a)} \qquad \sigma_r = \frac{pa^2}{b^2 - a^2}\left(1 - \frac{b^2}{r^2}\right),$$

$$\text{(b)} \qquad \sigma_\theta = \frac{pa^2}{b^2 - a^2}\left(1 + \frac{b^2}{r^2}\right). \tag{11.70}$$

In plane strain $\varepsilon_z = 0$ and Hooke's law gives

$$\sigma_z = \nu\,(\sigma_r + \sigma_\theta) = \frac{2\,\nu\, pa^2}{b^2 - a^2}. \tag{11.71}$$

In the case of a closed-end tube, eq. (11.71) must be replaced by

$$\sigma_z = \frac{pa^2}{b^2 - a^2} = \frac{1}{2}(\sigma_r + \sigma_\theta) . \tag{11.72}$$

In both situations σ_2 is the intermediate principal stress. With the Tresca yield condition, the maximum elastic pressure p_e will be given by the condition

$$(\sigma_\theta - \sigma_r)_{max} = \sigma_Y . \tag{11.73}$$

From eq. (11.70) it is readily seen that $\sigma_\theta - \sigma_r$ is maximum at r=a. We obtain

$$p_e = \frac{\sigma_Y}{2} \frac{b^2 - a^2}{b^2} , \tag{11.74}$$

an expression valid for both end conditions considered above. With the von Mises yield condition, the end conditions play a role because σ_Z enters the yield condition

$$\sigma_R^2 \equiv \sigma_r^2 + \sigma_\theta^2 + \sigma_z^2 - \sigma_r \sigma_\theta - \sigma_\theta \sigma_z - \sigma_z \sigma_r = \sigma_Y^2 . \tag{11.75}$$

It turns out that σ_R is maximum at r=a [3.12] and it is found that in the state of plane strain,

$$p_e = \frac{\left(\frac{\sigma_Y}{\sqrt{3}}\right)\left[\frac{(b^2 - a^2)}{b^2}\right]}{\left[1 + (1 - 2\nu)^2 \left(\frac{a^4}{3b^4}\right)\right]^{1/2}} , \tag{11.76}$$

and in the case of a closed end tube,

$$p_e = \frac{\sigma_Y}{\sqrt{3}} \frac{b^2 - a^2}{b^2} . \tag{11.77}$$

Note that, for an annulus in plane stress $[\sigma_z = 0, \ \varepsilon_z = \dfrac{-2\,\nu\,pa^2}{E\,(b^2 - a^2)}$ in elastic range],

one would have

$$p_e = \frac{\left(\dfrac{\sigma_Y}{\sqrt{3}}\right)\left[\dfrac{(b^2 - a^2)}{b^2}\right]}{\left[1 + \left(\dfrac{a^4}{3\,b^4}\right)\right]^{1/2}} . \tag{11.78}$$

If we now increase p beyond p_e, a plastic annulus spreads from the internal surface toward the external surface. The state of restricted plastic flow, due to the presence of the external elastic annulus, ceases when the elastic-plastic interface reaches the outer surface, and plastic collapse occurs. The corresponding limit pressure is obtained in the following manner, assuming that the Tresca condition holds and that σ_z remains the intermediate principal stress. The stresses σ_r and σ_θ (fig. 11.42) satisfy the equilibrium equation

$$\frac{d\,\sigma_r}{dr} + \frac{\sigma_r - \sigma_\theta}{r} = 0, \tag{11.79}$$

and the yield condition

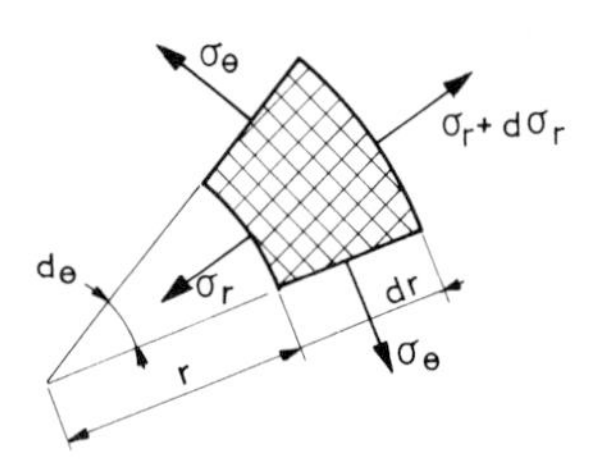

Fig. 11.42.

$$\sigma_\theta - \sigma_r = \sigma_Y \qquad \text{for} \qquad a \le r \le b. \tag{11.80}$$

Integration of eq. (11.79), after substitution of σ_Y for $\sigma_\theta - \sigma_r$, gives

$$\sigma_r = \sigma_Y \ln r + C_1. \tag{11.81}$$

The integration constant is obtained from the condition $\sigma_r = 0$ at $r = b$.

We obtain $C_1 = -\sigma_Y \ln b$, and eq. (11.81) is rewritten

$$\sigma_r = \sigma_Y \ln \frac{r}{b} . \tag{11.82}$$

The limit pressure is given by $\sigma_r = -p_l$ at $r = a$, a condition that yields

$$p_l = \sigma_Y \ln \frac{b}{a} . \tag{11.83}$$

From eqs. (11.80) and (11.82) it is deduced that

$$\sigma_\theta = \sigma_Y \left(1 + \ln \frac{r}{b} \right) \tag{11.84}$$

The reader will verify as an exercise that σ_z is actually intermediate between σ_θ and σ_r, and that a mechanism corresponding to the stress field, eqs. (11.82) and (11.84), can be found.

In the absence of work hardening, the limit pressure (11.83) is a real collapse pressure because, when changes of geometry are taken into account, the pressure decreases with continuing plastic flow.

From a detailed elastic-plastic analysis by Hill, Lee and Tupper [11.22, 11.23] where it is found that

$$\text{(a)} \qquad \sigma_\theta = \sigma_Y \left(\frac{1}{2} + \frac{\rho^2}{2\, b^2} + \ln \frac{r}{\rho} \right),$$

$$(b) \qquad \sigma_r = \sigma_Y \left(-\frac{1}{2} + \frac{\rho^2}{2\,b^2} + \ln \frac{r}{\rho} \right),$$

$$(c) \qquad p = \sigma_Y \left(\frac{1}{2} + \frac{\rho^2}{2\,b^2} + \ln \frac{\rho}{a} \right), \tag{11.85}$$

(ρ being the radius of the plastic region), it is also shown that σ_z tends very rapidly to the value $1/2\ (\sigma_r + \sigma_\theta)$. In a fully plastic state, it can thus be assumed with a good approximation (of the order of 1%) that $\sigma = 1/2\ (\sigma_r + \sigma_\theta)$, and the condition of von Mises becomes formally identical to that of Tresca, except that σ_Y is replaced by $(2/\sqrt{3})\ /\sigma_Y$. Eq. (11.83) thus simply becomes

$$p_l = \frac{2\,\sigma_Y}{\sqrt{3}} \ln \frac{b}{a}, \tag{11.86}$$

valid for both the closed-end tube and the tube in plane strain.

Bursting tests of approximately one hundred closed-end thick tubes have been performed by Faupel [11.19]. Various materials were used, among them mild-steel, brass and allied steels, with fracture stresses ranging from 466 N/mm to133 N/mm^2 and average strains at rupture from 12% to 83%. The general shape of the stress-versus-strain curves of the tested materials is shown in fig. 10.33. The conventional highest elastic stress σ_e is

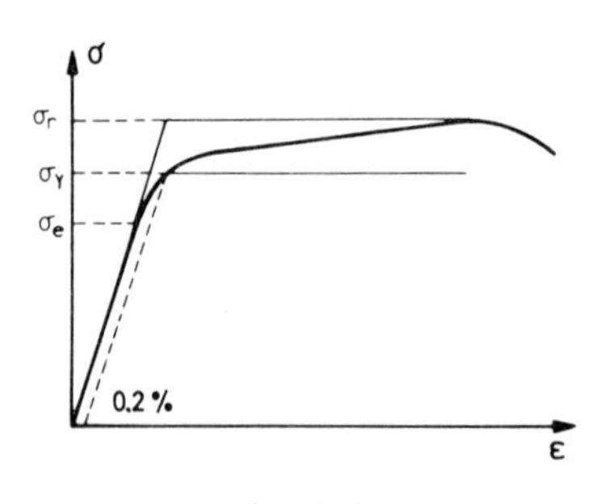

Fig. 11.43.

chosen as the stress producing a residual strain of 0.01%, whereas the conventional yield limit $\sigma_{Y,0.2}$ corresponds to a plastic strain of 0.2%. The rupture stress σ_r is the highest attainable stress.

Eq. (11.77), in which σ_e will be substituted for σ_Y, will give the highest elastic pressure. Because of work hardening of the material, eq. (11.86) must be used with a "weighted" yield stress that depends on $\sigma_{Y,0.2}$ and on σ_r, to give the pressure at which continued plastic flow occurs, and eventually burst under constant pressure. Faupel has suggested the formula

$$p_l = \frac{2\,\sigma_r}{\sqrt{2}}\left[2 - \frac{\sigma_r}{\sigma_{Y,0.2}}\right] \ln\frac{b}{a}, \tag{11.87}$$

which is supported by his experimental results for ninety tests among one hundred, with 15% maximum discrepancy.

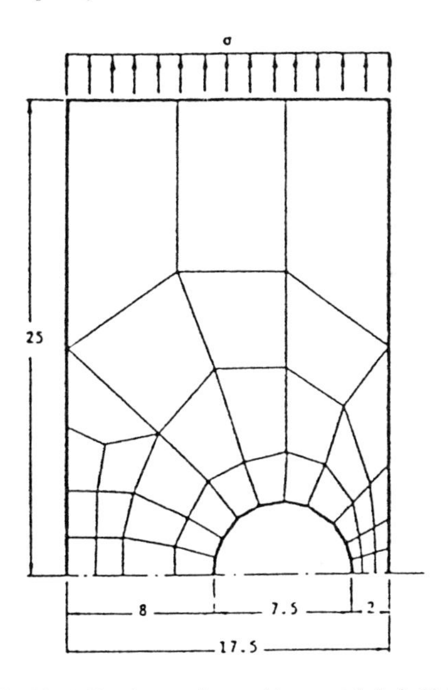

Fig. 11.44. Tension specimen with eccentric hole [6.47].

11.7. Shakedown problems.

11.7.1. *Strip with an eccentric circular hole.*

This example is computed by Nguyen Dang Hung and Palgen in [6.47]. Equilibrium finite elements are used.

The theoretical justification of the method is established in Section 6.8.

Fig. 11.44 represents the strip with an eccentric circular hole, considered in [6.47]. In ref [6.33], the proportional loading was assumed and this strip was computed by limit analysis. Now, we suppose the strip subjected to an uniform traction that varies between fixed limits :

$$\sigma_1 \leq \sigma \leq \sigma_2 . \tag{11.88}$$

The adopted finite element mesh is indicated in fig. 11.44. The elastic analysis was performed for a Poisson's ratio of 0.3. In ref. [6.47], the von Mises yield criterion is used in the sense of the mean (see Section 6.3.2).

In this example, the yield condition was examined in only 17 of the 37 elements of the structure ; the selected elements were those with the highest effective elastic stresses.

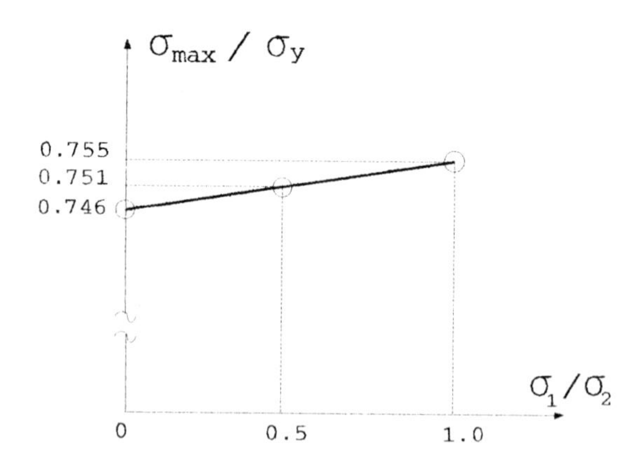

Fig. 11.45. Shakedown nominal stress for the strip of fig. 11.44. [6.47].

This simplification reduced the size of the nonlinear programming problem significantly. Its validity was established by checking *a posteriori* that the yield was not violated in the discarded elements. Elastic stresses corresponding to applied tractions $\alpha \, \sigma_1$ and $\alpha \, \sigma_2$ added to the residual stresses had to satisfy the yield condition in the 17 elements under consideration. This gave 34 constraints in the nonlinear programming problem. Let a nominal stress $\sigma_{nominal}$ be defined as the average stress acting on the net section when $\sigma = \alpha \, \sigma_2$. Referring to the dimensions of the strip indicated in fig. 11.45, one gets

$$\sigma_{nominal} = \frac{17.5}{10} \, \alpha \, \sigma_2 \, .$$

$\sigma_{nominal}$ reaches its maximum permissible value, σ_{max}, when $\alpha = \alpha_{max}$, the factor of safety with respect to shakedown. α_{max} was computed for different ratios σ_1/σ_2. A different shakedown problem corresponds to each value of this ratio. Results are shown in fig. 11.45. In all cases, incremental collapse (ratcheting) was responsible for failure.

The case $\sigma_1 = \sigma_2$ corresponds to limit analysis. The limit load is seen to be always a good approixmation to the shakedown load in the range of ratios σ_1/σ_2 that was investigated. Different results, including those of Rimawi and Dogan [11.26] for the limit analysis problem are compared in Table 11.3. A good agreement exists because the limit load computed by Nguyen Dang Hung ([6.33] and Section 11.5) is an upper bound, whereas the limit load evaluated by the present method is a lower bound.

Table 11.3. Limit nominal stress for the strip of fig. 11.44 [6.47].

Author	σ_{max}/σ_Y	Number of degrees of freedom	Method
Rimawi and Dogan [11.26]	0.77	---	Experimental
Nguyen [6.33]	0.793	267	Limit analysis (FEM)
Nguyen and Palgen [6.47]	0.755	178	Shakedown (FEM)

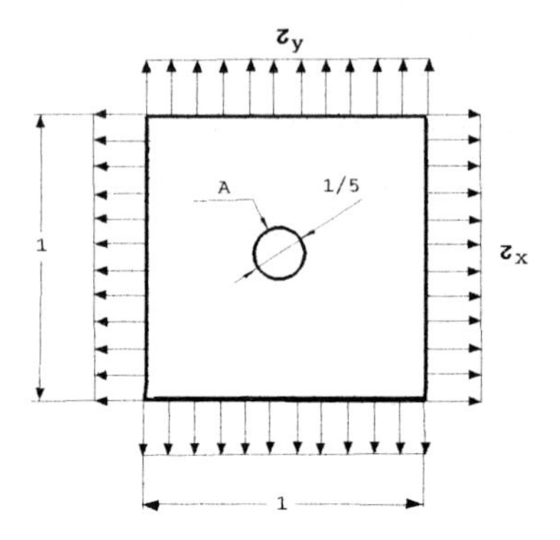

Fig. 11.46. Disk with circular hole [6.47].

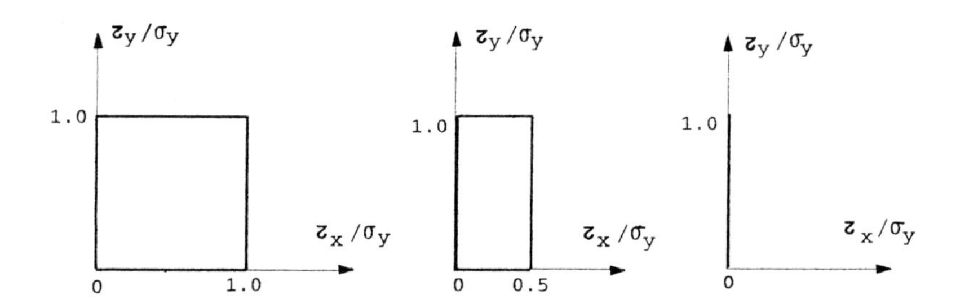

Fig. 11.47. Load domains [6.47].

11.7.2. *Square slab with a circular opening.*

The finite element method is applied to this problem by Belytschko [6.45], Corradi and Zavelani [6.46], Nguyen Dang Hung and Palgen [6.47]. A thin square slab with a circular opening is subjected to biaxial tension (fig. 11.46). The uniform tractions T_x and T_y vary between zero and an assigned value.

In [6.47], there are 24 equilibrium finite elements and 116 degrees of freedom.

In this example, the non-violation of the von Mises criterion in the mean sense (see Section 6.3.2) was required for all elements of the structure. Elastic stresses corresponding to the four vertices of the load domain had to be considered ; hence, the nonlinear programming problem contained 96 constraints. A different shakedown problem corresponds to each ratio of the upper limits of the applied tractions. The three load domains shown in fig. 11.47 were studied. The elastic and limit behaviours of the structure were also investigated. Computations were made for a Poisson'ratio of 0.3.

Table 11.4. Normalized shakedown traction $(T_y)_{max} / \sigma_Y$

$T_x / T_y 1$ / Authors	1	0.5	0
Belytschko [6.45]	0.431	0.501	
Corradi and Zavelani [6.46]	0.504		0.654
Nguyen and Palgen [6.47]	0.431	0.532	0.557

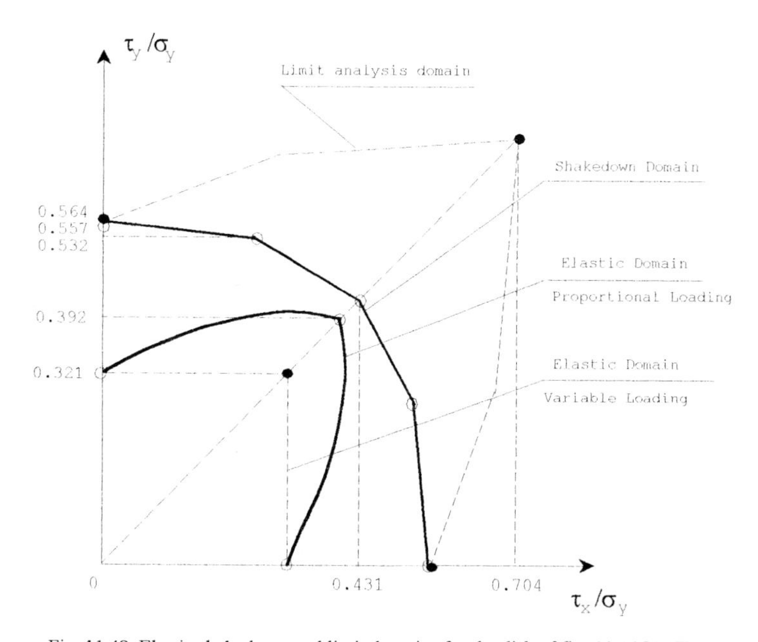

Fig. 11.48. Elastic shakedown and limit domains for the disk of fig. 11.46 [6.47].

Results are gathered in fig. 11.48. The solid line bounding the shakedown domain is the locus of shakedown tractions as the load domain varies. In the elastic domain with domain with variable loading they are allowed to vary independently between zero and their upper limits. The dotted line associated with the limit-analysis domain is the locus of tractions for which static collapse occurs. As seen from fig. 11.48, the purely elastic response domain is well inside the shakedown domain. If loading occurs only along the y axis, i.e. $T_x = 0$, limit analysis gives a good approximation to the shakedown load ; however it is very much on the unsafe side when tractions T_x, T_y have equal upper limits. Table 11.4 oportional loading the applied tractions are assumed to grow proportionally ; in the elastic compares the results of Belytschko [6.45], Corradi and Zavelani [6.46], Nguyen Dang Hung and Palgen [6.47], for different ratios of the upper limits of the applied tractions. The approaches of Belytschko and Nguyen Dang Hung use equilibrium finite elements and both methods provide therefore lower bounds of the shakedown multiplier.

The results are identical for equal axial loads whereas a high value is obtained in [6.47] for $T_x = T_y /2$. This may be due to the use of the mean criterion, which allows some local violation of the yield condition. The results of Corradi and Zavelani are the highest ; their procedure gives upper bounds to the shakedown load. Belytschko [6.45] obtained an analytical upper bound for the shakedown domain by assuming that failure was due to alternating plasticity at point A (fig. 11.46). On the basis of Howland's elastic solution [11.27], a shakedown traction of 0.470 σ_Y was found for equal axial loads.

References.

[11.1] F.A. GAYDON, "On the Yield-point Loading of a Square Plate with Concentric Circular Hole", J. Mech. Phys. Solids, **2** : 170, 1954.

[11.2] W.G. BRADY and D.C. DRUCKER, "An Experimental Investigation and Limit Analysis of Net Area in Tension", Transactions A.S.C.E., **120** : 1133, 1955.

[11.3] R. SCHLEICHER, Taschenbuch für Bauingenieure, vol. 1, Springer, Berlin, 1955.

[11.4] Manual of Steel Construction, pp. 5-32, A.S.C.I., New-York, 1964.

[11.5] P.G. HODGE Jr and N. PERRONE, "Yield loads of Slabs with Reinforced Cutouts", J. Appl. Mech., **24** : 85, 1957.

[11.6] D. VASARHELYI and R.A. HECHTMAN, "Welded Reinforcements of Openings in Structural Steel Members", Weld. J. Research, Suppl., **16** : 182s, 1951.

[11.7] D.C. DRUCKER, "On Obtaining Plane Strain or Plane Stress Conditions in Plasticity", Proc. 2nd U.S. Nat. Cong. Appl. Mech., pp. 485-488, A.S.M.E., 1954.

[11.8] E.H. LEE, "Plastic flow of a V-notched Bar Pulled in Tension", J. Appl. Mech., **19** : Trans. A.S.M.E., **74** : 331, 1952.

[11.9] W.N. FINDLEY and D.C. DRUCKER, "An Experimental Study of Plane Plastic Straining of Notched Bars", J. Appl. Mech., **32** : 493, 1965.

[11.10] W. SZCZEPINSKI and J. MIASTKOWSKI, "Experimental Limit Analysis of Flat Notched Bars Subjected to Tension" (in Polish), Rosprawy Inzynierskie, **13** : 637, 1965.

[11.11] L. DIETRICH, "Theoretical and Experimental Analysis of Load Carrying Capacity in Tension of Bars Weakened by Nonsymmetric Notches", Bull. Acad. Pol. Sci., Serie des Sc. Tech., **14** : 7, 363, 1966.

[11.12] W. SZCZEPINSKI, "A Survey of Papers dealing with the Problem of Notched Bars Pulled in Tension" (in Polish), Mechanika Teoretyczna i Stosowana, **3** : 51, 1965.

[11.13] J. HEYMAN, "Plastic Design of Rotating Disk", Proc. Inst. Mech. Eng., **172** : 531, 1958.

[11.14] J. HEYMAN, "Rotating Disks. Insensitivity of Design", Proc. 3rd U.S. Nat. Cong. Appl. Mech., Brown University, Providence, R.I., 1958.

[11.15] K. SZMODITS, "Scheibenmessung auf Grund der Traglastverfahrens", Act. Techn. Hung., **46** : 371, 1964.

[11.16] G. SACCHI and M. SAVE, "On the Evaluation of the Limit Load for Rigid Perfectly Plastic Continuum", Meccanica, **3** : 3, 1968.

[11.17] W. PRAGER, "Dimensionnement plastique et économie des matériaux", Bull. C.E.R.E.S., **X** : Institut du Génie Civil, Liège, 1959.

[11.18] T.C. HU and R.T. SHIELD, "Minimum Volume of Disks", Z.A.M.P., **XII** : 5, 414, 1961.

[11.19] J.H. FAUPEL, “Yield and Bursting Characteristics of Heavy-wall Cylinders”, Trans. A.S.M.E., **78** : 1031, 1956.

[11.20] P.G. HODGE Jr and G.N. WHITE Jr, “A Quantitative Comparison of Flow and Deformation Theories in Plasticity”, J. Appl. Mech., **17** : 180, 1950.

[11.21] D.N. de G. ALLEN and D.G. SOPWITH, “The Stresses and Strains in a Partly Plastic Thick Tube under Internal Pressure and End Load”, Proc. 7th Int. Cong. Appl. Mech., 403, Cambridge, 1958.

[11.22] R. HILL, E.H. LEE and S.J. TUPPER, “The Theory of Combined Plastic and Elastic Deformation with Particular Reference to a Thick Tube under Internal Pressure”, Proc. Roy. Soc., **A191** : 278, 1947.

[11.23] R. HILL, E.H. LEE and S.J. TUPPER, “Plastic Flow in a Closed End Tube with Internal Pressure”, Proc. Ist. U.S. Nat. Cong. Appl. Mech., Chicago, 1951, 561, A.S.M.E., ed., 1952.

[11.24] L.M. KACHANOV, “Foundation of the Theory of Plasticity”, American Elsevier, New-York, 1971.

[11.25] J.G. MARTIN, “Plasticity”, M.I.T. Press, Cambridge, Mass. 1975.

[11.26] W.H. RIMAWI and E. DOGAN, “Experiments on Yielding of Tension Specimens with Notches and Holes”, Exper. Mech., 10, pp. 427-432, 1970.

[11.27] R.C.J. HOWLAND, Trans. R. Soc., **A229** : 49, 1930.